Readings from *Conservation Biology*

EDITED BY DAVID EHRENFELD

The Landscape Perspective

A Joint Publication of
The Society for Conservation Biology
and Blackwell Science, Inc.

BLACKWELL SCIENCE

EDITORIAL OFFICES:
238 Main Street, Cambridge, Massachusetts 02142, USA
Osney Mead, Oxford OX2 0EL, England
25 John Street, London WC1N 2BL, England
23 Ainslie Place, Edinburgh EH3 6AJ, Scotland
54 University Street, Carlton, Victoria 3053, Australia
Arnette Blackwell SA, 1 rue de Lille, 75007 Paris, France
Blackwell Wissenschafts-Verlag GmbH, Kurfürstendam 57, 10707 Berlin, Germany
Blackwell MZV, Feldgasse 13, A-1238 Vienna, Austria

DISTRIBUTORS:

USA
Blackwell Science, Inc.
238 Main Street
Cambridge, Massachusetts 02142
(Telephone orders: 800-215-1000 or 617-876-7000)

Canada
Oxford University Press
70 Wynford Drive
Don Mills, Ontario M3C 1J9
(Telephone orders: 416-441-2941)

Australia
Blackwell Science Pty Ltd
54 University Street
Carlton, Victoria 3053
(Telephone orders: 03-347-5552)

Outside North America and Australia
Blackwell Science, Ltd.
c/o Marston Book Services, Ltd.
P.O. Box 87
Oxford OX2 0DT
England
(Telephone orders: 44-865-791155)

Printed and bound by The Sheridan Press, Hanover, PA, on acid-free recycled stock.

95 96 97 98 5 4 3 2 1

Library of Congress Cataloging in Publication Data

The landscape perspective/edited by David Ehrenfeld. p.
cm.—(Readings from Conservation biology)
Includes bibliographical references.
ISBN 0-86542-453-5
1. Habitat conservation. 2. Wildlife conservation. 3. Landscape ecology. I. Ehrenfeld, David. II. Society for Conservation Biology. III. Series.
QH75.L29 1994
333.95'16—dc20 94-46714
CIP

Contents

Preface

Landscape ecology and conservation biology are both young disciplines, and their interaction has been unusually fertile. By the 1980s, the theory of island biogeography had brought landscape-scale processes into the main stream of conservation thinking, especially concerning the design of nature reserves. In the 1990s, this occasionally abstract subject was increasingly supplemented by studies of the biological effects of habitat fragmentation and the benefits and costs of corridors connecting discrete areas of protected land. New technologies of geographic data analysis and mapping have also helped to make the landscape approach to conservation productive and popular.

I have selected 29 articles from the 30 issues of Conservation Biology published at the time of this writing, to provide a representative sample of the best of landscape-oriented conservation. My objective was to choose papers that are interesting, that make general points of lasting value, and that will be of use in teaching and in the libraries of workers in the field. I hope that this collection will make the landscape perspective accessible to all conservation researchers and managers, so that they can incorporate landscape thinking in their planning and decisions. In keeping with the topic, I have aimed for geographic diversity among the papers—many of them are North American in locale, but they also range from Australia and New Zealand to South and Central America to Europe to Africa.

The volume begins with a review by Saunders, Hobbs, and Margules of the effects of fragmentation on ecosystems. The changes that occur are physical as well as biological, and they are related to the size, shape, and position in the landscape of the habitat fragments. The authors call for an integrated approach to conservation that treats the landscape as a whole, rather than as a collection of biotic and legal entities. Lenore Fahrig and Gray Merriam argue that the effects of fragmentation on population survival cannot be understood without knowledge of landscape spatial structure—the spatial relationships among habitat patches and their matrix. Metapopulation models should include these spatial details of landscape structure.

No collection of papers on the interface between landscape ecology and conservation biology can ignore the great SLOSS (Single Large Or Several Small) controversy about whether large nature reserves are better than equivalent-sized clusters of small ones. Although this question—in its original form—seems less pressing than it did in the 1980s, the exchange still contains valuable lessons. James Quinn and Alan Hastings model habitat fragmentation and discover that some fragmentation may actually increase the ability of a threatened species to persist. But Michael Gilpin challenges the validity of their model and questions its use as a basis for management action. Quinn and Hastings consider Gilpin's criticisms, reassert their prior general conclusions, and suggest that more complex models might help with future management of real-world metapopulations.

Another healthy and fascinating debate has arisen about the need for corridors connecting discrete patches of conservation habitat. Daniel Simberloff and James Cox fire the first salvo, claiming that there are management alternatives to corridors that are frequently cheaper and biologically safer than the purchase of connecting routes. The Florida panther provides a case in point. Reed Noss counters that island analogies do not apply directly to conservation planning: corridors are often the most cost-effective solution to the problem of habitat fragmentation, and their disadvantages can be dealt with in various ways. Simberloff, et al. then review the history of the idea of conservation corridors and consider the possible advantages and disadvantages in greater detail. They conclude that connection of isolated patches is not always desirable: many considerations should govern the choice of corridors or their alternatives in the management of entire landscapes.

Some species, such as desert-dwelling mountain sheep, occupy a naturally fragmented habitat. Bleich, Wehausen, and Holl telemetered mountain sheep in California, and advocate the use of corridors and "stepping stones" (small areas of mountainous habitat that are not permanently occupied) in the conservation of these large, vagile animals. David Lindenmayer and Henry Nix observed arboreal marsupials in 49 wildlife corridors in southeastern Australia. They emphasize the importance of developing ecological principles for corridor design, and suggest from their experience that an understanding of the ecological principles could reveal those species that will be poorly conserved by wildlife corridors.

Bruce Paterson takes a close look at the land and freshwater bird species of New Zealand, with its many islands, and finds that species inhabiting fragments of once-continuous distributions constitute nested subsets of the species in richer, more intact biotas. The species preserved in such fragments tend to be those abundant generalists least in need of protection. John Blake applies the nested subset theory to bird species distributions in isolated woodlots in the central United States, and confirms that small habitat patches do not preserve many species.

Current thinking has it that a high ratio of habitat edge to interior area is detrimental to many species of interior-dwelling wildlife. Peter Paton reviews the literature on the effect of edge on avian nest success: he concludes that edge effects are real important, but that they are less convincing when sampling is more than 50 m from the edge boundary. Reuven Yosef studied Loggerhead Shrikes nesting in linear fencelines, and found that they had much lower reproductive success than shrikes nesting in nonlinear habitats. Some nest-predators use the fencelines as travel corridors, and thus encounter the nests frequently. Alverson, Waller, and Solheim claim that forest logging practices leading to fragmentation have greatly increased forest edge in northern Wisconsin, thus boosting deer populations, which in turn has led to the decline of browsed species such as Canada yew, eastern hemlock, and white cedar. But David Mladenoff and Forest Stearns disagree: they state that the abundance of deer is only one factor in the decline of eastern hemlock—the interaction of climate, hemlock life history, disturbance, ecosystem processes, and historical land use is more important than deer browsing in preventing hemlock regeneration.

Much of the work on the effects of forest fragmentation has involved the study of bird populations. This is because birds are comparatively easy to census and good historical distribution and abundance data are available for many sites. James Karr looked at selective extinction on Barro Colorado Island, in Panama, a protected "landbridge" island formed in 1914 by construction of the Panama Canal. The methods he used might help to identify extinction-prone species in other areas undergoing fragmentation. Kattan, Alvarez-López, and Giraldo studied a fragmented cloud forest site in the western Andes of Colombia and found that species at the upper limits of their altitudinal range, understory insectivores, and large canopy frugivores were the most at risk of extinction. Brian Maurer and Greg Heywood used data from the North American Breeding Bird Survey to examine the effect of fragmentation of breeding habitat on the declining populations of neotropical migrant landbirds. They learned that neotropical migrants and resident bird species are distributed differently across the continent. The migrants turn out to be much more vulnerable than the residents to fragmentation-caused extinction. The authors recommend that conservation management activities be concentrated on the maintenance and improvement of current migrant breeding habitat, rather than on the acquisition of new reserves.

The study of bird distribution and abundance in large, undisturbed tracts can yield information about the potential effects of fragmentation. J. M. Thiollay surveyed raptor populations in huge tracts of unbroken rain forest in French Guiana. He estimates that the smallest Guianan rain forest reserve that would include a complete bird community would be between 1 and 10 million hectares in size. The battle over the conservation of the Northern Spotted Owl of the U.S. Pacific northwest has provided a practical testing ground for theories of forest fragmentation and extinction. Harrison, Stahl, and Doak look at the models that have gone to court and question whether the U.S. Forest Service's conservation strategy based on them will save the owl.

Landscape-scale phenomena affect most species of plants and animals—not just birds. Michael Samways demonstrates that topography is a critical factor in the conservation of montane grasshoppers in South Africa. John Litvaitis shows that historic changes in land use and land ownership patterns in the northeastern U.S. have had a marked effect on populations of early successional vertebrates. Ola Jennersten finds that habitat fragmentation greatly reduces the reproductive success of a perennial, butterfly-pollinated herb in Swedish meadows. According to Timothy Brothers and Arthur Spingarn, forest fragments in agricultural central Indiana are relatively resistant to invasion by alien plant species. But this, they claim, is not a finding that can safely be generalized.

Principles derived from island biogeography have been used to develop design criteria for nature reserves, but as Jonathan Ambrose and Susan Bratton point out, a lack of congruence between the legal and ecological boundaries of a reserve limits the utility of the island model.

Geographic information system (GIS) technology can be very useful in organizing and analyzing ecological information for conservation. Wright, MacCracken, and Hall applied this technology at the ecoregion scale to evaluate vegetation types in four sites in Idaho proposed as U.S. national parks. None of the sites meet predetermined preservation criteria, but additions of relatively few hectares would allow each to reach the protection goal. J. B. Kirkpatrick and M. J. Brown, studying forest sites in Tasmania, compared two methods of evaluating proposed nature reserves: use of environmental classifications versus direct biological data. Neither method was adequate by itself, but a combination of the two holds promise for selecting a representative system of nature reserves.

In the concluding paper, Dale, et al. develop a comprehensive land use model that incorporates social and biological variables to predict the extent of forest clearing, carbon release, and farm survival in the Brazilian Amazonian settlement of Rondonia. The results of the model suggest that socio-economic and ecologic factors must be considered together in the planning of any long-term management strategy.

This is the fifth of six volumes currently planned for this series. The first was *To Preserve Biodiversity—An Overview,* a general survey of the field of conservation biology. The second, third, and fourth were: *The*

Social Dimension—Ethics, Policy, Law, Management, Development, Economics, Education; Wildlife and Forests; and *Plant Conservation.* The sixth volume will be *Genes, Populations, and Species.* Overlap of papers between volumes is small—approximately 10 to 20 percent. Each volume is intended to stand by itself with a minimum of duplication. I have made every effort to select papers that will not age rapidly; nevertheless, the contents of each volume will be reviewed and revised at regular intervals.

I thank Jim Krosschell, Jane Humphreys, and Kathleen Grimes, of Blackwell Science, Inc., for their assistance and for their effort to produce attractive, durable volumes printed on acid-free, recycled paper and published at an affordable price. I am also grateful to the officers of the Society for Conservation Biology, especially Peter Brussard, Steven Humphrey, and Reed Noss, for their support.

David Ehrenfeld

Review

Biological Consequences of Ecosystem Fragmentation: A Review

DENIS A. SAUNDERS

CSIRO Division of Wildlife and Ecology
LMB 4
P.O. Midland
Western Australia 6056, Australia

RICHARD J. HOBBS

CSIRO Division of Wildlife and Ecology
LMB 4
P.O. Midland
Western Australia 6056, Australia

CHRIS R. MARGULES

CSIRO Division of Wildlife and Ecology
P.O. Box 84
Lyneham
Australian Capital Territory 2602, Australia

Abstract: *Research on fragmented ecosystems has focused mostly on the biogeographic consequences of the creation of habitat "islands" of different sizes, and has provided little of practical value to managers. However, ecosystem fragmentation causes large changes in the physical environment as well as biogeographic changes. Fragmentation generally results in a landscape that consists of remnant areas of native vegetation surrounded by a matrix of agricultural or other developed land. As a result, fluxes of radiation, momentum (i.e., wind), water, and nutrients across the landscape are altered significantly. These in turn can have important influences on the biota within remnant areas, especially at or near the edge between the remnant and the surrounding matrix. The isolation of remnant areas by clearing also has important consequences for the biota. These consequences vary with the time since isolation, distance from other remnants, and degree of connectivity with other remnants. The influences of physical and biogeographic changes are modified by the size, shape, and position in the landscape of individual remnants, with larger remnants being less adversely affected by the fragmentation process. The dynamics*

Paper submitted May 2, 1990; revised manuscript accepted September 20, 1990.

Resumen: *La investigación sobre los ecosistemas fragmentados se ha enfocado principalmente en las consecuencias biogeográficas de la creación de "islas" de hábitat de diferentes tamaños y ha sido de muy poco valor práctico para los manejadores del recurso. Como quiera que sea, la fragmentación de los ecosistemas causa grandes cambios en el medio ambiente físico así como en el ámbito biogeográfico. La fragmentación resulta generalmente en terrenos que consisten de áreas remanentes de vegetación nativa rodeada de una matriz de tierras agrícolas u otras formas de uso de la tierra. Como un resultado de esto, el flujo de la radiación, del momentum (ej. el viento), del agua y de los nutrientes a través de la tierra son alterados significativamente. Esto en su turno, puede influenciar e la biota dentro de las áreas remanentes, especialmente en o cerca de los límites entre los remanentes y la matriz que los rodea. El aislamiento de las áreas remanentes por la tala también tiene importantes consecuencias para la biota y estas consecuencias varian con, el tiempo desde el momento del aislamiento, la distancia hasta los otros remanentes y el grado de conección entre ellos. La influencia de los cambios físicos y biogeográficos es modificada por el tamaño, la forma y la posición en el terreno de remanentes individuales siendo los remanentes grandes los menos afectados adversamante por el proceso de fragmentación. La dinámica de las áreas remanentes son dirigidas*

of remnant areas are predominantly driven by factors arising in the surrounding landscape. Management of, and research on, fragmented ecosystems should be directed at understanding and controlling these external influences as much as at the biota of the remnants themselves. There is a strong need to develop an integrated approach to landscape management that places conservation reserves in the context of the overall landscape.

predominantemente por factores que surgen en el terreno circundante. El manejo y la investigación de los ecosistemas fragmentados debería de dirigirse tanto al entendimiento y control de estas influencias externas como a las biotas remanentes en sí. Hay una fuerte necesidad para el desarrollo de un enfoque integrado en el manejo de tierras que coloca a las reservas para la conservación en el contexto del terreno en general.

Introduction

Since the development of agriculture, the natural vegetation cover of every continent except Antarctica has been extensively modified. A cycle of agricultural development followed by overexploitation of the land has been repeated throughout recorded history. Forman (1987) quotes Plato's (ca. 2350 BP) description of how an area of ancient Greece was stripped of its soil following clearing and grazing, leaving "the mere skeleton of the land." Overexploitation and the use of inappropriate agricultural practices have led to desert encroachment, as in the Sahel region in North Africa (Le Houérou & Gillet 1986; Ehrlich & Ehrlich 1987), and to extensive loss of soil, often with disastrous and dramatic consequences, as in the Dust Bowl of the United States (Hudson 1981).

The process of land clearing and the consequent environmental degradation is continuing rapidly in many regions such as Southeast Asia and South America, particularly in areas of tropical rain forest (Myers 1988). Australia provides an example of recent agricultural development, with vast areas being cleared over the last 100 years for cereal cropping and stock grazing (Saunders & Hobbs 1989; Hobbs & Hopkins 1990; Saunders et al. 1990). In some regions over 93% of the native vegetation has been removed and the agricultural land that has replaced it is subject to extensive wind and water erosion or soil salinization, with consequent pollution of water supplies for drinking and irrigation.

One legacy of the extensive removal of native vegetation is that the remaining vegetation is usually in fragmented patches across the landscape. These patches or remnants are situated in different positions in the landscape and on different soil types, possess different vegetation types, and vary in their size, shape, isolation, and type of ownership. Over much of the world, conservation of regional biotas depends entirely on the retention and management of these remnants. Conservation managers are therefore faced with the dual issues of whether the remnants have any practical conservation values, and if they do, of how they must be managed to retain these values (Saunders et al. 1987*a*).

In this paper we use the word *remnant* to define any patch of native vegetation around which most or all of the original vegetation has been removed (Saunders et al. 1987*a*). Remnants often have been called habitat islands and the changes that result from the isolation of these islands have been the subject of considerable debate in the literature. This debate has centered mostly on the equilibrium theory of island biogeography (MacArthur & Wilson 1963, 1967) and its applicability to conservation. In particular, the importance of size, shape, and design of single reserves and reserve systems, extinction and colonization rates, and species-area relationships have been much discussed (Wilson & Willis 1975; Diamond 1975, 1976; May 1975; Terborgh 1976; Whitcomb et al. 1976; Simberloff & Abele 1976*a*, 1982; Pickett & Thompson 1978; Game 1980; Margules et al. 1982; Boecklen & Gotelli 1984). Particular attention has been paid to the question of whether one large reserve could preserve more species than several small reserves adding up to the equivalent area of the larger reserve (the so-called SLOSS, or "single large or several small" debate; Simberloff & Abele 1976*b*, 1984; Gilpin & Diamond 1980; Higgs & Usher 1980; Järvinen 1982; Willis 1984).

These questions have been reviewed elsewhere (Simberloff 1988) and are not examined in detail here. While of theoretical interest, most of these issues are of little practical value in managing fragmented systems (Zimmerman & Bierragaard 1986; Margules 1987; Hobbs 1988; Margules & Stein 1989). The species-area equation, for example, may give a manager a rough idea of how many species will be maintained on a remnant of a given area, but will yield absolutely no information on the practical issue of which habitats contribute most to species richness or on which species are most likely to be lost from the remnant. Simberloff (1986) stated, "It is also sad that unwarranted focus on the supposed lessons of island biogeography theory has detracted from the main task of refuge planners, determining what habitats are important and how to maintain them." Margules (1986), in a discussion about two conservation evaluation exercises, notes that, "no panel members in either exercise considered the species-area relationship or the equilibrium theory of island biogeography in their evaluations; at least not explicitly." In addition, the debates about reserve design and SLOSS are of limited relevance because, with very few exceptions, managers of conservation reserves are faced with a fait accompli. Conservation considerations have rarely been taken into account during the development of areas for agriculture, mining, forestry, or other such uses. Conservation managers must work with the remnants left following these developments and virtually never have the opportunity to design a reserve network before an area is fragmented. There is an increasing need to utilize design criteria to improve conservation networks in already fragmented areas, but this requires a clear understanding of the problems created by fragmentation in the first place.

Hence we believe that research and discussion should focus on practical issues relating to the impact of fragmentation on natural ecosystems and managing remnants for conservation (Saunders et al. 1987*b*). We share the fear of Noss and Harris (1986) that conservation agencies have not realized the important biological consequences of ecosystem fragmentation and have therefore not developed policies to manage their remnants to maintain conservation values. The aim of this paper is to point out the physical effects of fragmentation, the biological consequences for natural ecosystems of these effects, and the options available for conservation research and management.

Characteristics of Fragmented Ecosystems

Fragmentation of the landscape produces a series of remnant vegetation patches surrounded by a matrix of different vegetation and/or land use. Two primary effects of this are an alteration of the *microclimate* within and surrounding the remnant and the *isolation* of each area from other remnant patches in the surrounding landscape. Thus, in a fragmented landscape there are changes in the physical environment as well as biogeographic changes. Most discussions of habitat fragmentation have concentrated on the biogeographic aspects, and the physical changes have received little attention. All remnants are exposed to these physical and biogeographic changes to a greater or lesser degree, but their effects are modified by the *size, shape,* and *position in the landscape* of individual remnants. We examine first the physical effects of fragmentation and then discuss the operation of the modifying factors.

Changes in Microclimate

Fragmentation of the landscape results in changes in the physical fluxes across the landscape. Alterations in fluxes of radiation, wind, and water can all have important effects on remnants of native vegetation.

1. RADIATION FLUXES

The energy balance of a fragmented landscape will differ markedly from one with a complete cover of native vegetation, especially where the native vegetation was dense before clearing. Removing native vegetation and replacing it with crop species with differing architecture and phenology alters the radiation balance by increasing the solar radiation reaching the ground surface during the day, changing the albedo, and increasing reradiation at night. These features vary depending on time of year; ploughing, crop growth, and harvesting produce an alternation of bare ground and varying degrees of vegetation cover (Geiger 1965; Milthorpe & Moorby 1974). In cleared areas, in general, daytime temperatures are higher and night temperatures lower than in naturally vegetated areas. This leads to greater temperature ranges both at the surface and in the upper layers of the soil, and an increased incidence of frost (Geiger 1965).

These changes in the cleared parts of the landscape impinge on the remnant native vegetation, especially at the edge between the two. Except near the equator, the orientation of the edge affects the degree to which solar radiation increases within a remnant at different times of year (Geiger 1965; Wales 1972; Ranney et al. 1981). Latitude also affects radiation input, and at high latitudes especially, where solar angles are low, a remnant edge can receive significantly more solar radiation than unfragmented areas receive (De Walle & McGuire 1973; Hutchinson & Matt 1976, 1977). Air temperatures at the edge of a forest remnant can be significantly higher than those found in either the interior of the remnant or the surrounding agricultural land (Geiger 1965; Kapos 1989).

The consequences of increasing solar radiation at the edges of remnants are not clear. The indications are, however, that a different suite of species may come to occupy this altered habitat. Lovejoy et al. (1986) report that fragmentation in tropical forests results in the rapid growth of vines and other secondary vegetation in a 10–25 m strip around the remnant edge. This may effectively seal off the remnant and maintain an environment within the remnant similar to that that existed prior to fragmentation (F. Crome, personal communication). This also occurs to some extent in temperate regions (Gysel 1951; Trimble & Tryon 1966; Ranney et al. 1981).

Shade-tolerant species may become restricted to the interior parts of the remnant, with different species requiring different distances from the edge. The composition of remnant edges may be affected by edge aspect (Wales 1972; Palik & Murphy 1990). Distinct sets of "interior" and "edge" species have been recognized in landscapes that have been fragmented for a long time, for instance in the eastern United States (Ranney et al. 1981).

Nutrient cycling processes may be affected by increased soil heating and its effect on soil microorganism and invertebrate numbers and activity (Klein 1989; Parker 1989), on litter decomposition, and on soil moisture retention. Changes in the radiation balance also affect larger fauna both directly and indirectly through altering resource availability (via changes in plant growth and phenology). Lovejoy et al. (1986) attribute changes in butterfly community composition in tropical forest fragments partly to increased insolation within small remnants. Increased radiation load and dessication rates may lead to reduced foraging opportunities. Saunders (1982) considered that elevated temperatures in fragmented landscapes reduced the foraging time available to adult Carnaby's cockatoos (*Calyptorhynchus funereus latirostrus*) and contributed to their local extinction. Alternatively, some species dependent on temperature thresholds may be able to forage for longer periods. For instance, the strongly dominant ants of the *Iridomyrmex* genus are known to forage only when insolation and temperatures are high, and other ant species forage only when *Iridomyrmex* is absent (Greenslade & Halliday 1983; Andersen 1987). Increased insolation and ambient temperatures could then increase the foraging time available to *Iridomyrmex* but reduce it for subordinate species.

Altered temperature regimes can also have the effect of destabilizing competitive, predator-prey, and parasitic interactions. Geiger (1965) quotes the example of the timber pest *Ocneria monacha*, which lays eggs on tree trunks. Elevated springtime temperatures on trunks at the forest edge allow larvae to emerge before their parasites, which emerge from the cooler forest floor. This gives the pest a head start and results in population buildups.

2. WIND

With the removal of surrounding vegetation, the entire pattern of momentum transfer over the landscape is altered. As air flows from one vegetation type to another, the wind profile adjusts to the new roughness characteristics. When air flowing over one vegetation type comes to a boundary with a new vegetation type, the upper part of the wind profile initially retains the characteristics of the previous vegetation type, but the lower part takes on new characteristics reflecting the roughness of the new vegetation type. The wind profile does not fully equilibrate with the new vegetation for some distance. Rules of thumb for wind profile measurements give the minimum "fetch" (i.e., the minimum distance from the vegetation boundary that will ensure that the profile has taken on the characteristics of the vegetation under study) as 100–200 times the height of the vegetation under study (Monteith 1975; Grace 1983; Jones 1983). Turbulent transfer is important for the transport of atmospheric gases, and gas fluxes above vegetation are controlled by these processes. It follows, then, that these processes must be significantly modified in remnant vegetation areas, where the wind profile will not be in equilibrium with the remnant vegetation (Jarvis & McNaughton 1986). Given the fetch requirements, a woodland with 20-m-tall trees would need to be at least 2–4 km wide before wind profiles would resemble those in an unfragmented situation. The implications of this for plant gas exchange and growth have not been examined, but could be significant.

A more obvious effect of landscape fragmentation is that remnants are subjected to increased exposure to wind. This may result in damage to the vegetation, either through direct physical damage (Moen 1974; Grace 1977), or by increasing evapotranspiration with reduced humidity and increased dessication (Tranquillini 1979; Lovejoy et al. 1986). Direct physical damage can take the form of wind pruning (Caborn 1957) or windthrow of trees. Distinct edge structures have been found to develop at the edges of tree plantations (Fraser 1972), and this is likely to be the case for remnant areas also (see Ranney et al. 1981 for a discussion of edge structures). Trees near the edge of recently isolated remnants are particularly at risk of windthrow since they have matured within a closed canopy and have therefore developed in the absence of strong winds and lack the necessary support mechanisms to deal with such winds. Windthrow of dominant trees results in changes in the vegetation structure and increased availability of regeneration gaps, allowing recruitment, particularly of pioneer or light-demanding species. Increased litter fall through tree damage is likely to alter soil surface characteristics and hence the habitat of ground-dwelling fauna. Similarly, increased exposure to wind may remove loose bark and reduce the substrate available for bark-dwelling invertebrates, and hence also reduce their availability as a food resource. Increased wind turbulence due to clearing has been shown to affect the breeding success of birds by creating difficulties in landing due to wind shear and vigorous canopy movement (Brett 1989; Reville et al. 1990). In the case of tropical forests, fragmentation can result in hot, dry winds blowing into remnant areas from the surrounding cleared areas, with the probable result of increased tree mortality (Lovejoy et al. 1986) and prevention of regeneration of species whose successful establishment requires humid conditions or persistent soil moisture (Janzen 1986). This may also be important in the regeneration of species in other areas, such as those with mediterranean climates, where successful establishment requires adequate soil moisture (e.g., Gordon et al. 1989; Williams & Hobbs 1989). This effect will be lessened in cases where the edges of remnants are sealed off by rapid secondary growth.

Increased wind speeds at remnant edges have the secondary effect of increasing the transfer of material such as dust and seeds in from the surrounding matrix. Geiger (1965) gives an example where particulate matter deposition at the edge of a forest remnant increased by 40% over that in the open. Transfer of nutrients by saltation of surface soil particles is also possible, and strong gradients in soil nutrient levels have been found at the edges of remnant areas (Muir 1979; Cale & Hobbs 1991). Wind can deposit seeds of nonnative species over considerable distances into remnant areas (Hobbs & Atkins 1988). Transfer of insects and disease organisms into remnant areas may also be increased.

3. WATER FLUX

Fragmentation of the landscape results in modification of the local water regime by altering the various components of the hydrological cycle. Removal of native vegetation changes the rates of rainfall interception and evapotranspiration, and hence changes soil moisture levels (Kapos 1989). The pathways by which water penetrates the soil may also be markedly altered (Bormann & Likens 1979; Nulsen et al. 1986; Peck & Williamson 1987; Sharma et al. 1987; Bell 1988). Replacement of deep-rooted perennial species with herbaceous crop or pasture species leads to greatly reduced evapotranspiration and increased surface- and groundwater flows. The hydrological system in general becomes much less buffered with more extreme run-off events (Hornbeck 1973; Simons 1989). Increased surface water flows lead to increased erosion and transport of particulate matter (e.g., Bormann et al. 1974). Topsoil removed from high in the catchment ends up as sediment in the river system. Transport of nutrients into streams also increases (Likens et al. 1970). Rises in water tables can bring stored salts to the surface and cause secondary salinity, with considerable impacts on both remnant vegetation and the surrounding agricultural matrix (Peck 1978; Williamson et al. 1987). Movement of stored salts, nutrients, and pesticides washed from cleared land can have significant impacts on river systems (Kendrick 1976; Karr & Schlosser 1978).

The impact of these changes on an individual remnant depends greatly on its position in the landscape (Swanson et al. 1988). Remnants at the top of a catchment can be expected to be relatively little affected by changes in water flux, whereas remnants on midslopes and valley bottoms will be more affected. Remnants in run-off areas can be expected to experience more erosion, while those in run-on areas will experience more soil deposition, especially on the up-slope edge.

Further impacts on remnant areas can be expected following management operations in the surrounding matrix that alter hydrological processes. Thus irrigation, water storage, or drainage may affect remnant areas. An extreme example of this is found in the fens of eastern England, where drainage has led to peat shrinkage and a drop of 4 m in land level in 130 years. Remnant areas of natural wetland now require pumping systems to retain adequate water levels (Hutchinson 1980; Rowell 1986).

Changes in water fluxes and associated particulate and nutrient fluxes can have important influences on the biota of remnants. Altered patterns of erosion lead to changes in drainage patterns and the production of new substrates for plant colonization. Of particular importance is the deposition of nutrient-rich material in run-on areas, which can act as a focus for invasion by species requiring disturbance and/or nutrient enrichment for successful establishment (Hobbs & Atkins 1988). Changes in surface and soil moisture levels could also lead to changes in decomposition rates, altered seedbed characteristics, and changes in habitat for ground-dwelling fauna.

Isolation

Landscape fragmentation has two important consequences for the biota. First, there is a reduction in the total area of habitat available, with possible increased densities of surviving fauna in the remnants, and second, the habitat that is left is broken up into remnants that are isolated to varying degrees (Lovejoy et al. 1984, 1986; Haila & Hanski 1984; Wilcove et al. 1986). The time since isolation, the distance between adjacent remnants, and the degree of connectivity between them are all important determinants of the biotic response to fragmentation.

1. TIME SINCE ISOLATION

Upon isolation, a remnant is likely to have more species than it will be capable of maintaining, and species will be lost as the changes brought about by fragmentation take effect (Miller & Harris 1977; Miller 1978; Wilcox 1980; Harris 1984). This process of "species relaxation" is considered an inevitable consequence of area reduction and isolation, on the basis of island biogeographical predictions. However, the various mechanisms by which local extinctions occur will result from the physical changes discussed above and resulting changes in biotic interactions. The rate of species relaxation will differ among different taxa. The most rapid extinctions are likely in species that depend entirely on native vegetation, those that require large territories, and those that exist at low densities. Dispersal behavior and demography will determine the response of individual species to fragmentation (Karieva 1987).

Populations that are too small to be viable may persist for long periods simply because of the longevity of individuals. For example, in remnants in the Western Australian wheatbelt, female trapdoor spiders *Anidiops villosus* can live for at least 23 years (Main 1987), and small Australian passerines of about 25 g may live over 20 years (Australian Bird Banding Scheme records). It may take several hundred years to lose some species such as long-lived trees, especially since adult plants are often less sensitive to changed environmental conditions than seedling and juvenile stages. Alterations in disturbance regime in remnant areas may prevent successful regeneration (Hobbs 1987; Bond et al. 1988). Presence of a species in a remnant is thus no guarantee of its continued existence there; successful reproduction and recruitment are required. Managers therefore may need to examine the age structure of species on remnants to identify vulnerable species to be targeted for special management.

Time since isolation will therefore determine how far down the "relaxation track" any given remnant has traveled. Recently isolated remnants can be expected to continue losing species; this process may continue for relatively long periods in the absence of interventive management (Soulé et al. 1988; Saunders 1989). Long-isolated remnants can be expected to have lost a proportion of the species originally present, and gained an additional component of invading species that are capable of establishing in the fragmented system. It is thus wrong to consider only species numbers and not species composition when discussing species diversity in remnant areas: species numbers can potentially increase in fragmented systems where invasive and edge species can establish, but the numbers of species originally found in the area may continue to decline (Verner 1986; Murphy 1989; Webb 1989; Harris & Scheck 1991).

It is commonly assumed that at some stage the remnant will reequilibrate with the surrounding landscape. It is, however, questionable whether a new stable equilibrium will be reached since the equilibration process is liable to be disrupted by changing fluxes from the surrounding matrix, disturbances, and influx of new invasive species. The final equilibrium can be likened to an idealized endpoint that is never likely to be reached, in much the same fashion as the climatic climax is now conceptualized in succession theory. Management of remnant areas will thus be an adaptive process directed at minimizing potential future species losses.

2. DISTANCE FROM OTHER REMNANTS

The ability of species to colonize a remnant depends to some extent on the distance of the remnant from other areas of native vegetation, be they other remnants or nearby uncleared areas. Colonizing ability is related to dispersal mode, with wind-dispersed and vagile species more likely to arrive at isolated remnants. However, whether such species become successful colonists depends on physical and biotic factors such as nutrient availability and competitive interactions (Vepsäläinen & Pisarski 1982). Animal species may have the physical ability to disperse long distances, but lack the behavioral repertoire to traverse the matrix surrounding the remnant; the matrix becomes an effective barrier to movement. Organism size is also important, and 100 m over agricultural fields may be a complete barrier to dispersal for small organisms such as invertebrates (Mader 1984) and some species of bird (Saunders & de Rebeira 1991). The persistence of such species on a remnant would then depend entirely on the retention of enough suitable habitat to maintain sufficient numbers to withstand the risks of extinction (Soulé 1987*a*; Ewens et al. 1987). Some evidence exists that fragmentation of large populations into subpopulations may decrease the risk of overall species extinction even though local extinctions may occur (Higgs 1981; Quinn & Hastings 1987). It seems likely that different species will respond differently to the creation of subpopulations and that knowledge of the details of an organism's behavior will be necessary to predict its response (Karieva 1987; Merriam 1991).

3. CONNECTIVITY

Associated with the effects of distance is the degree to which individual remnants are connected in some way to adjacent areas. The issue of connectivity and the usefulness of corridors connecting remnants has received increasing attention in the literature (MacClintock et al. 1977; Wegner & Merriam 1979; Baudry 1984; Forman & Baudry 1984; Merriam 1984; Harris 1984, 1985; Fahrig & Merriam 1985; Noss & Harris 1986; Bridgewater 1987; Simberloff & Cox 1987; Noss 1987; Soulé et al. 1988), and was the subject of a recent symposium (Hobbs et al. 1990; Saunders & Hobbs 1991*a*). Corridors are generally believed to provide benefits such as enhanced biotic movement, extra foraging areas, refuges during disturbances, and enhancement of the aesthetic appeal of the landscape. In some areas they significantly add to the area of native vegetation left following fragmentation.

Simberloff & Cox (1987) pointed out that most of the work on the value of corridors has not been sufficiently controlled to demonstrate an unequivocal role in increasing immigration and/or decreasing extinctions. An increasing number of studies, however, now indicate that corridors are of value for movement, at least for a subset of the biota (papers in Saunders & Hobbs, 1991*a*). Simberloff and Cox (1987) also discussed potential disadvantages of corridors, which include facilitation of the spread of disease, pests, and fire and other disturbances, increased predation, and high costs of maintaining linear remnants with high edge:area ratios. The relative merits of corridors and their required characteristics (i.e., width, composition, etc.) will vary from place to place and will depend on the target species. Detailed predictions of corridor value in reducing isolation of remnant areas are not possible without information on the movement of individual species across the landscape. Such information is, however, difficult and time-consuming to collect (Saunders & de Rebeira 1991). Nevertheless, while such data are being gathered, we need to take the approach that corridors do have value for biotic movement and attempt to retain a good corridor network wherever possible (Harris & Scheck 1991; Saunders & Hobbs 1991*b*).

4. CHANGES IN THE SURROUNDING LANDSCAPE

Removal of the vegetation from the area surrounding a remnant leads to the remnant becoming the only area of suitable habitat remaining for biota displaced by clearing. This may lead to the concentration of mobile elements of the biota in the remnants (Lovejoy et al. 1986). This concentration or crowding effect may be rapid and result in supersaturation of the remnant by some species. Crowding can alter intra- and interspecific interactions. Competition and predation, for example, can be increased, resulting in changes in fecundity and the potential ultimate collapse of the population. Resource availability is also affected by overexploitation; for instance, increased herbivory by large herbivores such as the elephant in African reserves can lead to quite dramatic changes in habitat (Spinage & Guiness 1971; Laws et al. 1975; Walker 1981).

Supersaturation results from the influx of species native to the area, but there are also potential influxes of new suites of species that have increased in abundance or established in the surrounding landscape following fragmentation. Such species include those that have been introduced in the process of agricultural development (pasture and crop plant species and livestock), other deliberate and accidental introductions, and native or migrant species that can take advantage of the new habitat conditions caused by fragmentation.

Natural communities vary in their susceptibility to invasion, although there is still debate over which characteristics render one community more invasible than another (Fox & Fox 1986; MacDonald et al. 1986; Crawley 1987; Usher 1988; Rejmánek 1989). For vegetation, establishment of nonnative species seems to be enhanced by some form of disturbance, especially if this increases the availability of a limiting resource (Hobbs 1989; Panetta & Hopkins 1991). Thus the opening of light gaps in dense forests where light was limiting could enhance invasion, whereas in nutrient-limited systems, nutrient input significantly increases the performance of nonnatives, especially in conjunction with soil disturbance (Hobbs & Atkins 1988). Invasion may be restricted to the edge of remnants if disturbance factors decline with distance from edge (Cale & Hobbs 1991; Panetta & Hopkins 1991), but species with wind- or animal-dispersed seeds can establish in suitable areas within a remnant, away from the edge. Invading species can establish, for example, from seeds carried in by, or deposited in feces of, animals that feed in the area surrounding the remnant but use the remnant for shelter. An example from Western Australia is the Wedge-tailed Eagle (*Aquila audax*), which breeds or roosts in remnants but forages on carrion in the surrounding farmland, bringing parts of sheep carcasses back to the roost to consume them. Wool from carcasses carries seed that is dropped under the roost tree. Nutrient input from eagle droppings and the disturbance caused by other scavanging animals provides a focus for the establishment of nonnative species (Saunders, personal observation). Thus, even in the absence of deliberate disturbance within remnant areas, invasions may occur.

Invasive species can have significant impacts on the plant communities within remnants; for instance, invading plant species can significantly alter the fuel structure and hence fire regime, and can inhibit the regeneration of native species (Wycherly 1984; Macdonald et al. 1989; Panetta & Hopkins 1991). Nonnative herbivores, including stock, can also dramatically change vegetation structure and prevent regeneration. Species that increase because of landscape modification can also have significant impacts on the rest of the biota. For instance, in North America increased pressure from nest predators and parasites such as the Brown-headed Cowbird (*Molothrus ater*) has affected passerine bird populations in fragmented systems (Brittingham & Temple 1983; Wilcove 1985; Andren & Angelstam 1988). Similarly, the Galah (*Cacatua roseicapilla*) has moved into all agricultural areas in Australia as a result of the development of cereal cropping and the provision of watering points for stock (Saunders et al. 1985). It now roosts in remnant woodland areas, often competing with other indigenous hole nesters (Saunders & Ingram 1987; Saunders 1990), and can damage tree foliage and bark, in extreme examples causing tree mortality.

Modifying Influences

1. REMNANT SIZE

The smaller a remnant is, the greater the influence external factors are likely to have. In small remnants, ecosystem dynamics are probably driven predominantly by external rather than internal forces. Of importance here are "edge effects" (Williamson 1975; Janzen 1983; Harris 1988; Yahner 1988). Larger remnants have a bigger core area that is unaffected by the environmental and biotic changes associated with edges. Here, we regard edge effects as mainly detrimental; this is opposite to the traditional view that edges and ecotones are beneficial to wildlife (Harris 1988). The difference is that remnant edges are created by removal of the surrounding vegetation and place the remainder in juxtaposition with a completely altered surrounding matrix.

Noss and Harris (1986) have pointed out that we do not know the minimum critical size an ecosystem needs to be to preserve its characteristic species diversity and species composition (Lovejoy & Oren 1981). In fact the "minimum dynamic area" of Pickett and Thompson (1978) or "the smallest area with a natural disturbance regime which maintains internal recolonization sources" would probably exist only in the largest conservation parks.

Larger remnants usually contain greater habitat diversity than smaller ones. A collection of smaller reserves may, however, cover a greater array of habitats than a single large one simply because a single large reserve will not contain all of the habitats likely to occur in a region. These arguments have been discussed in detail elsewhere and will not be pursued here. It is important, however, to recognize that the process of fragmentation is generally not random (Usher 1987). Land clearance usually occurs on a selective basis, with the best soil types being cleared first. For example, in southwest Western Australia, settlers selectively cleared woodlands because they occurred on the heavier soils best suited for agriculture. As a result, woodland communities are now poorly represented in conservation reserves and most woodland remaining on farms is in a highly degraded state (Saunders & Hobbs 1989). Few reserves in the area are large enough to contain representative samples of all preexisting vegetation types. Kitchener et al. (1980*a, b,* 1982) found, however, that even relatively small reserves (i.e., 30 ha) could be rich in some groups of fauna, but whether these populations are viable in the long term is debatable (see Saunders 1989).

Remnant size determines the potential size of populations of component species. Clearly, the number of individuals of any particular species that a given remnant will support depends on organism size and requirements. The larger the remnant, the more likely it is that populations will be large and more likely to resist chance extinctions (Gilpin & Soulé 1986; Soulé 1987*a*). Retaining populations in the long term may require large population sizes (of the order of hundreds or very much greater; Shaffer 1987), although the actual numbers required will depend on the life history and population growth rate of the species involved. Pimm et al. (1988) confirm that over a few decades extinction risk does decrease with population size, but they found no extinctions among British island birds numbering over 18 pairs. The issue of minimum population size has been discussed extensively in the literature (see Soulé 1987*b*), but there is still no real resolution as to what constitutes a minimum viable population. There has been extensive modeling of the concept, but little experimental work.

Larger populations tend to have higher levels of heterozygosity than small isolated populations. Current thinking is that heterozygosity is beneficial. Species that have gone through genetic "bottlenecks" are likely to suffer a reduction in heterozygosity with consequent loss of ability to adapt to changing conditions. Species isolated on remnants may go through such genetic bottlenecks because of small population sizes, and deliberate transfer of individuals between populations may be required (Boecklen & Bell 1987). However, the general assumption that heterozygosity is essential for long term population viability is still open to debate.

Large reserves may have some disadvantages; in particular, the possibility of disease spreading through entire populations on a large reserve has been discussed. However, there is a wide range of species' life histories and distribution patterns in the biota, and the effects of reserve size are largely species-specific. Species that have large area requirements or that require combinations of different habitats are likely to survive only in relatively large areas, whereas organisms with small, localized populations and simple habitat requirements can survive on smaller remnants. However, all will be affected by the disruption of physical ecosystem processes that result from fragmentation and were discussed earlier.

2. SHAPE

The shape of a remnant is important only for relatively small areas; there is some size beyond which shape does not really matter. However, for small remnants, shape determines the perimeter:core (or edge:interior) ratio. Long, thin remnants have proportionally much more edge than square or round remnants (Diamond 1975; Wilson & Willis 1975), and are more open to detrimental edge effects. However, some vegetation types, such as riparian strips, are naturally thin, and corridors are by definition generally long thin remnants (although the wider they are the more useful they may be to aid movement of biota; Saunders & Hobbs 1991*b*). Long, thin remnants may also, depending on their orientation, lie along environmental gradients and thus contain more vegetation types and habitats than a square reserve of similar area. Linear features are thus part of the natural and fragmented landscape, and there is no point in trying to develop optimal design principles that do not take them into account; the most important question is how to manage the remnants, whatever shape they are, so as to minimize external effects.

3. POSITION IN LANDSCAPE

The position of a remnant in the landscape has important influences on all the features so far discussed. It affects prefragmentation patterns of geomorphology, soil, and vegetation, and hence determines the structure and vegetation composition of any given remnant area. It also significantly affects postfragmentation processes.

For instance, there is an important distinction between remnants that are predominantly run-on versus those that are predominantly run-off. This influences not only the hydrological regime of the remnant, but also the movement of soil, nutrients, and seeds into and out of the remnant area.

Lessons for Management

Management of fragmented ecosystems has two basic components: (1) management of the natural system, or the internal dynamics of remnant areas, and (2) Management of the external influences on the natural system. For large remnant areas, the emphasis should be on managing the internal dynamics, including, for instance, the disturbance regime and population dynamics of key organisms. For small remnants, on the other hand, management should be directed primarily at controlling the external influences. Janzen (1983, 1986), however, has pointed out that external influences can be important whatever the remnant size.

Since most impacts on remnant areas originate from the surrounding landscape, there is clearly a need to depart from the traditional notions of reserve management, and look instead toward integrated landscape management. It will become increasingly difficult to maintain remnants of native vegetation if the management practices in the surrounding matrix have continuous adverse impacts on them. Traditional reserve management stops at the reserve boundary; fluxes of water, particulates, and organisms do not. Placing the conservation reserves firmly within the context of the surrounding landscape and attempting to develop complementary management strategies seems to be the only way to ensure the long term viability of remnant areas (Hobbs & Saunders 1991). This has important implications for land managers since it involves a radically new way of viewing management and requires that neighboring land uses, and hence neighboring landowners, interact in a positive way. This is difficult, but not impossible, and there are encouraging examples of attempts at this type of integrated management (e.g., Fitzgerald Biosphere Project 1989; Bradby 1991).

The landscape approach to management is also essential since several remnant areas taken together may represent a system over which components of the biota travel to meet habitat and food requirements. The loss of a single component of such a network could severely affect the capability of the remaining remnants to carry out the same functions, if for instance a particular species or habitat was lost. Such a network consists not only of the designated reserves, but also other remnant areas and linkages between them.

The goal of conservation management usually is to maintain species diversity, and the method of achieving this is to attempt to maintain representative examples of each ecosystem or community type present before fragmentation. To do this, we need to know the distributions of species and communities and then select areas that represent them. In general, there are two possible scenarios. In the first, we have a system that is about to be fragmented and we have to design the ideal set of reserves for the area. In the second, we have an already-fragmented system that we need to make the most of. Most theories in conservation biology, including virtually all the discussions of island biogeography in relation to reserves, have dealt with the first scenario, whereas it is the second scenario that we most frequently have to confront. Here we present a series of guidelines for management in this situation.

1. The initial step must be to determine the minimum subset of the existing remnants that are required to represent the diversity of a given region (Margules et al. 1988; Margules & Stein 1989). This requires that we have some knowledge of the distribution of species or ecosystem types. Clearly, it would be desirable to have all existing remnants available for this purpose, but in many cases this is not achievable, and there must be priorities for reserve retention or acquisition.

2. The system must then be managed to maintain the diversity of species or ecosystems. The question of whether management should be for individual species or whole ecosystems is largely irrelevant, because individual species require functioning ecosystems to survive. Management guidelines will be area-specific, but the need to manage external influences is universal.

3. Priorities for management must be established. Clearly there are many problems to be tackled, and usually there are limited resources available for the job. There must therefore be a clear priority ranking to ensure that resources are deployed optimally. Problems that are likely to disrupt ecosystem processes and hence threaten the viability of a remnant area should be given high priority for treatment.

4. Continuous management is needed to maintain remnant areas in their current state, due to the constant pressure of altered internal dynamics and external influences. Here again, the allocation of scarce management resources must be considered. Effort should go into maintaining some remnant areas in as near a "natural" state as possible, but it will not be feasible to do this for all remnants. There is a strong case to be made for letting some areas degrade so that they become less natural but are easier to manage and still retain some conservation value (Bridgewater 1990). This is not as radical as it may sound, since the process is ongoing in many remnant areas anyway. Once we accept that many remnants now contain "synthetic" communities that are not likely ever to return to their pristine state, management priorities become easier to set.

Research Requirements

Research to date on fragmented ecosystems has provided few answers on the issues of practical importance to management. It is just as important to set priorities for research as for management, and in the same way, research costs must be taken into account since resources for research are also limited. For instance, is it better to concentrate on single-species studies that can produce results with direct practical application but are very costly and time-consuming, or should we concentrate on the community/ecosystem approach that is cheaper but may yield more equivocal results? Clearly, balanced use of both approaches is needed. We have identified a number of priority areas that require research effort (see also Soulé & Kohm 1989).

1. A major priority is to understand the effects of external factors. Comparisons of pre- and postfragmentation systems will be particularly useful (Lovejoy et al. 1986; Margules 1987). Effects of changes in radiation and water fluxes are particularly important, as is the biotic invasion process.
2. Changes in internal processes since fragmentation also require further investigation. In particular, the interaction between internal and external processes is likely to be of critical importance.
3. Isolation factors need to be better understood. In particular, rates of genetic change in isolated populations require study, as does the question of whether reduction in genetic variability is important. We also require more and better data on the role of corridors in allowing biotic (and hence genetic) movement in fragmented landscapes.
4. While theoretical studies have their place, there is an urgent need for field experimentation in both management and restoration. While such experiments are costly to set up, it is possible to make use of many situations that offer ready-made experiments. Our understanding of succession has benefited greatly from the study of abandoned old fields, and in the same way, we can use current or past management activities as large-scale experiments; there is plenty of experimental material around (McNab 1983; Hopkins & Saunders 1987; Pimm 1986; Jordan et al. 1987; Hobbs & Hopkins 1990). Research has much to gain from a close liaison with management, especially if management operations actually can be carried out as designed experiments and the results suitably monitored.

Conclusions

Emphasis in the literature has been on the design of nature reserves, but we are usually too late to do anything except try to maintain the remnants left following fragmentation. Emphasis also has been on biogeographic explanations for the patterns of species loss after fragmentation, whereas a whole suite of physical and biotic parameters are significantly altered in the fragmented system and have significant impacts on remnant biota. In particular, the switch from predominantly internally driven to predominantly externally driven dynamics is a key factor in the fragmented system. Management and research should focus on this factor. There is a pressing need for an integrated approach that treats the landscape as a whole instead of as a collection of separate biotic and legal entities.

Acknowledgments

We are grateful to Francis Crome, Harry Recher, Michael Soulé, and two anonymous referees for their constructive comments on the manuscript.

Literature Cited

Andersen, A. N. 1987. Ant community organisation and environmental assessment. Pages 43–53 in J. D. Majer, editor. The role of invertebrates in conservation and biological survey. Department of Conservation and Land Management, Perth, Australia.

Andren, H., and P. Angelstam. 1988. Elevated predation rates as an edge effect in habitat islands: experimental evidence. Ecology 69:544–547.

Baudry, J. 1984. Effects of landscape structure on biological communities: the case of hedgerow network landscapes. Pages 55–65 in J. Brandt and P. Agger, editors. Proceedings of the first international seminar on methodology in landscape ecological research and planning. Volume 1. International Association for Landscape Ecology, Rosskilde University Center, Rosskilde, Denmark.

Bell, A. 1988. Trees, water and salt—a fine balance. Ecos 58: 2–8.

Boecklen, W. J., and C. W. Bell. 1987. Consequences of faunal collapse and genetic drift for the design of nature reserves. Pages 141–149 in D. A. Saunders, G. W. Arnold, A. A. Burbidge, and A. J. M. Hopkins, editors. Nature conservation: the role of remnants of native vegetation. Surrey Beatty and Sons, Chipping Norton, Australia.

Boecklen, W. J., and N. J. Gottelli. 1984. Island biogeographic theory and conservation practice: species-area or specious-area relationships? Biological Conservation 29:63–80.

Bond, W. J., J. Midgley, and J. Vlok. 1988. When is an island not an island? Insular effects and their causes in fynbos shrublands. Oecologia (Berlin) 77:515–521.

Bormann, F. H., and G. E. Likens. 1979. Pattern and process in a forested ecosystem. Springer, New York.

Bormann, F. H., G. E. Likens, R. S. Siccama, R. S. Pierce, and J. S. Eaton. 1974. The export of nutrients and recovery of stable conditions following deforestation at Hubbard Brook. Ecological Monographs 44:255–277.

Bradby, K. 1991. A data bank is never enough—the local approach to landcare. Pages 377–385 in D. A. Saunders and R. J. Hobbs, editors. Nature conservation: the role of corridors. Surrey Beatty and Sons, Chipping Norton, Australia. In press.

Brett, D. 1989. Sea birds in the trees. Ecos 61:4–8.

Bridgewater, P. B. 1987. Connectivity: an Australian perspective. Pages 195–200 in D. A. Saunders, G. W. Arnold, A. A. Burbidge, and A. J. M. Hopkins, editors. Nature conservation: the role of remnants of native vegetation. Surrey Beatty and Sons, Chipping Norton, Australia.

Bridgewater, P. B. 1990. The role of synthetic vegetation in present and future landscapes of Australia. Proceedings, Ecological Society of Australia 16:129–134.

Brittingham, M. C., and S. A. Temple. 1983. Have cowbirds caused forest songbirds to decline? BioScience 33:31–35.

Caborn, J. M. 1957. Shelterbelts and microclimate. Forestry Commission Bulletin 29. Her Majesty's Stationary Office, Edinburgh, Scotland.

Cale, P., and R. J. Hobbs. 1991. Condition of roadside vegetation in relation to nutrient status. Pages 353–362 in D. A. Saunders and R. J. Hobbs, editors. Nature conservation the role of corridors. Surrey Beatty and Sons, Chipping Norton, Australia.

Crawley, M. J. 1987. What makes a community invasible? Pages 429–453 in M. J. Crawley, P. J. Edwards, and A. J. Gray, editors. Colonization, succession and stability. Blackwell Scientific Publications, Oxford, England.

De Walle, D. R., and S. G. McGuire. 1973. Albedo variations of an oak forest in Pennsylvania. Agricultural Meteorology 11:107–113.

Diamond, J. M. 1975. The island dilemma: lessons of modern biogeographic studies for the design of nature reserves. Biological Conservation 7:129–146.

Diamond, J. M. 1976. Island biogeography and conservation: strategy and limitations. Science 193:1027–1029.

Ehrlich, A. H., and P. R. Ehrlich. 1987. Earth. Franklin Watts, New York.

Ewens, W. J., P. J. Brockwell, J. M. Gani, and S. I. Resnick. 1987. Minimum viable population size in the presence of catastrophes. Pages 59–68 in M. E. Soulé, editor. Viable populations for conservation. Cambridge University Press, Cambridge, England.

Fahrig, L., and G. Merriam. 1985. Habitat patch connectivity and population survival. Ecology 66:1762–1768.

Fitzgerald Biosphere Project. 1989. The bush comes to the city. Fitzgerald Biosphere Project, Perth, Australia.

Forman, R. T. T. 1987. The ethics of isolation, the spread of disturbance, and landscape ecology. Pages 213–229 in M. G. Turner, editor. Landscape heterogeneity and disturbance. Springer, New York.

Forman, R. T. T., and J. Baudry. 1984. Hedgerows and hedgerow networks in landscape ecology. Environmental Management 8:495–510.

Fox, M. D., and B. J. Fox. 1986. The susceptibility of natural communities to invasion. Pages 57–66 in R. H. Groves and J. J. Burdon, editors. Ecology of biological invasions: an Australian perspective. Australian Academy of Science, Canberra, Australia.

Fraser, A. I. 1972. The effect of climate factors on the development of plantation forest structure. Review Papers in Forest Meteorology 1972:109–125.

Game, M. 1980. Best shape for nature reserves. Nature 287:630–632.

Geiger, R. 1965. The climate near the ground. Harvard University Press, Cambridge, Massachusetts.

Gilpin, M. E., and J. M. Diamond. 1980. Subdivision of nature reserves and the maintenance of species diversity. Nature 285:567–568.

Gilpin, M. E., and M. E. Soulé. 1986. Minimum viable populations: processes of species extinctions. Pages 19–34 in M. E. Soulé, editor. Conservation biology. The science of scarcity and diversity. Sinauer Associates, Sunderland, Massachusetts.

Gordon, D. R., J. M. Welker, J. W. Menke, and K. J. Rice. 1989. Competition for soil water between annual plants and blue oak (*Quercus douglasii*) seedlings. Oecologia (Berlin) 79:533–541.

Grace, J. 1977. Plant response to wind. Academic Press, London, England.

Grace, J. 1983. Plant-atmosphere relationships. Chapman and Hall, London, England.

Greenslade, P. J. M., and R. B. Halliday. 1983. Colony dispersion and relationships of meat ants *Iridomyrmex purpureus* and allies in an arid locality in South Australia. Insectes Sociaux 30:82–99.

Gysel, L. W. 1951. Borders and openings of beech-maple woodlands in southern Michigan. Journal of Forestry 49:13–19.

Haila, Y., and I. Hanski. 1984. Methodology for studying the effect of fragmentation on land birds. Annals Zoologica Fennica 21:393–397.

Harris, L. D. 1984. The fragmented forest. Island biogeographic theory and the preservation of biotic diversity. University of Chicago Press, Chicago, Illinois.

Harris, L. D. 1985. Conservation corridors: a highway system for wildlife. ENFO Report 85-5. Environmental Information Center of the Florida Conservation Foundation, Winter Park, Florida.

Harris, L. D. 1988. Edge effects and conservation of biotic diversity. Conservation Biology 2:330–332.

Harris, L. D., and J. Scheck. 1991. From implications to applications: the dispersal corridor principle applied to conservation of biological diversity. Pages 189–220 in D. A. Saunders and R. J. Hobbs, editors. Nature conservation the role of corridors. Surrey Beatty and Sons, Chipping Norton, Australia.

Higgs, A. J. 1981. Island biogeographic theory and nature reserve design. Journal of Biogeography **8**:117–124.

Higgs, A. J., and M. B. Usher. 1980. Should nature reserves be large or small? Nature **285**:568.

Hobbs, R. J. 1987. Disturbance regimes in remnants of natural vegetation. Pages 233–240 in D. A. Saunders, G. W. Arnold, A. A. Burbidge, and A. J. M. Hopkins, editors. Nature conservation: the role of remnants of native vegetation. Surrey Beatty and Sons, Chipping Norton, Australia.

Hobbs, R. J. 1988. What is ecological theory and is it of any use to managers? Pages 15–27 in D. A. Saunders and A. A. Burbidge, editors. Ecological theory and biological management of ecosystems. CALM Occasional Paper 1/88. Department of Conservation and Land Management, Perth, Australia.

Hobbs, R. J. 1989. The nature and effects of disturbance relative to invasions. Pages 389–405 in J. A. Drake, H. A. Mooney, F. di Castri, et al., editors. Biological invasions. A global perspective. John Wiley & Sons, New York.

Hobbs, R. J., and L. Atkins. 1988. The effect of disturbance and nutrient addition on native and introduced annuals in the Western Australian wheatbelt. Australian Journal of Ecology **13**:171–179.

Hobbs, R. J., and A. J. M. Hopkins. 1990. From frontier to fragments: European impact on Australia's vegetation. Proceedings, Ecological Society of Australia **16**:93–114.

Hobbs, R. J., and D. A. Saunders. 1991. Reintegrating fragmented landscapes: a proposed framework for the Western Australian wheatbelt. Proc. V Aust. Soil Cons. Conf. In press.

Hobbs, R. J., D. A. Saunders, and B. M. T. Hussey. 1990. Nature conservation: the role of corridors. Ambio **19**:94–95.

Hopkins, A. J. M., and D. A. Saunders. 1987. Ecological studies as the basis for management. Pages 15–28 in D. A. Saunders, G. W. Arnold, A. A. Burbidge, and A. J. M. Hopkins, editors. Nature conservation: the role of remnants of native vegetation. Surrey Beatty and Sons, Chipping Norton, Australia.

Hornbeck, J. W. 1973. Storm flow from hardwood-forested and cleared watersheds in New Hampshire. Water Resources Research **9**:346–354.

Hudson, N. 1981. Soil conservation. Cornell University Press, Ithaca, New York.

Hutchinson, B. A., and D. R. Matt. 1976. Beam enrichment of diffuse radiation in a deciduous forest. Agricultural Meteorology **17**:93–110.

Hutchinson, B. A., and D. R. Matt. 1977. The distribution of solar radiation within a deciduous forest. Ecological Monographs **47**:185–207.

Hutchinson, J. N. 1980. The record of peat wastage in the East Anglian fenlands at Holme Post, 1848–1978 AD. Journal of Ecology **68**:229–49.

Janzen, D. H. 1983. No park is an island: increase in interference from outside as park size decreases. Oikos **41**:402–410.

Janzen, D. H. 1986. The eternal external threat. Pages 286–303 in M. E. Soulé, editor. Conservation biology. The science of scarcity and diversity. Sinauer Associates, Sunderland, Massachusetts.

Järvinen, O. 1982. Conservation of endangered plant populations: single large or several small reserves? Oikos **38**:301–7.

Jarvis, P. G., and K. G. McNaughton. 1986. Stomatal control of transpiration: scaling up from leaf to region. Advances in Ecological Research **15**:1–4.

Jones, H. G. 1983. Plants and microclimate. Cambridge University Press, Cambridge, England.

Jordan, W. R., III, M. E. Gilpin, and J. D. Aber. 1987. Restoration ecology: ecological restoration as a technique for basic research. Pages 3–21 in W. R. Jordan III, M. E. Gilpin, and J. D. Aber, editors. Restoration ecology: a synthetic approach to ecological research. Cambridge University Press, Cambridge, England.

Kapos, V. 1989. Effects of isolation on the water status of forest patches in the Brazilian Amazon. Journal of Tropical Ecology **5**:173–185.

Karieva, P. 1987. Habitat fragmentation and the stability of predator-prey interactions. Nature **326**:388–90.

Karr, J. R., and I. J. Schlosser. 1978. Water resources and the land-water interface. Science **201**:229–34.

Kendrick, G. W. 1976. The Avon: faunal and other notes on a dying river in south-western Australia. West Australian Naturalist **13**:97–114.

Kitchener, D. J., A. Chapman, J. Dell, B. G. Muir, and M. Palmer. 1980*a*. Lizard species assemblage and reserve size and structure in the Western Australian wheatbelt—some implications for conservation. Biological Conservation **17**:25–62.

Kitchener, D. J., A. Chapman, B. G. Muir, and M. Palmer. 1980*b*. The conservation value for mammals of reserves in the Western Australian wheatbelt. Biological Conservation **18**:179–207.

Kitchener, D. J., J. Dell, B. G. Muir, and M. Palmer. 1982. Birds in Western Australian wheatbelt reserves—implications for conservation. Biological Conservation **22**:127–163.

Klein, B. C. 1989. Effects of forest fragmentation on dung and carrion beetle communities in Central Amazonia. Ecology **70**:1715–1725.

Laws, R. M., I. S. C. Parker, and R. C. B. Johnstone. 1975. Elephants and their habitat. Clarendon Press, Oxford, England.

Le Houérou, H. N., and H. Gillet. 1986. Desertization in African arid lands. Pages 44–461 in M. E. Soulé, editor. Conservation biology. The science of scarcity and diversity. Sinauer Associates, Sunderland, Massachusetts.

Likens, G. E., F. H. Bormann, N. M. Johnson, D. W. Fisher, and R. S. Pierce. 1970. Effects of forest cutting and herbicide treatment on nutrient budgets in the Hubbard Brook watershed-ecosystem. Ecological Monographs **40**:23–47.

Lovejoy, T. E., R. O. Bierregaard, K. S. Brown, L. H. Emmons, and M. E. Van der Voort. 1984. Ecosystem decay of Amazon forest fragments. Pages 295–325 in M. H. Niteki, editor. Extinctions. University of Chicago Press, Chicago, Illinois.

Lovejoy, T. E., B. O. Bierregaard, Jr., A. B. Rylands, et al. 1986. Edge and other effects of isolation on Amazon forest fragments. Pages 257–285 in M. E. Soulé, editor. Conservation biology. The science of scarcity and diversity. Sinauer Associates, Sunderland, Massachusetts.

Lovejoy, T. E., and D. C. Oren. 1981. The minimum critical size of ecosystems. Pages 7–12 in R. L. Burgess and D. M. Sharpe, editors. Forest island dynamics in man-dominated landscapes. Springer, New York.

MacArthur, R. H., and E. O. Wilson. 1963. An equilibrium theory of insular biogeography. Evolution **17**:373–387.

MacArthur, R. H., and E. O. Wilson. 1967. The theory of island biogeography. Princeton University Press, Princeton, New Jersey.

MacClintock, L., R. F. Whitcomb, and B. L. Whitcomb. 1977. Island biogeography and the "habitat islands" of eastern forest II. Evidence for the value of corridors and minimization of isolation in preservation of biotic diversity. American Birds **31**:6–16.

Macdonald, I. A. W., L. L. Loope, M. B. Usher, and O. Hamann. 1989. Wildlife conservation and the invasion of nature reserves by introduced species: a global perspective. Pages 215–255 in J. A. Drake, H. A. Mooney, F. di Castri, et al., editors. Biological invasions. A global perspective. John Wiley & Sons, New York.

Macdonald, I. A. W., F. J. Powrie, and W. R. Siegfried. 1986. The differential invasion of southern Africa's biomes and ecosystems by alien plants and animals. Pages 209–225 in I. A. W. Macdonald, F. J. Kruger, and A. A. Ferrar, editors. The ecology and management of biological invasions in Southern Africa. Oxford University Press, Cape Town, South Africa.

McNab, J. 1983. Wildlife management as scientific experimentation. The Wildlife Society Bulletin **11**:397–401.

Mader, H.-J. 1984. Animal habitat isolation by roads and agricultural fields. Biological Conservation **29**:81–96.

Main, B. Y. 1987. Persistence of invertebrates in small areas: case studies of trapdoor spiders in Western Australia. Pages 29–39 in D. A. Saunders, G. W. Arnold, A. A. Burbidge, and A. J. M. Hopkins, editors. Nature conservation: the role of remnants of native vegetation. Surrey Beatty and Sons, Chipping Norton, Australia.

Margules, C. R. 1986. Conservation evaluation in practice. Pages 297–314 in M. B. Usher, editor. Wildlife conservation evaluation. Chapman and Hall, London, England.

Margules, C. R. 1987. The Wog Wog habitat patch experiment: Background, objectives, experimental design and sample strategy. Technical Memorandum 85/18. CSIRO, Division of Water and Land Resources, Canberra, Australia.

Margules, C. R., A. J. Higgs, and R. W. Rafe. 1982. Modern biogeographic theory: are there any lessons for nature reserve design? Biological Conservation **24**:115–128.

Margules, C. R., A. O. Nicholls, and R. L. Pressey. 1988. Selecting networks of reserves to maximise biological diversity. Biological Conservation **43**:63–76.

Margules, C. R., and J. L. Stein. 1989. Patterns and distributions of species and the selection of nature reserves: an example from *Eucalyptus* forests in south-eastern New South Wales. Biological Conservation **50**:219–238.

May, R. M. 1975. Island biogeography and the design of wildlife preserves. Nature **245**:177–178.

Merriam, G. 1984. Connectivity: a fundamental ecological characteristic of landscape pattern. Pages 5–15 in J. Brandt and P. Agger, editors. Proceedings of the First International Seminar on Methodology in Landscape Ecological Research and Planning. International Association for Landscape Ecology, Rosskilde University Center, Rosskilde, Denmark.

Merriam, G. 1991. Corridors for connectivity: animal populations in heterogeneous environments. Pages 133–142 D. A. Saunders and R. J. Hobbs, editors. Nature conservation: the role of corridors. Surrey Beatty and Sons, Chipping Norton, Australia.

Miller, R. I. 1978. Applying island biogeographic theory to an East African reserve. Environmental Conservation **5**:191–195.

Miller, R. I., and L. D. Harris. 1977. Isolation and extirpations in wildlife reserves. Biological Conservation **12**:311–315.

Milthorpe, F. L., and J. Moorby. 1974. An introduction to crop physiology. Cambridge University Press, Cambridge, England.

Moen, A. N. 1974. Turbulence and the visualization of wind flow. Ecology **55**:1420–1424.

Monteith, J. L. 1975. Vegetation and the atmosphere. Volume 1. Principles. Academic Press, London, England.

Muir, B. G. 1979. Observations of wind-blown superphosphate in native vegetation. West Australian Naturalist **14**:128–130.

Murphy, D. D. 1989. Conservation and confusion: wrong species, wrong scale, wrong conclusions. Conservation Biology **3**:82–84.

Myers, N. 1988. Tropical forests and their species: going, going . . . ? Pages 28–35 in E. O. Wilson and F. M. Peter, editors. Biodiversity. National Academy Press, Washington, D.C.

Noss, R. F. 1987. Corridors in real landscapes: a reply to Simberloff and Cox. Conservation Biology **1**:159–164.

Noss, R. F., and L. D. Harris. 1986. Nodes, networks and MUMs: preserving diversity at all scales. Environmental Management **10**:299–309.

Nulsen, R. A., K. J. Bligh, I. N. Baxter, E. J. Solin, and D. H. Imrie. 1986. The fate of rainfall in a mallee and heath vegetated catchment in southern Western Australia. Australian Journal of Ecology **11**:361–371.

Palik, B. J., and P. G. Murphy. 1990. Disturbance versus edge effects in sugar-maple/beech forest fragments. Forest Ecology and Management **32**:187–202.

Panetta, D., and A. J. M. Hopkins. 1991. Weeds in corridors: invasion and management. Pages 341–351 in D. A. Saunders and R. J. Hobbs, editors. Nature conservation the role of corridors. Surrey Beatty and Sons, Chipping Norton, Australia.

Parker, C. A. 1989. Soil biota and plants in the rehabilitation of degraded agricultural soils. Pages 423–438 in J. D. Majer, editor. Animals in primary succession. The role of fauna in reclaimed lands. Cambridge University Press, Cambridge, England.

Peck, A. J. 1978. Salinisation of non-irrigated soils and associated streams: a review. Australian Journal of Soil Research **16**:157–168.

Peck, A. J., and D. R. Williamson. 1987. Effects of forest clearing on groundwater. Journal of Hydrology **94**:47–66.

Pickett, S. T. A., and J. N. Thompson. 1978. Patch dynamics and the size of nature reserves. Biological Conservation **13**:27–37.

Pimm, S. A. 1986. Community stability and structure. Pages 309–329 in M. E. Soulé, editor. Conservation biology. The science of scarcity and diversity. Sinauer Associates, Sunderland, Massachusetts.

Pimm, S. A., H. L. Jones, and J. Diamond. 1988. On the risk of extinction. American Naturalist **132**:757–785.

Quinn, J. F., and A. Hastings. 1987. Extinction in subdivided habitats. Conservation Biology **1**:198–208.

Ranney, J. W., M. C. Bruner, and J. B. Levenson. 1981. The importance of edge in the structure and dynamics of forest islands. Pages 67–95 in R. L. Burgess and D. M. Sharpe, editors. Forest island dynamics in man-dominated landscapes. Springer, New York.

Rejmánek, M. 1989. Invasibility of plant communities. Pages 369–388 in J. A. Drake, H. A. Mooney, F. di Castri, et al. editors. Biological invasions. A global perspective. John Wiley & Sons, New York.

Reville, B. J., J. D. Tranter, and H. D. Yorkston. 1990. Impact of forest clearing on the endangered seabird, *Sula abbotti*. Biological Conservation **51**:23–38.

Rowell, T. A. 1986. The history of drainage at Wicken Fen, Cambridgeshire, England, and its relevance to conservation. Biological Conservation **35**:111–142.

Saunders, D. A. 1982. The breeding behaviour and biology of the short-billed form of the White-tailed Black Cockatoo *Calyptorhynchus funereus*. Ibis **124**:422–455.

Saunders, D. A. 1989. Changes in the avifauna of a region, district and remnant as a result of fragmentation of native vegetation: the wheatbelt of Western Australia. A case study. Biological Conservation **50**:99–135.

Saunders, D. A. 1990. Problems of survival in an extensively cultivated landscape: the case of Carnaby's Cockatoo *Calyptorhynchus funereus latirostris*. Biological Conservation **54**:111–124.

Saunders, D. A., G. W. Arnold, A. A. Burbidge, and A. J. M. Hopkins. 1987*a*. The role of remnants of native vegetation in nature conservation: future directions. Pages 387–392 in D. A. Saunders, G. W. Arnold, A. A. Burbidge, and A. J. M. Hopkins, editors. Nature conservation: the role of remnants of native vegetation. Surrey Beatty and Sons, Chipping Norton, Australia.

Saunders, D. A., G. W. Arnold, A. A. Burbidge, and A. J. M. Hopkins, editors. 1987*b*. Nature conservation: the role of remnants of native vegetation. Surrey Beatty and Sons, Chipping Norton, Australia.

Saunders, D., and R. Hobbs. 1989. Corridors for conservation. New Scientist **1642**:63–68.

Saunders, D. A., and R. J. Hobbs, editors. 1991*a*. Nature conservation: the role of corridors. Surrey Beatty and Sons, Chipping Norton, Australia. In press.

Saunders, D. A., and R. J. Hobbs. 1991*b*. The role of corridors in nature conservation: what do we know and where do we go? Pages 421–427 in D. A. Saunders and R. J. Hobbs, editors. Nature conservation: the role of corridors. Surrey Beatty and Sons, Chipping Norton, Australia.

Saunders, D. A., A. J. M. Hopkins, and R. A. How, editors. 1990. Australian ecosystems: 200 years of utilisation, degradation and reconstruction. Proceedings of the Ecological Society of Australia **16.**

Saunders, D. A., and J. A. Ingram. 1987. Factors affecting survival of breeding populations of Carnaby's cockatoo *Calyptorhynchus funereus latirostris* in remnants of native vegetation. Pages 249–258 in D. A. Saunders, G. W. Arnold, A. A. Burbidge, and A. J. M. Hopkins, editors. Nature conservation: the role of remnants of native vegetation. Surrey Beatty and Sons, Chipping Norton, Australia.

Saunders, D. A., and C. P. de Rebeira. 1991. Values of corridors to avian populations in a fragmented landscape. Pages 221–244 in D. A. Saunders and R. J. Hobbs, editors. Nature conservation the Role of Corridors. Surrey Beatty and Sons, Chipping Norton, Australia.

Saunders, D. A., I. Rowley, and G. T. Smith. 1985. The effects of clearing for agriculture on the distribution of cockatoos in the south west of Western Australia. Pages 309–321 in A. Keast, H. F. Recher, H. Ford, and D. Saunders, editors. Birds of eucalypt forests and woodlands: ecology, conservation, management. Surrey Beatty and Sons, Chipping Norton, Australia.

Shaffer, M. 1987. Minimum viable populations: coping with uncertainty. Pages 69–86 in M. E. Soulé, editor. Viable populations for conservation. Cambridge University Press, Cambridge, England.

Sharma, M. L., R. J. W. Barron, and D. R. Williamson. 1987. Soil water dynamics of lateritic catchments as affected by forest clearing for pasture. Journal of Hydrology **94**:109–127.

Simberloff, D. 1986. Design of nature reserves. Pages 315–37 in M. B. Usher, editor. Wildlife conservation evaluation. Chapman and Hall, London, England.

Simberloff, D. 1988. The contribution of population and community biology to conservation science. Annual Review of Ecology and Systematics **19**:473–511.

Simberloff, D. S., and L. G. Abele. 1976*a*. Island biogeography and conservation: strategy and limitations. Science **193**:1032.

Simberloff, D. S., and L. G. Abele. 1976*b*. Island biogeography theory and conservation practice. Science **191**:285–286.

Simberloff, D., and L. G. Abele. 1982. Refuge design and island biogeographic theory: effects of fragmentation. American Naturalist **120**:41–50.

Simberloff, D., and L. G. Abele. 1984. Conservation and obfuscations: subdivision of reserves. Oikos **42**:399–401.

Simberloff, D., and J. Cox. 1987. Consequences and costs of conservation corridors. Conservation Biology **1**:63–71.

Simons, P. 1989. Nobody loves a canal with no water. New Scientist **1685**:30–34.

Soulé, M. E. 1987*a*. Introduction. Pages 1–10 in M. E. Soulé, editor. Viable populations for conservation. Cambridge University Press, Cambridge, England.

Soulé, M. E., editor. 1987*b*. Viable populations for conservation. Cambridge University Press, Cambridge, England.

Soulé, M. E., D. T. Bolger, A. C. Alberts, J. Wright, M. Sorice, and S. Hill. 1988. Reconstructed dynamics of rapid extinctions of chaparral-requiring birds in urban habitat islands. Conservation Biology **2**:75–92.

Soulé, M. E., and K. A. Kohm. 1989. Research priorities for conservation biology. Island Press, Washington, D.C.

Spinage, C. A., and F. E. Guiness. 1971. Tree survival in the absence of elephants in the Akagera National Park, Rwanda. Journal of Applied Ecology **8**:723–728.

Swanson, F. J., T. K. Kratz, N. Caine, and R. G. Woodmansee. 1988. Landform effects on ecosystem patterns and processes. BioScience **38**:92–98.

Terborgh, J. 1976. Island biogeography and conservation: strategy and limitations. Science **193**:1029–1030.

Tranquillini, W. 1979. Physiological ecology of the alpine timberline; tree existence at high altitudes with special reference to the European Alps. Springer, New York.

Trimble, G. R., and E. H. Tryon. 1966. Crown encroachment into openings cut in Appalachian hardwood stands. Journal of Forestry **64**:104–108.

Usher, M. B. 1987. Effects of fragmentation on communities and populations: a review with applications to wildlife conservation. Pages 103–121 in D. A. Saunders, G. W. Arnold, A. A. Burbidge, and A. J. M. Hopkins, editors. Nature conservation: the role of remnants of native vegetation. Surrey Beatty and Sons, Chipping Norton, Australia.

Usher, M. B. 1988. Biological invasions of nature reserves: a search for generalisations. Biological Conservation **44**:119–135.

Vepsäläilnen, K., and B. Pisarski. 1982. Assembly of island ant communities. Annals Zoologica Fennici **19**:327–335.

Verner, J. 1986. Predicting effects of habitat patchiness and fragmentation—the researcher's viewpoint. Pages 327–329 in J. Verner, M. L. Morrison, and C. J. Ralph, editors. Wildlife 2000: modeling habitat relationships of terrestrial vertebrates. University of Wisconsin Press, Madison, Wisconsin.

Wales, B. A. 1972. Vegetation analysis of north and south edges in a mature oak-hickory forest. Ecological Monographs **42**:451–471.

Walker, B. H. 1981. Is succession a viable concept in African savannah ecosystems? Pages 431–447 in D. C. West, H. H. Shugart, and D. B. Botkin, editors. Forest succession: concepts and application. Springer, New York.

Webb, N. R. 1989. Studies on the invertebrate fauna of fragmented heathland in Dorset, U.K., and the implications for conservation. Biological Conservation **47**:153–165.

Wegner, J. F., and G. Merriam. 1979. Movements by birds and small mammals between a wood and adjoining farmland habitats. Journal of Applied Ecology **16**:349–357.

Whitcomb, R. F., J. F. Lynch, P. A. Opler, and C. S. Chandler. 1976. Island biogeography and conservation: strategy and limitation. Science **193**:1030–1032.

Wilcove, D. S. 1985. Nest predation in forest tracts and the decline of migratory songbirds. Ecology **66**:1211–1214.

Wilcove, D. S., C. H. McLellan, and A. P. Dobson. 1986. Habitat fragmentation in the temperate zone. Pages 273–256 in M. E. Soulé, editor. Conservation biology. The science of scarcity and diversity. Sinauer Associates, Sunderland, Massachusetts.

Wilcox, B. A. 1980. Insular ecology and conservation. Pages 95–117 in M. E. Soulé and B. A. Wilcox, editors. Conservation Biology an Evolutionary-Ecological Perspective. Sinauer, Sunderland, Massachusetts.

Williams, K., and R. J. Hobbs. 1989. Control of shrub establishment by springtime soil water availability in an annual grassland. Oecologia (Berlin) **81**:62–66.

Williamson, D. R., R. A. Stokes, and J. K. Ruprecht. 1987. Response of input and output of water and chloride to clearing for agriculture. Journal of Hydrology **94**:1–28.

Williamson, M. 1975. The design of nature preserves. Nature (London) **256**:519.

Willis, E. O. 1984. Conservation, subdivision of reserves and the anti-dismemberment hypothesis. Oikos **42**:396–398.

Wilson, E. O., and E. O. Willis. 1975. Applied biogeography. Pages 522–34 in M. L. Cody and J. M. Diamond, editors. Ecology and evolution of communities. Belknap Press, Cambridge, Massachusetts.

Wycherly, P. 1984. People, fire and weeds: can the vicious spiral be broken? Pages 11–17 in S. A. Moore, editor. The management of small bush areas in the Perth metropolitan region. Department of Fisheries and Wildlife, Perth, Australia.

Yahner, R. H. 1988. Changes in wildlife communities near edges. Conservation Biology **2**:333–339.

Zimmerman, B. L., and R. O. Bierragaard. 1986. Relevance of the equilibrium theory of island biogeography and species area relations to conservation with a case from Amazonia. Journal of Biogeography **13**:133–143.

Conservation of Fragmented Populations

LENORE FAHRIG
GRAY MERRIAM
Ottawa-Carleton Institute of Biology
Department of Biology
Carleton University
Ottawa, Canada K1S 5B6

Abstract: *In this paper we argue that landscape spatial structure is of central importance in understanding the effects of fragmentation on population survival. Landscape spatial structure is the spatial relationships among habitat patches and the matrix in which they are embedded. Many general models of subdivided populations make the assumptions that (1) all habitat patches are equivalent in size and quality and (2) all local populations (in the patches) are equally accessible by dispersers. Models that gloss over spatial details of landscape structure can be useful for theoretical developments but will almost always be misleading when applied to real-world conservation problems. We show that local extinctions of fragmented populations are common. From this it follows that recolonization of local extinctions is critical for regional survival of fragmented populations. The probability of recolonization depends on (1) spatial relationships among landscape elements used by the population, including habitat patches for breeding and elements of the inter-patch matrix through which dispersers move, (2) dispersal characteristics of the organism of interest, and (3) temporal changes in the landscape structure. For endangered species, which are typically restricted in their dispersal range and in the kinds of habitat through which they can disperse, these factors are of primary importance and must be explicitly considered in management decisions.*

Conservación de poblaciones fragmentadas

Resumen: *En este trabajo nosotros argumentamos que la estructura espacial del paisaje es de central importancia para la comprensión de los efectos de la fragmentación sobre la superviencia de las poblaciones. La estructura espacial del paisaje consiste en la relación espacial entre parches de hábitas y la matriz en la cual están incluidos. Muchos modelos de poblaciones subdivididas asumen que (1) todos los parches de hábitat son equivalentes en tamaño y calidad y (2) todas las poblaciones locales (en los parches) son igualmente accesibles a los dispersores. Modelos que trivializan los detalles espaciales de la estructura del paisaje pueden ser útiles par desarrollos teóricos pero casi siempre serán engañosos cuando se los aplique a problemas reales de conservación. Nosotros demostramos que las extinciones locales de poblaciones fragmentadas son comunes. De esto se deduce que la recolonización de extinciones locales es crítica para la supervivencia regional de las poblaciones fragmentadas. La probabilidad de recolonización depende de (1) relaciones espaciales entre los elementos del paisaje usados por las poblaciones, incluyendo parches de hábitats para cría y elementos de la matriz inter-parches a través de los cuales los dispersores se movilizan, (2) características de dispersión del organismo en cuestión y (3) cambios temporales en la estructura del paisaje. Estos factores son de primordial importancia y deben ser considerados explícitamente en las decisiones de manejo para especies en peligro de extinción, las cuales están tipicamente restringidas en su rango de dispersión y en los tipos de hábitats a través de los cuales se dispesan.*

Paper submitted August 31, 1992; revised manuscript accepted January 29, 1993.

Introduction

The habitat of all species is heterogeneous on many scales due to both natural processes and human activities (Lord & Norton 1990). This results in heterogeneous distributions of populations at different spatial scales (Wiens 1989). Andrewartha and Birch (1984) simplify this spatial structuring to two relevant scales for understanding the population dynamics of any species: the "local population" scale (small scale) and the "natural population" scale (large or regional scale). The local population is the unit within which the classical assumption of random mating holds (see Wright 1977). Because of disturbances and demographic variability, the local population is often prone to extinction. For many populations, survival at the larger regional scale depends on recolonization of these local extinctions from other areas through dispersal (Addicott 1978; Henderson et al. 1985; Harrison et al. 1988; Paine 1988; Wegner & Merriam 1990; Merriam & Wegner 1992; Villard et al. 1992). The natural or regional population is "the sum of a large number of interacting local populations." The regional population is persistent even though the local populations are ephemeral. Andrewartha and Birch (1984:184) assert the general importance of this spatial structuring of populations in their statement that "No general theory about the distribution and abundance of animals should have a chance of being accepted as realistic unless it takes full cognizance of the patchy dispersion of animals in natural populations." Note that the scales of local and regional populations depend on the size and dispersal capability of the organism. The regional population of a small immobile organism may cover a smaller area than the local population of a large mobile organism (Kotliar & Wiens 1990).

Several general theoretical constructs have been proposed for subdivided populations. Den Boer (1977, 1979) envisioned the local population or "interaction group" from which offspring are spread to other asynchronously fluctuating interaction groups, forming a "multipartite population." Levins (1970) coined the term "metapopulation" for such a "population of populations." Wilson (1980) proposed the "winking patches" model for the extinction and recolonization dynamic of local populations. Hastings and Wolin (1989) and Hastings (1991) developed a model in which patches are grouped in "classes" defined by their population sizes. Pulliam (1988) developed the "source-sink" model of interaction among local populations, in which some patch populations act as sources of colonists for other patches that depend on these sources for persistence.

In "core-satellite" or "island-mainland" models (Boorman & Levitt 1973), precipitated by the work of MacArthur and Wilson (1967), one patch, the mainland, acts as the ultimate source of immigrants for the local island populations. Although the spatial structure of some patchy populations conforms to the island-mainland analogue, we focus here on the increasingly common spatial structure in which there is no mainland equivalent. Because any patch can experience an extinction, and because recolonization is not assured by the presence of a large mainland source (see Karr 1990), spatial configuration of the habitat patches is particularly important in this situation.

Models of subdivided populations have been useful for development of theoretical insights, but their application to real problems of conservation biology is dangerous for the following reasons: (1) many of the models, especially the metapopulation and winking patches models assume that all patches are equivalent in size and quality; and (2) the models frequently assume that all local populations are equally accessible by dispersers—in other words, the models are not spatially explicit. There are some exceptions (for example, Harrison & Quinn 1989), which are reviewed in Kareiva (1990). In a recent paper, Adler and Nürnberger (in press) conducted a spatially explicit simulation of patchy populations and compared the regional population size to that predicted by an approximation that ignores spatial structure. They concluded that the approximation failed to predict the spatially explicit simulation results.

In this paper we present arguments and evidence for the view that, if the goal is species conservation, the spatial structure of the landscape in which the species is found must be explicitly considered. The following factors must be examined: (1) differences among the patch populations in terms of habitat area and quality; (2) spatial relationships among landscape elements; (3) dispersal characteristics of the organism of interest, and (4) temporal changes in the landscape structure. Parallel issues arise when one investigates genetic questions, but in this paper ecological concerns will be the focus.

Static Landscape Structure

Persistent Populations Extend across Several Habitat Patches

A habitat patch is defined as any discrete area that is used by a species for breeding or for obtaining other resources. Note that continuous habitat, such as forest, is internally heterogeneous (Freemark & Merriam 1986). For a particular species, extensive habitat may actually contain several habitat patches with functionally separate, local populations (Krohne et al. 1984; Tomialojc et al. 1984; Wilcove 1988; Krohne & Burgin 1990).

Fluctuations in local abundance mean that there is always some possibility that the local population in a habitat patch will become extinct. For regional survival, it is important that the fluctuations of the local populations are not synchronous, since this would result in a high chance of simultaneous extinction of all local populations (Gilpin 1990; Hanski 1991). The number of habitat patches and the spatial scale over which they are distributed must be large enough for recolonizations to balance local extinctions.

In Figure 1 we present recorded annual extinction rates of local populations for a variety of organisms in habitat patches. Since scales of observation vary from study to study, the extinction and recolonization rates in these studies should not be directly compared. They do serve to demonstrate, however, that local extinctions and recolonizations are common, often in the range of 10–20% of local populations becoming extinct per year. Other authors have reported frequent local extinctions without presenting annual rates. For example, in a study of four species of aphids on fireweeds, Addicott (1978) found that local populations became extinct and colonizations occurred throughout the summer. Henderson et al. (1985) found that local extinctions of chipmunks (*Tamias striatus*) were recolonized within three to nine weeks. Many measures of turnover rates also imply high rates of local extinctions and colonizations (see Williamson 1983; Schoener & Spiller 1987).

Landscape Spatial Structure Constrains Regional Survival

Because local extinctions occur, population survival (both locally and over the landscape) depends on recolonization of habitat patches that have experienced extinctions. Whether or not patches can be recolonized depends on the availability of dispersing individuals and the ease with which these individuals can move about within the landscape. Both of these depend on landscape spatial structure. If the landscape spatial structure restricts movement between patches, the area (number of habitat patches) required for population survival is large. For example, Duelli (1990) studied movements of 97 species of arthropods across the edges of agricultural fields. Depending on the species, the edge could restrict movement, enhance movement, or have no effect on movement. Also, Mader et al. (1990) found that movements of arthropods are restricted by some kinds of linear barriers (tracks and roads) but not by others. If movement is enhanced by landscape spatial structure, only a small number of patches is needed for regional survival.

Landscape spatial structure is defined as the spatial relationships among the landscape components. Landscape components include both habitat patches and components of the matrix in which the patches are imbedded. They are areas in the landscape that are defined functionally for a particular species by the way in which they are used by that organism. Examples are patches of breeding habitat or feeding habitat, areas of inhospitable habitat, and areas that can be used as part or all of a dispersal route. The characteristics of landscape spatial structure are:

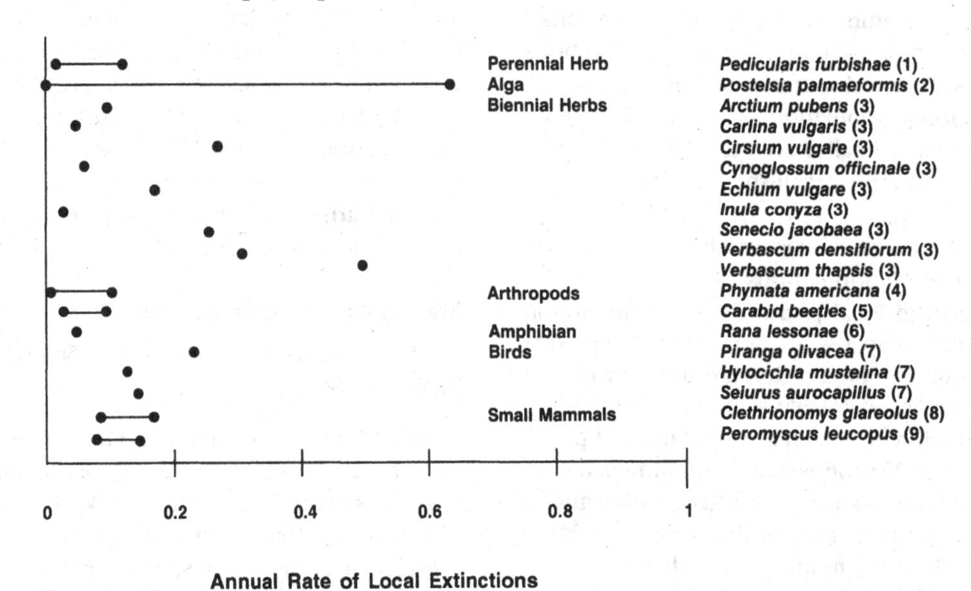

Figure 1. Rate of local extinctions per year (the fraction of local populations becoming extinct per year) for different organisms. Data from true islands are not included. Numbers to the right of the organism correspond to the following habitats and sources: (1) river bank (Menges 1990), (2) rocky intertidal (Paine 1988), (3) coastal sand-dunes (van der Meijden et al. 1985), (4) goldenrod patches (Mason 1977), (5) grassy sites (den Boer 1985), (6) ponds (Sjögren 1991), (7) forest fragments (Villard et al. 1992), (8) forest fragments (R. van Apeldoorn, Research Institute Nature Management, The Netherlands [personal communication]), and (9) forest fragments (Merriam & Wegner 1992).

(1) SIZE, SHAPE, AND QUALITY OF PATCHES

The size of a patch has been shown to influence local population persistence. Verboom et al. (1991) have shown a relationship between habitat patch size and persistence of local populations of the European nuthatch (*Sitta europea*). Also, the positive relationships found between patch size and local population size (see Lynch & Whigam 1984) and between local population size and population persistence (see Paine 1988; Berger 1990) imply a positive relationship between patch size and population persistence.

One way that patch shape can influence population survival is if the population dynamics change in relation to the distance from the edge of the patch. For example, Gates and Gysel (1978) found that the number of passerine birds increased at the field-forest edge, but this was accompanied by increased rates of predation and nest parasitism. Ambuel and Temple (1983) suggest that forest-edge and farmland bird species exclude certain forest-dwelling species, and that this exclusion has a greater impact than changes in patch area or isolation. Two patches of the same area but with different amounts of edge may therefore have different population dynamics.

Patch quality can also influence the survival and abundance of populations. For example, Saunders et al. (1985) and Saunders and Ingram (1987) have shown that the availability of trees with nesting holes within range of food resources can influence the population survival of cockatoos. Dingle (1991) showed that 36% of the variability among patch populations of the large milkweed bug was due to patch quality.

(2) THE PRESENCE OF DISPERSAL ROUTES THROUGH THE LANDSCAPE

Dispersal routes are comprised of components of the matrix through which the organism can move. A habitat patch can also form part of a dispersal route between two other patches. White-footed mice in a landscape of woodlots and farmland move mainly through fencerows, as shown in a radiotagging study of Merriam and Lanoue (1990). Fahrig and Merriam (1985) showed that the function of these fencerows as dispersal routes is important for regional population abundance of white-footed mice in this landscape. A shrub strip between a forest and a littoral zone was shown to act as a dispersal route for birds (Dmowski & Kozakiewicz 1990). Bennett (1990) studied small mammal populations in a fragmented landscape in Australia. Survival of regional populations was found to be facilitated by movements between remnant forest patches along vegetated dispersal routes. The presence of red squirrel (*Sciuris vulgaris*) in wooded fragments was shown to be positively related to the amount of hedgerow surrounding the fragments (Verboom & van Apeldoorn 1990). Note that "corridors" connecting habitat patches can either enhance or reduce regional population survival (Hobbs 1992; Simberloff et al. 1992).

(3) THE QUALITY OF DISPERSAL ROUTES

Quality can affect the likelihood of dispersers using the route and/or the probability that dispersers using the route will survive. Henein and Merriam (1990) showed in a simulation study that the quality of dispersal routes can be important for the abundance of the regional population. If an additional habitat patch is added to a region but is connected to other patches by a dispersal route of low quality (in terms of the survival probability of dispersers), then adding the patch will create a sink and lower the regional population abundance. Quality of dispersal routes can refer to more than just quality for movement. Availability of dispersal routes with qualities permitting both breeding and movement between patches is also important to population survival (Bennett 1987).

At the extreme of low quality, an element of the matrix having zero quality for movement is a movement barrier. For example, Duelli (1990) showed that edges of agricultural fields can be barriers to movement through the landscape by some farmland arthropods. Mader et al. (1990) showed that minor roads stopped some forest arthropods. Merriam et al. (1989) demonstrated the differential effects of roads and cropland as barriers to *Peromyscus leucopus*.

(4) THE SPATIAL CONFIGURATION OF THE COMPONENTS OF THE LANDSCAPE

Landscape components include both habitat patches and the matrix in which the patches are imbedded. Lefkovitch and Fahrig (1985) showed in a simulation study that, for population survival and abundance, the total number of dispersal routes in the landscape may be less important than their configuration relative to the habitat patches. In particular, they found that it is the overall shape and size of the geometric figure formed by interconnected patches that is most important; large, closed figures produce the highest probability of regional persistence. In a model of a one-dimensional patchy population in which one patch was different from the others in terms of population demography and/or dispersal, Seno (1988) showed that the spatial location of this patch affected the regional population dynamics: the more central the patch, the greater its impact. The degree of clumping of substrate patches was shown in a modeling study to affect persistence of moss populations (Herben et al. 1991). The positive effect of clumping on population persistence has been shown in a general model by Adler and Nürnberger (in press). The effect of spatial configuration on regional population survival also depends on the spatial configuration of the high- versus the low-quality patches and connections.

Many field studies have demonstrated the importance of the spatial configuration of the landscape, particularly the importance of patch isolation on local population abundance and persistence. Carnaby's Cockatoo is more likely to persist in agricultural landscapes that have nest sites linked to feeding areas by well-vegetated strips (Saunders & Ingram 1987). Lawton and Woodroffe (1991) showed that breeding water voles were less likely to be present in isolated sites. The large milkweed bug is less likely to be found in host patches that are far from major rivers, possibly due to the fact that the bugs follow water courses during migration (Dingle 1991). Potter (1990) found that degree of isolation of forest remnants affected the probability of use by the Brown Kiwi in New Zealand. If large remnants are interspersed with small ones, the kiwis can move between the large remnants by using the small ones as "stepping stones." Similar use of stepping stones by forest carabid beetles in Brittany farmland was reported by Burel (1989).

Effect of Landscape Spatial Structure on Regional Survival is Determined by Dispersal and Its Underlying Behaviour

Even if the spatial structure of the landscape is static, local population abundance fluctuates, which can lead to local extinctions. Recolonization of locally extinct patch populations depends on dispersal. If the landscape structure restricts dispersal, extinctions will cover larger areas and these areas will remain extinct for longer. Therefore, the population must cover a larger area, including more patches, to ensure regional survival (Merriam 1984; Hansson 1991). Our use of the term dispersal refers to a change in location of organisms, accompanied by reproduction. Reproduction can occur after the move to the new area, or the organisms may reproduce in the original area and their offspring disperse to a new one (Lidicker 1975).

Fahrig (1990) conducted a simulation study in which local populations in patches could become extinct with a given probability. The dispersal rate resulting in the highest probability of regional survival depends on the probability of local extinction. Populations that typically experience high rates of local extinction are expected to have relatively high dispersal rates, while those that typically experience low rates of local extinction are expected to have low dispersal rates. In the former case, the positive effect of recolonization on regional survival outweighs the negative effect of dispersal mortality. In the latter case, where recolonization is less critical for regional survival, the negative effect of dispersal mortality outweighs the positive effect of recolonization. It must be noted, however, that this result depends on the assumption that dispersers experience higher mortality rates than nondispersers; evidence for this is equivocal at best (see, for example, Dhont 1979; Greenwood et al. 1979; Johnson & Gaines 1987; Merriam & Lanoue 1990).

The components of dispersal that are important for population response to the constraint of landscape structure are as follows:

(1) Dispersal probability, or the probability of individuals leaving patches per unit time.
(2) Dispersal distance, or the probability of individuals successfully reaching a range of distances or landscape components. If the dispersers do not travel far enough to move between habitat patches, they will not be able to recolonize local extinctions.
(3) Temporal pattern of dispersal. For example, dispersal may be seasonal. If local extinctions are more likely to occur at some times of the year than others, then the timing of movement that results in recolonization will affect its demographic impact.
(4) Dispersal behavior that improves the probability of dispersers finding habitat patches and/or that decreases the risk of disperser mortality. For example, some herbivorous insects, including bark beetles (McMullen & Atkins 1962), desert locusts (Wallace 1958), and cabbage root fly (Prokopy et al. 1983), locate and orient toward host-plant patches from a distance.

Fahrig and Paloheimo (1988*a*) conducted a simulation experiment in which they examined the relative importance of various components of demography and dispersal on regional abundance in patchy environments. In general, dispersal was shown to be more important than demographics (such as birth rate) in determining regional population abundance. This result is supported by a study of small rodent populations in which immigration was found to be more important than local demographics in affecting local population persistence (Blaustein 1981). Fahrig and Paloheimo (1988*a*) also found that the most important determinant of regional population abundance was the probability that dispersers successfully locate new patches. The exact spatial pattern of habitat patches was found to be most important when dispersal distances are small relative to the distance between patches.

It is important to note that the spatial pattern of interconnectance among habitat patches can not be described in isolation from the nature of the landscape matrix or the dispersal behavior of the particular species under study. For example, Fahrig and Paloheimo (1987, 1988*b*) studied the effect of the spatial pattern of habitat patches (patches of cabbages) on the local abundance of the cabbage butterfly. In this case, patches that are very close together (less than about 100 m) are in fact more isolated from each other for cabbage butterfly dispersal than are patches somewhat farther apart (about 200–400 m) because of the dispersal behavior of the butterflies.

Dynamic Landscape Structure

Landscape spatial structure is not constant but changes over time. The higher the rate of landscape change, the lower the probability of regional population survival.

Habitats are clearly variable over time on the scale of the local population (seasonal changes, for example). However, changes also occur at larger (regional or landscape) scales. These we refer to as changes in landscape spatial structure; they normally occur on longer time scales than changes at the local scale (Urban et al. 1987). Examples of changes in landscape spatial structure include (1) fragmentation and/or habitat removal, (2) increase or decrease in the number and/or quality of dispersal routes, and (3) spatial reorganization in which the proportions of various patch types remain constant but the patches change location. These changes can be gradual or abrupt, natural or anthropogenic. If the landscape structure changes, the constraint imposed by the landscape structure on regional survival also changes.

For example, Chew (1981) attributes local extinction of the butterfly *Pieris oleracea* near Boston to changes in landscape structure resulting in reduction of the host-plant distribution. Saunders and Ingram (1987) and Saunders (1990) found that in southwest Western Australia the rate of change of the landscape due to clearing of natural vegetation was so high that local extinctions of Carnaby's Cockatoo (*Calyptorhyncus funereus latirostris*) accumulated into regional extinctions. Decreasing population survival was associated with decreased landscape connectivity. Den Boer (1990) showed that habitat change due to cultivation in the Netherlands has meant that carabid beetle species with low powers of dispersal can no longer compensate for local population extinctions by recolonization.

Fahrig (1992) showed in a simulation study that in a dynamic landscape the rate of change in landscape structure is more important than the degree of patch isolation in determining population survival and abundance. If the changes in landscape structure occur at an "unnaturally" high rate (as in some anthropogenic changes), dispersal may not be able to keep up with the high rates of local extinction. In this case the regional population will become extinct. Figure 2 shows simulation results from Fahrig's (1992) model for the regional sustainable (long-term average) population size of a hypothetical forest-floor plant having a generation time of

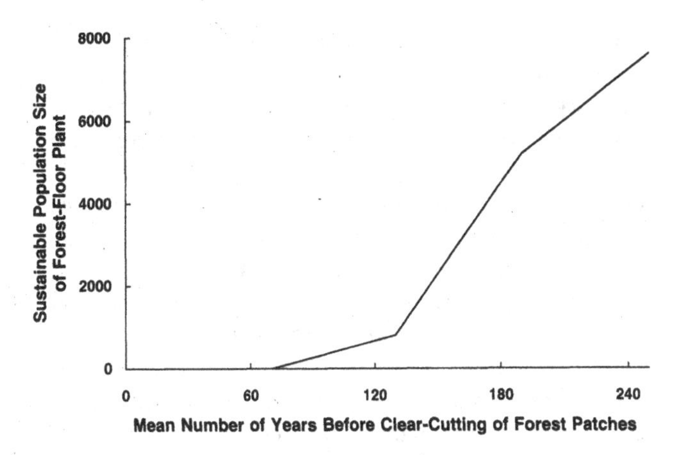

Figure 2. Simulation results from the model described in Fahrig (1992) for the regional sustainable (long-term average) population size of a hypothetical forest-floor plant species in a fragmented landscape with 10% cover by forest patches, versus mean number of years before clear-cutting of patches. Note that new patches of forest grow in so that the landscape is always comprised of 10% forest; the amount of habitat available to the forest-floor plant is therefore constant among and within simulations over time.

10 years. Ten percent of the landscape is assumed to be comprised of forest fragments; these fragments are the habitat patches for the forest-floor plant. When a forest fragment is clear-cut, the local population of the forest-floor plant dies out. Cutting of different fragments is not synchronous but is spread out approximately evenly over some time period, so that all of the forest present at the beginning of the period has been cut by the end. New patches of forest grow in over the same period, so that the landscape is always comprised of 10% forest. The simulations show that there is a lower limit of approximately 80 years for the mean number of years before clear-cutting, below which the regional population of the forest-floor plant cannot persist. Above this limit, population size increases with increasing time before clear-cutting. This result is due completely to the rate of change in landscape structure, since the amount of habitat available to the forest-floor plant (10% of the landscape) is constant among and within simulations over time.

Rapidly changing landscape structure can result in increased distance and rate of dispersal. If the rate of change in dispersal is not as fast as the rate of change in the landscape, the regional population will not survive.

If the regional population does survive in the face of a rapidly changing landscape structure, it is due to changes in dispersal behaviour (Merriam 1991). The changes can be either learned or genetically based, resulting from strong selection pressure (Pease et al.

1989; Olivieri et al. 1990). This pressure favors increasingly flexible and exploratory dispersal behavior. Exploratory behavior is favored in the short term because individuals need access to resources over larger areas; the animal is not able to count on certain locations for certain resources. An increase in exploratory behavior leads to an increased probability of dispersers finding new habitat patches (see Wegner & Merriam 1990). This leads to an increase in dispersal scale (Hansson 1991). For example, in its original habitat of woods or brush, *Peromyscus leucopus* used home ranges on the scale of less than 0.5 ha. Where agricultural clearing has fragmented wooded habitat, this mouse uses up to tens of hectares and may move hundreds of meters in a night's activity (Merriam & Lanoue 1990; Wegner & Merriam 1990). Also, Capman et al. (1990) found that different spacings of host-plant patches can alter the dispersal behavior of the Common Sooty-Wing Skipper. Finally, dispersal distances of the red fox (*Vulpes vulpes*) are greater in urban environments than in boreal forest habitats (Lindström 1989; Hansson 1991).

The interaction between the rate of change of landscape spatial structure and the rate of change in dispersal behavior determines the probability of a species' regional survival. As long as the rate of change in dispersal behavior is greater than the rate of change in landscape spatial structure, it is possible for the organism to survive in the changing landscape by moving around in it and integrating the resources over space. However, there will be a maximum possible rate of change in dispersal behavior. If the landscape structure is changing at a rate higher than this, the organism will be unable to recolonize local extinctions at a sufficient rate and the regional population will become extinct.

Discussion

The aim of this paper is to argue that landscape structure is of central importance in understanding the effects of population subdivision on population survival for conservation. Models that gloss over spatial details of landscape structure can be useful for theoretical developments (Hastings 1991) but will almost always be misleading when applied to real-world conservation problems. This is particularly important for endangered species and populations. For example, the inaccuracies due to the assumption of equal accessibility are largest when the proportion of occupied patches is small (Gurney & Nisbet 1978; Hanski 1991), which is often the case for endangered species. Endangered species are also often poor dispersers (see Cappucino & Kareiva 1985), and these are the conditions under which the landscape spatial structure has the greatest effect (Fahrig & Paloheimo 1988*a*). As stated by Hanski (1991), "When quantitative predictions are needed it may be necessary to assume the more realistic stepping-stone dispersal and to resort to simulation studies.... There is an urgent need to develop metapopulation models that include variation in habitat patch size and quality."

A related point concerns the definition of the "metapopulation." Some authors focus on the dynamic of extinction and recolonization (such as Hanski & Gilpin 1991), while others focus on the relative rates of movement within versus between local populations (such as Harrison 1991). The fact that populations are subdivided on many scales means that the delimitation of the local population is often subjective (Hanski & Gilpin 1991). We believe that, for application of the metapopulation concept to real problems of conservation, it is most profitable to recognize that all populations are spatially structured, and to focus on the spatial structuring and dynamics of the population at hand.

Acknowledgments

We are grateful to many colleagues at Carleton University and to Jana Verboom for useful comments on earlier versions. We are also grateful to Peter Kareiva, James Karr, and two anonymous reviewers for their helpful suggestions and comments. This work was supported by grants to both authors from the National Sciences and Engineering Research Council of Canada.

Literature Cited

Addicott, J. F. 1978. The population dynamics of aphids on fireweed: A comparison of local populations and metapopulations. Canadian Journal of Zoology **56**:2554–2564.

Adler, F. R., and B. Nürnberger. 1993. Persistence in patchy irregular landscapes. Theoretical Population Biology.

Ambuel, B., and S. A. Temple. 1984. Area-dependent changes in the bird communities and vegetation of southern Wisconsin forests. Ecology **64**:1057–1068.

Andrewartha, H. G., and L. C. Birch. 1984. The ecological web. University of Chicago Press, Chicago, Illinois.

Bennett, A F. 1987. Conservation of mammals within a fragmented forest environment: The contributions of insular biogeography and autecology. Pages 41–52 in D. A. Saunders, G. W. Arnold, A. A. Burbidge, and A. J. M. Hopkins, editors. Nature conservation: The role of remnants of native vegetation. Surrey Beatty and Sons, Chipping Norton, Australia.

Bennett, A. F. 1990. Habitat corridors and the conservation of small mammals in a fragmented forest environment. Landscape Ecology **4**:109–122.

Berger, J. 1990. Persistence of different-sized populations: An empirical assessment of rapid extinctions in bighorn sheep. Conservation Biology **4**:91–98.

Blaustein, A. R. 1981. Population fluctuations and extinctions of small rodents in coastal southern California. Oecologia **48**:71–78.

Boorman, S. A., and P. R. Levitt. 1973. Group selection on the boundary of a stable population. Theoretical Population Biology **4**:85–128.

Burel, F. 1989. Landscape structure effects on carabid beetles spatial patterns in western France. Landscape Ecology **2**:215–226.

Capman, W. C., G. O. Batzli, and L. E. Simms. 1990. Response of the common sooty wing skipper to patches of host plants. Ecology **71**:1430–1440.

Cappuccino, N., and P. Kareiva. 1985. Coping with a capricious environment: A population study of a rare pierid butterfly. Ecology **66**:152–161.

Chew, F. S. 1981. Coexistence and local extinction in two pierid butterflies. American Naturalist **118**:655–672.

den Boer, P. J. 1977. Dispersal power and survival: Carabids in a cultivated countryside. Landbouwhogeschool Wageningen, Miscellaneous Papers **14**:1–190.

den Boer, P. J. 1979. The significance of dispersal power for the survival of species, with special reference to the carabid beetles in a cultivated countryside. Fortschritte der Zoologie **25**:79–94.

den Boer, P. J. 1985. Fluctuations of density and survival of carabid populations. Oecologia **67**:322–330.

den Boer, P. J. 1990. Density limits and survival of local populations in 64 carabid species with different powers of dispersal. Journal of Evolutionary Biology **3**:19–48.

Dhont, A. A. 1979. Summer dispersal and survival of juvenile great tits in southern Sweden. Oecologia **42**:139–157.

Dingle, H. 1991. Factors influencing spatial and temporal variation in abundance of the large milkweed bug (Hemiptera: Lygaeidae). Annals of the Entomological Society of America **84**:47–51.

Dmowski, K., and M. Kozakiewicz. 1990. Influence of shrub corridor on movements of passerine birds to a lake littoral zone. Landscape Ecology **4**:99–108.

Duelli, P. 1990. Population movements of arthropods between natural and cultivated areas. Biological Conservation **54**:193–207.

Fahrig, L. 1990. Interacting effects of disturbance and dispersal on individual selection and population stability. Comments on Theoretical Biology **1**:275–297.

Fahrig, L. 1992. Relative importance of spatial and temporal scales in a patchy environment. Theoretical Population Biology **41**:300–314.

Fahrig, L., and G. Merriam. 1985. Habitat patch connectivity and population survival. Ecology **66**:1762–1768.

Fahrig, L., and J. Paloheimo. 1987. Interpatch dispersal of the cabbage butterfly. Canadian Journal of Zoology **65**:616–622.

Fahrig, L., and J. Paloheimo. 1988*a*. Determinants of local population size in patchy habitats. Theoretical Population Biology **34**:194–213.

Fahrig, L., and J. Paloheimo. 1988*b*. Effect of spatial arrangement of habitat patches on local population size. Ecology **69**:468–475.

Freemark, K. E., and G. Merriam. 1986. Importance of area and habitat heterogeneity to bird assemblages in temperate forest fragments. Biological Conservation **36**:115–141.

Gates, J. E., and L. W. Gysel. 1978. Avian nest dispersion and fledgling success in field-forest ecotones. Ecology **59**:871–883.

Gilpin, M. E. 1990. Extinction of finite metapopulations in correlated environments. Pages 177–186 in B. Shorrocks and I. R. Swingland, editors. Living in a patchy environment. Oxford Science Publications, Oxford, England.

Greenwood, P. J., P. H. Harvey, and C.-M. Perrins. 1979. The role of dispersal in the Great Tit (*Parus major*): The causes, consequences and heritability of natal dispersal. Journal of Animal Ecology **48**:123–142.

Gurney, W. S. C., and R. M. Nisbet. 1978. Single-species population fluctuations in patchy environments. American Naturalist **112**:1075–1090.

Hanski, I. 1991. Single-species metapopulation dynamics: Concepts, models and observations. Biological Journal of the Linnean Society **42**:17–38.

Hanski, I., and M. Gilpin. 1991. Metapopulation dynamics: Brief history and conceptual domain. Biological Journal of the Linnean Society **42**:3–16.

Hansson, L. 1991. Dispersal and connectivity in metapopulations. Biological Journal of the Linnean Society **42**:89–103.

Harrison, S. 1991. Local extinction in a metapopulation context: An empirical evaluation. Biological Journal of the Linnean Society **42**:73–88.

Harrison, S., and J. F. Quinn. 1989. Correlated environments and the persistence of metapopulations. Oikos **56**:293–298.

Harrison, S., D. D. Murphy, and P. R. Ehrlich. 1988. Distribution of the Bay checkerspot butterfly, *Euphydryas editha bayensis*: Evidence for a metapopulation model. American Naturalist **132**:360–382.

Hastings, A. 1991. Structured models of metapopulation dynamics. Biological Journal of the Linnean Society **42**:57–71.

Hastings, W., and C. Wolin. 1989. Within-patch dynamics in a metapopulation. Ecology **70**:1261–1266.

Henderson, M. T., G. Merriam, and J. Wegner. 1985. Patchy environments and species survival: Chipmunks in an agricultural mosaic. Biological Conservation **31**:95–105.

Henein, K., and G. Merriam. 1990. The elements of connectivity where corridor quality is variable. Landscape Ecology **4**:147–170.

Herben, T., H. Rydin, and L. Söderström. 1991. Spore establishment probability and the persistence of the fugitive invading moss, *Orthodontium lineare*: A spatial simulation model. Oikos **60**:215–221.

Hobbs, R. J. 1992. The role of corridors in conservation: Solution or bandwagon? Trends in Ecology and Evolution **7**:389–392.

Johnson, M. L., and M. S. Gaines. 1987. The selective basis for dispersal of the prairie vole, *Microtus ochrogaster*. Ecology **68**:684–694.

Kareiva, P. 1990. Population dynamics in spatially complex environments: Theory and data. Philosophical Transactions of the Royal Society of London **B330**:175–190.

Karr, J. R. 1990. Avian survival rates and the extinction process on Barro Colorado Island, Panama. Conservation Biology **4**:391–397.

Kotliar, N. B., and J. A. Wiens. 1990. Multiple scales of patchiness and patch structure: A hierarchical framework for the study of heterogeneity. Oikos **59**:253–260.

Krohne, D. T., and A. B. Burgin. 1990. The scale of demographic heterogeneity in a population of *Peromyscus leucopus*. Oecologia **82**:97–101.

Krohne, D. T., B. A. Dubbs, and R. Baccus. 1984. Analysis of dispersal in an unmanipulated population of *Peromyscus leucopus*. American Midland Naturalist **112**:146–156.

Lawton, J. H., and G. L. Woodroffe. 1991. Habitat and the distribution of water voles: Why are there gaps in a species' range? Journal of Animal Ecology **60**:79–91.

Lefkovitch, L. P., and L. Fahrig. 1985. Spatial characteristics of habitat patches and population survival. Ecological Modelling **30**:297–308.

Levins, R. 1970. Extinction. Pages 77–107 in M. Gerstenhaber, editor. Lectures on mathematics in the life sciences. Vol. 2. American Mathematical Society, Providence, Rhode Island.

Lidicker, W. Z., Jr. 1975. The role of dispersal in the demography of small mammals. Pages 103–128 in F. B. Golley, K. Petrusewicz, and L. Ryskowski, editors. Small mammals: Their productivity and population dynamics. Cambridge University Press, New York, New York.

Lindström, E. 1989. Food limitation and social regulation in a red fox population. Holarctic Ecology **12**:70–79.

Lord, J. M., and D. A. Norton. 1990. Scale and the spatial concept of fragmentation. Conservation Biology **4**:197–202.

Lynch, J. E., and D. F. Whigham. 1984. Effects of forest fragmentation on breeding bird communities in Maryland, USA. Biological Conservation. **28**:287–324.

MacArthur, R. H., and W. O. Wilson. 1967. The theory of island biogeography. Princeton University Press, Princeton, New Jersey.

Mader, H.-J., C. Schell, and P. Kornacker. 1990. Linear barriers to arthropod movements in the landscape. Biological Conservation **54**:209–222.

Mason, L. G. 1977. Extinction, reproduction and population size in natural populations of ambushbugs. Evolution **31**:445–447.

McMullen, L. H., and M. D. Atkins. 1962. On the flight and host selection of the Douglas-fir beetle, *Dendroctonus pseudotsugae* Hopk. (Coleoptera: Scolytidae). Canadian Entomologist **94**:1309–1325.

Menges, E. S. 1990. Population viability analysis for an endangered plant. Conservation Biology **4**:52–70.

Merriam, G. 1984. Connectivity: A fundamental ecological characteristic of landscape pattern. Pages 5–15, Theme 1, in Proceedings of the first International Seminar on Methodology in Landscape Ecological Research and Planning. International Association for Landscape Ecology, Roskilde, Denmark.

Merriam, G. 1991. Corridors and connectivity: Animal populations in heterogeneous environments. Pages 134–142 in D. Saunders and R. Hobbs, editors. The role of corridors in nature conservation. Surrey Beatty & Sons, Chipping Norton, Australia.

Merriam, G., and A. Lanoue. 1990. Corridor use by small mammals: Field measurements for three experimental types of *Peromyscus leucopus*. Landscape Ecology **4**:123–131.

Merriam, G., and J. Wegner. 1992. Local extinctions, habitat fragmentation and ecotones. Pages 150–169 in A. J. Hansen and F. Di Castri, editors. Landscape boundaries. Springer Verlag, New York, New York.

Merriam, G., M. Kosakiewicz, E. Tsuchiya, and K. Hawley. 1989. Barriers as boundaries for metapopulations and demes. Landscape Ecology **2**:227–235.

Olivieri, I., D. Couvet, and P.-H. Gouyon. 1990. The genetics of transient populations: Research at the metapopulation level. Trends in Ecology and Evolution **5**:207–210.

Paine, R. T. 1988. Habitat suitability and local population persistence of the sea palm *Postelsia palmaeformis*. Ecology **69**:1787–1794.

Pease, C. M., R. Lande, and J. J. Bull. 1989. A model of population growth, dispersal and evolution in a changing environment. Ecology **70**:1657–1664.

Potter, M. A. 1990. Movement of North Island brown kiwi (*Apteryx australis mantelli*) between forest fragments. New Zealand Journal of Ecology **14**:17–24.

Prokopy, R. J., R. H. Collier, and S. Finch. 1983. Visual detection of host plants by cabbage root flies. Entomologica Experimentalis et Applicata **34**:85–89.

Pulliam, H. R. 1988. Sources, sinks, and population regulation. American Naturalist **132**:652–661.

Saunders, D. A. 1990. Problems of survival in an extensively cultivated landscape: The case of Carnaby's cockatoo *Calyp-*

torhyncus funereus latirostris. Biological Conservation **54**:277–290.

Saunders, D. A., and J. A. Ingram. 1987. Factors affecting survival of breeding populations of Carnaby's cockatoo *Calyptorhyncus funereus latirostris* in remnants of native vegetation. Pages 249–258 in D. A. Saunders, G. W. Arnold, A. A. Burbidge, and A. J. M. Hopkins, editors. Nature conservation: The role of remnants of native vegetation. Surrey Beatty and Sons, Chipping Norton, Australia.

Saunders, D. A., I. Rowley, and C. T. Smith. 1985. The effects of clearing for agriculture on the distribution of cockatoos in the southwest of Australia. Pages 309–321 in A. Keast, H. D. Recher, H. Ford, and D. Saunders, editors. Birds of eucalypt forests and woodlands: Ecology, conservation, management. Royal Australian Ornithologists Union, Surrey Beatty and Sons, Chipping Norton, Australia.

Schoener, T. W., and D. A. Spiller. 1987. High population persistence in a system with high turnover. Nature **330**:474–477.

Seno, H. 1988. Effect of a singular patch on population persistence in a multi-patch system. Ecological Modelling **43**:271–286.

Simberloff, D., J. A. Karr, J. Cox, and D. W. Mehlman. 1992. Movement corridors: Conservation bargains or poor investments? Conservation Biology **6**:493–504.

Sjögren, P. 1991. Extinction and isolation gradients in metapopulations: The case of the pool frog (*Rana lessonae*). Biological Journal of the Linnean Society **42**:135–147.

Tomialojc, L., T. Wesolowski, and W. Walankiewicz. 1984. Breeding bird community of a primaeval temperate forest (Bialowieza National Park, Poland). Acta Ornithologica **20**:251–311.

Tsuchiya, E. 1990. Genetic differentiation of mitochondrial DNA at the landscape scale in patchy populations of *Peromyscus leucopus.* M.Sc. thesis. Carleton University, Ottawa, Canada.

Urban, D. L., R. V. O'Neill, and H. H. Shugart. 1987. Landscape ecology. Bioscience **37**:119–127.

van der Meijden, E., T. J. de Jong, P. G. L. Klinkhamer, and R. E. Kooi. 1985. Temporal and spatial dynamics of biennial plants. Pages 91–103 in J. Haeck and J. W. Woldendorp, editors. Structure and functioning of plant populations 2. North-Holland, Amsterdam, The Netherlands.

Verboom, B., and R. van Apeldoorn. 1990. Effects of habitat fragmentation of the red squirrel. *Sciurus vulgaris* L. Landscape Ecology **4**:171–176.

Verboom, J., A. Schotman, P. Opdam, and J. A. J. Metz. 1991. European nuthatch metapopulations in a fragmented agricultural landscape. Oikos **61**:149–156.

Villard, M.-A., K. E. Freemark, and G. Merriam. 1992. Metapopulation dynamics as a conceptual model for neotropical migrant birds: An empirical investigation. Pages 474–482 in J. M. Hagan and D. W. Johnston, editors. Ecology and conservation of neotropical migrant landbirds. Smithsonian Institution Press, Washington, D.C.

Wallace, G. K. 1958. Some experiments on form perception in the nymphs of the desert locust, *Schisterocerca gregaria* Forskal. Journal of Experimental Biology **35**:765–775.

Wegner, J., and G. Merriam. 1990. Use of spatial elements in a farmland mosaic by a woodland rodent. Biological Conservation **54**:263–276.

Wiens, J. A. 1989. Spatial scaling in ecology. Functional Ecology **3**:385–397.

Wilcove, D. S. 1988. Changes in the avifauna of the Great Smoky Mountains: 1947–1983. Wilson Bulletin **100**:256–271.

Williamson, M. 1983. The land-bird community of Stockholm: Ordination and turnover. Oikos **41**:378–384.

Wilson, D. S. 1980. The natural selection of populations and communities. Benjamin/Cummings, Menlo Park, California.

Wright, S. 1977. Evolution and genetics of populations. Vol. 3. Experimental results and evolutionary deductions. University of Chicago Press, Chicago, Illinois.

Extinction in Subdivided Habitats

JAMES F. QUINN*
Division of Environmental Studies
University of California
Davis, California 95616

ALAN HASTINGS
Division of Environmental Studies
University of California
Davis, California 95616

Abstract: *The effect of the spatial structure of populations on extinction rates is a central question in conservation biology, ecology, and evolution. We show that the effect of subdivision on the mean time to extinction in independently varying habitat patches depends upon the relationship between extinction probabilities and habitat area. Models of extinction by demographic stochasticity alone predict that subdivision should decrease the mean time to extinction. If environmental stochasticity is incorporated into the model, subdivision may increase the mean time to extinction. Empirical species persistence times estimated from island biogeography studies show no systematic relationship to the degree of subdivision. However, subdivision always increases the probability of survival over a sufficiently short time interval. These results suggest that over ecologically relevant timescales, subdivision into a number of independent subpopulations may frequently act to decrease the probability of overall extinction in rare species. To the degree that these considerations may be applied to rare and endangered species, maintaining sufficient numbers of nature reserves is likely to be crucial to effective conservation strategies.*

Resumen: *El efecto de la estructura espacial de las poblaciones en las tasas de extinción es una interrogante fundamental en conservación, ecología y evolución. Este trabajo de muestra que el efecto de la subdivisión de poblaciones sobre el tiempo promedio de extinción, en habitats fragmentados que varían independientemente, depende de la relación entre la probabilidad de extinción y la extensión del habitat. Los modelos de extinción basados exclusivamente en demografía estocástica, predicen que la subdivisión debe reducir el tiempo promedio de extinción. Si se incorporan eventos aleatorias del ambiente (medio ambiente estocástico) en el modelo, la subdivisión puede incrementar el tiempo promedio de extinción. Las estimaciones empíricas del tiempo de persistencia de las especies, provenientes de los estudios sobre biogeografía de islas, no demuestran una relación sistemática con el grado de subdivisión. No obstante, la subdivisión siempre aumenta la probabilidad de sobrevivencia en un período de tiempo suficientemente corto. Estos resultados sugieren que sobre escalas de tiempo ecológicamente pertinentes, la subdivisión en varias subpoblaciones independientes frecuentemente puede producir una reducción en la probabilidad de extinción de las especies raras. Será crucial para la efectividad de las estrategias de conservación, el grado en que éstas consideraciones se apliquen a las especies raras y en peligro de extinctión, manteniendo un número suficiente de reservas naturales.*

Introduction

One promising application of theoretical population biology is to the problem of establishing new parks and nature reserves. A central focus in the last 10 years has been the tradeoff between the number of reserves and the average size of reserves dictated by limited resources available for land acquisition. One major goal has been the determination of strategies for maximizing species diversity, however measured, within the refuge system.

**Correspondence and requests for reprints should be addressed tothis author.*

A predominant view has been that very large contiguous areas should be preserved, even if this results in a small number of new parks (Diamond 1975, Shaffer & Sullivan 1975, Whitcomb et al. 1976, Lovejoy et al. 1984, Wilcox & Murphy 1985). Other authors have observed that collections of small natural islands or habitat patches ["fragmented" habitats in the terminology of Simberloff and Abele (1982)] often harbor more species than comparable contiguous areas on large islands or patches. Hence, by analogy, large numbers of modest-sized parks might well be more effective in maximizing total species numbers than would be single very large parks (Simberloff & Abele 1976, 1982; Higgs & Usher 1980; Gilpin & Diamond 1980; Järvinen 1982, Simberloff & Gotelli 1984; McLellan et al. 1986; Quinn & Robinson 1987).

However, maximizing total species diversity is rarely if ever the principal objective of conservation strategies (Soulé & Simberloff 1986). Other aesthetic, resource preservation, and recreational values are often more important. Even within the realm of species conservation, decisions are often made on the basis of the requirements of particular target species, such as large vertebrates, because of particular ecological or human values, and because many of the species contributing to total species counts are common or not particularly threatened by human activities (Wilcox & Murphy 1985). As a result, it is perfectly possible that the optimum size and geographical distribution of reserves designed to protect particularly threatened target species might be very different from that suggested by considerations of total species numbers alone (Shaffer & Samson 1985, Soulé & Simberloff 1986, Goodman 1987). In particular, several authors have suggested that very large reserves are needed to maintain viably large populations of target species, such as large carnivores, which require large habitats (Soulé et al. 1979, Shaffer 1981, Shaffer & Samson 1985).

There are at least two important components of population viability. One is the need to maintain genetic variation, both over the entire species range, and, if inbreeding depression is important, within local populations (Frankel 1974, Frankel & Soulé 1981, Beardmore 1983, Soulé 1986). We will consider some genetic models in a future paper. The second is demographic processes. Presumably effective conservation efforts should minimize the probability of the total extinction of target species over some relevant time frame (Shaffer 1981). Obviously a practical tradeoff occurs between minimizing the probability of population extinction in each individual reserve, which would suggest a small number of large reserves, and spreading the risk over a larger number of protected populations. As a result, optimum conservation strategies for protecting any given species depend upon how the probability per unit time of extinction varies with habitat area and initial population size (Simberloff & Abele 1982, Shaffer & Samson 1985, Wilcox & Murphy 1985).

Here, we use a simple modeling approach to examine the tradeoff between refuge size and refuge number. We study a mosaic of independently varying isolated patches and examine the effect of subdivision on the extinction probability of a single species. The problem of the distribution over time of extinctions in multiple reserves is analogous to failure problems in mechanical or electronic systems. Results may be borrowed from reliability theory (Polovko 1968), which gives the distribution of failure time for systems with redundant subunits, each with known failure probabilities. Local extinction in a reserve corresponds to a subunit failure, whereas global extinction over all reserves is analogous to a total (irrecoverable) system failure. Multiple reserves represent redundancy in the conservation system. As with mechanical systems, it may be more effective to increase either subunit reliability or redundancy, depending upon the relative costs of improving subunit reliability versus replicating the subunits.

Even in the simplest of cases, the optimum choice between larger or more numerous reserves depends upon the nature of the extinction process. In this paper we examine the simplest model that captures the essential features of reserve systems. We assume identical reserves with no immigration between reserves. All reserves are subject to the same extinction processes, which are assumed to act independently in the different reserves. This model thus best approximates species with limited dispersal abilities living in highly insularized habitats. However, the qualitative behavior of the model is likely to be much more general.

We observe that under some of the simplest models, assuming multiple reserves without recolonization, there may be an optimum intermediate size-number combination that maximizes expected species survival time. However, examination of overall per-species extinction rates estimated from several empirical studies of extinction versus island or habitat size shows no clear evidence that expected survival time varies in any systematic way with the size-number tradeoff in these cases.

Extinction in Multiple, Independently Varying Populations

Let us assume that a species population is subdivided into m identical subpopulations. Under a wide variety of random environment models, if the expected time to extinction of each subpopulation is finite (as necessarily is the case for any realistic model with density dependence), extinction in individual subpopulations after some period of time resembles a Poisson process, with each extant subpopulation having a roughly constant probability of disappearing in the next small time interval (Keilson 1979).

There are two intuitive possibilities for explaining why extinction should resemble a Poisson process. The first possibility is if the extinction probability was relatively independent of population size (i.e., epidemics and environmental catastrophes). Here, extinction clearly could be approximately Poisson. The other possibility is if there were an equilibrium for each population that would be stable in the absence of stochastic factors and if extinction occurred as the result of a number of small stochastic events. In this case, the systems would rapidly approach the deterministic equilibrium in most cases. These systems thus could be modelled as a random walk described by a finite Markov chain. Keilson (1979) shows that in this case, extinction would approximate a Poisson process.

If λ is the probability per unit time of extinction of a single subpopulation (or the mean extinction rate), the probability $P(t)$ that one or more of the m subpopulations survives until time t is derived as such:

Equation 1

$$P(t) = 1 - (1-e^{-\lambda t})^m$$

Most published discussions of the effects of habitat fragmentation on extinction rate (e.g., Wilcox & Murphy 1985) use the mean time to extinction, T_e, as the measure of species persistence (although, as we will discuss below, T_e may not be a particularly good metric for considering short-term risk). The main result for independently varying reserves may be taken from reliability theory. With m subpopulations, each with a probability of extinction of λ_m per population per unit time, the expected time to extinction of each subpopulation is $1/\lambda_m$. A standard result from reliability theory is that the mean time to extinction for the entire population is

Equation 2

$$T_e = (1/\lambda_m) \sum_{i=1}^{m} 1/i$$
$$\approx (1/\lambda_m)\, ln(m).$$

The expected persistence time of the species somewhere in the reserve system increases roughly with the logarithm of the number of reserves (Polovko 1968).

The typical problem facing conservation organizations is that for any given set of land acquisition strategies, the attainable per-reserve extinction rates (e.g., λ_m's) and the maximum number of reserves that can be protected (m) are interrelated. The problem may be caricatured as follows.

Let us assume the target species has an initial density (or perhaps a carrying capacity) D over its original range. Resources exist to set aside an area A in some m equal-size reserves, each of which is assumed to be independently susceptible to random extinction. The number of individuals incorporated into the reserve system is then $N = DA$ with N/m in each individual reserve. Suppose the probability per unit time of extinction $\lambda(x)$ expected on a reserve with an initial population of x individuals (or, alternatively, a carrying capacity of x) is known. The mean time to extinction is then

Equation 3

$$T_e = 1/[(\lambda(N/m)]\sum_{i=1}^{m} 1/i.$$

If extinction is a Poisson process, $1/\lambda(x)$ is the expected time to extinction for a subpopulation of size x. Replication of equal-size subpopulations increases the expected survival time, relative to that of a single population of the same size, by a multiple of the summation term.

Subdivision will tend to increase persistence time by replication, but decrease persistence time by reducing the size of each subpopulation. Thus, the value of m for which T_e is maximized depends upon the form of the function $\lambda(N/m)$. Depending upon the population model assumed, rather different expressions might be proposed for the dependence of extinction probability (or, equivalently, mean time to extinction per subpopulation). Shaffer (1981) divides the processes resulting in extinction into four categories: demographic stochasticity, chronic environmental variation, environmental catastrophes, and genetic effects (in particular inbreeding). The applicability of models of these processes depends upon the circumstances envisioned. Both demographic and environmental stochasticity affect all real populations, with the variability attributable to demographic effects only becoming important in comparison with the environmentally driven variation when populations are very small.

Several rather different stochastic extinction models have been proposed.

1. *Pure demographic stochasticity.* These models assume that probabilities per unit time of birth or death depend only on population size, with stochasticity entering only in the order of births and deaths. Thus population growth is viewed as a stationary Markov process with absorption at the boundary of a population of zero.

Richter-Dyn and Goel (1972) model pure demographic stochasticity in a small population with an expected positive growth rate. When the population size is N, the expected time to extinction is

Equation 4

$$T_e = \sum_{i=1}^{N} \sum_{j=i}^{K} (1/d(1)) \prod_{n=i}^{j-1} (b(j)/d(j)).$$

In this equation K is the upper limit to the population, and in a population with n individuals $d(n)$ is the death and $b(n)$ is the birth rate. (Note that these are total rates, not per capita rates.) Roughly, the time to extinction increases approximately as the *N*th power of the birth rate divided by the death rate. The net result is that once the population reaches a small threshold size (e.g., 20), the probability of extinction in ecologically relevant time periods becomes vanishingly small.

Wright and Hubbell (1983) consider a related model in which the total number of individuals of all species is fixed, as might be expected in a space-limited environment. Individuals die at random, and are equally likely to be replaced by the offspring of any other individual in the environment (Hubbell 1979). This model is mathematically equivalent to the genetic drift model of Moran (1958), with fixation as a reflecting barrier, which yields a random walk in gene frequencies. In this and similar random walk models, the expected number of generations to extinction is approximately proportional to the population size (Ewens 1979, Hubbell 1979, Leigh 1981). Hubbell (1979) has proposed this kind of dynamics as a model for "nonequilibrium coexistence" of tropical forest trees, as the expected time to extinction quickly exceeds the age of the forests as carrying capacities and initial population sizes become moderate. Leigh (1981) analyzes a similar model.

The Wright and Hubbell model may be viewed as a limiting case for models incorporating only demographic stochasticity, in which individuals just replace themselves (on the average) when rare. In models like those of Richter-Dyn and Goel, in which populations tend to increase when rare, extinction probability declines more rapidly with increasing population size. Rare species in which individuals do not on the average replace themselves are likely to disappear in short order, regardless of the spatial structure of the populations.

2. *Population growth in a continuously varying environment.* Ludwig (1976) and Leigh (1981) examine models in which environmental quality is assumed to vary rapidly, as roughly a white noise process. Both find that the expected time to extinction varies approximately with the logarithm of carrying capacity.

3. *Catastrophes—logistic population growth punctuated by environmental disasters.* In a model that perhaps better captures the likely mechanisms of extinction in many populations, Hanson and Tuckwell (1981) consider a case in which deterministic logistic growth is punctuated by randomly occurring disasters. They analyze the specific case in which the disasters decrease the population size by a fixed fraction. The population never reaches zero in this model. The time until the population falls below some lower threshold value appears to resemble a gamma distribution for the parameters simulated by Hanson and Tuckwell, but they were unable to attain an analytic expression for the mean or distribution of extinction times. More recent results on this model are contained in Gripenberg (1985). A future paper will report on the behavior of subdivided populations with within-patch dynamics following this model.

Thus the relationship between the expected time to extinction in the ensemble and the number of units into which the reserve system is divided varies with the extinction dynamics assumed, as these determine the form of the function $\lambda(N/m)$.

Subpopulation Extinction Rate (λ) Determined by Demographic Stochasticity

In the pure demographic extinction models, the expected time to extinction in each subpopulation, $(1/\lambda)$, rises sufficiently rapidly with subpopulation size that the time to global extinction (T_e) is maximized at $m = 1$ (i.e., a single large reserve; see Fig. 1, curve d). Strictly speaking, Equation 3 cannot be applied to these models because the waiting time to extinction is not exponentially distributed as assumed in a Poisson extinction process. However, this does not greatly affect the calculation of mean extinction time with various degrees of subdivision, though it may modify the expected variance.

The effect of subdivision on species persistence is most easily seen in the limiting random walk case (Leigh 1981, Wright & Hubbell 1983) in which individuals are just able to replace themselves on the average, and the expected persistence time of each subpopulation is proportional to its size. Wright and Hubbell (1983) have simulated extinction times for $m = 1$ and $m = 2$, with no migration between reserves, and shown that the population in a single larger reserve on the average persists longer. More generally, if mean persistence of a subpopulation $(1/\lambda)$ is proportional to N/m, Equation 3 may easily be shown to have a single maximum at $m = 1$.

Although it is algebraically messy, using persistence time in the Richter-Dyn and Goel model (Equation 4) to approximate the subpopulation persistence time $(1/\lambda)$ maximizes the mean time to global extinction (T_e) at $m = 1$. This is not unexpected. If expected growth rates are positive, the probability of short-term extinction decreases much more rapidly with size than in the random walk case (mean growth rate of zero) considered by Wright and Hubbell. Thus the tradeoff between larger versus more subpopulations will favor large subpopulations even more than in the neutral random walk case (Fig. 1). More generally, mean time to extinction should be maximized by a single large reserve whenever the expected persistence time of a single population increases more rapidly than linearly with initial population size.

To some extent, the apparent superiority of a single-large-reserve strategy results from the choice of T_e (mean time to extinction) as the persistence metric. As noted below, if the criterion is altered to be the probability of survival over a fixed time, different results are possible. In any reasonable model, if the time chosen is short enough, it will always be the case that the survival prob-

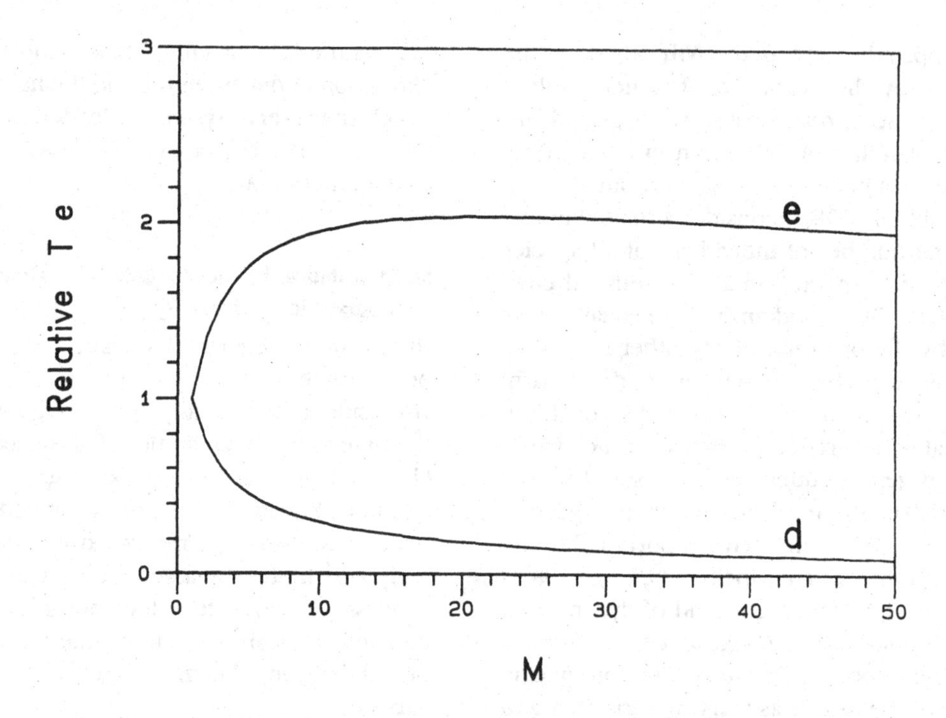

Figure 1. The relative change in mean time to extinction, T_e, with increasing subdivision, assuming N=1000 individuals subdivided into m subpopulations. Curve d = demographic stochasticity. Here extinction probability for a subpopulation, λ, is proportional to the inverse of subpopulation size (N/m), as in the pure demographic stochasticity model of Wright and Hubbell (1983). The relative T_e decreases with m. Curve e = environmental stochasticity. Here extinction probability for a subpopulation, λ, is proportional to the inverse of the natural log of subpopulation size as in the models of Ludwig (1976) and Leigh (1981). For this model T_e increases to a maximum when m is approximately the square root of N. T_e = 1 for m=1 in both models.

ability is greater in a two-patch ($m=2$) system than in a one-patch system (see Fig. 2).

As noted by Shaffer and Samson (1985) and Wilcox and Murphy (1985), the pure demographic stochasticity models probably have limited applicability to conservation decisions, except for populations of a few individuals, such as might occur in some captive breeding or species introduction programs. The absence of environmental variability leads to expected persistence times for even small populations (e.g., 50 to several hundred) that are much longer than the usual time scale for conservation planning, whereas such populations are often recognized as severely threatened in reality.

Subpopulation Extinction Rate (λ) Determined by a Continuously Varying Environment

The continuously variable environment models behave differently. If, following Larkin (1976) and Leigh (1981), we assume that the mean time to extinction of each subpopulation varies as *ln(N/m)*, we get the following:

Equation 5

$$T_e = k\, ln(N/m) \sum_{i=1} 1/i$$

The summation term may be approximated as *ln(m)*. T_e is then maximized when $dT_e/dm = 0$, or the following:

Equation 6

$$0 = k(m/N)\, ln(m)\,(-N/m^2) + k\, ln(N/m)/m$$

When solved for *m* this equation yields:

Equation 7

$$m = N^{1/2}.$$

In this model, the mean time until a population disappears completely is minimized when the reserve system is highly subdivided—with the number of reserves on the order of the square root of the total number of individuals potentially inhabiting them (see Fig. 1, curve e). Equation 7 reinforces the argument that the optimal reserve configuration may depend upon the target spe-

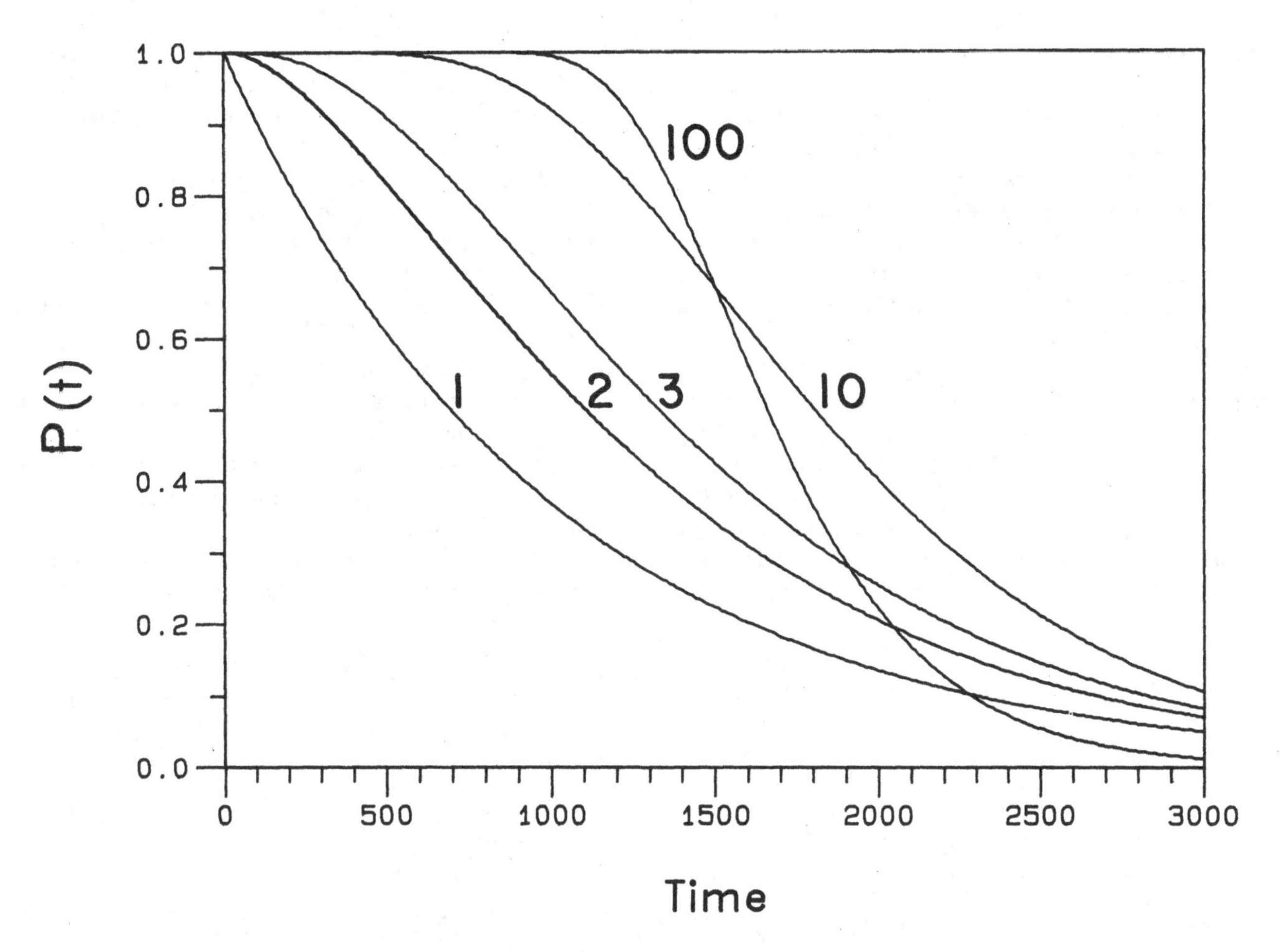

Figure 2. P(t), the probability of survival of one or more populations to time t, versus time (t), under the environmental stochasticity model of Ludwig (1976), for m = 1, 2, 3, 10, and 100 subpopulations. λ(1) = 0.001, N = 1000.

cies considered (Schaffer 1981, Belovsky 1987). For species with low maximum population densities, such as many vertebrate predators, the maximum population size (N) will be smaller than for other coexisting species, and the value of m maximizing global persistence will be correspondingly reduced. Belovsky (1987) reaches a similar conclusion from a model based on energetics. Obviously, the exact form of Equation 7 should not be taken seriously—the important point is that considerable redundancies (m much greater than 1) in reserve systems may be effective in preventing overall extinction.

As we will explain below, if a criterion of maximizing the survival over a short time is used instead, the advantage of subdivision would be further increased.

Subpopulation Extinction Rate (λ) Determined by Environmental Disasters

The general behavior of the random disasters model is difficult to analyze. It may be expected to be similar to the continuously varying environment model when the disasters are frequent and small, but may differ considerably when disasters are infrequent but intense.

Figure 2 shows the distribution of survival time for different numbers of subpopulations under the continuously varying environment model, assuming 1000 individuals broken into 1, 2, 3, 10, or 100 subpopulations, and a probability of 0.001 per year of extinction for the nonsubdivided population ($m = 1$). It is worth noting that the probability of extinction over very short time periods, as well as the mean time to extinction, is minimized at the moderate to large values of m.

Insofar as the assumptions of independence between the events affecting the isolated subpopulations are met, reserves are isolated from one another, and population growth or decline behaves approximately as envisioned by a continuous environmental noise model, this analysis suggests that substantial replication of even rather small reserves might be helpful in minimizing chances of extinction of target species.

Empirical Evidence

The preferred tradeoff between refuge size and numbers depends critically upon the way that extinction probabilities for target species decline with refuge area. Empirical evidence on this relationship for any actual rare or endangered species is necessarily poor, and not subject to experimental investigation. Most of the evidence for a dependence of extinction rates on area comes from island studies (MacArthur & Wilson 1967; reviewed by Diamond 1984). The quality of these data vary—in the majority of studies, pre-extinction occurrences are inferred from present distributions rather than from historical or fossil records. We have identified six island biogeography studies on the relationship between extinction (or species turnover) and area in which earlier occurrences of species were determined from published census data, and in which the main causes of extinction are presumed to have been natural population processes rather than direct human intervention. Although similar data are no doubt available from other studies, this group is probably representative.

There are a variety of potential sources of error in these data. Extinctions may be missed if species disappeared and then reinvaded between censuses (Simberloff 1976). "Pseudoextinctions" may also be recorded if rare species are missed in the final census (Nilsson & Nilsson 1985). Extinction rates are averaged over all species present, and therefore certainly greatly underestimate the risk to the smaller number of species (and perhaps the more important conservation targets) particularly susceptible to local extinction. How these considerations might bias the estimated dependence of λ on area is not obvious.

The change in average per-species extinction rate with increasing area is shown in Figure 3. While all studies

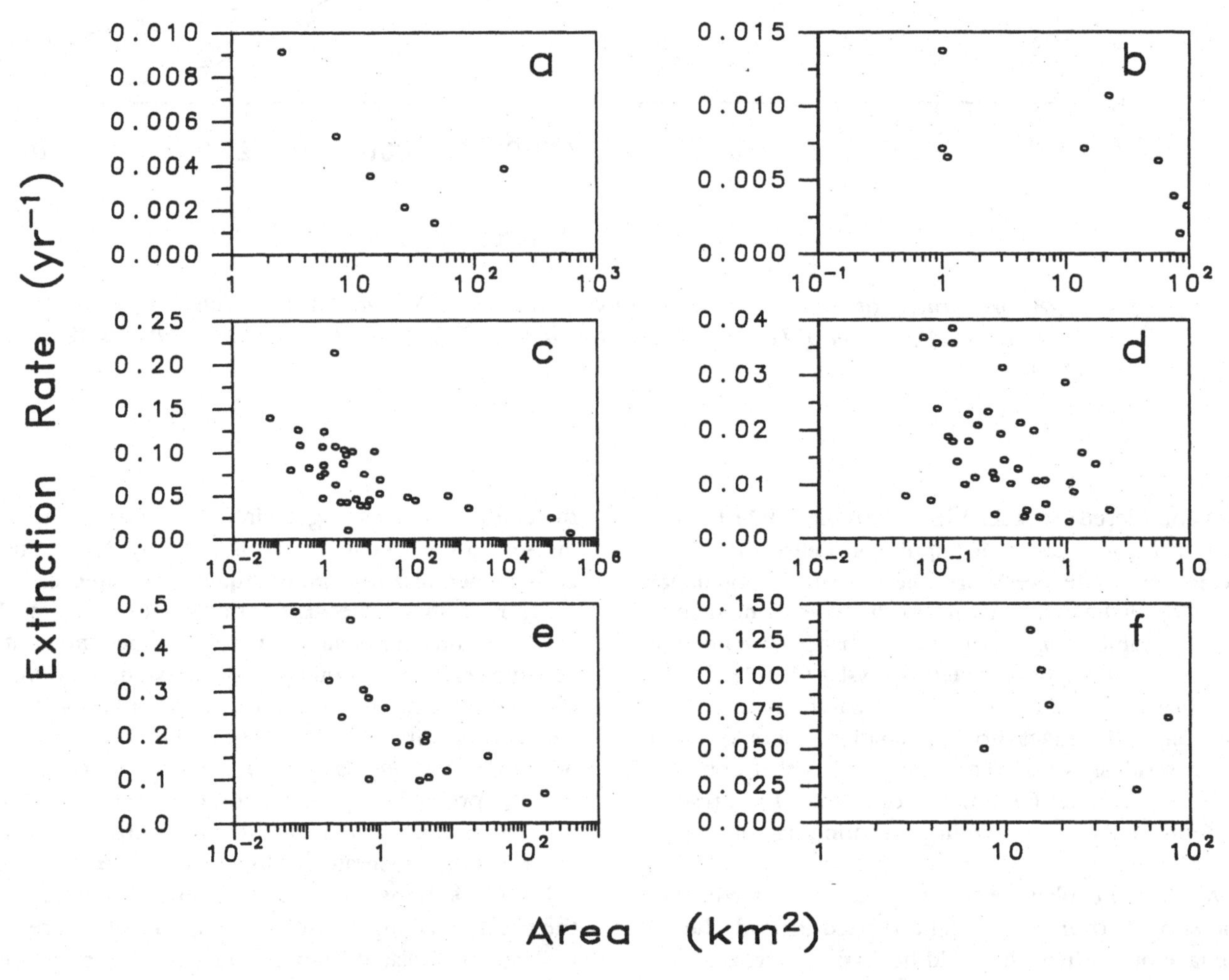

Figure 3. Mean per-species extinction rates as a function of habitat area, from the studies of (A) zooplankton in lakes in the northeastern United States (Browne 1981); (B) birds of the California Channel Islands (Diamond 1969); (C) birds on northern European islands (Diamond 1984); (D) vascular plants in southern Sweden (Nilsson and Nilsson 1982); (E) birds on Finnish islands (Vaisanen and Järvinen 1977); and (F) birds on islands in Gatun Lake, Panama (Wright 1985).

show a significant decline, the scatter appears too great to distinguish between models reliably.

In the absence of better information, one could assume that the extinction risk for any particular species is some species-specific multiple of the mean extinction rate averaged over all species. In this case, Equation 5 can be used with the data in Figure 3 to estimate the relative time to extinction for a hypothetical system of reserves, each the size of each actual island, and replicated sufficient times to equal the area of the entire real archipelago. (That is, if a_i is the area of each island, and A (the total area in the study) is the sum of the a_i, then m_i is the nearest integer to A/a_i. λ_i is estimated as the mean extinction rate actually observed on island i.) In the hypothetical reserve system consisting of m independent identical copies of island i, the expected time to extinction for any given species will then be proportional to

Equation 9

$$(1/\lambda_i)\sum_{i=1}^{m_i} 1/i.$$

Figure 4 shows the results of this calculation. Two cases in which no extinction was observed were eliminated, since they would otherwise yield estimates of infinite survival. There seems no uniform tendency for the hypothetical expected time to extinction either to increase or decrease with size versus number of islands. None of the regression lines differs significantly from zero in slope, and none explains more than 15 percent of the variance. Only Figure 4*a* suggests that combina-

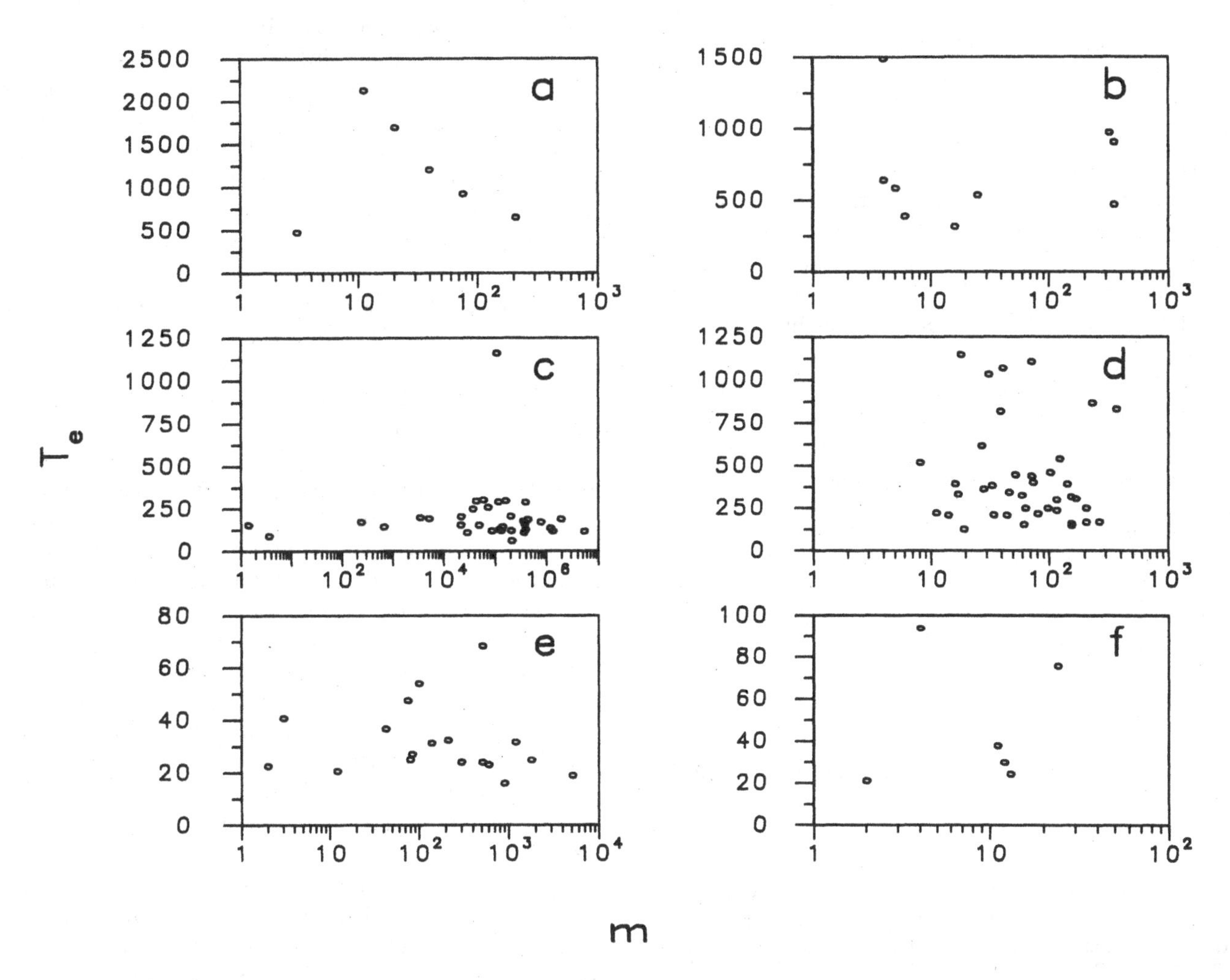

Figure 4. Mean time to extinction, T_e versus number of independent subpopulations, m, estimated from eq. (5) using the per island extinction rates in Figure 3. (A) zooplankton in lakes in the northeastern United States (Browne 1981); (B) birds of the California Channel Islands (Diamond 1969); (C) birds on northern European islands (Diamond 1984); (D) vascular plants in southern Sweden (Nilsson and Nilsson 1982); (E) birds on Finnish islands (Vaisanen and Järvinen 1977); and (F) birds on islands in Gatun Lake, Panama (Wright 1985).

tions of intermediate size areas might lead to the longest expected persistence, as suggested by the environmental noise model, and the apparent decrease in persistence time at small m is due to a single point.

Short-Term Population Survival

As noted above, the mean time to extinction may be a poor choice as a measure of the viability of a threatened population. Mean time may be strongly influenced by probabilities of extinctions over time scales far longer than ecologically relevant time scales—time scales over which our simple model would apply. As suggested by Shaffer (1981) and Shaffer and Samson (1985), it is probably more appropriate to attempt to minimize the probability of extinction (e.g., to less than 1%) over some period of time where present policy decisions and technology may have some impact (perhaps a century?). Table 1 illustrates the relationship between the estimated risk of extinction under different refuge size/number tradeoffs and the time period considered. Three hypothetical populations, subdivided into $m = 1$, 2, or 10 subpopulations, are assumed to have equal mean times to extinction of 10,000 years. If extinction occurs as a Poisson process, the most fragmented population ($m = 10$) has a 10-billion-fold lower probability of disappearing in the first century than does the single contiguous population ($m = 1$), even though it is slightly less likely to last 10,000 years.

To emphasize the importance of time scale, we now turn to properties of the full distribution of extinction times. The rate at which extinctions occur in the entire reserve system, where λ_m is the rate in one reserve, is

Equation 10

$$\lambda_m m e^{-\lambda_m t}(1 - e^{-\lambda_m t})^{(m-1)}$$

If $\lambda_m t$ is sufficiently small (say 0.2 or less), Equation 10 can be approximated by

Equation 11

$$\lambda_m^m m(1 - \lambda_m t) t^{m-1}$$

From this one can see that much of the improvement of a system of many patches over short time scales is a general property of this kind of model.

Following in this vein, we can express the viability of a threatened population as the survival probability over some fixed time, T. If λT is sufficiently small, this survival probability (Equation 1) can be approximated as:

Equation 12

$$P(m) \approx 1 - (\lambda_m T)^m$$

Note that no matter how the extinction probabilities vary with m, for any particular m, one can always choose T short enough so that $P(m+1) > P(m)$.

Discussion and Conclusions

The problem of how to best minimize the extinction rates of threatened species is perhaps the central problem in conservation biology. The problem is particularly acute in rich tropical rainforest areas, where the data needed to tailor conservation reserves to the ecological needs of threatened species are particularly sparse. Clearly, where data are available, planners are well advised to acquire reserve lands that harbor concentrations of rare species (Terborgh & Winter 1983), incorporate specific habitat and resource requirements, and provide for space for the movement and dispersal of the particular target populations (Gilbert 1980, Margules et al. 1982, Boecklen & Gotelli 1984). However, with the incomplete information generally available to make such decisions, general ecological principles may prove of some value in planning.

Extinction rates have long been believed to decrease with increasing habitat area (MacArthur & Wilson 1967), and empirical evidence (e.g., Fig. 3) generally appears to conform to this expectation (see Diamond 1984 for a review). However, this observation does not unto itself demonstrate that the optimum reserve size is large. An increase in the number of reserves, at the expense of average reserve size, may, in some circumstances, decrease the overall probability of a species disappearing from the system because the number of locations in which the species might escape otherwise catastrophic diseases, competitors, predators, or environmental stresses would be maximized (Simberloff & Abele 1976, 1982).

Our analysis presents a simple case in which a certain degree of fragmentation would appear to increase the ability of threatened species to persist. However, as the reserves become too small, the average time to extinction throughout the reserve system would again fall. Our results suggest extreme caution toward the conventional wisdom that concentrating reserve lands into a small number of very large parks is necessarily the most effective way to preserve threatened species. Of course, real reserve systems will be more complex, and we do

Table 1. Extinction probabilities over time for three hypothetical populations with identical 10,000-year expected times of extinction (T_e). Extinction is assumed to be a Poisson process.

Degree of fragmentation	T = 100	T = 1000	T = 10,000
m = 1	9.95×10^{-3}	9.52×10^{-2}	0.632
m = 2	2.22×10^{-4}	1.94×10^{-2}	0.604
m = 10	4.03×10^{-16}	1.12×10^{-6}	0.577

T = time in years

not expect these simple models to have much predictive power. Factors not in the model, such as recolonizations, inbreeding, or species' social and reproductive requirements, may complicate the scenario. We have analyzed the model assuming that the subpopulations are independent—extinction in one does not affect the extinction probability in any other subpopulation. Environmental correlation between reserves will tend to decrease the risk-spreading advantages of multiple reserves. Deviations from independence would be particularly important in the analysis of models where reintroductions can occur. It seems apparent that the ability to reintroduce species to reserves from which they have been lost increases the effectiveness of spreading this risk among multiple reserves.

Ideally, this issue is best tested experimentally by establishing alternative refuge systems, each with the same total area, but with different degrees of fragmentation into subunits. This experiment is clearly infeasible with actual parks or islands, or with most threatened species, although it may be conducted on a smaller scale with more common species (Quinn et al. 1985, Quinn & Robinson 1987).

Without being able to replicate the smaller areas sufficiently to test the overall effect of habitat fragmentation directly, it may still be possible to examine experimentally the extinction probabilities of particular target species over a range of experimental refuge sizes (Lovejoy et al. 1984). Equation 5 provides a mechanism for applying the findings of such experiments to testing the overall effect of fragmentation.

Unfortunately, applying Equation 5 to existing studies relating extinction rate to areas of natural islands and habitat isolates yields no clear patterns. Presumably more replicates, and greater attention to the dynamics of particular species will be needed to test these ideas.

Figure 2 and Table 1 illustrate an important consideration in analyzing the effects of fragmentation on species persistence. The value of m (the number of reserves) that minimizes the probability of overall extinction over a fixed time, T, varies with the time, T. At the initial stages, a large value of m, representing considerable fragmentation, increases short-term survival, since it is very unlikely that the species will disappear in a large number of independent subpopulations simultaneously. Over a longer term, somewhat smaller values of m may be favored. Thus, effectiveness in crisis management may predict somewhat different strategies than the greatest long-term effectiveness.

More work, particularly in the empirical area, is needed to gain satisfactory insight into the best allocation of lands and resources to conservation. Initial analyses suggest, however, that increasing the numbers of conservation areas, as well as requiring them to be of a viable size, is likely to be essential for effective conservation strategies.

Acknowledgment

This paper was supported by the Hewlett Foundation through a grant to the Public Service Research and Dissemination Program at the University of California, Davis.

Literature Cited

Beardmore, J. A. 1983. Extinction, survival, and genetic variation. Pages 125–151 in C. M. Schonewald-Cox, S. M. Chambers, B. MacBryde, and W. L. Thomas, editors. Genetics and Conservation. Menlo Park, California: The Benjamin/Cummings Publ. Co.

Belovsky, G. 1987. Extinction models and mammalian persistence. In M. E. Soulé, editor. Minimum Viable Populations. Cambridge: Cambridge University Press. In press.

Boecklen, W. J., and N. J. Gotelli. 1984. Island biogeographical theory and conservation practice: Species-area or specious–area relationships? Biol. Conserv. 29:63–80.

Browne, R. A. 1981. Lakes as islands: biogeographic distribution, turnover rates, and species composition of the lakes of central New York. J. Biogeogr. 8:75–83.

Diamond, J. M. 1969. Avifaunal equilibria and species turnover rates on the Channel Islands of California. Proc. Natl. Acad. Sci. USA 64:57–62.

Diamond, J. M. 1975. The island dilemma: lessons of modern biogeographic studies for the design of natural reserves. Biol. Conserv. 7:129–146.

Diamond, J. M. 1976. Island biogeography and conservation: strategy and limitations. Science 193:1027–1029.

Diamond, J. M. 1984. "Normal" extinction of isolated populations. Pages 191–196 in M. H. Nitecki, editor. Extinctions. Chicago: University of Chicago Press.

Diamond, J. M., and R. M. May. 1976. Island biogeography and the design of natural reserves. Pages 228–252 in R. M. May, editor. Theoretical Ecology. Philadelphia: W. B. Saunders Co.

Ewens, W. J. 1979. Mathematical Population Genetics. Berlin: Springer-Verlag.

Frankel, O. H. 1974. Genetic conservation: our evolutionary responsibility. Genetics 78:53–65.

Frankel, O. H., and M. E. Soulé. 1981. Conservation and Evolution. Cambridge: Cambridge University Press.

Gilbert, F. S. 1980. The equilibrium theory of island biogeography: fact or fiction? J. Biogeogr.7:209–235.

Gilpin, M. E., and J. M. Diamond. 1980. Subdivision of nature reserves and the maintenance of species diversity. Nature 285:567–568.

Goodman, D. 1987. How do any species persist? Lessons for conservation biology. Conservation Biology 1:59–62.

Gripenberg, G. 1985. Extinction in a model for the growth of a population subject to catastrophes. Stochastics 14:149–163.

Hanson, F. B., and H. C. Tuckwell. 1981. Logistic growth with random density independent disasters. Theor. Pop. Biol. 19:1–18.

Harris, L. D. 1984. The Fragmented Forest. Chicago: University of Chicago Press.

Higgs, A. J., and M. B. Usher. 1980. Should nature reserves be large or small? Nature 285:568–569.

Hubbell, S. P. 1979. Tree dispersion, abundance and diversity in a tropical dry forest. Science 203:1299–1309.

Järvinen, O. 1982. Conservation of endangered plant populations: single large or several small reserves? Oikos 38:301–307.

Keilson, J. 1979. Markov Chain Models–Rarity and Exponentiality. New York: Springer-Verlag.

Lovejoy, T. E., J. M. Rankin, R. O. Bieregaard, Jr., K. S. Brown, Jr., L. H. Emmons, and M. E. Van der Voort. 1984. Ecosystem decay of Amazon forest remnants. Pages 295–326 in M. H. Nitecki, editor. Extinctions. Chicago: University of Chicago Press.

Leigh, E. G. Jr. 1981. The average lifetime of a population in a varying environment. J. Theor. Biol. 90:213–239.

Ludwig, D. 1976. A singular perturbation problem in the theory of population extinction. SIAM-AMS Proc. 10:87–104.

MacArthur, R. H., and E. O. Wilson. 1967. The theory of island biogeography. Princeton, NJ: Princeton University Press.

Margules, C., A. J. Higgs, and R. W. Rafe. 1982. Modern biogeographic theory: are there any lessons for nature reserve design? Biol. Conserv. 24:115–128.

McLellan, C. H., A. P. Dobson, D. S. Wilcove, and J. F. Lynch. 1986. Effects of forest fragmentation on New and Old World bird communities: empirical observations and theoretical implications. In J. Verner, M. L. Morrison, C. J. Ralph, editors. Modelling Habitat Relationships of Terrestrial Vertebrates. Madison: University of Wisconsin Press.

Moran, P. A. P. 1958. Random processes in genetics. Proc. Camb. Phil. Soc. 54:60–71.

Nilsson, I. N., and S. G. Nilsson. 1982. Turnover of vascular plant species on small islands in Lake Mockeln, South Sweden 1976–1980. Oecologia 53:128–133.

Nilsson, I. N., and S. G. Nilsson. 1985. Experimental estimates of census efficiency and pseudoturnover on islands: error trend and between-observer variation when recording vascular plants. J. Ecol. 73:65–70.

Polovko, A. M. 1968. Fundamentals of Reliability Theory. New York: Academic Press.

Quinn, J. F., and G. R. Robinson. 1987. The effects of experimental subdivision on flowering plant diversity in a California annual grassland. J. Ecology 75:837–856.

Quinn, J. F., C. Van Riper III, C. R. Karban, G. R. Robinson, and S. P. Harrison. 1985. Conservation strategies: island biogeography and the design of nature reserves. Park Science 6:6–9.

Richter-Dyn, N., and N. S. Goel. 1972. On the extinction of a colonizing species. Theor. Pop. Biol. 3:406–433.

Shaffer, M. L. 1981. Minimum population sizes for species conservation. BioScience 31:131–134.

Shaffer, M. L., and F. B. Samson. 1985. Population size and extinction: a note on determining critical population size. Amer. Nat. 125:144–152.

Simberloff, D. S. 1976. Species turnover and equilibrium island biogeography. Science 194:572–578.

Simberloff, D. S., and L. G. Abele. 1976. Island biogeography theory and conservation practice. Science 191:285–286.

Simberloff, D. S., and L. G. Abele. 1982. Refuge design and island biogeographic theory effects and fragmentation. Amer. Nat. 120:41–50.

Simberloff, D. S., and N. Gotelli. 1984. Effects of insularization on plant species richness in the prairie–forest ecotone. Biol. Conserv. 29:27–46.

Soule, M. E., B. A. Wilcox, and C. Holtby. 1979. Benign neglect: a model of faunal collapse in the game reserves of East Africa. Biol. Conserv. 15:259–272.

Sullivan, A. L., and M. L. Shaffer. 1975. Biogeography of the megazoo. Science 189:13–17.

Terborgh, J. 1976. Island biogeography and conservation: strategy and limitations. Science 193:1029–1030.

Terborgh, J., and B. Winter. 1983. A method for siting parks and reserves with special references to Colombia and Ecuador. Biol. Conserv. 27:45–58.

Vaisanen, R., and O. Järvinen. 1977. Dynamics of protected bird communities in a Finnish archipelago. J. Anim Ecol. 46:891–908.

Whitcomb, R. F., J. F. Lynch, P. A. Opler, and C. S. Robbins. 1976. Island biogeography and conservation: strategy and limitations. Science 193:1030–1032.

Wilcox, B. A., and D. D. Murphy. 1985. Conservation strategy: the effects of fragmentation on extinction. Amer. Nat. 125:879–887.

Wright, S. J. 1985. How isolation affects rates of turnover of species on islands. Oikos 44:331–340.

Wright, J. S., and S. P. Hubbell. 1983. Stochastic extinction and reserve size: a focal species approach. Oikos 41:466–476.

A Comment on Quinn and Hastings: Extinction in Subdivided Habitats

MICHAEL E. GILPIN
Department of Biology (C-016)
University of California at San Diego
La Jolla, CA 92093, U.S.A.

In a review of the question of habitat subdivision for the conservation of a target species, Quinn and Hastings (*Conservation Biology* 1:198–208) review the SLOSS (Single Large Or Several Small) controversy and add a new view of their own. By likening reserve design to production line engineering and species to failure-prone components, they conclude that redundancy is the best policy in planning reliably for the persistence of a species. Following models of stochastic population growth developed by Leigh (1981), they derive a formula that specifies that the number of equally sized subdivided habitat patches be equal to the square root of the numeric carrying capacity of the target species. For example, given sufficient habitat to support 1000 individuals, it would be best to divide this habitat into 32 small patches of 32 individuals each. For black rhinos, down to about 8000 individuals, this works out to 90 reserves of 90 rhinos each. Thus, they push the old controversy of SLOSS to a more extreme dichotomy, SLOPP (Single Large Or Plentifully Patchy). While they do caution that their square root result should not be taken too seriously, they add that "the important point is that considerable redundancy in reserve systems may be effective in preventing overall extinction." I challenge the foundations of this theoretical finding. For the purposes of this discussion, I define Single Large as 1 or 2 patches (10^0), Several Small as 3 to 10 patches (10^1), and Plentifully Patchy as 10 to 100 patches (10^2).

Quinn and Hastings' support for Plentifully Patchy is based on the view that each population is isolated and dynamically independent. That is, there is no movement of individuals, which would allow gene flow, rescue (in the sense described by Brown & Kodric-Brown [1971]), and recolonization; nor are their populations assumed to share in any way a common environment. Further, Quinn & Hastings ignore genetics, which would affect isolated small populations earlier and more seriously than large populations, and demographic stochasticity, which would also have a disproportionately larger effect on small populations. With regard to demographic stochasticity, it must be understood that it is always a factor in extinction, for it applies whenever a population becomes small, which it must do at least once on the way to extinction. Together, these omissions mean that their Population Vulnerability Analysis (Gilpin & Soulé 1986) is incomplete — probably seriously so — and that it cannot, therefore, be taken as an unalloyed management recommendation.

There might, however, be a small subset of species for which all of the considerations mentioned above are of secondary importance. For these possible cases, I wish to criticize Quinn and Hastings' model in the context of their own assumptions. For concreteness, I refer to a situation where the habitat will support 256 individuals of the species of concern. Thus, the Plentifully Patchy strategy is for 16 habitat subdivisions of 16 individuals each. Figure 1a shows the Single Large configuration, while Figures 1b and 1c show two differently spaced versions of Plentifully Patchy, which I term Close Separated and Far Separated, respectively. Because they assume isolation and independence, Quinn and Hastings accept the two separated cases (Figs. 1b and 1c) as equivalent, and both, according to their reasoning, as superior to the Single Large.

Quinn and Hastings have reached their conclusion based on the ideas of reliability theory, which is a successful and important branch of engineering and economics. Reliability theory speaks to the failure time of a system fashioned of components each of which has its own characteristic probabilities of failure. This theory assumes that it is possible to place components in parallel. For several identical components in parallel, the system functions so long as one of these components works; it does not function any better when more than one component is working. The environment of the system in which these components function is constant. Reliability theory is of commercial importance since it is often possible to replace one expensive part with several cheaper and inferior parts placed in parallel, or vice versa. Superficially, it seems that reliability theory might be equally applicable to conservation biology, where scarce resources must also be invested wisely.

My first criticism is of their idea of "success." If at some point in time, say after 10 generations, 15 of the 16 patches of one of the separated systems have gone extinct, the system is taken to be just as "successful" as the single large system, which is assumed to be extant. Yet there are 256 individuals in the large system, representing virtually all of the initial genetic variability of the species, while the patch population of 16 individuals will have lost 30% of its heterozygosity ($H_{10} = H_0 * (1 - 1/(2*16))^{10}$). Further, the Single Large still allows much scope for management action, e.g., the establishment of a captive population, while this is greatly limited with the final patch population.

The second criticism concerns component independence. Even in production line machinery, the failure of components is not an independent Poisson process, for parts can become senescent or can fail together in response to a common shock. The same applies at least as strongly for biological components. In Quinn and Hastings' analysis, the persistence probability of the species declines dramatically just as the habitat patches (components) become contiguous, i.e., at the transition from Close Separated to Single Large. This happens because the Single Large is supposed to suddenly become a new kind of component. Think of 16 copper-plated circuit boards replaced by a single gold-plated version. Biologically, however, the material of the habitat patch components does not change; there is simply a juxtaposition of the patches. In Quinn and Hastings' model, this juxtaposition is assumed to produce two sharp changes in behavior. They assume first that the population becomes a cohesive unit in the Single Large, and second, that the totally uncorrelated environments of the patches in the Close Separated become perfectly correlated in the Single Large.

Both of these assumptions are questionable. If the species is sedentary to the extent that the Close Separated patches, Figure 1b, are totally isolated from one another, then it is unlikely that the juxtaposition pictured in Figure 1a will produce a homogeneous, well-mixed population. It could take substantial time for diffusional movements of individuals to even out the population distribution. To the extent that this is true, the Single Large will resemble the Close Separated with regard to independent population movement.

The more important deficiency involves their assumptions concerning the spatial correlation of the environment. They assume that the upper left patch in Figure 1a experiences the same environment as the lower right patch, whereas patches exactly the same distance apart in Figure 1b experience totally different environments.

This last point can be viewed in a different way. Suppose one were concerned with preserving a fish species distributed in the headwaters of the Madison, Galletin, and Yellowstone Rivers of Yellowstone National Park. As long as the park exists the fish are jeopardized by being in a Single Large. According to Quinn & Hastings, however, the dismemberment of Yellowstone National Park into three wilderness areas surrounding each of the rivers would break the environmental correlation and increase the species survival probability. This is, of course, absurd.

Different kinds of environmental variability have different spatial covariances. Extreme temperatures may span continents, whereas landslide will be restricted to single watersheds. This is an important aspect of reserve design that is not adequately understood by conservation biologists.

Epidemic diseases, especially those spread directly from host to host, may behave in the way envisioned by

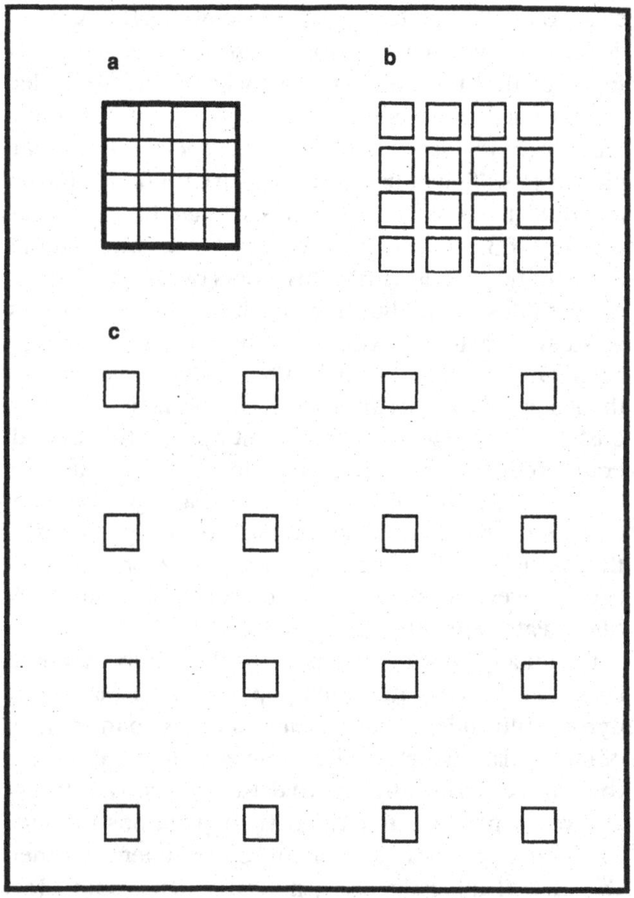

Figure 1.

Quinn and Hastings, that is, be correlated in the Single Large and uncorrelated in the separated systems. But this consideration for reserve design has been adequately explored (see Soulé & Simberloff 1986) and does not require the large degree of redundancy stipulated by reliability theory.

Even granting Quinn and Hastings' dismissal of genetics, demographic stochasticity, and island biogeographic dynamics, their model is not logically supportable. I have argued they were misguided in transferring the objects of reliability theory to biological conservation. In evolutionary ecology, where many conservationists have earlier worked, the play of contending ideas is the source of selectable intellectual variance, and it has almost always been rewarding to go into a sister science, such as economics, and borrow some of its more developed theory. In conservation biology, which is more applied, there is a different charge. Conservation biologists must make accurate predictions and build reliable structures. While Quinn and Hastings have addressed an important question, they make a specific recommendation, based on reliability theory, for subdivisions "much greater than 1." This recommendation is currently too weakly supported to be used as the basis for management action.

Literature Cited

Brown, J. H., and A. Kodric-Brown. 1977. Turnover rates in insular biogeography: Effects of immigration on extinction. Ecology **58**:445–449.

Gilpin, M. E., and M. E. Soulé. 1986. Population vulnerability analysis. Pages 19–34 *in* M. E. Soulé, editor. Conservation Biology, Sinauer, Sunderland, Massachusetts.

Leigh, E. G. 1981. The average lifetime of a population in a varying environment. Journal of Theoretical Biology **90**:213–239.

Soulé, M. E., and D. Simberloff. 1986. What do genetics and ecology tell us about the design of nature reserves? Biological Conservation **35**:19–40.

Comment

Extinction in Subdivided Habitats: Reply to Gilpin

JAMES F. QUINN
Division of Environmental Studies
University of California
Davis, CA 95616, U.S.A.

ALAN HASTINGS
Department of Mathematics and
Division of Environmental Studies
University of California
Davis, CA 95616, U.S.A.

The purpose of our article (Quinn & Hastings 1987) was to provoke discussion of the apparently widely held belief that subdivided populations are necessarily more susceptible to regional extinction than are comparably large panmictic populations. Gilpin's thoughtful comment suggests that we have been at least partially successful, and expands on some important issues that bear further investigation.

Our approach was to construct the simplest model of extinction in populations subdivided into multiple habitat patches (which might represent isolated nature reserves) in which the essential feature of both patch-size-dependent extinction rates and spreading extinction risks in space and time can be captured. Not surprisingly, the effects of subdivision on regional extinction do not follow merely from the degree of habitat fragmentation, but also depend critically on the relationship between fragment size and extinction probability. If there is a relatively abrupt threshold size above which extinction becomes very unlikely, as is the case with demographic stochasticity models (Richter-Dyn & Goel 1972) and a few empirical studies (Schoener & Schoener 1983), subdivision increases the chance of regional extinction in even the simplest of models. In essence, subdivision can push population sizes below the persistence threshold. However, if the decrease in extinction probability with increasing population size is fairly gradual, as in most environmental stochasticity models (Leigh 1981; Hanson & Tuckwell 1981; Gripenberg 1985) and a number of empirical studies (reviewed by Diamond 1984; Quinn & Hastings 1987; Quinn & Robinson 1988), then the model predicts that multiple independent populations of moderate size are likely to be more persistent than single large ones.

Consequently, if one wishes to argue as a matter of ecological principle that subdivided habitats, including dispersed reserves, increase the risk of regional extinction, ancillary assumptions are needed. Gilpin offers several candidates.

1. Demographic Stochasticity

As local populations become very small, the sampling variation inherent in the discrete nature of births and deaths can contribute to eventual extinction, even if average trends would predict persistence. As we discussed at some length (Quinn & Hastings 1987, pp. 200–201 and Fig. 1), our model behaves as Gilpin suggests — when the variation due to demographic stochasticity is much greater than that due to the environment, subdivision should increase regional extinction probabilities. Wright & Hubbell (1983) reach similar conclusions. This scenario is likely when local populations drop below roughly 20 individuals (Richter-Dyn & Goel 1972; Shaffer & Samson 1985). In the absence of more detailed analysis, it seems prudent to suggest that threatened populations should not be subdivided into local populations smaller than about 20 breeding females.

2. Genetic Effects

As populations become more subdivided, the average number of alleles in any subpopulation declines, the level of inbreeding may increase, and average homozygosity is likely to rise (See Soulé & Simberloff 1986 for a review). The effects of these processes on overall population size and extinction risk obviously vary from species to species, depending upon breeding system, susceptibility to inbreeding (and outbreeding) depression, and the extent of gene flow between subpopulations. Traditional population genetics models hold that approximately one migrant per generation is sufficient to maintain genetic variability in local populations (Ewens 1979).

From the perspective of minimizing the risk of regional extinction, increased mortality due to inbreeding effects would have to swamp the inherent variability in mortality from environmental causes to have a substantial effect on the behavior of our model. While such effects are plausible, convincing cases in natural populations are few.

From the perspective of maintaining genetic variability within threatened populations, subdivision poses inevitable tradeoffs. As the degree of subdivision increases, local subpopulations typically become more homozygous and less genetically diverse. At the species level, however, subdivision may be expected to increase overall genetic variability. This is most obvious for pure genetic drift in isolated subpopulations. Small subpopulations are likely to drift to fixation more quickly than larger subpopulations. However, it is likely that different subpopulations will fix different alleles, whereas all population-wide polymorphism is lost when fixation eventually occurs in a single large population. Similar arguments can be made for the maintenance of variability in the form of local adaptations in multiple subpopulations (see Slatkin 1985 for a review).

3. Migration

Our model assumed that migration between subpopulations can be ignored. For many species in parks and reserves, this may not be an unreasonable assumption. However, as Gilpin observes, migration tends to increase regional persistence, both through a "rescue effect" (Brown & Kodric-Brown 1977) and through re-immigration to uninhabited patches. In general, migration between patches might be expected to *increase* the advantages of multiple reserves, because, as the number of subpopulations grows, it becomes increasingly unlikely that all will disappear simultaneously. If migration rates are high, even a few surviving subpopulations can re-establish the entire metapopulation, effectively buffering it against overall extinction.

More formally, we can add immigration to our model by supposing that local immigration of empty patches, like local extinction from occupied patches, occurs with some specified probability per unit time. Presumably the per-empty-patch immigration probability will be roughly proportional to the number of inhabited patches. In this scenario, changes in the number of occupied patches will approximate those predicted by a stationary branching process analogous to those analyzed by Richter-Dyn & Goel (1972) and Leigh (1981) for populations of individuals. These models have the general feature that the expected persistence time increases approximately exponentially with the number of individuals present (where, in this case, the "individuals" are occupied patches). Assuming that the expected number of births (immigration events) exceeds the expected number of deaths (local extinctions) as populations become small, these models generally predict that populations in excess of roughly 20 (patches) will persist essentially indefinitely.

For these reasons, it is far from clear that ignoring immigration invalidates our general conclusions. The "reliability theory" model, without migration, predicts that overall persistence times should increase with something like the logarithm of the number of occupied reserves. Under many reasonable assumptions, this suggests that maintaining considerable redundancy (multiple populations) is the prudent course. If migration is important, persistence time should increase essentially exponentially, rather than logarithmically, with the number of reserves. This can only serve to increase the value of multiple populations.

4. Environmental Correlation

Gilpin correctly observes that our model assumed that the environment fluctuates independently among the various reserves. Clearly some kinds of environmental variation, e.g., weather, are likely to be somewhat correlated among reserves. We noted in passing that such correlation reduces the value of redundant subpopulations, but presented no analysis.

In the extreme case that all individuals experience identical environments—i.e., the environments in the reserves are perfectly correlated—the probability of losing the entire population to an environmental catastrophe is roughly the product of the probabilities of losing each individual, irrespective of distributions of individuals among reserves. We did not suggest and do not believe that purposeful fragmentation *per se* will decrease the probability of extinction of highly correlated subpopulations. By the same token, there is no logical reason to accept Diamond & May's (1976) suggestion that the establishment of a barrier (in their example, a road) dividing a population into two subpopulations should increase the probability of overall extinction. Any large effects would have to be a result of changes in mortality or fecundity due to the barrier itself.

A barrier could increase mortality or reduce fecundity, for example, by disrupting seasonal migration. Gilpin raises the example of stream fish in Yellowstone as a case where establishment of a barrier might promote extinction. Secondary effects of barriers (e.g., slash-and-burn agriculture following roadbuilding in tropical forests) can certainly have substantial and deleterious consequences.

However, barriers may also decrease mortality by preventing the spread of predators, competitors, or disease. One example of a barrier protecting species in correlated environments was once provided by Niagara Falls, which acted to protect native fishes in the western Great Lakes from decimation by lamprey predation (Smith 1968). Barriers to the spread of disease could be similarly valuable, as the black-footed ferret experience has shown. We certainly do not share Gilpin's confidence that the problem of extinction caused by epidemics "has been adequately explored."

If the correlation among patches is not complete, the results for the uncorrelated case may provide a reasonable approximation of the behavior of a correlated patch model. Chesson (1981) has shown that in many models, the dynamics of populations in patchy environments with environmental variability and correlation are similar to the uncorrelated case unless the correlation is almost complete. Explicit results depended upon the details of the model.

Elsewhere, Gilpin (1988) observes that persistence times of subdivided populations with migration and considerable environmental correlation should increase roughly in proportion to the logarithm of the number of subpopulations. Harrison & Quinn (1988) find similar results in computer simulations. Thus a model with both migration and correlated environments behaves qualitatively the same as the "reliability theory" model with neither.

In sum, we are not persuaded that genetics, migration, or environmental correlation invalidates our general approach. Nevertheless, we agree with Gilpin that the simple reliability theory models are too general to provide good estimates of the optimum number of reserves (in his terms, to distinguish between "several small" and "plentifully patchy"). As in mechanical systems, the most dramatic benefits of redundancy occur when moving from a single failure-prone unit to several levels of backup (see Quinn & Hastings 1987, Fig. 1).

Our main point is that having too few reserves may lead to long-term failure as surely as having reserves that are too small. Therefore we suggest that more reserves are needed, and that targeted small-scale acquisition efforts, such as those of the Nature Conservancy, can make very important contributions to long-term persistence of threatened species. However, as a practical matter, location, habitat diversity, and a variety of management and political considerations are likely to override simple reserve size/number optimizations in establishing the preferred reserve acquisition policy.

Finally, we do not accept Gilpin's prescription that models in conservation biology must be specifically predictive to be useful. Models in many applied fields (e.g., economics) are used for heuristic purposes (to clarify assumptions and verbal arguments) and for identifying particularly important uncertainties for further study (sensitivity analysis), as well as to make numerical predictions. Typically, somewhat different formulations are needed to achieve these varying purposes (see Levins 1966). Simple heuristic models generally do not incorporate enough detail to make precise predictions. Conversely, detailed management models frequently cannot be understood intuitively, analyzed for general behavior, or easily applied to different species or communities. Sensitivity analyses may concentrate on effects that are thought to be important, but that are too poorly understood to be useful for either heuristic or predictive purposes.

The intent of our paper was primarily heuristic. However, we believe the analysis is of potential value to conservation biologists in:

1. suggesting that the specific relationship between patch (reserve) size and extinction probability needs to be better understood to make informed reserve acquisition decisions;

2. showing that the mean persistence time of populations can be a misleading statistic; and

3. demonstrating that the common intuition that habitat subdivision *per se* elevates rates requires additional ancillary assumptions that may or may not be satisfied in particular cases.

Gilpin provides candidates for several kinds of ancillary biology that might make larger single populations persist longer than multiple moderate-sized populations. At this stage, none seems to invalidate our general conclusions, but all bear further investigation, and should be incorporated into more complex models to tease apart their potential relative contributions to effective management of real-world threatened metapopulations.

Literature Cited

Brown, J. H., and A. Kodric-Brown. 1977. Turnover rates in insular biogeography: Effect of immigration on extinction. Ecology **58**:445–449.

Chesson, P. L. 1981. Models for spatially distributed populations: The effect of within-patch variability. Theoretical Population Biology **19**:288–325.

Diamond, J. M. 1984. "Normal" extinctions of island populations. Pages 191–196 *in* M. H. Nitecki, editor. Extinctions. University of Chicago Press, Chicago.

Diamond, J. M., and R. M. May. 1976. Island biogeography and the design of nature reserves. Pages 163–186 *in* R. M. May, editor. Theoretical Ecology: Principles and Applications. Saunders, Philadelphia.

Ewens, W. J. 1979. Mathematical Population Genetics. Springer-Verlag, New York.

Gilpin, M. E. 1988. Extinction of finite metapopulations in correlated environments. Unpublished Manuscript. University of California at San Diego, La Jolla, California.

Gripenberg, G. 1985. Extinction in a model for the growth of a population subject to catastrophes. Stochastics **14**:149–163.

Hanson, F. B., and H. C. Tuckwell. 1981. Logistic growth with random density independent disasters. Theoretical Population Biology **19**:1–18.

Harrison, S., and J. F. Quinn. Correlated environments and the persistence of metapopulations. Unpublished material.

Leigh, E. G., Jr. 1981. The average lifetime of a population in a varying environment. Journal of Theoretical Biology **90**:213–239.

Levins, R. 1966. The strategy of model building in population biology. American Scientist 54:421–431.

Quinn, J. F., and A. Hastings. 1987. Extinction in subdivided habitats. Conservation Biology **1**:198–208.

Quinn, J. F., and G. R. Robinson. 1988. Habitat fragmentation, species diversity, extinction, and the design of nature reserves. In S. Jain and L. Botsford, editors. Applied Population Biology. Dr. W. Junk, Dordrecht, Netherlands, *in press.*

Richter-Dyn, N., and N. S. Goel. 1972. On the extinction of a colonizing species. Theoretical Population Biology **3**:406–433.

Schoener, T. W., and A. Schoener. 1983. The time to extinction of a colonizing propagule of lizards increases with island area. Nature **302**:332–334.

Shaffer, M. L., and F. B. Samson, 1985. Population size and extinction: A note on determining critical population size. American Naturalist **125**:144–152.

Slatkin, M. 1985. Gene flow in natural populations. Annual Review of Ecology and Systematics **16**:393–430.

Smith, S. H. 1968. Species succession and fishery exploitation in the Great Lakes. Journal of the Fisheries Research Board of Canada **25**:667–693.

Soulé, M. E., and D. S. Simberloff. 1986. What do genetics and ecology tell us about the design of nature reserves? Biological Conservation **35**:19–40.

Wright, J. S., and S. P. Hubbell. 1983. Stochastic extinction and reserve size: A focal species approach. Oikos **41**:466–476.

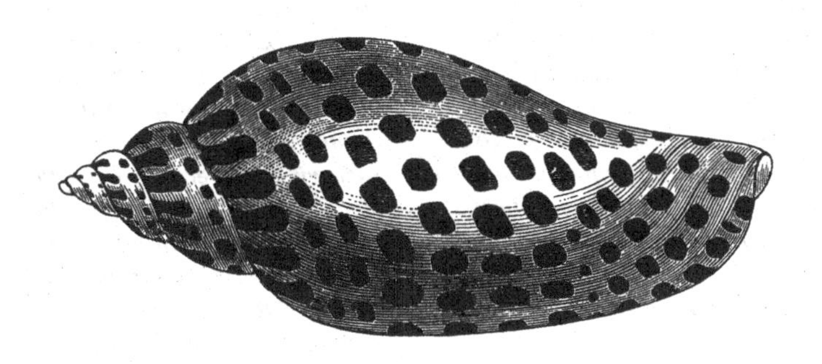

Consequences and Costs of Conservation Corridors

DANIEL SIMBERLOFF*

Department of Biological Science
Florida State University
Tallahassee, FL 32306

JAMES COX

Nongame Wildlife Program
Florida Game and Fresh Water Fish Commission
Tallahassee, FL 32301

Abstract: *There are few controlled data with which to assess the conservation role of corridors connecting refuges. If corridors were used sufficiently, they could alleviate threats from inbreeding depression and demographic stochasticity. For species that require more resources than are available in single refuges, a network of refuges connected by corridors may allow persistence. Finally, a corridor, such as a riparian forest, may constitute an important habitat in its own right. A dearth of information on the degree to which different species use corridors makes it difficult to tell which of these potential advantages will be realized in any particular case. Some experimental field studies suggest that certain species will use corridors, although lack of controls usually precludes a firm statement that corridors will prevent extinction.*

Corridors may have costs as well as potential benefits. They may transmit contagious diseases, fires, and other catastrophes, and they may increase exposure of animals to predators, domestic animals, and poachers. Corridors also bear economic costs. For example, a bridge that would maintain a riparian corridor costs about 13 times as much per lane-mile as would a road that would sever the corridor. Also, per-unit-area management costs may be larger for corridors than for refuges. It may be cheaper to manage some species by moving individuals between refuges rather than by buying and maintaining corridors.

Each case must be judged on its own merits because species–environment interactions differ. As an example, we used the case of the Florida panther (Felis concolor coryi), of which there remain about 30. The Florida panther's potential inbreeding problems could possibly be stemmed somewhat by a corridor system, but it is far from certain that even an extensive system will save this animal, and the cost of such a system would lessen the resources that could be devoted to land acquisition and other means of aiding many other threatened species.

Resumen: *Existen pocos datos probados con los cuales evaluar el rol para con la conservación, de los corredores que conectan los refugios naturales. Si los corredores se usaran suficientemente, esto reduciría las amenazas de depresión endogamica y la estocasticidad demográfica. Para las especies que requieren mas recursos que los disponibles en un solo refugio, una red de refugios conectados por corredores asegurara la persistencia de estas. Finalmente, un corredor, como por ejemplo un bosque ripario (de galeria) puede constituir un importante región por sí mismo. La escasez de información acerca de hasta qué punto las especies utilizan estos recursos dificulta la determinación de las ventajas potenciales para cada caso. Algunos estudios de campo indican que ciertas especies utilizan los corredores, no obstante que la falta de controles experimentales precluyen afirmaciones que los corredores previenen la extincion.*

Aun cuando los corredores proveen beneficios potenciales, tambien pueden incurrir costos. Las enfermedades contagiosas, los incendios y otras catástrofes pueden ser transmitidas por los corredores. Estos pueden aumentar la vulnerabilidad de los animales y sus predatores, los animales domésticos y los cazadores furtivos. Los corredores pueden incurrir costos económicos. Por ejemplo, un puente que mantiene un corredor ribereño puede costar trece veces mas por carril/milla que una carretera que cortaria el corredor, y el manejo por unidad/area puede ser mayor que para los refugios. Puede ser mas costeable el manejar ciertas especies trasladándolas entre refugios que adquiriendo y manejando corredores.

Cada caso debe juzgarse por sus propios méritos porque las relaciones especie-ambiente son tan diferentes. Queremos usar el caso de la pantera de Florida como ejemplo (Felis concolor coryi). Quedan aproximadamente 30 individuos de esta especie y es posible que los problemas potenciales de depresión endogamica puedan ser disminuidos con un sistema de corredores. Sin embargo, no es seguro que un sistema extensivo de corredores pueda salvar a esta especie y el costo reduciría los recursos disponibles destinados a la adquisición de tierras y otras formas de asistencia para otras especies amenazadas.

Correspondence and requests for reprints should be mailed to this author.

Introduction

The Seychelles islands in the Indian Ocean contained 14 endemic land bird species when Europeans first arrived there in 1770. In the two centuries since then, land clearing, a series of fires, and introduction of predators such as rats and cats have devastated the archipelago. However, only the green parakeet *(Psittacula eupatria wardi)* and the chestnut-flanked white-eye *(Zosterops mayottensis semiflava)* have been extinguished (Penny 1974). Losses were limited partly because the Seychelles consists of several separate, small islands (fires and introduced predators were unable to reach all the islands). For example, the Seychelles magpie robin *(Copsychus sechellarum)* remains only on Frigate Island; feral cats have destroyed it elsewhere but were controlled on Frigate and cannot reinvade because of the water barrier.

We describe this example simply to point out that corridors connecting refuges are not always an unmitigated blessing. The idea that corridors should be maintained between refuges whenever possible was suggested by Wilson and Willis (1975) as a logical consequence of the equilibrium theory of island biogeography (MacArthur and Wilson 1967). This suggestion is the subject of our article.

Possible Advantages

The corridor recommendation is an automatic consequence of the equilibrium theory. This theory states that the number of species in an insular site (like a refuge) is a dynamic equilibrium between local, on-site extinction of resident species and occasional stochastic immigration to the site by species not currently resident. Thus the composition changes but the number of species stays approximately constant. According to the original equilibrium theory, corridors would act by increasing the immigration rate. Once a species is locally extinguished on a refuge, the expected time to the next reimmigration is lowered by the availability of corridors. On average, then, there would be more species present any time one censused the site. Wilson and Willis (1975) add that corridors will also maintain higher numbers of species in refuges because species that would otherwise have become extinct will be maintained by continuing reciprocal immigration from other sites (Fig. 1). This is a version of the phenomenon subsequently termed the "rescue effect" by Brown and Kodric-Brown (1977).

Harris (1984, 1985) suggested two other reasons why corridors should be part of conservation plans: 1) individuals of some species, especially large mammals, must

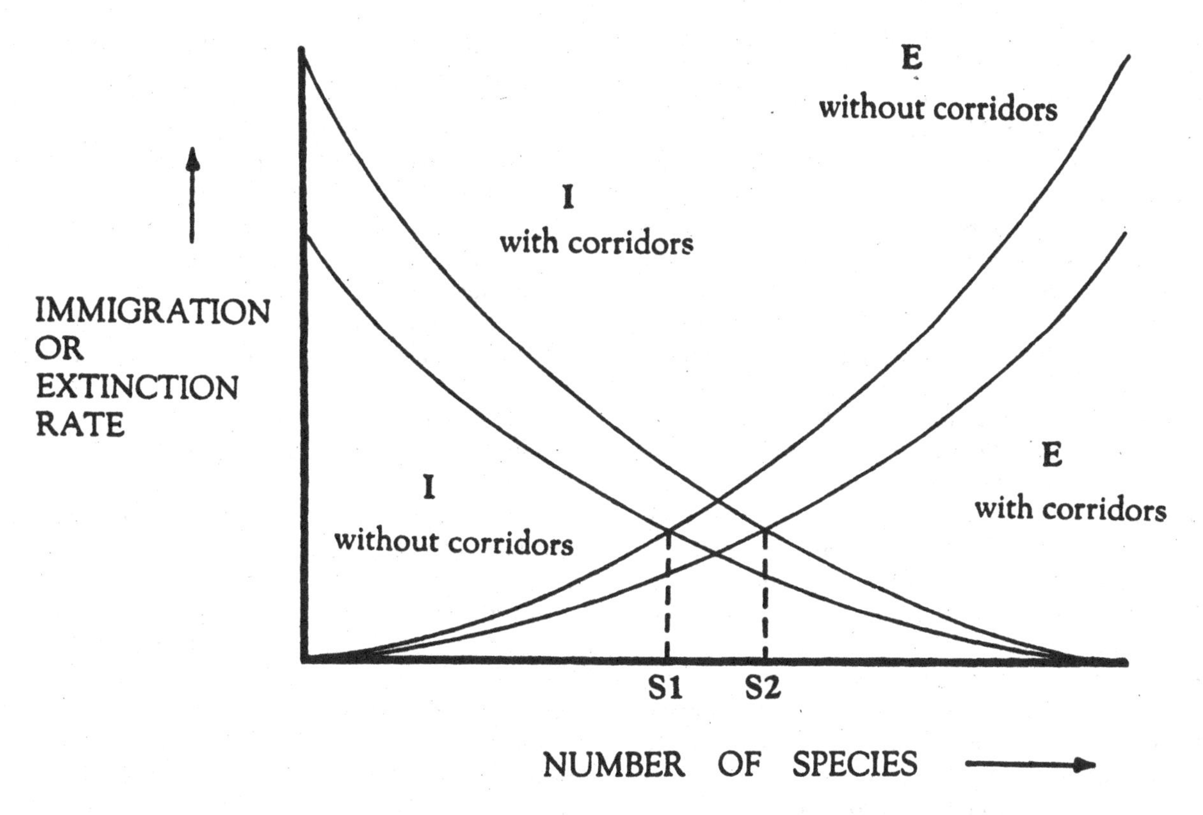

Figure 1. Effect of corridors on immigration rate (I), extinction rate (E), and resulting number of species in equilibrium island biogeographic model. S1 is equilibrium number of species without corridors. S2 is equilibrium number of species with corridors.

range widely in order to meet their food requirements (single, small refuges will not contain enough food), and 2) if population sizes within single refuges are too small, inbreeding depression will ensue and lead to extinction.

With respect to 1), if each individual of a species requires X calories of some food per year and a refuge produces only 20X calories per year, it is likely that the species cannot persist in this refuge alone even if all the food were available and the individuals were able to find all of it. This is the question of the minimum viable population (MVP) size consistent with long-term viability of the population. There are several stochastic reasons, such as temporal variation in demographic parameters, why very small populations are at much greater risk than larger ones (Shaffer 1981, Soulé and Simberloff 1986). If food within a refuge were limiting and if individuals within a population actually used the corridors to feed in other areas, a network of refuges might support a population where no one component within the network would have sufficed.

Inbreeding depression is more of a threat to small populations than large ones and could be alleviated by corridors. Although current consensus (May 1973, Shaffer 1981) is that demographic stochasticity is probably a much more important short-term danger to most threatened populations (e.g., National Research Council 1986, Dawson et al. 1986), inbreeding depression has often been demonstrated in small populations (e.g., Ralls and Ballou 1982*a*, 1982*b*, 1983) and must be considered as a possible threat. However, different species appear to tolerate inbreeding to different degrees (Ralls and Ballou 1983), and the sort of research that establishes the effect of inbreeding on fitness and even the existing degree of inbreeding is extremely difficult. The threat of extinction from demographic stochasticity could also be alleviated by corridors. Certainly it is alleviated by migration into populations via the "rescue effect" mentioned above. Fahring and Merriam (1985) modeled the performance of a white-footed mouse *(Peromyscus leucopus)* metapopulation with a range of deterministic birth and death rates and migration rates between populations in different simulated woodlots. Populations within the metapopulation that received no immigrants were more likely to become extinct than those that received immigrants. Fahring and Merriam also suggested that, as a matter of common sense, isolated populations that became extinct would be replaced more slowly than would connected populations that did so. Because local *Peromyscus leucopus* populations occasionally become extinct and because they are known to move occasionally between woodlots, this modeling approach may help to pinpoint the contribution of migration to population persistence.

One can easily envision the incorporation of corridors into such a model, although the results would be useful only for realistic values of corridor use: Do individuals move through the corridors and at what rates? The key additional feature required by such a model would be to substitute stochastic demography (birth rates, death rates, sex ratio, etc.) for the deterministic treatment, as suggested by Shaffer (1985) for the spotted owl *(Strix occidentalis)* metapopulation in the Pacific Northwest. Such stochastic variation would probably tend to increase the rate of local population extinction and thus enhance the role of migration. However, without data on use of migration corridors, even a simulation enlarged in this way would not be able to indicate how important corridors are.

Finally, if a corridor constitutes an important and dwindling habitat in its own right, as do some riparian corridors (Noss and Harris 1986), then the corridor can function as a refuge, and the benefits from this source must be taken into account. Forman (1983) argues that most North American transmission line corridors would constitute unique habitats for the species residing in them only if the rest of the land were forested, while hedgerows are unique habitats only when no large woods are present. Henderson et al. (1985) found that chipmunks *(Tamius striatus)* bred within fencerows even when nearby woods supported breeding populations. Whether a corridor is an important habitat in its own right is obviously a matter that depends on the target biota and surrounding habitat matrix.

Does Extinction Occur? To What Extent Are Corridors Used?

If the equilibrium theory model is to represent validly even part of the role of corridors as proposed by Wilson and Willis (1975), two criteria must be met. First, stochastic population extinction must be continual in the refuges at measurable rates in the short term. Second, there must be immigration through the corridors. In short, there must be "turnover." It is widely agreed that extinction will occur for refuges (Soulé and Simberloff 1986), although the probable rates of such extinction are in dispute (e.g., Soulé et al. 1979, East 1983; vs. Western and Ssemakula 1981, Boecklen and Gotelli 1984, Boecklen and Simberloff 1986). Part of the problem in estimating these rates is the lack of controlled experimental data, a lacuna that is being redressed (e.g., Lovejoy et al. 1983).

Whether particular kinds of corridors will actually be used by the organisms for which they are intended and whether they will be used at rates high enough to forestall extinction have generally not been considered as frequently as the question of extinction rates. Data are slight, and no consensus has emerged (Forman 1983). Rivers and riparian habitats are often suggested as corridors, but many deep forest species may not venture into these habitats (Frankel and Soulé 1981). For example, Harris (1984) recommends riparian corridors between stands of mature Douglas fir forest but the red tree vole *(Arborimus longicaudus)* and the California red-backed vole *(Clethrionomys californicus)* would probably avoid such corridors (Soulé and Simberloff 1986). Forman (1983) and Forman and Baudry (1984) report that plants and small mammals do not move along hedgerow corridors efficiently while some birds and larger mammals do. Harris (1985) argues that corridors such as the Bering land bridge and the Panamanian Isthmus have, over geological periods of time, served as conduits for entire biotae. However, it is difficult to see how this is relevant to present conservation needs. These historical land bridges are tens or hundreds of miles across and have persisted for thousands of years. Such corridors were much larger than most present or projected refuges! Many species have used these land bridges, but this does not imply they would use the kind of corridors currently envisioned in conservation planning.

Studies cited to support the contention that corridors actually increase immigration and/or decrease extinction and thus raise the number of species in a site are, upon close examination, often insufficiently controlled to demonstrate this point. For example, MacClintock et al. (1977) entitled their paper "Island biogeography and the habitat islands of eastern forest. II. Evidence for the value of corridors and minimization of isolation in preservation of biotic diversity." One would expect to see a design approximating one or more habitat islands a given distance from a source area and connected to the source by corridors, compared to one or more habitat islands equidistant from the same (or a similar) source but not connected by corridors. Instead, there are two forest islands connected to a large forest by a corridor, plus two quadrats within the large forest. There is no way to determine from data on these sites whether the corridor facilitated movement to the habitat islands.

In several studies, Merriam and coworkers have attempted to provide more control on whether small mammals use corridors (Wegner and Merriam 1979, Middleton and Merriam 1981, Henderson et al. 1985). They concluded that both chipmunks and white-footed mice probably use fencerows to move between woods and that this movement is important because populations in small woods probably are extinguished occasionally. The importance of the fencerows was indicated, particularly in the recolonization of woods from which chipmunks had been experimentally removed, but there is still some ambiguity in the results. For example, Henderson et al. (1985) found some colonizations to have required crossing 60 m of open pasture and some marked individuals to have moved about 1000 m and crossed a two-lane gravel road. In no instance was the same wood experimentally defaunated more than once and then recolonized with and without corridors. Nor were connected and unconnected woods defaunated. However, of all recorded movements between woods, fewer than 10 percent were not known to have required fencerows. It therefore seems likely that the fencerows facilitate both recolonization of areas in which populations have become extinct and the operation of the rescue effect to forestall such extinction.

Lee Harper (personal communication) has studied the effect of a corridor on ant birds (Formicariidae) in an experimental forest island of Brazilian rain forest (Lovejoy et al. 1983). The island, of approximately 100 ha, was connected by a 2 km corridor at least 100 m wide along a stream to a large rain forest. The island had been surveyed for at least a year before 300 m of the corridor was destroyed in August 1984. Three species of ant birds disappeared within four weeks even though army ants were still present. There was no control island for which the corridor was *not* severed, but using the same site over again as a temporal control is possible. Harper (personal communication) reports that, after a year of second growth in the corridor, one of the three ant bird species is beginning to recolonize.

Possible Disadvantages

So far we have spoken as if a set of corridors has few potential costs, only potential benefits. The equilibrium theory of island biogeography treats only a limited range of phenomena that might bear on extinction. It does not, for example, deal explicitly with genetics. We have already discussed potential benefits from corridors in addition to those predicted by the equilibrium theory. Several possible detrimental effects should also be considered.

First and foremost, the catastrophic contagious effects such as fires and introduced predators in the Seychelles have attracted much attention (Simberloff and Abele 1982). For example, the Solomon Islands' brown tree snake *(Bioga irregularis),* introduced to Guam during World War II, is eating its way through the island's avifauna (Jaffe 1985, Savidge 1985). This nocturnal predator has already eliminated the bridle white-eye *(Zosterops conspicillata)* and the rufous fantail flycatcher *(Rhipidura rufifrons)* from the island and threatens at least three other species. As in the Seychelles example cited earlier, one can imagine that if Guam consisted of several isolated, small islands instead of one large one, the snake's effect would be contained or at least greatly retarded.

Contagious diseases are another such agent of extinction; for example, 20 of the remaining 60 Javan rhinoceroses *(Rhinoceros sondiacus)* died of an unknown contagious disease in 1984 (*Oryx* 1984), and a key factor in the extinction of the heath hen *(Tympanuchus cupido cupido)* in the eastern United States was blackhead, a disease contagious among poultry, that ravaged the population, which had been gathered entirely in a single refuge on an island off Martha's Vineyard, Massachusetts

(Simberloff 1986). Needless to say, a corridor can transmit a disease, fire, or predator.

Second, corridors increase the exposure of animals to humans (facilitating poaching), domestic animals (facilitating diseases), and predators (Soulé and Simberloff 1986). If a corridor were thin enough, it would not be surprising if hunters used it, legally or illegally, by stationing themselves at appropriate locations. Both mule deer *(Odocoileus hemionus)* and Virginia white-tailed deer *(Odocoileus virginianus)*, for example, tend to follow specific routes, and hunters exploit this behavior by monitoring these routes. Bobcats tend to follow established routes quite closely. In Wyoming, during one 90-day period, 39 bobcats were trapped at one location through a catnip bait placed near a trail (Young 1978). Animal predators also patrol particular travel routes. For example, the eastern diamondback rattlesnake *(Crotalus adamanteus)* typically stations itself along frequently used mammal runs (D.B. Means, personal communication). Whether such threats should be a major concern would, of course, depend on the sorts of animals expected to use a corridor, the width of the corridor, potential predators, potential vectors of diseases, and a host of other factors.

Third, corridors can provide an entrée for weedy and opportunistic species into a forested habitat that might otherwise be closed to them (Noss and Harris 1986). In Florida, Maehr and Brady (1984) warn that black bears *(Ursus americanus)* may disperse Brazilian pepper *(Schinus terebinthifolius)*, a noxious exotic whose fruits are readily consumed by bears. Corridors proposed by Noss and Harris (1986) for bears and other large vertebrates may help to spread this plant. Janzen (1983) describes how facilitating movement of secondary successional species can lead to the gradual degradation of pristine tropical forest.

Fourth, highway right-of-way and median strip corridors can lead to lead poisoning (O'Neill et al. 1983), depending on the volume of traffic and the amount of time animals spend in such corridors, not to mention the potential death from collisions (Harris 1985).

Fifth, we must address the economic component of this issue. It costs something to purchase, maintain, and protect corridors. As one example, Harris (1985) argued that high bridges over rivers and highways will allow safe passage through the resultant corridors while planar intersections contribute to highway carnage. This contention is certainly true, but in Florida the average cost of a road is $500,000 per lane-mile, while the average cost of a bridge is $6.86 million per lane-mile, which is greater than a tenfold increase (Florida Department of Transportation, S. Burnett, personal communication). Other management costs are probably greater for corridors than for an equivalent area of refuge if only because corridors have more edge so their habitat is more likely to be influenced by the surrounding habitats. Just the cost of fencing alone would be far greater for a corridor than for an equal-size compact refuge. Maintaining plants can be another cost (e.g., Nilson 1977, Ericson 1979).

As Harris (1985) notes, "We can never buy, own, possess, or totally control enough land to preserve everything." Thus current conservation strategies are fraught with important, perhaps paramount, cost/benefit considerations. We must therefore ask how corridor acquisition and management complements or conflicts with the acquisition and management of isolated areas that have intrinsic biological values beyond their connective values. We should also consider the suggestion that the accretion of human development may soon hinder our abilities to acquire large reserves (Soulé and Simberloff 1986).

Corridors and Gene Flow

The facilitation of gene flow by corridors need not be automatically desirable. If separate populations of a species have characteristically different genotypes, such differences will tend to be broken down. Another example from the Seychelles avifauna (Penny 1974, Cade 1983) makes this point. The Seychelles turtledove *(Streptopelia picturata rostrata)* is a morphologically distinct subspecies of a widely ranging species. It differs greatly in color and size from the Madagascar subspecies *S.p. picturata.* The latter subspecies was introduced from Mauritius or Madagascar at least by the mid-nineteenth century (Long 1981) and has either replaced or diluted (through interbreeding) the native turtledove on all but two or three small, remote islands: Cousin, Cousine and possibly Frigate (Penny 1974). On these islands the genetic swamping was greatly slowed, but even here there has been sufficient recent introgression that the pure form of the turtledove can be found only in museums (Penny 1974). It is highly likely that, if the Seychelles had been connected by corridors of land, the destruction of its turtledove would have been much quicker.

The preservation of local genetic variants, which may be adapted to the local environment or simply the fortuitous product of the founder effect, genetic drift, or random mutation, may or may not be a paramount goal. Sometimes such preservation is of great symbolic value. Certainly the struggle of Disney World and a number of ornithologists to save at least a substantial fraction of the racial characteristics of the dusky seaside sparrow *(Ammodramus maritimus nigrescens)* falls in this category. As long as there was a chance of finding more individuals, the U.S. Fish and Wildlife Service would not sanction breeding the five remaining males with females of other subspecies in order to protect the genetic purity of the subspecies (Cade 1983, Wilford 1986). In the debate over the importance and cost of maintaining or resurrecting this form, the adaptive role of the threatened genotypes was also adduced. Even when the adaptive value of geographic variation has not been ascertained (e.g., the Florida panther, Belden 1985, Cristoffer and Eisenberg 1985, R. Noss, personal communication), the interest in unusual morphology alone may be sufficient to compel a wish to preserve a variant.

A concern of conservation genetics is whether loss of genetic variability in threatened species will hinder the ability of the species to evolve in response to future environmental change (Soulé and Simberloff 1986). Chesser (1983) points out that one possible strategy for maintaining genetic variability is to maintain a species as a collection of isolated populations, each containing a high frequency of a different allele. In this approach, the subdivision of a larger population allows the effects of genetic drift to assist in the maintenance of genetic variability in the species. The within-population variation decreases more quickly than it would in the single large population, but the among-population variation decreases much more slowly.

One problem with this strategy, of course, is that smaller, isolated populations can suffer inbreeding depression more quickly than would a united, larger population and can eventually disappear. Chesser et al. (1980) and Chesser (1983) suggest that by deliberate movement of small numbers of individuals from population to population the dangers of inbreeding can be greatly lessened. Just how many individuals should be moved, and in what direction, is a complicated matter resting on effective population sizes, selection coefficients, and the specific effects of inbreeding. For particular sets of these variables the number of individuals that must be transferred among populations can be surprisingly small—so small that it might be genetically advantageous and far less expensive to move individuals manually rather than to provide corridors. Translocation also allows managers to monitor closely the gene flow between populations. Transferring bird eggs from population to population, for example, is a relatively straightforward and inexpensive procedure for many species (James 1983, Logan and Nesbitt 1986). Even large vertebrates can be efficiently and inexpensively translocated under certain circumstances. For example, Weise et al. (1975) successfully moved timber wolves *(Canis lupus)* from Minnesota to Michigan, although illegal killings and collisions with cars eliminated all four transplanted individuals.

Another problem with balkanizing a threatened population is that each smaller population is more susceptible to extinction by stochastic processes and certain catastrophic events. As mentioned, if corridors increase the chances of the "rescue effect," then populations connected by corridors would persist longer than separated populations.

Deciding to split or to lump populations for these reasons involves the type of "risk analysis" that is becoming part of many conservation and management programs (Salwasser et al. 1984). At the heart of this analysis is estimating the relative chances of success under different conservation strategies before implementing a single course of action. As an example, by including genetical parameters in probabilistic demographic models (e.g., Shaffer 1985), a manager can estimate whether several small populations could maintain higher levels of genetic variability without experiencing many localized extinctions, or whether a single larger population could maintain reproductive vigor under reduced levels of genetic variability in future generations. It seems to us that this type of analysis is required before corridors are established.

An Example: The Florida Panther

A discussion of current conservation strategies for the Florida panther, which include a possible extensive corridor system (Cristoffer and Eisenberg 1985, Noss and Harris 1986), helps to illustrate the need for a careful assessment of all costs and benefits associated with corridor establishment.

The population of panthers in Florida has been reduced from a presettlement population of approximately 1400 (Cristoffer and Eisenberg 1985) to a vestigial population of perhaps as few as 30 isolated in as yet undeveloped areas of South Florida (Florida Game and Fresh Water Fish Commission Annual Research Report, 1985). The population was once contiguous with other North American populations and the adaptive significance of subspecific characteristics has not been established (Noss, personal communication; Belden 1985; Cristoffer and Eisenberg 1985).

One corridor system planned for the Florida panther and other wide-ranging mammals (Cristoffer and Eisenberg 1985, Noss and Harris 1986) envisions establishing a captive breeding program and releasing individuals into an integrated refuge system in North Florida. Two national forests, three national wildlife refuges, and several state and private land-holdings may be united in this effort, connected by riverine and coastal corridors of an undetermined width, some stretching nearly 130 km through unprotected areas (Noss and Harris 1986). Harris (1985) contends that this corridor system is essential in safeguarding the survival of many wide-ranging mammals, such as the panther and the black bear. Discussions on establishing corridors between panthers translocated to North Florida and panthers in South Florida have also begun (Noss, personal communication).

Many unresolved biological and monetary questions linger over this proposition. As Harris (1985) notes, "Virtually no research has been done to establish the necessary dispersal corridor widths [for mammals]," which leaves the value of any proposed corridors in question until considerable additional research is funded.

That panthers, bears, and many other wide-ranging mammals will move 50–130 km between refuges along riverine corridors seems especially uncertain without specifying extremely broad widths.

The cost of the captive breeding program for panthers has been estimated at over $2 million for 15 years (Grow 1984), and unexpected costs may be associated with preventing the spread of contagious diseases among connected populations (e.g., feline distemper is prevalent in the South Florida population) (Florida Game and Fresh Water Fish Commission Annual Research Report, 1985). It will also cost tens of millions of dollars to construct bridges and manage corridors to prevent road and illegal killings—the major sources of known mortality (Florida Game and Fresh Water Fish Commission Annual Research Report, 1985).

Expanding the current population size and range of Florida panthers seems essential to the animal's survival, but whether corridors are needed among translocated populations to offset problems of inbreeding depression is less certain. There is evidence that the South Florida population already suffers from genetic problems (as indicated by a very large percentage of abnormal sperm in panther semen) (Florida Game and Fresh Water Fish Commission Annual Research Report 1985), but a better method of preserving extant levels of genetic variability might be to isolate new translocated populations. Given the home range sizes for panthers presented by Harris (1985), two isolated populations of approximately 25 to 40 might be established on the two largest tracts of protected lands involved in the proposed system (which have already been acquired). Three populations of approximately 25 to 40 panthers could then be maintained and cross-bred selectively through a recurring captive breeding program. Without analysis of the various risks associated with these integrated or divided panther populations, it is difficult to determine the better strategy.

Finally, the cost of this proposed corridor system may detract from other valuable conservation and management efforts in Florida. Many areas in South Florida where panthers range are not currently protected and perhaps could be acquired with money spent on proposed corridors. The degree of human growth projected to occur in South Florida (Fernald 1981) certainly will threaten chances of purchasing large tracts later. Because panthers apparently prefer deciduous habitats, it has been recommended that panther reintroduction areas in North Florida be managed for "phasing out mature pines" (Cristoffer and Eisenberg 1985), although mature pines are the primary nesting habitat of the endangered red-cockaded woodpecker *(Picoides borealis).* Also, Florida harbors hundreds of endemic species whose preservation may hinge upon the acquisition of isolated areas. Approximately 220 plant taxa (species and subspecies), seven freshwater fish species, seven amphibian and reptile species, seven subspecies of birds, three mammalian species, and hundreds of invertebrate species are endemic to Florida (Caire et al., in preparation). Little has been done to prepare conservation strategies or acquire isolated areas for many of these endemics, although they are very important in terms of regional or national conservation efforts (Kushlan 1979).

Conclusion

As we have stressed throughout the foregoing discussion, each potential corridor must be considered on its own merits; generalizations made from theoretical considerations cannot be universally applied. In some situations corridors may be of great use, and in others one can reasonably argue that they will be irrelevant or even detrimental. Unfortunately, much of the current literature concerning corridors fails to consider potential disadvantages and often assumes potential benefits without the support of sufficient biological data, or even explicit recognition that such data are needed. Costs seem often to have been ignored. It is very important, though, that decisions on these choices and other conservation matters be based on data or well-founded inference, not on overarching generalities. In particular, the matter of whether the same money and effort might be spent better in other ways deserves more consideration.

Acknowledgments

We thank Lee Harper for access to unpublished data and an anonymous referee for suggestions on corridor use.

References and Notes

Belden, C. A discriminant analysis of skeletal characteristics in the Florida panther. Annual Research Report. Tallahassee, FL: Florida Game and Fresh Water Fish Commission; 1985.

Boecklen, W.J.; Gotelli, N.J. Island biogeographic theory and conservation practice: Species-area or specious-area relationships? *Biological Conservation 29*:63-80;1984.

Boecklen, W.J.; Simberloff, D. Area-based extinction models in conservation. In: D.K. Elliott, ed. *Dynamics of Extinction.* New York: John Wiley & Sons; 1986:247–276.

Brown, J.H.; Kodric-Brown, A. Turnover rates in insular biogeography: effect of immigration on extinction. *Ecology 58*:445–449;1977.

Cade, T.J. Hybridization and gene exchange among birds in relation to conservation. In: C.M. Schonewald-Cox, S.M. Chambers, B. MacBryde, L. Thomas, eds. *Genetics and Conservation.* Menlo Park, CA: Benjamin/Cummings; 1983:288–309.

Caire, N.; Gatewood, S.E.; Hardin, E.D.; Jackson, D.R.; Muller, J.W. Summary report on the vascular plants, animals, and natural communities endemic to Florida. Nongame Wildlife Technical Report. Tallahassee, FL: Florida Game and Fresh Water Fish Commission; 1986.

Chesser, R.K. Isolation by distance: relationship to the management of genetic resources. In: C.M. Schonewald-Cox, S.M. Chambers, B. MacBryde, L. Thomas, eds. *Genetics and Conservation.* Menlo Park, CA: Benjamin/Cummings; 1983:66–77.

Chesser, R.K.; Smith, M.H.; Brisbin, Jr., I.L. Management and maintenance of genetic variability in endangered species. *Intern. Zoo Yearbook* 20:146–154;1980.

Cristoffer, C.; Eisenberg, J. On the captive breeding and reintroduction of the Florida panther in suitable habitats. Task #1, Report #2. Tallahassee, FL: Florida Game and Fresh Water Fish Commission and Panther Technical Advisory Committee; 1985.

Dawson, W.R.; Ligon, J.D.; Murphy, J.R.; Myers, J.P.; Simberloff, D.; Verner, J. Report of the Advisory Panel on the Spotted Owl. New York: National Audubon Society;1986.

East, R. Application of species-area curves to African savannah reserves. *Afr. J. Ecol.* 21:123–128;1983.

Ericson, J. Rights-of-way. *North Dakota Outdoors* 42:18–21;1979.

Fahrig, L.; Merriam, G. Habitat patch connectivity and population survival. *Ecology* 66:1762–1768;1985.

Fernald, E., editor. *Atlas of Florida.* Tallahassee, FL: Florida State University Foundation, Inc.; 1981.

Florida's endangered wildlife/Panther health and reproduction. Study No. E-1-09. Annual Research Report. Tallahassee, FL: Florida Game and Fresh Water Fish Commission; 1985.

Forman, R.T.T. Corridors in a landscape: their ecological structure and function. *Ekologia* (C.S.S.R.) 2:375–387;1983.

Forman, R.T.T.; Baudry, J. Hedgerows and hedgerow networks in landscape ecology. *Environmental Management* 8:495–510;1984.

Frankel, O.H.; Soulé, M.E. *Conservation and Evolution.* Cambridge: Cambridge University Press; 1981.

Grow, G. New threats to the Florida panther. ENFO 84-6. Winter Park, FL: Environmental Information Center of Florida Conservation Foundation, Inc.; 1984.

Harris, L.D. *The Fragmented Forest: Island Biogeographic Theory and the Preservation of Biotic Diversity.* Chicago: University of Chicago Press; 1984.

Harris, L.D. Conservation corridors: A highway system for wildlife. ENFO Report 85-5. Winter Park, FL: Environmental Information Center of the Florida Conservation Foundation, Inc.; 1985.

Henderson, M.T.; Merriam, G.; Wegner, J. Patchy environments and species survival: chipmunks in an agricultural mosaic. *Biol. Conserv.* 31:95–105;1985.

Jaffe, M. Snakes look guilty in the case of Guam's disappearing birds. *Tallahassee Democrat,* June 23, 1985, pp. 1G,5G.

James, F.C. Environmental component of morphological differentiation in birds. *Science* 221:184–186;1983.

Janzen, D.H. No park is an island: increase in interference from outside as park size decreases. *Oikos* 41:402-410;1983.

Kushlan, J.A. Design and management of continental wildlife reserves: lessons from the Everglades. *Biol. Conserv.* 15:281–290;1979.

Logan, T.H.; Nesbitt, S.A. Status of sandhill/whooping crane studies in Florida. Proc. 1985 Crane Workshop. Grand Island, NB; in press.

Long, J.L. *Introduced Birds of the World.* New York: Universe Books; 1981.

Lovejoy, T.E.; Bierregaard, R.O.; Rankin, J.; Schubart, H.O.R. Ecological dynamics of tropical forest fragments. In: S.L. Sutton, T.C. Whitmore, A.C. Chadwick, eds. *Tropical Rain Forest: Ecology and Management.* Oxford: Blackwell Scientific Publications; 1983:377–384.

MacArthur, R.H.; Wilson, E.O. *The Theory of Island Biogeography.* Princeton, NJ: Princeton University Press; 1967.

MacClintock, L.; Whitcomb, R.F.; Whitcomb, B.L. Island biogeography and the "habitat islands" of eastern forest. II. Evidence for the value of corridors and minimization of isolation in preservation of biotic diversity. *Am. Birds.* 31:6–12;1977.

Maehr, D.; Brady, J. Florida Status report. In: D. Maehr, J. Brady, eds. *Seventh Eastern Workshop on Black Bear Research.* Florida Game and Fresh Water Fish Commission; 1984:2–3.

May, R.M. *Stability and Complexity in Model Ecosystems.* Princeton, NJ: Princeton University Press; 1973.

Middleton, J.; Merriam, G. Woodland mice in a farmland mosaic. *J. Anim. Ecol.* 18:703–710;1981.

National Research Council. *Ecological Knowledge and Environmental Problem-Solving: Concepts and Case Studies.* Washington, D.C.: National Academy Press; 1986.

Nilson, D. Roadside management and wetland development along North Dakota highways. *North Dakota Outdoors* 40:23–25;1977.

Noss, R.F.; Harris, L.D. Nodes, networks, and MUMs: preserving diversity at all scales. *Environ. Manage.* 10; in press.

Notes and news. *Oryx 16*:299;1984.

O'Neill, D.H.; Robel, R.J.; Dayton, A.D. Lead contamination near Kansas highways: implications for wildlife enhancement programs. *The Wildlife Society Bulletin* 11:152–160;1983.

Penny, M. *The Birds of Seychelles and the Outlying Islands.* New York: Taplinger; 1974.

Ralls, K.; Ballou, J. Effects of inbreeding on infant mortality in captive primates. *Int. J. Primatol.* 3:491–505;1982*a*.

Ralls, K.; Ballou, J. Effects of inbreeding on juvenile mortality in some small mammal species. *Lab. Anim.* 16:159–166;1982*b*.

Ralls, K.; Ballou, J. Extinction: lessons from zoos. In: C.M. Schonewald-Cox, S.M. Chambers, B. MacBryde, L. Thomas, eds.

Genetics and Conservation. Menlo Park, CA: Benjamin/Cummings; 1983:164–184.

Salwasser, H.; Mealey, S.; Johnson, K. Wildlife population viability—a question of risk. *Trans. N. Amer. Wildl. Nat. Res. Conf.* 49:421–439;1984.

Savidge, J. Population declines and extinctions of birds on Guam: the havoc an introduced predator can wreak. Abstract *81.* 103rd Stated Meeting of the American Ornithologists' Union. Tempe, AR; 1985.

Shaffer, M.L. Minimum population sizes for species conservation. *BioScience* 31:131–134;1981.

Shaffer, M.L. The metapopulation and species conservation: the special case of the northern spotted owl. In: R.J. Gutiérrez, A.B. Carey, eds. *Ecology and Management of the Spotted Owl in the Pacific Northwest.* Portland, OR: U.S.D.A. Forest Service; 1985;86–99.

Simberloff, D. The proximate causes of extinction. In: D. Raup, D. Jablonski, eds. *Patterns and Processes in the History of Life.* Berlin: Springer-Verlag; 1986:259–276.

Simberloff, D.; Abele, L.G. Refuge design and island biogeographic theory: effects of fragmentation. *Amer. Natur.* 120:41–50;1982.

Soulé, M.E.; Wilcox, B.A.; Holtby, C. Benign neglect: a model of faunal collapse in the game reserves of East Africa. *Biol. Conserv.* 15:259–272;1979.

Soulé, M.E.; Simberloff, D. What do genetics and ecology tell us about the design of nature reserves? *Biol. Conserv.* 35:19–40;1986.

Wegner, J.F.; Merriam, G. Movements by birds and small mammals between a wood and adjoining farmland habitats. *J. Appl. Ecol.* 16:349–357;1979.

Weise, T.F.; Robinson, W.L.; Hook, R.A.; Mech, L.D. An experimental translocation of the Eastern timber wolf *(Canus lupus lycaon).* Audubon Conservation Report Number 5. National Audubon Society; 1975.

Western, D.; Ssemakula, J. The future of savannah ecosystems: ecological islands or faunal enclaves? *Afr. J. Ecol.* 19:7–19; 1981.

Wilford, J.N. Last dusky sparrow struggles on. *New York Times,* April 29, p. 15–16;1986.

Wilson, E.O.; Willis, E.O. Applied biogeography. In: M.L. Cody, J.M. Diamond, eds. *Ecology and Evolution of Communities.* Cambridge, MA: Harvard University Press; 1975:522–534.

World Conservation Strategy. Gland, Switzerland: International Union for the Conservation of Nature and Natural Resources; 1980.

Young, S.P. *The Bobcat of North America.* Omaha, NB: University of Nebraska Press; 1978.

Corridors in Real Landscapes: A Reply to Simberloff and Cox

REED F. NOSS
Department of Wildlife and Range Sciences
School of Forest Resources and Conservation
University of Florida
Gainesville, Florida 32611

Abstract: *Habitat corridors have become popular in land-use plans and conservation strategies, yet few data are available to either support or refute their value. Simberloff and Cox (1987) have criticized what they consider an uncritical acceptance of corridors in conservation planning.*

Any reasonable conservation strategy must address the overwhelming problem of habitat fragmentation. Although Simberloff and Cox use island analogies to illustrate advantages of isolation, these analogies do not apply directly to problems in landscape planning. Genetics also does not offer unequivocal advice, but the life histories of wide-ranging animals (e.g., the Florida panther) suggest that the maintenance or restoration of connectivity in the landscape is a prudent strategy. Translocation of individuals among reserves—considered by Simberloff and Cox a viable alternative to natural dispersal—is impractical for whole communities of species that are likely to suffer from problems related to fragmentation.

Many of the potential disadvantages of corridors could be avoided or mitigated by enlarging corridor width or by applying ecologically sound zoning regulations. Corridors are not the solution to all of our conservation problems, nor should they be used as a justification for small reserves. But corridors can be a cost-effective complement to the strategy of large and multiple reserves in real-life landscapes.

Resumen: *Los corredores naturales se han hecho muy comunes en proyectos de uso de terreno y estrategias de conservación, pero hay muy pocos datos disponibles que puedan apoyar o rechazar su valor. Simberloff and Cox (1987) critican lo que ellos consideran es una aceptación poco crítica de los corredores naturales en la planificación de estrategias de conservación.*

Cualquier estrategia razonable tiene que dirigirse al problema preponderante de fragmentación de hábitat. Aunque Simberloff y Cox usan el ejemplo de las islas como analogía para ilustrar las ventajas del aislamiento, estas analogías no aplican directamente a problemas en la planificación del uso de la tierra. La genética tampoco ofrece consejo inequívoco, pero la historia natural de animales con ámbitos extensos (ejem. Florida panther) sugiere que el mantenimiento o restauración de una conexión entre áreas naturales es una estrategia prudente. El movimiento de individuos de una u otra especie entre las reservas, considerado por Simberloff y Cox como una alternativa viable a la dispersión natural, es impráctico para comunidades enteras de especies que son vulnerables a la fragmentación.

Muchas de las desventajas potenciales de corredores naturales se pueden evitar o mitigar ensanchando el corredor, o estableciendo reglamentos de zonificación que sean congruentes con principios ecológicos establecidos. Los corredores naturales no son la solución a todos los problemas de la conservación natural, ni tampoco deberían ser usados como una justificación para reservas pequeñas. Pero los corredores naturales pueden ser un complemento de bajo costo y rendimiento efectivo a la estrategia de establecer muchas reservas de gran extensión.

Corridors are a hot topic, perhaps even a fad, in conservation planning these days. Planners and environmentalists from county to federal levels are busy drawing "greenbelts" and other habitat corridors into their designs, sometimes with only a vague awareness of the biological issues underlying the corridor strategy. Re-

Paper submitted 12/9/86; revised manuscript accepted 3/31/87.

cently, some biologists have expressed concern that the corridor idea has been thrown into the political arena prematurely, without adequate field research or discussion among conservation biologists.

Soulé and Simberloff (1986), briefly, and Simberloff and Cox (1987), in more depth, have discussed some possible advantages and disadvantages of the corridor strategy. Simberloff and Cox (1987) place particular emphasis on disadvantages, because they believe that "much of the current literature concerning corridors fails to consider potential disadvantages and often assumes potential benefits without the support of sufficient biological data, or even explicit recognition that such data are needed." No doubt more research is needed to develop optimal connectivity strategies, but the continuing severance of natural linkages in many landscapes suggests that active strategies to combat the process and the consequences of fragmentation must proceed quickly, with or without "sufficient" data.

Many conservation biologists agree with Wilcox and Murphy (1985) that "habitat fragmentation is the most serious threat to biological diversity and is the primary cause of the present extinction crisis." Conservation strategies, therefore, might be evaluated on the basis of how well they counter the effects of fragmentation in real landscapes. The fragmentation problem has essentially two components: 1) a decrease in total habitat area, and 2) an apportionment of the remaining area into ever more isolated pieces (Wilcove et al. 1986). The two ways to counter fragmentation, then, are 1) increase effective habitat area, and 2) increase connectivity. My purpose in this note is to offer an alternative viewpoint to Simberloff and Cox (1987) and to evaluate potential advantages and disadvantages of corridors in the context of an integrated landscape conservation strategy.

Potential Advantages and Disadvantages of Corridors in Human-Dominated Landscapes

Some potential advantages and disadvantages of corridors in human-dominated landscapes, with particular reference to conservation of terrestrial species and habitats, are listed in Figure 1. These lists are not comprehensive, as many important functional attributes of corridors will not be discussed here. For example, vegetated riparian corridors are important in maintaining water quality in streams (Karr & Schlosser 1978, Schlosser & Karr 1981), and hedgerows and shelterbelts have well-known advantages in inhibiting soil erosion (Forman & Baudry 1984).

Simberloff and Cox (1987) propose that corridors be evaluated individually on their own merits, and that theoretical considerations cannot be applied universally. Few ecologists would quarrel with that statement. But although Simberloff and Cox (1987) criticize the use of biogeographic analogies in pro-corridor arguments, much of their argument against corridors (or in favor of habitat subdivision) is based on island analogies. The common goal in this debate—conservation of biodiversity—might be served best if all parties, whenever possible, refrain from arguments based on theory and analogy, and devote their efforts to solving concrete problems in real-world landscapes.

The extent to which a habitat corridor might facilitate dispersal and thus increase immigration rates to reserves is strictly an empirical matter, and would depend upon habitat structure within the corridor, corridor width and length, and the autecologies of the particular organisms in question (Forman 1983, Harris 1984, Forman & Godron 1986, Noss & Harris 1986). If we determine that immigration rate is enhanced, we still do not know

Potential Advantages of Corridors

1. Increase immigration rate to a reserve, which could
   A increase or maintain species richness and diversity (as predicted by island biogeography theory);
   B increase population sizes of particular species and decrease probability of extinction (provide a "rescue effect") or permit re-establishment of extinct local populations;
   C prevent inbreeding depression and maintain genetic variation within populations.
2. Provide increased foraging area for wide-ranging species.
3. Provide predator-escape cover for movements between patches.
4. Provide a mix of habitats and successional stages accessible to species that require a variety of habitats for different activities or stages of their life-cycles.
5. Provide alternative refugia from large disturbances (a "fire escape").
6. Provide "greenbelts" to limit urban sprawl, abate pollution, provide recreational opportunities, and enhance scenery and land values.

Potential Disadvantages of Corridors

1. Increase immigration rate to a reserve, which could
   A facilitate the spread of epidemic diseases, insect pests, exotic species, weeds, and other undesirable species into reserves and across the landscape;
   B decrease the level of genetic variation among population or subpopulations, or disrupt local adaptations and coadapted gene complexes ("outbreeding depression").
2. Facilitate spread of fire and other abiotic disturbances ("contagious catastrophes").
3. Increase exposure of wildlife to hunters, poachers, and other predators.
4. Riparian strips, often recommended as corridor sites, might not enhance dispersal or survival of upland species.
5. Cost, and conflicts with conventional land preservation strategy to preserve endangered species habitat (when inherent quality of corridor habitat is low).

Figure 1. Potential advantages and disadvantages of conservation corridors.

whether the net effect of this increased immigration is good or bad for conservation. (Fig. 1; item number 1). According to island biogeography theory (MacArthur & Wilson 1967), increased immigration should result in a higher equilibrium species number. But higher species richness at the local scale may not be a goal in conservation, particularly if the species that invade are alien to the landscape or not in need of reserves for survival (Diamond 1976, Noss 1983). Augmentation of local population size with immigrants from the same species ("rescue effect"; Brown & Kodric-Brown 1977) and re-establishment of extinct local populations (Fahrig & Merriam 1985) might be considered an advantage of increased connectivity, but not if the organism so benefitted is a competitor, predator, parasite, or pathogen of a species of greater conservation concern.

The genetic consequences of increased immigration rate are also controversial, as discussed by Simberloff and Cox (1987). On the one hand, inbreeding depression and genetic drift might be minimized with the influx of genetically different individuals, resulting in increased fitness for the average individual and increased genetic variation in the population. Evidence for immigrants contributing much to heterozygosity or fitness in a small population is meager, however (Frankel & Soulé 1981). Furthermore, increased gene flow between demes might lead to genetic swamping and homogenization of the gene pool. Corridors have never been implicated in this problem, although the possibility worries Simberloff and Cox (1987).

Gene flow might also disrupt local adaptation and coadapted gene complexes, resulting in outbreeding depression. Although outbreeding depression is likely to be a temporary phenomenon, rapidly eliminated by selection and sometimes replaced by a superior coadapted gene complex, it could greatly increase the chances of extinction in small populations (Templeton 1986). Outbreeding depression is a potential problem in captive breeding programs, when animals from distinct populations are mated (e.g., de Boer 1983, Templeton et al. 1986). On the other hand, maintaining or restoring natural landscape connectivity never has been shown to cause outbreeding depression in populations.

On the genetics issue, Simberloff and Cox (1987) suggest that "the subdivision of a larger population allows the effects of genetic drift to assist in the maintenance of genetic variability in the species" (*see* Chesser 1983). When inbreeding becomes a problem, manual translocation of individuals between isolated populations is considered preferable ("genetically advantageous and much less expensive"; Simberloff & Cox 1987) to protecting corridors for natural dispersal. The esthetic and perhaps ethical question of whether shipping animals around in crates is a satisfactory substitute for natural movements warrants philosophical discussion.

Although translocation might fulfill genetic management objectives for a few focal species, moving all the species of a community among reserves is clearly impractical. Fragmentation thus far has been documented to have deleterious effects in a relatively few sensitive species, but many secondary extinctions may be forthcoming (Wilcove et al. 1986). Small populations of organisms that today seem relatively secure, such as long-lived forest herbs with limited dispersal capabilities and slow responses to potentially significant forest changes, may prove vulnerable in the long term (Middleton & Merriam 1983). The same may be true for many forest songbirds (D. Wilcove, personal communication). Furthermore, because of mutualistic dependencies among certain species and the natural shifting of habitat patches over time and space, levels of biological organization above the population may depend on habitat connectivity for long-term persistence (Noss & Harris 1986). In the face of scientific ignorance about these phenomena, maintenance of habitat connectivity would seem to be the prudent course.

Simberloff and Cox (1987) choose the Florida panther (*Felis concolor coryi*) as an example of a species where proposed corridor strategies (Cristoffer & Eisenberg 1985, Noss 1985*a*, Noss & Harris 1986) are inappropriate. Unfortunately, they fail to consider the most critical component of that strategy: the re-creation of wilderness landscapes by closing roads and limiting access in large core areas of public lands in the present range of the panther in south Florida, and in proposed reintroduction sites in north Florida, as detailed in Noss (1985*a*, 1985*b*) and summarized in Noss and Harris (1986). In this admittedly idealistic strategy, large roadless areas would be surrounded by multiple-use buffer zones and connected by broad habitat corridors (revised map in Noss 1987*a*). Thus, corridors are envisioned as one critical component of an integrated landscape conservation strategy, not as panaceas. Wilderness restoration may be the only hope for the Florida panther, an animal demonstrably intolerant of habitat fragmentation. Like other pumas, it does not prosper in human-dominated landscapes (Noss 1985*a*, USFWS 1986, Van Dyke et al. 1986).

Simberloff and Cox (1987) propose that "a better method of preserving extant levels of genetic variability might be to isolate new translocated populations" of panthers, but their reasoning is unconvincing in light of *Felis concolor*'s natural history. *F. c. coryi* was originally distributed continuously from Florida northeast to South Carolina, and west to Arkansas and east Texas or west Louisiana (Goldman 1946, Hall 1981). Habitat fragmentation and direct persecution subsequently eliminated this subspecies from all of its range except south Florida, which until recently was remote wilderness (Belden, in press; USFWS 1986). Although the current level of genetic variation in the 20 to 50 remaining panthers and how this relates to fitness is unknown, this may be a classic case of inbreeding depression. All five males examined to date have had greater than 93 percent abnormal sperm (Roelke 1986, USFWS 1986). The Florida panther as we know it today is an "inbred subset" of the *F. c. coryi* recognized by Goldman (1946), and strategies to enhance genetic variation are being discussed (Eisenberg 1986).

Given what is known about the former continuity of panther distribution and the long-distance movements of individual *Felis concolor* (often over 100 km; Young 1946, Hornocker 1970, Dewar 1976, Hemker et al. 1984, Logan et al. 1986), it is possible that a single deme may have occupied all of peninsular Florida. Documented dispersal of juveniles and immigration of transients suggest that *F. concolor* cannot adequately be managed site-by-site, but instead requires management on a regional basis (Logan et al. 1986). Because several individuals often travel common corridors that are influenced by topography (e.g., Young 1946), the safest strategy may be to provide numerous, carefully selected, and well-protected swaths of habitat (including highway underpasses) for this movement. The alternative of isolated reserves would virtually assure that individuals will continue to be shot and run over in the mortality sink of the developed landscape.

Evidence that corridors on a finer scale can help animals avoid predation is accumulating, contrary to the suggestion of Simberloff and Cox (1987) that thin corridors may increase the exposure of animals to predators. Studies of fall movements of blue jays in Wisconsin have demonstrated that jays usually follow wooded fencerows in crossing open farmland. Apparently this is a response to predation from numerous migrating hawks, for jays frequently dive into fencerow cover when hawks approach (Johnson & Adkisson 1985). Experimental and radio-telemetry studies of *Peromyscus* movements along fencerows in Ontario farmland are providing data on what type of cover is preferred by these mice (Merriam 1986). Simberloff and Cox (1987) correctly note that the question of whether a corridor represents safety or a threat to an animal can be answered only by considering ecological factors specific to the organism and the site.

Most of the other potential advantages and disadvantages of corridors listed in Figure 1 are self-explanatory. Two, however, warrant further comment. Although biologists seldom consider the anthropocentric functions of "greenbelts" or "open space" in developed landscapes, these quality-of-life factors are of utmost importance to landscape architects and planners. Scenery, recreation, pollution abatement, and land value enhancement are what usually motivate planners to draw corridors into their designs (various human uses of corridors are mentioned in Forman & Godron 1986). And many of these corridors are being drawn. It would be auspicious for biologists and planners to work together to develop corridor designs that can optimize the quality of both the human and the nonhuman environment.

Finally, a major concern of Simberloff and Cox (1987) is that the cost of corridors will conflict with conventional conservation objectives to preserve endangered species habitat. I prefer to think of the corridor strategy as a complement to efforts to save "the last of the least and the best of the rest," as in The Nature Conservancy's heritage approach (Jenkins 1985). Whereas the heritage approach generally focuses on relatively small, discrete sites chosen for their occurrences of endangered elements (often plant species), corridors are an element of a landscape-level approach designed to restore and protect intact ecosystems and wide-ranging animals, many of which are critically endangered (Noss 1983, 1987*a*, *b*). I share Soulé's concern (personal communication) that corridors might be prescribed as an answer to every problem, or as a justification for preserves that are too small. But the fact remains that almost all existing reserves are far too small to maintain natural ecological processes and viable populations of the species with the largest home ranges (Pickett & Thompson 1978, White & Bratton 1980, Lovejoy & Oren 1981, Schonewald-Cox et al. 1983, Harris 1984, Noss & Harris 1986).

The corridor strategy can be an important complement to the strategy of large and multiple reserves (cf. Soulé & Simberloff 1986). Although money for conservation is never easy to come by, conservationists probably have not made strong enough demands for funds. For example, assuming that 26,000 ha of Florida wildland can be purchased for $20 million (i.e., a recent state acquisition of a critical coastal corridor), then over 4.5 million ha could be purchased for the cost of one $3.5 billion space shuttle. Furthermore, many corridors can be protected by conservation easements, tax incentives, management agreements, registry programs, and other less-than-fee negotiations (e.g., Noss & Harris 1986), although these options all involve their own complications (J. Cox, personal communication).

Conclusions

Perhaps the best argument for corridors is that the original landscape was interconnected. This is not to deny that dispersal barriers such as rivers and mountain ranges have been important in biogeography and evolution, or that naturally isolated habitats such as lakes, caves, mountain tops, and edaphic patches are important features of natural landscapes. But as can be observed readily in aerial photographs of undeveloped land, presettlement landscapes in general are interdigitating mosaics with high connectivity of similar habitats. Connectivity declines with human modification of the landscape (Godron & Forman 1983). Hence, wide-ranging animals such as large predators that once were distributed almost continuously over entire continents are now confined to the few remaining pockets of unfragmented land. Corridors are simply an attempt to maintain or restore some of the natural landscape connectivity. No one, to my knowledge, is suggesting that we build corridors or other connections between naturally isolated habitats.

Certainly, humans and associated disturbances will impinge on corridors, just as they impinge on small nature reserves. For this reason, corridors should generally be as wide as possible. Planners and conservationists often ask how wide corridors need to be, and corridor widths (especially for riparian corridors) are often specified in land-use plans. In reality, the necessary width will vary depending on habitat structure and quality within the corridor, the nature of the surrounding habitat, human use patterns, and the particular species that we expect to use the corridor. Narrow fencerows might suffice for many farmland species, but wilderness species such as large carnivores may require corridors many miles wide to travel safely among reserves. A wide enough swath, of course, effectively creates one large reserve out of two or more smaller reserves.

Corridors are not *the* answer to our conservation problems. Undoubtedly, in many situations acquiring marginal habitat for corridors should be of lower priority than preserving isolated sites for endemic or endangered species. The major area of common ground among the various conservation biologists involved in this debate is that we are all interested in maintaining biodiversity, and most of us are wary of facile generalizations and analogies as guides to conservation. We furthermore agree that conservation actions must be based on the autecologies of the species concerned, and on other site-specific attributes. But a holistic, "top-down" framework can be useful in providing context to autecology. When money is limited, as it always will be, alternative actions should be weighed carefully. Weighing alternatives is made easier by evaluating their potential contributions to a landscape conservation strategy that addresses the overwhelming problem of habitat fragmentation.

Acknowledgments

My thinking on corridors has benefitted from discussions and correspondence with Jim Cox, Richard Forman, Larry Harris, Michael Soulé, Jack Stout, David Wilcove, and many others. I particularly thank Jim Cox, David Ehrenfeld, Robert May, and an anonymous referee for their helpful comments on an earlier draft of this paper.

Literature Cited

Belden, R.C. Florida panther recovery plan implementation: A 1983 progress report. *Proceedings of the International Cat Symposium.* Kingsville, TX (in press).

Brown, J.H.; Kodric-Brown, A. Turnover rates in insular biogeography: Effect of immigration on extinction. *Ecology 58:* 445–449;1977.

Chesser, R.K. Isolation by distance: Relationship to the management of genetic resources. In: C.M. Schonewald-Cox, S.M. Chambers, B. MacBryde, W.L. Thomas, eds. *Genetics and Conservation: A Reference for Managing Wild Animal and Plant Populations.* Menlo Park, CA: Benjamin/Cummings; 1983:66–77.

Cristoffer, C.; Eisenberg, J. On the captive breeding and reintroduction of the Florida panther in suitable habitats. Task #1, Report #2. Report for the Florida Game and Fresh Water Fish Commission and Florida Panther Technical Advisory Council. Tallahassee, FL; 1985:23pp.

De Boer, L.E.M. Karyological problems in breeding owl monkeys, *Aotus trivirgatus. International Zoo Yearbook 22:* 119–124;1982.

Dewar, P. Comments. In: G.C. Christensen, R.J. Fischer, cochairmen. *Transactions of the Mountain Lion Workshop.* U.S. Fish and Wildlife Service, Portland, OR, and Nevada Fish and Game Department, Reno; 1976:65.

Diamond, J.M. Island biogeography and conservation: Strategy and limitations. *Science 193*:1027–1029;1976.

Eisenberg, J. Taxonomic status of the Florida panther. In: W.V. Branan, ed. *Survival of the Florida Panther: A Discussion of Issues and Accomplishments.* Tallahassee, FL: Florida Defenders of the Environment; 1986:14–18.

Fahrig, L.; Merriam, G. Habitat patch connectivity and population survival. *Ecology 66*:1762–1768;1985.

Forman, R.T.T. Corridors in a landscape: Their ecological structure and function. *Ekologiya* (CSSR) *2*:375–387;1983.

Forman, R.T.T.; Baudry, J. Hedgerows and hedgerow networks in landscape ecology. *Environmental Management 8*:495–510;1984.

Forman, R.T.T.; Godron, M. *Landscape Ecology.* New York: John Wiley & Sons; 1986.

Frankel, O.H.; Soulé, M.E. *Conservation and Evolution.* Cambridge: Cambridge University Press; 1981.

Godron, M.; Forman, R.T.T. Landscape modification and changing ecological characteristics. In: H.A. Mooney, M. Godron, eds. *Disturbance and Ecosystems.* Berlin: Springer-Verlag; 1983:12–28.

Goldman, E.A. Classification of the races of the puma, Part 2. In: S.P. Young, E.A. Goldman, eds. *The Puma, Mysterious American Cat.* Washington, D.C.: The American Wildlife Institute; 1946:177–302.

Hall, E.R. *The Mammals of North America,* 2nd ed. New York: John Wiley & Sons; 1981.

Harris, L.D. *The Fragmented Forest: Island Biogeography Theory and the Preservation of Biotic Diversity.* Chicago: University of Chicago Press; 1984.

Harris, L.D. Conservation corridors: A highway system for wildlife. ENFO Report 85-5. Environmental Information Center of the Florida Conservation Foundation, Inc., Winter Park, FL; 1985.

Hemker, T.P.; Lindzey, F.G.; Ackerman, B.B. Population characteristics and movement patterns of cougars in southern Utah. *Journal of Wildlife Management 48*:1275–1284;1984.

Hornocker, M.G. An analysis of mountain lion predation upon mule deer and elk in the Idaho Primitive Area. *Wildlife Monographs 21*:1–39;1970.

Jenkins, R.E. Information methods: Why the heritage programs work. *Nature Conservancy News 35*(6):21–23;1985.

Johnson, W.C.; Adkisson, C.S. Dispersal of beech nuts by blue jays in fragmented landscapes. *American Midland Naturalist 113*:319–324;1985.

Karr, J.R.: Schlosser, I.J. Water resources and the land-water interface. *Science 201*:229–234;1978.

Logan, K.A.; Irwin, L.L.; Skinner, R. Characteristics of a hunted mountain lion population in Wyoming. *Journal of Wildlife Management 50*:648–654;1986.

Lovejoy, T.E.; Oren, D.C. The minimum critical size of ecosystems. In: R.L. Burgess, D.M. Sharpe, eds. *Forest Island Dynamics in Man-Dominated Landscapes.* New York: Springer-Verlag; 1981:7–12.

MacArthur, R.H.; Wilson, E.O. *The Theory of Island Biogeography.* Princeton, NJ: Princeton University Press; 1967.

Merriam, G. Ecological processes in farmland mosaics. Abstract. In: Program of the IV International Congress of Ecology. Syracuse, NY: State University of New York and Syracuse University; 1986:238.

Middleton, J.; Merriam, G. Distribution of woodland species in farmland woods. *Journal of Applied Ecology 20*:623–644;1983.

Noss, R.F. A regional landscape approach to maintain diversity. *BioScience 33*:700–706;1983.

Noss, R.F. Landscape considerations in reintroducing and maintaining the Florida panther: Design of appropriate preserve networks. Report for the Florida Panther Technical Advisory Council, Gainesville, FL; 1985*a*:32 pp.

Noss, R.F. Wilderness recovery and ecological restoration: An example for Florida. *Earth First! 5*(8):18–19;1985*b*.

Noss, R.F. Protecting natural areas in fragmented landscapes. *Natural Areas Journal 7;* 1987*a* (in press).

Noss, R.F. From plant communities to landscapes in conservation inventories: A look at The Nature Conservancy (USA). *Biological Conservation* 1987*b* (in press).

Noss, R.F.; Harris, L.D. Nodes, networks, and MUMs: Preserving diversity at all scales. *Environmental Management 10*:299–309;1986.

Pickett, S.T.A.; Thompson, J.N. Patch dynamics and the size of nature reserves. *Biological Conservation 13*:27–37;1978.

Roelke, M.E. Medical management, biomedical findings, and research techniques for the Florida panther. In: W.V. Branan, ed. *Survival of the Florida Panther: A Discussion of Issues and Accomplishments.* Tallahassee, FL: Florida Defenders of the Environment; 1986:7–14.

Schlosser, I.J.; Karr, J.R. Water quality in agricultural watersheds: Impact of riparian vegetation during base flow. *Water Resources Bulletin 17*:233–240;1981.

Schonewald-Cox, C.M.; Chambers, S.M.; MacBryde, B.; Thomas, W.L. *Genetics and Conservation: A Reference for Managing Wild Animal and Plant Populations.* Menlo Park, CA: Benjamin/Cummings; 1983.

Simberloff, D.; Cox, J. Consequences and costs of conservation corridors. *Conservation Biology 1*: 63–71;1987.

Soulé, M.E.; Simberloff, D. What do genetics and ecology tell us about the design of nature reserves? *Biological Conservation 35*: 19–40;1986.

Templeton, A.R. Coadaptation and outbreeding depression. In: M.E. Soulé, ed. *Conservation Biology: The Science of Scarcity and Diversity.* Sunderland, MA: Sinauer Associates; 1986: 105–116.

Templeton, A.R.; Hemmer, H.; Mace, G.; Seal, U.S.; Shields, W.M.; Woodruff, D.S. Local adaptation, coadaptation, and population boundaries. *Zoo Biology 5*:115–125;1986.

U.S. Fish and Wildlife Service (USFWS). Florida panther *(Felis concolor coryi)* recovery plan. Revised technical draft. Prepared by the Florida Panther Interagency Committee for the U.S. Fish and Wildlife Service, Atlanta; 1986.

Van Dyke, F.G.; Brocke, R.H.; Shaw, H.G.; Ackerman, B.A.; Hemker, T.H.; Lindzey, F.G. Reactions of mountain lions to logging and human activity. *Journal of Wildlife Management 50*:95–102;1986.

White, P.S.; Bratton, S.P. After preservation: Philosophical and practical problems of change. *Biological Conservation 18*:241–255;1980.

Wilcove, D.S.; McLellan, C.H.; Dobson, A.P. Habitat fragmentation in the temperate zone. In: M.E. Soulé, ed. *Conservation Biology: The Science of Scarcity and Diversity.* Sunderland, MA: Sinauer Associates; 1986:237–256.

Wilcox, B.A.; Murphy, D.D. Conservation strategy: The effects of fragmentation on extinction. *American Naturalist 125*:879–887;1985.

Young, S.P. History, life habits, economic status, and control, Part 1. In: S.P. Young, E.A. Goldman, eds. *The Puma, Mysterious American Cat.* Washington, D.C.: The American Wildlife Institute;1946:1–173.

Essay

Movement Corridors: Conservation Bargains or Poor Investments?

DANIEL SIMBERLOFF
JAMES A. FARR
Department of Biological Science
Florida State University
Tallahassee, FL 32306, U.S.A.

JAMES COX
Florida Game and Fresh Water Fish Commission
620 S. Meridian Street
Tallahassee, FL 32399, U.S.A.

DAVID W. MEHLMAN
Department of Biological Science
Florida State University
Tallahassee, FL 32306, U.S.A.
and
Department of Biology
University of New Mexico
Albuquerque, NM 87131

Abstract: *Corridors for movement of organisms between refuges are confounded with corridors designed for other functions, obscuring an assessment of cost-effectiveness. The rationales for movement corridors are (1) to lower extinction rate in the sense of the equilibrium theory, (2) to lessen demographic stochasticity, (3) to stem inbreeding depression, and (4) to fulfill an inherent need for movement. There is a paucity of data showing how corridors are used and whether this use lessens extinction by solving these problems. Small, isolated populations need not be doomed to quick extinction from endogenous forces such as inbreeding depression or demographic stochasticity, if their habitats are protected from humans. In specific instances, corridors could have biological disadvantages. Corridor proposals cannot be adequately judged generically. In spite of weak theoretical and empirical bases, numerous movement corridor projects are planned. In the State of Florida, multi-million-dollar corridor proposals are unsupported by data on which species might use the corridors and to what effect. Similarly, plans for massive corridor networks to counter extinction caused*

Paper submitted August 13, 1991; revised manuscript accepted February 27, 1992.

Corredores para el movimiento: ¿Gangas de la conservación o malas inversiones?

Resumen: *Los corredores para el movimiento de organismos entre refugios son confundidos con corredores designados para otras funciones obscureciendo una evaluación sobre costo-efectividad. Las funciones atribuidas a los corredores para movimiento son (1) disminuir la tasa de extinción definida en términos de la teoría de equilibrio, (2) disminuir la estocasticidad demográfica, (3) contrarrestar la depresión endogámica y (4) satisfacer una necesidad innata de movimiento. Existe una carencia de datos que demuestren como son usados estos corredores y si este uso minimiza las extinciones al resolver estos problemas. Poblaciones pequeñas y aisladas no están necesariamente condenadas a una rápida extinción causada por fuerzas endógenas, como depresión endogámica o estocasticidad demográfica, si sus ábitats están protejidos de los humanos. En instancias particulares, los corredores pueden tener desventajas biológicas. Las propuestas sobre corredores no pueden ser juzgadas apropiadamente en forma genérica. Numerosos corredores para movimiento están proyectados a pesar de fundamentos teóricos y empiricos débiles. En el*

by global warming are weakly supported. Alternative approaches not mutually exclusive of corridors might be more effective, but such a judgment cannot be made without a cost-benefit analysis.

estado de Florida, propuestas multimillionarias para corredores no están fundamentadas por datos que indiquen que especies usarian los corredores y con que propósito lo harían. En forma semejante, los proyectos de redes masivas de corredores para contrarrestar la extinción causada por el calentamiento global están pobremente fundamentados. Estrategias alternativas, que no son mutuamente excluyentes con los corredores, podrían ser mas efectivas, pero tales juicios no pueden ser hechos sin un análisis de costo-beneficio.

Introduction

A remarkable publicity campaign, much of it outside the bounds of mainstream science, has promoted corridors for conservation. Wilson and Willis (1975) originally proposed corridors based on the equilibrium theory of island biogeography (MacArthur & Wilson 1967); the suggestion was reprinted in *World Conservation Strategy* (International Union for the Conservation of Nature and Natural Resources [IUCN] 1980). With the imprimatur of the IUCN, the United Nations Environmental Program, and the World Wildlife Fund, the idea was widely accepted. The popular promotion of corridors entails lead articles in lay magazines (such as Arnold 1990), a videocasette (Suchy & Harris 1988), a pamphlet (Anonymous 1990*a*), the entire December 1986 issue of ENFO, and a special publication by Defenders of Wildlife (Mackintosh 1989). In Montana, a court has ruled that corridors are scientifically established as important (Breen 1991; Pace 1991). Keith Hay of the Conservation Fund argued that corridors "hold more promise for the conservation of the diversity of life than any other management factor except stabilization of the human population" (Chadwick 1990).

This hype is occurring in spite of a dearth of evidence of whether corridors will be useful in specific situations. "The workshops that discussed the values of corridors were certain that they were positive features despite the lack of supporting research" (Dendy 1987). That they facilitate movement "is now almost an article of faith" (Hobbs & Hopkins 1991). Saunders & Hobbs (1991*a*) believe that we do not have time to test the efficacy of corridors, while Noss (1992) argues that, in the face of uncertainty, it is prudent to maintain or restore "natural" kinds of corridors.

Discussion of corridors is confused (Anonymous 1986). We are concerned in this article with corridors for movement, but at least six senses of "corridor" appear in the conservation literature.

First, some habitats constitute corridors. A corridor may deserve protection as a distinct habitat, whether or not it aids movement (Simberloff & Cox 1987). For example, riparian communities are very threatened in some regions (Johnson 1989), but their value as habitats is independent of whether they allow movement between other habitats. To embed a discussion of riparian habitats in an argument about movement corridors (see Johnson 1989; Ames 1990) confounds assessment of strategies for land acquisition. Some linear artificial habitats, such as rights-of-way for highways (Adams & Dove 1989; Wilcox 1989), railroads (Noss 1992), and transmission lines (Anderson et al. 1977; Kroodsma 1982; Forman 1983), are also called "corridors," and can bolster animal populations and enhance urban and suburban green space (Adams & Dove 1989). Again, their utility as habitats is separate from their importance for movement.

Second, greenbelts and buffers are occasionally called "corridors," and their function of ameliorating the human environment is cited in discussions of movement corridors (see Gilbrook 1986; Budd et al. 1987; Adams & Dove 1989). Although such areas might aid dispersal, some envisioned uses (such as hiking and horse trails, boating, outfall for storm sewers) could impede their utility as either dispersal routes or habitats. In any event, the value of such constructs as aesthetic amenities is independent of their value for movement.

Third, Harris (1985) and Suchy and Harris (1988) call biogeographic landbridges such as the Isthmus of Panama "corridors." Such large regions have aided intercontinental movement of entire communities (Brown & Gibson 1983), but the relevance of this fact to maintaining viable populations in refuges is obscure (Simberloff & Cox 1987).

Fourth, a series of discrete refuges for migratory waterfowl is occasionally called a corridor (see Harris 1985; Anonymous 1991). Below we will suggest that such a system can be construed as an alternative to corridors or complementary to a corridor network.

Fifth, underpasses and tunnels are often now called "corridors." They are commonly used to allow animals to cross highways (reviewed by Bennett 1990*b*; Noss 1992), primarily to keep individuals from being killed on the road rather than to decrease demographic stochasticity, prevent inbreeding depression, or serve other population-level functions. Costs of such structures should be construed as part of road construction rather than as land acquisition.

Sixth, and our concern here, are strips of land intended to facilitate movement between larger habitats. Wilson and Willis (1975) and Harris and Scheck (1991), in the context of equilibrium theory, envision such movement as increasing immigration rates, thus raising the equilibrium number of species at each site. Harris (1984, 1985) suggested two other rationales for corridors. First, individuals of some species typically range widely; second, inbreeding depression will lead to extinction in small refuges. Recently, alleviation of demographic stochasticity has been a suggested benefit of corridors (see Merriam 1991; Thomas 1991; Noss 1992).

We are particularly concerned with the economic cost of corridors. Hobbs and Hopkins (1991) talk of corridors as adding an option to a sparse conservation repertoire, but options may be foreclosed if a particular corridor is reserved. Noss (1992 and personal communication) believes it is prudent to retain existing corridors, conceding that establishing new ones may be cost-ineffective. But even an existing corridor is not necessarily free, and prudence when all options cannot be pursued requires some sort of cost-benefit analysis of each.

Rationales for Movement Corridors

The Equilibrium Theory of Island Biogeography

The equilibrium theory of island biogeography states that species number is constant, but that local turnover changes composition. Thus, corridors are useful by virtue of maintaining more species. An odd aspect of the rush to found a technology of refuge design based on island biogeographic theory is that, exactly when the IUCN and others were popularizing refuge design based on equilibrium theory, the theory was increasingly heavily criticized (Gilbert 1980; Williamson 1981, 1989; Williams 1984, 1986) as inapplicable to most of nature, largely because local population extinction was not demonstrated.

Demographic Stochasticity and the Metapopulation

Although not cited in the original recommendations for corridors, demographic stochasticity has been widely recognized as a potential threat to small populations (references in Simberloff 1988), and its diminution is now viewed as a rationale for corridors (see Merriam 1991; Thomas 1991; Noss 1992). No unified theory combines genetic, demographic, and other forces threatening small populations, nor is there accord on the relative importance of these threats. Lande (1988) believes that demography will usually be more important than genetics to very small populations, while Goodman (1987*a*) sees demographic stochasticity as important only when just a "handful" of individuals remain.

The metapopulation paradigm (Levins 1970) has replaced equilibrium theory for habitat islands (Merriam 1991), partly because of absence of evidence for turnover. The metapopulation is seen as lessening or redressing extinction by demographic stochasticity of the component populations. As with the equilibrium theory, however, there is now a tendency to take the metapopulation paradigm as broadly representative of nature, rather than as a new and untested hypothesis. Such statements as, "Many or most species are distributed as 'metapopulations'" (Noss 1992) are simply unwarranted. There are few empirical data, no specification of the range of dispersal rates that qualifies a group of populations as a metapopulation, and a variety of untested metapopulation models (references in Hanski & Gilpin 1991), none of which has been demonstrated to represent many situations in nature. Even if the metapopulation model should be shown to apply to a particular situation, the model requires movement, not corridors (Merriam 1991).

Inbreeding Depression

A genetic argument for corridors extends the concept of movement to genes as well as individuals. Harris (1984, 1985) argues that gene flow is required to prevent inbreeding depression from causing extinction and that corridors are needed for this gene flow.

Although a degree of inbreeding depression is usually found in captive animal populations, its threat must be established empirically. Some species with little genetic variation suffer no inbreeding depression (such as Pere David's deer and the European bison [Frankel & Soulé 1981]). It is often said that such species were lucky—their populations probably shrank gradually, and natural selection removed deleterious alleles that cause inbreeding depression. This line of reasoning implies that inbreeding depression is likely if normally outbred species suddenly inbreed (Frankel & Soulé 1981), as might be induced by rapid habitat fragmentation. However, in these cases most of the population decline occurred in a few generations. Therefore it remains mysterious why some species have little or no inbreeding depression. In nature, evidence from animal populations is extremely scarce (Charlesworth & Charlesworth 1987). The oft-cited evolution of behavior that reduces inbreeding need not have been selected by inbreeding depression (Charlesworth & Charlesworth 1987). In higher plants in nature, a measure of inbreeding depression has almost always been found when sought (Charlesworth & Charlesworth 1987). However, adequate data have been collected from far too few species.

Finally, finding that small populations are threatened by inbreeding depression is different from demonstrating its existence (Lande 1988). To evaluate the threat from inbreeding depression, it is important to bear in mind that a loss in genetic fitness need not endanger a population. "Inbreeding depression" means that more inbred individuals are less fit than less inbred ones. Most populations of most species, however, generation after generation produce some individuals less fit than others, yet are not endangered. So it is not axiomatic that inbreeding, even if it should lead to inbreeding depression, is a major threat to small populations, relative to other threats (Lande 1988). In any event, even if inbreeding depression is found to threaten a population, the key question is whether corridors are the best way to stem it; we return to this point below.

Need of Individual Animals for Movement

The Northern Spotted Owl (*Strix occidentalis caurina*) has a median pair home range in most areas of 1200 to 2000 ha (Thomas et al. 1990). Home ranges are larger in areas with less old-growth forest habitat (references in Thomas et al. [1990]), suggesting that the crucial requirement is enough small mammal prey inhabiting this habitat. There is an indication that owls that do not get enough favored prey experience lowered reproduction rates, but there is no suggestion that failure of space for movement per se is limiting. Mortality of dispersing juveniles is severe, but Thomas et al. (1990) argue against even wide corridors specifically for owl movement on the grounds that predators may thrive in them. Spotted Owls appear to disperse in random directions, which also argues against a corridor strategy (Noss 1992). Rather, Thomas et al. (1990) suggest management of the entire matrix surrounding owl habitat conservation areas to make it suitable for owl dispersal.

Harris (1985) and Noss and Harris (1986) have adduced the large home ranges of Florida panthers (*Felis concolor coryi*) and black bears (*Ursus americanus*) as a rationale for corridors. Most conservation areas in Florida are too small to provide the approximately 80 km^2 covered by a male black bear (Lindzey & Meslow 1977; Wooding & Hardisky 1990) or the approximately 400 km^2 covered by a male panther (Belden 1989) each year, and Maehr and Harris (1986) believe that areas as large as 800 km^2 may not support viable populations of either species. However, black bear populations have persisted despite confinement to small areas (Lindzey & Meslow 1977). Corridors might allow some species to avoid potentially fatal intraspecific encounters. Territorial battles between male panthers and between male black bears are sometimes lethal (Kemp 1976; Belden 1989). There are instances of black bear cannibalism (Tietje et al. 1986), and young black bears establish territories more readily at low densities (Kemp 1976; Rogers 1987). Therefore if corridors actually aided dispersal, they might be beneficial, but there is little evidence that dispersal is mediated by such social interactions (Rogers 1987) or that bears use corridors to disperse.

No Florida panther restricts its movements to one protected area, and panthers use a "hardwood strand" corridor linking the Big Cypress National Preserve to unprotected habitat (Maehr 1990). Hardwood strands follow depressions and slow-moving watercourses in Florida (Maehr & Cox, unpublished data). This particular strand is less than 5 km long—much less than the daily ambit of a panther (Belden 1989)—and bounded by agricultural areas, which panthers avoid. How well panthers would use a much longer corridor is unknown. Long-distance dispersal by bears and panthers is not nearly as well documented. Maehr et al. (1988) describe one black bear dispersal of 120 km, but they also contend that well-defined corridors are not necessary for bears so long as habitat does not fully impede movement.

In sum, the need for movement per se is doubtful, though animals may be driven to move for specific purposes, such as finding food or avoiding conspecifics. The key questions are whether such movement is necessary in a specific refuge system and, if it is, whether corridors of specified characteristics are the best strategy to facilitate such movement.

How Inevitable is Quick Extinction of Small Populations?

Given the attention paid in recent conservation literature to threats to small populations, it might seem that any small, isolated population is doomed in the short term. Many go extinct (see Petterson 1985). But many small populations persist and cannot easily be dismissed as recently-reduced populations en route to extinction. Numerous endemic species of small islands are endangered primarily by either habitat destruction or introduced species (references in Simberloff 1986; see Craig 1991) but apparently thrived for millennia in the absence of humans, though many probably numbered at most in the low hundreds. Species in the highest trophic levels are particularly striking. For example, the Red-Tailed Hawk (*Buteo jamaicensis socorroensis*) of Socorro Island (southwest of Baja California) has had a stable population of about 20 pairs for at least the last few decades, and there is no reason to think the population was larger in antiquity (Walter 1990). Of course, we do not even know the fates of many isolated populations that disappeared in the absence of human interference, but rapid extinction is not automatic.

Evidence for Corridor Use

Simberloff & Cox (1987) found few empirical data on corridor use and discovered that most reported observations were ambiguous. These problems persist (Hobbs & Hopkins 1991; Nicholls & Margules 1991; Saunders & Hobbs 1991*a*). There are still few data, and many widely cited reports are unconvincing.

Probably the first advocacy of corridors based on specific data was for birds of forest patches in the northeastern U.S. (MacClintock et al. 1977). This study is often cited (see Greenberg 1990) as showing that corridors increase diversity. The study was uncontrolled; there were no isolated sites *not* connected by a corridor to a larger forest, and no data on corridor use for sites that were connected to a larger forest. A frequently cited (see Noss 1992) recent paper (Saunders & Ingram 1987) contends that "Comparison of breeding results from all five populations studied and the amount of native vegetation remaining in each area showed that Carnaby's Cockatoo [*Calyptorhynchus funereus latirostris*] can breed successfully in areas which have been extensively cleared provided there are corridors of native vegetation connecting patches of remnant vegetation." In fact, data from this paper do not show that corridors are more important than amount of vegetation in the remnant (Saunders, personal communication 1987). However, additional uncited data (Saunders 1980) showed that, at the one site where the cockatoo disappeared, feeding areas near railway and road verges were used but did not connect to a nearby reserve.

Bennett (1987*a*), studying the long-nosed potoroo (*Potorous tridactylus*) in Australian forest patches, concluded that forest strips along roads and creeks aid dispersal and allow persistence in a fragmented landscape. No data in this paper addressed this question, but the paper cited a dissertation (Bennett 1987*b*). There the conclusion rested solely on two animals trapped in forested corridors, while there were no traps outside of corridors. Bennett (1990*a*) similarly felt that narrow forested corridors "facilitate continuity between populations" of eight mammals in Australia but did not study movement outside of corridors. Suckling (1984) argued that forested roadside strips connecting woodland patches prevent local extinction of the sugar glider (*Petaurus breviceps*) in Australia. However, he did not trap outside corridors and forest patches, and the only unconnected patch of the three studied was less than half the size of the others. Thus there are no data on increased dispersal with corridors or increased extinction without them. Bennett (1990*b*) argued that "all known dispersal movements" of the sugar glider involved corridors; in fact, none were sought otherwise.

Assessing these claims was difficult, entailing correspondence and examination of an unpublished document (Bennett 1987*b*) in Australia that could not be procured by interlibrary loan in Florida. One cannot expect management personnel to expend such effort, yet without this examination one would be left with a misleading impression. For example, Ogle (1989) noted that a 1986 corridor proposal in New Zealand was questioned on the grounds of lack of evidence on corridor use, "notwithstanding the plausibility of the concept and a growing body of supportive data from overseas." There were, at that time, virtually no supportive empirical data, only a plethora of statements advocating the concept (Simberloff & Cox 1987). Ogle (1989) went on to cite new supporting data:

> Biological corridors have been the subject of considerable field research in Australia, and their importance for movements of fauna between otherwise discrete patches of habitat has been established on, for example, small marsupials (Bennett 1987) and Cockatoos (Saunders & Ingram 1987). Bennett showed that narrow strips of forest along road verges and water courses provide routes for the dispersal of adult long-nosed potoroo (*Potorous*) between remnants of native vegetation in agricultural lands of south-west Victoria. Saunders & Ingram have shown that Carnaby's Cockatoos (*Calyptorhynchus funereus latirostris*) in Western Australia nest in isolated forest remnants in agricultural land, and feed in remnants of native heathland. Native vegetation of road verges provides some feeding habitat and the routes (corridors) between nesting and feeding areas.

Ogle was at pains to show that corridors are critical for these species, but neither study cited demonstrated their importance.

Nicholls and Margules (1991) detail statistical difficulties of experiments to show that corridors enhance movement. Largely because of problems achieving sufficient experimental sample sizes, they acknowledge that observational studies could be useful. As the examples above show, however, many such studies are inconclusive, particularly because of failure to examine movement without corridors. That an animal uses corridors when these are present need not mean movement without them is impossible, or even less frequent. Of 36 papers in "The Role of Corridors" (Saunders & Hobbs 1991*b*), five present new data on animal movement (Arnold et al. 1991; Catterall et al. 1991; Date et al. 1991; Prevett 1991; Saunders & de Rebeira 1991). Of these, only Arnold et al. (1991) gathered data on movement between habitat patches without corridors. Only three (Arnold et al. 1991; Date et al. 1991; Prevett 1991) concluded that corridors have a very small role in conserving a particular taxon.

Regional Corridors to Alleviate the Effects of Global Warming

The interaction of fragmentation with global warming has led to a different sort of corridor proposal. Global warming can cause a species' present sites to change more quickly than the species can evolve (Peters & Darling 1985). During earlier climatic changes many species could shift their geographic ranges to remain in their favored habitat. As habitats have become fragmented, routes are increasingly blocked (Peters 1988).

Graham (1988), Hunter et al. (1988), and Harris and Gallagher (1989) suggest a vast network of corridors over North America. Hunter et al. (1988) feel that even a 300-meter-wide corridor stretching thousands of kilometers could be very useful. Few data are available to assess this plan. Whether a strip 300 meters wide would allow the range shifts that typified the Pleistocene has barely been discussed. Noss (1992) worries that the anticipated speed of temperature increase may render such corridors of little use for many species. Hobbs and Hopkins (1991) admit that the utility of corridors in counteracting effects of global warming is very uncertain, but they argue for going ahead anyway.

If a corridor is intended to preserve an entire community, as in the global warming recommendation, it is particularly important that it be wide enough to permit breeding as well as movement. One need only consider the limited mobility of many soil invertebrates and plants to realize that a range shift would be painfully slow and require many generations.

A corridor to permit survival and breeding is serving as more than a corridor for movement; it is a habitat in its own right. Such a requirement raises the stakes considerably. One must know if the interior of a corridor functions as the intended habitat rather than as edge. Both micrometeorological effects and biotic intrusions can propagate an edge effect far into the forest (see Levenson 1981; Janzen 1983, 1986; Wilcove et al. 1986; Kapos 1989). Nest predation studies (such as Wilcove 1985; Andren & Angelstam 1988) also point to a substantial edge effect. The modified habitat in thin corridors is the motivation for the Klamath Corridor Proposal (Pace 1991), which envisions a corridor 5.5 km wide, connecting two wilderness areas 26 km apart in the Klamath National Forest. This proposal specifically aims at a corridor that is a habitat, not just a travel route.

Potential Biological Disadvantages of Corridors

Genetic and Demographic Costs and Benefits

Although the possibility of inbreeding depression would likely be lowered if corridors were used, the possibility of loss of alleles to drift in an ensemble of refuges would increase (references in Simberloff [1988]). This trade-off is inevitable; which problem should be of more concern depends on the severity of inbreeding depression (an empirical matter) and how pronounced and dangerous the slowing of future evolution by drift will be. There is no unanimity on this matter (Simberloff 1988). Goodman (1987*b*) has argued that effects of demographic stochasticity can be greatly lessened by a modest amount of migration among refuges, though he gives no figure. Whether a corridor would be the only or even the best way to provide whatever movement is necessary would depend on information on how much movement would occur with and without corridors. Depending on metapopulation structure, as little as one breeding migrant per generation can render a population effectively panmictic (Lande & Barrowclough 1987), which raises the possibility that even if a corridor increased movement it might be unnecessary genetically.

Spread of Catastrophes

Phenomena such as fires, diseases, or introduced species can spread through a corridor (Simberloff & Cox 1987). For example, in New Zealand, introduced feral pigs may eliminate *Paryphanta* snails from large forest fragments but do not invade small, isolated ones (Ogle 1987). Similarly, introduced vertebrate browsers in New Zealand have almost eliminated mistletoe from large forest fragments, but not from isolated small groups of trees (Ogle & Wilson 1985). In most instances, however, the added risk of spreading catastrophes via corridors may be low because introduced predators or diseases can reach fragments without corridors.

Corridors as Reservoirs of Edge and Introduced Species

Several authors have suggested that corridors might be inimical as a habitat in their own right. For example, forest corridors have a high fraction of edge habitat and might attract edge-inhabiting predators (Ambuel & Temple 1983). The same concern is one reason for the disenchantment with corridors on the part of Thomas et al. (1990), and numerous other authors have pointed to this potential liability in particular systems (see Catterall et al. 1991). Others have warned that some corridors may favor movement by introduced species (see Forman 1991; Hobbs & Hopkins 1991; Panetta & Hopkins 1991).

Corridors as Traps or Sinks

Henein and Merriam (1990) and Soulé and Gilpin (1991) propose on the basis of simulation modeling that low-quality corridors could act as sinks, decreasing the size of a metapopulation. Field data on this proposition are nonexistent.

Discussion

The notion that corridors can't hurt, even if the possible biological costs could be discounted, is not necessarily always true. Much would depend on the relative costs and benefits of a proposed corridor and alternative uses of the funds (Simberloff & Cox 1987). Because there are so few data on the importance of movement through corridors, such an analysis will be very difficult. We believe no thorough analysis of this sort has ever been conducted. Possibly the most that can be done today is to say that some options are much less likely to be important than others, but even this attempt is rarely made. The enormous price of some corridor proposals surely implies that other options would not be pursued. For example, a 300-meter-wide corridor from south Florida to Canada, proposed by Hunter et al. (1988), would be approximately 720 km^2, enough for a very large refuge. If it were possible to buy conservation lands of this magnitude, would this corridor be the best possible purchase?

Florida provides examples of the uncritical advocacy of extremely expensive corridors. Florida's Conservation and Recreation Lands (CARL) acquisition program has an annual budget of approximately $50 million from the sale of real estate stamps and an additional $135 million from the Landmark Preservation 2000 bond program. Preservation 2000 bonds have been authorized for the first two years only, although the program is envisioned to extend ten years. Proposed acquisitions still far exceed available funds. For example, of 93 CARL proposals passing a stringent double review by 1991, a priority list of 60 had a tax-assessed value of $408,000,000 (fair market values are often twice tax-assessed ones). Thus, every acquisition means other land will not be acquired. Citizens of several counties have voted to tax themselves to acquire land. Many such counties apply for joint acquisition projects to the CARL program. County governments also face lists of projects far exceeding available funds.

The Florida Natural Areas Inventory submitted the Blackwater-Eglin Connector project to the CARL program in 1988 (Anonymous 1988). This 2400-ha corridor was intended to connect Eglin Air Force Base (187,500 ha) with the Blackwater River State Forest and the adjacent Conecuh National Forest (143,250 ha combined). Anticipated fair market value exceeded $5 million. The corridor was "to join three large publicly owned areas ... into a single, uninterrupted ecological unit of nearly one million acres; to assure perpetual opportunities for gene flow among all populations in these areas; to allow for direct movements of individuals of more vagile species ..." (Anonymous 1988). The application spoke of the "urgency and desirability of developing systems of interconnected reserves for maintaining the full scope of biotic diversity" and stated that "connections between reserves create buffers against stochastic changes in populations, including local extinctions, and provide suitable pathways for 'natural' inoculations of new genetic material...."

No data showed that any species needed to move from one site to the other. Black bears are mentioned as potential users of the corridor, but they do not occur regularly in the Blackwater River State Forest or north of it (Brady & Maehr 1985). The corridor is complicated by Interstate 10, which separates Eglin from Blackwater. While small culverts run under the highway, there is no evidence that bears or other species use them or that they would substantially decrease highway mortality. Black bears are loath to cross interstate highways (Brady & Pelton 1989).

The Blackwater-Eglin Connector was also proposed to aid the movement of Red-cockaded Woodpeckers (*Picoides borealis*) (Anonymous 1988). The woodpeckers move up to 90 km (Walters et al. 1988*a*), but we doubt that this corridor would be effective. Of breeding females, 19% move per year, fewer than 5% of these move more than 5 km, and about 70% survive the year (Walters et al. 1988*b*). Therefore, the chance that a female will move that far and survive is about 0.7% per year. For all life-history categories together, the annual chance that a bird will disperse and survive is 4–8%. For breeding, this percentage is at least halved, because over 60% of dispersal is by first-year birds that produce only 20–67% as many fledglings as older ones.

Probabilities must be further reduced depending on the chance that a bird finds and uses the corridor. The distance traversed by the corridor is about 6 km, but midpoints of the two refuges are 45 km apart. Much less than 1% of the population monitored by Walters et al. (1988*b*) moved this far, so the great majority of dispersers will probably stay within their original refuge. The corridor itself is unsuitable for the woodpecker. Greatly altered by logging, it consists of streamside hardwoods and commercial pinelands. Several streams through the site have been impounded to create amenity lakes. It would surprise us if even 1% of dispersing individuals used this corridor, but if that many did, 0.04% of the population might reach the adjacent refuge and contribute to future generations. About 300 colonies inhabit both refuges (Wood & Wenner 1983), and each colony produces 1.2–1.7 fledglings per year (Walters et al. 1988*b*). So one successful dispersal might occur every 5–7 years. The corridor would thus be a very expensive way to move very few woodpeckers. It is unfair to ascribe the entire cost of a corridor to one or two target species when others would surely use it, but an expenditure of this size merits discussion of the full roster of species that would benefit.

This proposal was not approved by the CARL program. This fate contrasts with that of the Pal-Mar project, submitted to the CARL program in 1990 as a 8860 ha site in Martin and Palm Beach counties (Anonymous 1990*b*). Its fair market value was estimated at $22 million. The site is a large, relatively intact system of mesic/wet flatwoods and depression marsh/wet prairies containing many listed plant and animal species—perhaps the biggest relatively undisturbed functional wetland in south Florida east of the Everglades.

After the original application had been submitted, staff of the Florida Department of Natural Resources suggested expanding the project (Timmerman et al. 1991). They stated that the main thrust of a plan they had been developing to protect biological diversity in southeast Florida was establishment of two large corridors, and they suggested adding a corridor connecting Pal-Mar to Jonathan Dickinson State Park, which contains approximately 4850 ha. This corridor, subsequently added to the proposed Pal-Mar project, is 1.6 km wide and 10 km long, with one other section of land. The corridor is thus 18 km^2. It is bordered to the north by orange groves and to the south by residences. No estimate of cost was provided, but the proposed corridor contains substantial developable uplands, and its tax-assessed value is $21,000,000. Not only did the proposal request addition of a corridor (without biological data on which species would use it or why it is necessary), but it added that the corridor was key to the whole project and should be purchased *before* the original proposed acquisitions. The letter failed to mention that this corridor is traversed by two major highways. The final project design with recommended acquisition phasing has not yet been completed.

The detailed assessment of the Pal-Mar project, prepared by staff of the CARL land acquisition selection body (Anonymous 1990*b*), stated: "Acquisition of the Pal-Mar proposal will secure a corridor connecting J. W. Corbett Wildlife Management Area to Jonathan Dickinson State Park, and prevent otherwise inevitable development that would isolate the state park and *lead to faunal extirpations*" (our italics). Indeed, despite the fact that acquisition of the site proposed initially would protect a vast wetland system containing many state- and federally-listed species, the project assessment said, "Acquisition of the Pal-Mar tract would serve several purposes of regional significance, *the primary one* being the protection of a large portion of a proposed wildlife corridor that would link Jonathan Dickinson State Park to J. W. Corbett . . ." (our italics). The assessment even acknowledged that "Perhaps the most significant hurdle that this plan must overcome is the fact that wildlife routes between the state park and Pal-Mar are impeded by I-95 (six lanes) and the Florida Turnpike (four lanes), which run side by side less than a mile west of the park." Thus, even though large mammals have never been documented on the site and would need to cross ten heavily traveled lanes, the State of Florida is considering buying an 18 km^2 corridor on very expensive developable south Florida land.

We detail these examples because we feel that no abstract discussion of corridors can adequately guide acquisition and management decisions. There may well be specific cases in which a proposed corridor would be more effective than an alternative. For example, in parts of the American West, where many potential high-quality corridors are already in public ownership, one might expect acquisition costs to be low (Noss, personal communication) and use high. Noss (personal communication) suggests that elimination of below-cost timber sales in such sites might even save public money. However, the same benefits might derive from other land that might be set aside in the same region; are these the acres we would most wish a public agency to sequester? Noss (1987) believes that even major corridor costs can be met by the right sort of publicity and/or by peace dividends. We hope he is correct, but if he is, could not other conservation expenditures similarly increase? Surely costs and benefits of alternative strategies should be considered.

It is important to realize that there *are* general alternative strategies to facilitate survival where refuges are insufficient. Franklin (1989) has recently propounded the "new forestry," a collection of methods by which even logged forests in the American Northwest might be more useful for conservation than they are now. The key is that the entire landscape be managed as a matrix supporting the entire biotic community. The specific costs entailed in such management have not been tallied. In the longleaf pine forests of the American Southeast, topography and aspects of the biology of key species suggest that such an approach would not be prohibitively expensive (Simberloff 1992). The tablelands of northern New South Wales can be managed so that the entire landscape, not just reserves, contributes heavily to the conservation of the community (McIntyre 1991).

Numerous authors have pointed to the importance of small *un*connected patches of forest or even single trees for the persistence of populations (see Date et al. 1991; Prevett 1991). McDowell et al. (1991) call for a network of such "stepping stones" connected by corridors to minimize extinctions in the South African fynbos community. If resources are insufficient for both, which would be more useful? How dense and how large must a network of stepping stones be to be useful in this system?

Of course these strategies are not incompatible; corridors and stepping stones could be part of an entire landscape managed for both extraction and conservation. However, limited funds will almost certainly prevent the simultaneous adoption of all possible approaches, and it is simply not useful in practice to advocate "strategically placed larger reserves complemented by local or regional networks of smaller reserves and short connecting links between them in as many directions as possible" (Blyth 1991). One must be willing to set priorities, and these should be based on relative costs and benefits. We do not agree with Saunders and Hobbs (1991*a*) that we should preserve all corridors now because we do not have the time to find out if they are useful. If, by preserving them, we fail to pursue other options, such an approach is not prudent.

Finally, to the extent that this trend is encouraged by the fact that the concept of corridors is easily understood by the public and legislators (Harris & Gallagher 1989) or that "people feel they are doing something for conservation" (Bennett, in Stolzenburg 1991), it is not a scientific phenomenon at all; but it still has costs and benefits, and these are not being addressed. Is it good conservation biology to sell legislators and the public on the easiest program for them to understand, in the absence of evidence that it is the most effective one? Is it beneficial for people to feel they are doing something important for conservation by preserving narrow roadside strips (Bennett's corridors) in the absence of evidence that they really are doing something? Even if they are, is preserving such corridors sufficient? Does it foster the belief that one has done enough and need not preserve larger tracts of valuable habitat? We cannot answer such questions, but we believe that it is important to raise them.

Acknowledgments

Bob Jenkins, Reed Noss, Jack Ward Thomas, and three anonymous referees provided numerous suggestions for improving this manuscript.

Literature Cited

Adams, L. W., and L. E. Dove. 1989. Wildlife Reserves and Corridors in the Urban Environment. National Institute for Urban Wildlife, Columbia, Maryland.

Ambuel, B., and S. A. Temple. 1983. Area-dependent changes in the bird communities and vegetation of southern Wisconsin forests. Ecology **64**:1057–1068.

Ames, L. 1990. The Otay River corridor. Coast and Ocean **6(2)**:13.

Anderson, S. H., K. Mann, and H. H. Shugart, Jr. 1977. The effect of transmission-line corridors on bird populations. American Midland Naturalist **97**:216–221.

Andren, H., and P. Angelstam. 1988. Elevated predation rates as an edge effect in habitat islands: experimental evidence. Ecology **69**:544–547.

Anonymous. 1986. What happened to "corridors"? ENFO **86(6)**:5.

Anonymous. 1988. Blackwater-Eglin Connector. File No. 880131-57-1, CARL Program. Florida Department of Natural Resources, Tallahassee, Florida.

Anonymous. 1990*a*. America's Greenways: Linking the Nation's Open Space. Conservation Fund, Arlington, Virginia.

Anonymous. 1990*b*. Pal-Mar. File No. 901203-43-1, CARL Program. Florida Department of Natural Resources, Tallahassee, Florida.

Anonymous. 1991. Lee County Wildlife Corridor System Plan. Lee County Division of Environmental Sciences, Ft. Myers, Florida.

Arnold, C. 1990. Wildlife corridors. Coast and Ocean **6(3)**:10–14,16,19–21.

Arnold, G. W., J. R. Weeldenberg, and D. E. Steven. 1991. Distribution and abundance of two species of kangaroo in remnants of native vegetation in the central wheatbelt of Western Australia and the role of vegetation along road verges and fence lines as linkages. Pages 273–280 in D. A. Saunders and R. J. Hobbs, editors. The role of corridors. Surrey Beatty, Chipping Norton, New South Wales, Australia.

Belden, C. 1989. The Florida panther. Pages 515–532 in Audubon 1988/1989 Wildlife Report. National Audubon Society, New York, New York.

Bennett, A. F. 1987*a*. Conservation of mammals within a fragmented forest environment: The contributions of insular biogeography and autecology. Pages 41–52 in D. A. Saunders, G. W. Arnold, A. A. Burbridge, and A. J. M. Hopkins, editors. Nature conservation: the role of remnants of native vegetation. Surrey Beatty, Chipping Norton, New South Wales, Australia.

Bennett, A. F. 1987*b*. Biogeography and conservation of mammals in a fragmented forest environment in south-western Victoria. Ph.D. dissertation. Department of Zoology, University of Melbourne, Melbourne, Australia.

Bennett, A. F. 1990*a*. Habitat corridors and the conservation of small mammals in a fragmented forest environment. Landscape Ecology **4**:109–122.

Bennett, A. F. 1990*b*. Habitat corridors: their role in wildlife management and conservation. Department of Conservation and Environment, Melbourne, Australia.

Blyth, J. D. 1991. The role of corridors in a changing climate. Pages 402–403 in D. A. Saunders and R. J. Hobbs, editors. The role of corridors. Surrey Beatty, Chipping Norton, New South Wales, Australia.

Brady, A. J., and M. R. Pelton. 1989. Effects of roads on black bear movements in western North Carolina. Wildlife Society Bulletin **17**:5–10.

Brady, J. R., and D. S. Maehr. 1985. Black bear distribution in Florida. Florida Field Naturalist **13**:1–7.

Breen, B. 1991. Animals win the right-of-way. Garbage **3(2)**:18–20.

Brown, J. H., and A. C. Gibson. 1983. Biogeography. C. V. Mosby, St. Louis, Missouri.

Budd, W. W., P. L. Cohen, P. R. Saunders, and F. R. Steiner. 1987. Stream corridor management in the Pacific Northwest. I. Determination of stream-corridor widths. Environmental Management **11**:587–597.

Catterall, C. P., R. J. Green, and D. N. Jones. 1991. Habitat use by birds across a forest-suburb interface in Brisbane: implications for corridors. Pages 247–258 in D. A. Saunders and R. J. Hobbs, editors. The role of corridors. Surrey Beatty, Chipping Norton, New South Wales, Australia.

Chadwick, D. H. 1990. The biodiversity challenge. Defenders Magazine **65** (May/June):19–30.

Charlesworth, D., and B. Charlesworth. 1987. Inbreeding depression and its evolutionary consequences. Annual Review of Ecology and Systematics **18**:237–268.

Craig, J. L. 1991. Are small populations viable? Pages 2546–2552 in Acta XX Congressus Internationalis Ornithologici, Vol. IV. New Zealand Ornithological Congress Trust Board, Wellington, New Zealand.

Date, E. M., H. A. Ford, and H. F. Recher. 1991. Frugivorous pigeons, stepping stones, and weeds in northern New South Wales. Pages 241–245 in D. A. Saunders and R. J. Hobbs, editors. The role of corridors. Surrey Beatty, Chipping Norton, New South Wales, Australia.

Dendy, T. 1987. The value of corridors (and design features of same) and small patches of habitat. Pages 357–359 in D. A. Saunders, G. W. Arnold, A. A. Burbidge, and A. J. M. Hopkins, editors. Nature conservation: the role of remnants of native vegetation. Surrey Beatty, Chipping Norton, New South Wales, Australia.

Forman, R. T. T. 1983. Corridors in a landscape: their ecological structure and function. Ekologiya (C.S.S.R.) **2**:375–387.

Forman, R. T. T. 1991. Landscape corridors: from theoretical foundations to public policy. Pages 71–84 in D. A. Saunders and R. J. Hobbs editors. The role of corridors. Surrey Beatty, Chipping Norton, New South Wales, Australia.

Frankel, O. H., and M. E. Soulé. 1981. Conservation and Evolution. Cambridge University Press, Cambridge, England.

Franklin, J. 1989. Toward a new forestry. American Forests November/December **1989**:1–8.

Gilbert, F. S. 1980. The equilibrium theory of island biogeography: fact or fiction? Journal of Biogeography **7**:209–235.

Gilbrook, M. J. 1986. Choosing preserves and connecting links. ENFO **86(6)**:9–10.

Goodman, D. 1987*a*. The demography of chance extinction. Pages 11–34 in M. E. Soulé, editor. Viable populations for conservation. Cambridge University Press, Cambridge, England.

Goodman, D. 1987*b*. Consideration of stochastic demography in the design and management of biological reserves. Natural Resource Modeling **1**:205–234.

Graham, R. W. 1988. The role of climatic change in the design of biological reserves: the paleoecological perspective for conservation biology. Conservation Biology **2**:391–394.

Greenberg, K. 1990. Florida highways and landscape linkages: considering wildlife in transportation planning. Unpublished report. Florida Department of Transportation, Tallahassee, Florida.

Hanski, I., and M. Gilpin. 1991. Metapopulation dynamics: brief history and conceptual domain. Biological Bulletin of the Linnaean Society **42**:3–16.

Harris, L. D. 1984. The Fragmented Forest: Island Biogeographic Theory and the Preservation of Biotic Diversity. University of Chicago Press, Chicago, Illinois.

Harris, L. D. 1985. Conservation corridors: a highway system for wildlife. ENFO Report 85-5. Environmental Information Center of the Florida Conservation Foundation, Inc., Winter Park, Florida.

Harris, L. D., and P. B. Gallagher. 1989. New initiatives for wildlife conservation: the need for movement corridors. Pages 11–34 in G. Mackintosh, editor. Preserving communities and corridors. Defenders of Wildlife, Washington, D.C.

Harris, L. B., and J. Scheck. 1991. From implications to applications: the dispersal corridor principle applied to the conservation of biological diversity. Pages 189–220 in D. A. Saunders and R. J. Hobbs, editors. The role of corridors. Surrey Beatty, Chipping Norton, New South Wales, Australia.

Henein, K. M., and G. Merriam. 1990. The elements of connectivity where corridor quality is variable. Landscape Ecology **4**:157–170.

Hobbs, R. J., and A. J. M. Hopkins. 1991. The role of conservation corridors in a changing climate. Pages 281–290 in D. A. Saunders and R. J. Hobbs, editors. The role of corridors. Chipping Norton, Surrey Beatty, New South Wales, Australia.

Hunter, M. L., G. L. Jacobson, Jr., and T. Webb III. 1988. Paleoecology and the coarse-filter approach to maintaining biological diversity. Conservation Biology **2**:375–385.

International Union for the Conservation of Nature and Natural Resources. 1980. World Conservation Strategy, Gland, Switzerland.

Janzen, D. H. 1983. No park is an island: increase in interference from outside as park size decreases. Oikos **41**:402–410.

Janzen, D. H. 1986. The external threat. Pages 286–303 in M. E. Soulé, editor. Conservation biology: the science of scarcity and diversity. Sinauer Associates, Sunderland, Massachusetts.

Johnson, A. S. 1989. The thin green line: riparian corridors and endangered species in Arizona and New Mexico. Pages 35–46 in G. Mackintosh, editor. Preserving communities and corridors. Defenders of Wildlife, Washington, D.C.

Kapos, V. 1989. Effects of isolation on the water status of forest patches in the Brazilian Amazon. Journal of Tropical Ecology **5**:173–185.

Kemp, G. A. 1976. The dynamics and regulation of black bear (*Ursus americanus*) populations in northern Alberta. Pages 191–197 in M. R. Pelton, J. W. Lentfer, and G. E. Polk, Jr., editors. Bears: their biology and management. International Union for the Conservation of Nature, Gland, Switzerland.

Kroodsma, R. L. 1982. Bird community ecology on power-line corridors in east Tennessee. Biological Conservation **23**:79–94.

Lacy, R. C. 1987. Loss of genetic diversity from managed populations: interacting effects of drift, mutation, immigration, selection, and population subdivision. Conservation Biology **1**:143–158.

Lande, R. 1988. Genetics and demography in biological conservation. Science **241**:1455–1460.

Lande, R., and G. F. Barrowclough. 1987. Effective population size, genetic variation, and their use in population management. Pages 87–123 in M. E. Soulé, editor. Viable populations for conservation. Cambridge University Press, Cambridge, England.

Levenson, J. B. 1981. Woodlots as biogeographic islands in southeastern Wisconsin. Pages 13–39 in R. L. Burgess and D. M. Sharpe, editors. Forest island dynamics in man-dominated landscapes. Springer-Verlag, New York, New York.

Levins, R. 1970. Extinction. Pages 77–107 in M. Gerstenhaber, editor. Some mathematical questions in biology. Lectures on mathematics in the life sciences, vol. 2. American Mathematical Society, Providence, Rhode Island.

Lindzey, F. G., and E. C. Meslow. 1977. Population characteristics of black bears on an island in Washington. Journal of Wildlife Management **41**:408–412.

MacArthur, R. H., and E. O. Wilson. 1967. The Theory of Island Biogeography. Princeton University Press, Princeton, New Jersey.

MacClintock, L., R. F. Whitcomb, and B. L. Whitcomb. 1977. Island biogeography and the "habitat islands" of eastern forest. II. Evidence for the value of corridors and minimization of isolation in preservation of biotic diversity. American Birds **31**:6–12.

Mackintosh, G. 1989. Preserving Communities and Corridors. Defenders of Wildlife, Washington, D.C.

Maehr, D. S. 1990. The Florida panther and private lands. Conservation Biology **4**:167–170.

Maehr, D. S., and L. D. Harris. 1986. Black bear distribution and conservation strategy in Florida, U.S.A. Abstract of paper presented at 7th International Conference on Bear Research and Management, Williamsburg, Virginia, February 21–26.

Maehr, D. S., J. N. Layne, E. D. Land, J. W. McCown, and J. Roof. 1988. Long distance movement of a Florida black bear. Florida Field Naturalist **16**:1–6.

McDowell, C. R., A. B. Low, and B. McKenzie. 1991. Natural remnants and corridors in Greater Cape Town: their role in threatened plant conservation. Pages 27–39 in D. A. Saunders and R. J. Hobbs, editors. The role of corridors. Surrey Beatty, Chipping Norton, New South Wales, Australia.

McIntyre, S. 1991. Habitat variegation: an alternative model to vegetation fragmentation in a pastoral landscape. Abstract in the program Conservation Biology in Australia and Oceania. Centre for Conservation Biology, Brisbane, Australia.

Merriam, G. 1991. Corridors and connectivity: animal populations in heterogeneous environments. Pages 133–142 in D. A. Saunders and R. J. Hobbs, editors. The role of corridors. Surrey Beatty, Chipping Norton, New South Wales, Australia.

Nicholls, A. O., and C. R. Margules. 1991. The design of studies to demonstrate the biological importance of corridors. Pages 49–61 in D. A. Saunders and R. J. Hobbs, editors. The role of corridors. Surrey Beatty, Chipping Norton, New South Wales, Australia.

Noss, R. F. 1987. Corridors in real landscapes: a reply to Simberloff and Cox. Conservation Biology **1**:159–164.

Noss, R. F. 1992. Wildlife corridors. In D. Smith and P. Hellmund, editors. Ecology of greenways. University of Minnesota Press, Minneapolis, Minnesota. In Press.

Noss, R. F., and L. D. Harris. 1986. Nodes, networks, and MUMs: preserving diversity at all scales. Environmental Management **10**:299–309.

Ogle, C. C. 1987. The incidence and conservation of animal and plant species in remnants of native vegetation within New Zealand. Pages 79–87 in D. A. Saunders, G. W. Arnold, A. A. Burbridge, and A. J. M. Hopkins, editors. Nature conservation: the role of remnants of native vegetation. Surrey Beatty, Chipping Norton, New South Wales, Australia.

Ogle, C. C. 1989. An overview of reserve design and location in New Zealand. Pages 11–18 in D. A. Norton, editor. Management of New Zealand's natural estate (Occasional Publication No. 1). New Zealand Ecological Society, Christchurch, New Zealand.

Ogle, C. C., and P. R. Wilson. 1985. Where have all the mistletoes gone? Forest and Bird **13**:8–15.

Pace, F. 1991. The Grider Creek story. Wild Earth **1(1)**:28–33.

Panetta, F. D., and A. J. M. Hopkins. 1991. Weeds in corridors: invasion and management. Pages 341–351 in D. A. Saunders and R. J. Hobbs, editors. The role of corridors. Surrey Beatty, Chipping Norton, New South Wales, Australia.

Peters, R. L. 1988. The effect of global climatic change on natural communities. Pages 450–461 in E. O. Wilson, editor. Biodiversity. National Academy Press, Washington, D.C.

Peters, R. L., and J. D. S. Darling. 1985. The greenhouse effect and nature reserves. BioScience **35**:707–717.

Petterson, B. 1985. Extinction of an isolated population of the middle spotted woodpecker *Dendrocopos medius* (L.) in Sweden and its relation to general theories on extinction. Biological Conservation **32**:335–353.

Prevett, P. T. 1991. Movement paths of koalas in the urban-rural fringes of Ballarat, Victoria: implications for management. Pages 259–272 in D. A. Saunders and R. J. Hobbs, editors. The role of corridors. Surrey Beatty, Chipping Norton, New South Wales, Australia.

Rogers, L. 1987. Effects of food supply and kinship on social behavior, movements, and population growth of black bears in northern Minnesota. Wildlife Monographs **97**.

Saunders, D. A. 1980. Food and movements of the short-billed form of the White-Tailed Black Cockatoo. Australian Wildlife Research **7**:257–269.

Saunders, D. A., and C. P. de Rebeira. 1991. Values of corridors to avian populations in a fragmented landscape. Pages 221–240 in D. A. Saunders and R. J. Hobbs, editors. The role of corridors. Surrey Beatty, Chipping Norton, New South Wales, Australia.

Saunders, D. A., and R. J. Hobbs. 1991*a.* The role of corridors in conservation: what do we know and where do we go? Pages 421–427 in D. A. Saunders and R. J. Hobbs, editors. The role of corridors. Surrey Beatty, Chipping Norton, New South Wales, Australia.

Saunders, D. A., and R. J. Hobbs, editors. 1991*b.* The Role of Corridors. Surrey Beatty, Chipping Norton, New South Wales, Australia.

Saunders, D. A., and J. A. Ingram. 1987. Factors affecting survival of breeding populations of Carnaby's Cockatoo *Calyptorhynchus funereus latirostris* in remnants of native vegetation. Pages 249–258 in D. A. Saunders, G. W. Arnold, A. A. Burbridge, and A. J. M. Hopkins, editors. Nature conservation: the role of remnants of native vegetation. Surrey Beatty, Chipping Norton, New South Wales, Australia.

Simberloff, D. 1986. The proximate causes of extinction. Pages 259–276 in D. M. Raup and D. Jablonski, editors. Patterns and processes in the history of life. Springer-Verlag, Berlin, Germany.

Simberloff, D. 1988. The contribution of population and community biology to conservation science. Annual Review of Ecological Systems 19:473–511.

Simberloff, D. 1992. Species-area and fragmentation effects on old-growth forests: prospects for longleaf pine communities. In S. Hermann, editor. Proceedings of the Conference on Longleaf Pine Forests. Tall Timbers, Inc., Tallahassee, Florida. In press.

Simberloff, D., and J. Cox. 1987. Consequences and costs of conservation corridors. Conservation Biology 1:63–71.

Soulé, M. E., and M. E. Gilpin. 1991. The theory of wildlife corridor capability. Pages 3–8 in D. A. Saunders and R. J. Hobbs, editors. The role of corridors in nature conservation. Surrey Beatty, Chipping Norton, New South Wales, Australia.

Stolzenburg, W. 1991. The fragment connection. Nature Conservancy 41(4):19–25.

Suchy, W., and L. Harris. 1988. Landscape Linkages. Florida Films, Gainesville, Florida.

Suckling, G. C. 1984. Population ecology of the sugar glider, *Petaurus breviceps,* in a system of fragmented habitats. Australian Wildlife Research 11:49–75.

Thomas, C. D. 1991. Ecological corridors: an assessment. Department of Conservation, Wellington, New Zealand.

Thomas, J. W., E. D. Forsman, J. B. Lint, E. C. Meslow, B. R. Noon, and J. Verner. 1990. A Conservation Strategy for the Northern Spotted Owl. U.S.D.A. Forest Service, U.S.D.I. Bureau of Land Management, Fish and Wildlife Service, and National Park Service, Portland, Oregon.

Tietje, W. D., B. O. Pelchat, and R. L. Ruff. 1986. Cannibalism of denned black bears. Journal of Mammalogy. 67:762–766.

Timmerman, W. W., C. Tamborski, and J. B. Miller. 1991. Letter dated June 5 to O. G. Brock, Florida Department of Natural Resources. Included in Anonymous 1990.

Walter, H. S. 1990. Small viable population: the Red-tailed Hawk of Socorro Island. Conservation Biology 4:441–443.

Walters, J. R., S. K. Hansen, J. H. Carter III, P. D. Manor, and R. J. Blue. 1988*a.* Long-distance dispersal of an adult Red-cockaded Woodpecker. Wilson Bulletin 100:494–496.

Walters, J. R., P. D. Doerr, and J. H. Carter, III. 1988*b.* The cooperative breeding system of the Red-cockaded Woodpecker. Ethology 78:275–305.

Wilcove, D. S. 1985. Nest predation in forest tracts and the decline of migratory songbirds. Ecology 66:1211–1214.

Wilcove, D. S., C. H. McLellan, and A. P. Dobson. 1986. Habitat fragmentation in the temperate zone. Pages 237–256 in M. E. Soulé, editor. Conservation biology: the science of scarcity and diversity. Sinauer Associates, Sunderland, Massachusetts.

Wilcox, D. A. 1989. Migration and control of purple loosestrife (*Lythrum salicaria* L.) along highway corridors. Environmental Management 13:365–370.

Williams, G. R. 1984. Has island biogeography theory any relevance to the design of nature reserves in New Zealand? Journal of the Royal Society of New Zealand 14:7–10.

Williams, G. R. 1986. Some criticisms of generally accepted island biogeographic theory. Pages 229–237 in A. E. Wright and R. E. Beever, editors. The offshore islands of northern New Zealand. New Zealand Department of Lands and Survey, Wellington, New Zealand.

Williamson, M. 1981. Island Populations. Oxford University Press, Oxford, England.

Williamson, M. 1989. The MacArthur and Wilson theory today: true but trivial. Journal of Biogeography 16:3–4.

Wilson, E. O., and E. O. Willis. 1975. Applied biogeography, Pages 523–534 in M. L. Cody and J. M. Diamond, editors. Ecology and evolution of communities. Harvard University Press, Cambridge, Massachusetts.

Wood, D. A., and A. S. Wenner. 1983. Status of the Red-cockaded Woodpecker in Florida: 1983 update. Pages 89–91 in D. A. Wood, editor. Proceedings of the Red-cockaded Woodpecker symposium, vol. 2. Florida Game and Freshwater Fish Commission, Tallahassee, Florida.

Wooding, J., and T. S. Hardisky. 1990. Black bear habitat study. W–41–35. Final performance report. Florida Game and Fresh Water Fish Commission, Tallahassee, Florida.

Desert-dwelling Mountain Sheep: Conservation Implications of a Naturally Fragmented Distribution

V. C. BLEICH*

California Department of Fish and Game
407 West Line Street
Bishop, CA 93514, U.S.A.
and
Institute of Arctic Biology and Department of Biology and Wildlife
University of Alaska Fairbanks
Fairbanks, AK 99775, U.S.A.

J. D. WEHAUSEN

University of California
White Mountain Research Station
Bishop, CA 93514, U.S.A.

S. A. HOLL†

U.S. Forest Service
San Bernardino National Forest
Mill Creek Ranger Station
Route 1, Box 264
Mentone, CA 92359, U.S.A.

Abstract: *Mountain sheep (*Ovis canadensis*) are closely associated with steep, mountainous, open terrain. Their habitat consequently occurs in a naturally fragmented pattern, often with substantial expanses of unsuitable habitat between suitable patches; the sheep have been noted to be slow colonizers of vacant suitable habitat. As a result, resource managers have focused on (1) conserving "traditional" mountainous habitats, and (2) forced colonization through reintroduction. Telemetry studies in desert habitats have recorded more intermountain movement by desert sheep than was previously thought to occur. Given the heretofore unrecognized vagility of mountain sheep, we argue that existing corridors of "nontraditional" habitat connecting mountain*

* *Correspondence should be addressed to this author.*
† *Present address: Jones and 2600 V Street Suite 100, Sacramento, CA 95818, U.S.A.*
Paper submitted January 25, 1989; revised manuscript accepted November 7, 1989.

Resumen: *Los borregos cimarrones (*Ovis canadensis*) exhiben una asociación cercana con terrenos montañosos escarpados y abiertos. En consecuencia, su hábitat ocurre en un patrón naturalmente fragmentado, frecuentemente con grandes extensiones de hábitat impropio separando las áreas apropiadas; y se ha notado que los borregos cimarrones son lentos en colonizar hábitat apropiado vacante. Como resultado de estes observaciones, gerentes de recursos naturales han enfocado su atención en (1) conservación de hábitat montañoso "tradicional" y (2) colonización forzado por reintroducción. Estudios telemétricos en hábitates desiertos han demostrado más movimiento entre sierras que antes se creía que ocurría. Dado la tendencia de vagar hasta ahora no reconocido de los borregos cimarrones, proponemos que corredores actuales de hábitat no tradicional que conectan sierras merecen consideración adecuada para conservación. Además, se debe reconocer la importancia para poblaciones relativamente aisladas de los borregos cimarrones de áreas pequeñas de hábitat montañoso que, aunque*

ranges be given adequate conservation consideration. Additionally, small areas of mountainous habitat that are not permanently occupied but that may serve as "stepping stones" within such corridors must be recognized for their potential importance to relatively isolated populations of mountain sheep. We discuss the potential importance of such corridors to other large, vagile species.

no ocupadas de manera permanente, pueden facilitar el movimiento de los borregos cimarrones dentro de dichos corredores. Se señala la importancia de estos corredores para otros especies errantes grandes.

Introduction

Wilcox and Murphy (1985:884) echoed an increasingly common concern when they stated, "That current ecological theory is inadequate for resolving many of the details should not detract from what is obvious and accepted by most ecologists: habitat fragmentation ... is the primary cause of the present extinction crisis." Indeed, fragmentation has been a central theme of much recent literature dealing with conservation biology (e.g., Soulé & Wilcox 1980; Schonewald-Cox et al. 1983; Harris 1984; Lehmkuhl 1984; Schwartz et al. 1986; Soulé 1986; and Chepko-Sade & Halpin 1987). Both community-level (e.g., Wilcove et al. 1986) and population-level (e.g., Ralls et al. 1986; Allendorf & Leary 1986) theory have been applied to current conservation problems. The former has been concerned with species diversity and the latter with the long-term integrity of gene pools. Two primary approaches are used to maintain adequate gene pools in fragmented situations (1) periodic induced migration (Frankel 1983); and (2) maintaining or creating corridors to connect fragments (Schonewald-Cox 1983; Simberloff & Cox 1987; Noss 1987). This paper addresses the long-term maintenance of genetically viable populations of desert-dwelling mountain sheep (*Ovis canadensis* ssp.) via the latter approach. In addition, we discuss the related topic of protecting islands of habitat that do not support permanent populations but may be used occasionally, serving as important "stepping stones" in migration corridors.

Philopatry in Desert-dwelling Mountain Sheep

Mountain sheep, in general, are closely associated with steep, mountainous, open terrain (Geist 1971), which results in naturally disjunct demes. This habitat preference reflects two basic adaptations of mountain sheep relative to predation (1) great agility on rocks; and (2) keen vision to detect predators at sufficient distances to make escape probable.

Following the early decimation of mountain sheep in North America, it became evident that this species was inherently slow to recolonize vacant habitat. Consequently, reintroductions became an important management technique, dating back as far as the 1930s. Geist (1967, 1971) was the first to propose a general theory on the conservative colonization behavior of mountain sheep. The result has been an emphasis on conservation of mountainous habitats for wild sheep, with little concern for intermountain areas. For example, the 21 specific plans to conserve habitat for desert-dwelling mountain sheep called for in the California Desert Conservation Area Plan (Bureau of Land Management 1980:35–36) are restricted to specific mountainous areas totaling only 4,800 km^2. It is not our purpose to criticize these attempts to protect and enhance habitat; instead, we cite that document as an example of "traditional" thinking with respect to the protection of "traditional" habitat.

In addition to the behavioral conservatism of mountain sheep that Geist (1971) emphasized relative to dispersal, he documented some interpopulation movements, mostly by rams, in which they crossed "nontraditional" sheep habitat. Desert ecosystems differ markedly from the northern systems studied by Geist (1971), in that the relatively flat terrain separating "traditional" habitat islands lacks dense vegetation. Such terrain should represent less of a barrier to dispersal than the forests of more northern ecosystems. Additionally, many desert mountain ranges lack large carnivores such as mountain lions (*Felis concolor*) and wolves (*Canis lupus*), which may be more common in intermountain habitats of the north. Consequently, one might expect less conservative dispersal behavior of mountain sheep in desert ecosystems compared with northern systems.

Early researchers (e.g., Russo 1956) were aware of intermountain movements by desert-dwelling mountain sheep. Recent technology has resulted in a vastly expanded knowledge of patterns of habitat utilization by these animals. In Arizona, occasional intermountain movements by ewes were documented, in addition to extensive intermountain movements by rams (Witham & Smith 1979; Cochran & Smith 1983; Ough & deVos 1984; Krausman & Leopold 1986). In Nevada, McQuivey (1978) noted the presence of rams and ewes in ranges not known to have resident populations. Similarly, Elenowitz (1982) and King & Workman (1983) documented movements of mountain sheep across highways, fences, and intermountain flats in New Mexico and Utah, respectively. Extensive ongoing telemetry studies in the Mojave Desert of California also confirm intermountain movement by both rams and ewes (Berbach 1987; V. C. Bleich, A. M. Pauli, R. L. Vernoy, J. D. Wehausen, unpublished data).

Wilson et al. (1980) noted that, "all areas utilized by desert bighorn are essential to their continued survival". This has become an increasingly common concept in recent years, as more and more investigators have considered the role of habitats separating desert mountain ranges (e.g., Ough & deVos 1984; Cooperrider 1985; Krausman & Leopold 1986; Schwartz et al. 1986). These authors also considered the importance of small populations and began to incorporate concepts of population genetics relative to questions of wild sheep management.

Genetic Considerations

The concern about genetic health of desert-dwelling mountain sheep arose from (1) a popular (Seton 1929; Buechner 1960; DeForge et al. 1979), but probably greatly exaggerated (Welles 1962; V. C. Bleich and S. A. Holl, unpublished data) assumption that mountain sheep in general have declined to approximately 2% of their historical population level in North America; (2) their relatively isolated natural habitat, the rugged peaks of desert mountain ranges (Hansen 1980); (3) their polygynous mating system (Geist 1971); and (4) the assumption that cultural features developed in the last century prevent dispersal across the relatively flat ground between desert mountain ranges (Bailey 1980). Geist (1975) raised the general question of genetic effects on mountain sheep populations when interpopulation movements could no longer occur.

Schwartz et al. (1986) looked at this question through applying population genetics theory to a "metapopulation" of mountain sheep in the Mojave Desert of California and Nevada that was bounded by two major fenced highways and the Colorado River, and included about 1,600 sheep distributed in 15 subpopulations (demes). Their analyses suggested that relatively low levels of gene migration were necessary to prevent loss of genetic diversity in small populations. While migration of genes is difficult to document, the increasing evidence of intermountain movement by rams in the breeding season suggests that the low levels of gene migration considered necessary probably are met. Their study area would satisfy the requirements for a preserve of a size consistent with (1) the long-term genetic health of populations, (2) the possibility of establishing additional subpopulations, and (3) the possibility of continued divergence and long-term evolution (level 7 or 8 preserve; Schonewald-Cox 1983). Also, mounting evidence (Festa-Bianchet 1986; Geist 1971; J. D. Wehausen, V. C. Bleich, A. M. Pauli, and R. L. Vernoy, unpublished data) indicates substructuring within traditionally defined populations that would minimize inbreeding. Many populations appear to consist of a number of distinct but overlapping female home ranges. Although female offspring generally appear to adopt the home range of their mother, mature males appear to spend the rut outside of their maternal home range.

Both dispersal and social structure potentially are important determinants of effective population size (Chepko-Sade et al. 1987). Dispersal, coupled with substructuring of populations, probably acts to maintain genetic variation within populations of mountain sheep. Maintaining such variation presumably is important in preserving the evolutionary potential of metapopulations and, as such, should be of concern to managers.

Ecological Considerations

In addition to corridors necessary to facilitate gene flow, the ecological value of mountainous habitats not permanently occupied should be recognized. Recent work in California has documented further the use of areas not traditionally considered to be mountain sheep habitat. For example, Cowhole Mountain, located approximately 5 km across a broad, sandy area west of Old Dad Peak, has been found to be a lambing area for the Old Dad Peak population and is used at other times of the year by different cohorts of the population as well. Similarly, in 1987 two telemetered ewes from the Old Woman Mountains visited the neighboring Iron Mountains in winter and the Ship Mountains in spring. One of these bore a lamb in the Iron Mountains and returned to the Old Woman Mountains three months later. Her disappearance from the Old Woman Mountains in the winter of 1986 suggests that this may be a regular pattern. This view is supported by a native of Milligan (personal communication 1987), a town at the southern tip of the Old Woman Mountains, who reported regularly seeing sheep tracks crossing between the Old Woman and Iron mountains in winter and spring. Both the Ship and Iron mountains are separated from the Old Woman Mountains by 6–8 km of desert flats and blow sand. The potential ecological importance of these and similar areas should not be underestimated. The sheep population in the Old Woman Mountains has been depressed during the 1980s, possibly because of a high prevalence of cattle diseases (Clark et al. 1985; Wehausen 1988). The observed intermountain movements by ewes may be remnants of movements that formerly occurred on a larger scale and that could be in danger of being lost as a regular pattern. No land management plan even considers the potential importance of the Ship and Iron mountains to the Old Woman Mountains population.

Although the Iron Mountains have been identified as

a potential reintroduction site (see below), such action has been delayed by the potential that animals moving from the Old Woman Mountains will transmit disease to the Iron Mountains. Dobson and May (1986) have cautioned against such scenarios. Indeed, intermountain movements are a double-edged sword — necessary for gene flow, but potentially deleterious due to disease transmission (Simberloff & Cox 1987). Such movements may have been a major factor in the current widespread distribution of parainfluenza-III virus in desert populations of mountain sheep (Clark et al. 1985; Wehausen 1987).

Conclusions

The notion that the habitat of desert-dwelling mountain sheep is restricted to those mountain masses that provide food, cover, and water and that support permanent populations of the species is no longer adequate. Although habitat within mountain ranges can be enhanced (e.g., Bleich et al. 1982*a*, 1982*b*; Werner 1985), such activities must be conducted with the awareness that all areas used by mountain sheep may be essential for their long-term survival. For viable populations of mountain sheep to persist, more than "mountain islands within desert seas" must be protected. Although natural forces such as precipitation may drive the dynamics of populations within these "islands" (Monson 1960; Bleich 1986; Douglas & Leslie 1986; Wehausen et al. 1987), and disjunct populations may simultaneously experience "boom" or "bust" phenomena, the actions of humans will determine the ultimate fate of this species.

Wilcox and Murphy (1985) concluded that the risk of fragmentation is threefold (1) demographic units may be destroyed outright, reduced in size, or subdivided; (2) potential sources of emigrants may be lost; and (3) immigration may be impeded by conversion of natural habitat. All of these are applicable to the conservation of mountain sheep in desert ecosystems. Nonetheless, it is important to recognize that a *naturally* fragmented distribution, as found among populations of desert-dwelling mountain sheep, can minimize the probability of extinction where catastrophic population losses are a factor (Quinn & Hastings 1987). This is the fundamental concept underlying the Recovery and Conservation Plan for mountain sheep in the Sierra Nevada of California (Sierra Bighorn Interagency Advisory Group 1984). The history of mountain sheep is replete with examples of decimation and extinction of local populations due to diseases, mostly contracted from domestic livestock (Buechner 1960; Robinson et al. 1967; Stelfox 1971; Sandoval 1980; Foreyt & Jessup 1982; Goodson 1982; Onderka & Wishart 1984; Jessup 1985). Although such demographic impacts may far outweigh long-term genetic considerations from a conservation standpoint (Lande 1988), migration between disjunct subpopulations remains critically important, not only for genetic reasons, but also for natural recolonization of habitat that may become vacant. Berger (1990) has recently demonstrated the high probability of extinctions of small populations of mountain sheep in this century. If even a fraction of these extinctions would have occurred in the absence of influences related to the white man, natural extinction and recolonization may be considerably more common than previously thought.

To ensure the long-term conservation of these animals in a wild state, future management strategies for mountain sheep in the desert must take more factors into account in a larger-scale approach. Management documents should begin to seriously consider intermountain travel corridors for sheep, taking steps to minimize potential barriers such as range fences and motorized recreational activities. Managers should also recognize that if domestic livestock graze along such corridors, diseases may be transmitted to mountain sheep populations via migrating animals. Domestic sheep are particularly dangerous in this regard because they carry fatal respiratory bacterial strains (Onderka & Wishart 1988; Onderka et al. 1988; Foreyt 1989). Small, isolated tracts of "traditional" habitat that is not permanently occupied should be recognized as potential seasonal habitat and as "stepping stones" within migration corridors. Translocation programs should give priority to reestablishing populations on ranges that will decrease interdeme distances so as to facilitate gene migration.

The Bureau of Land Management recently prepared a management plan for mountain sheep on all applicable desert ranges in the southwestern United States. The plan incorporates the concept of metapopulations (BLM 1988). It sets as its goal the recovery of 115 "populations" to "viable" status (≥100 sheep). However, there remains a need to map all potential metapopulations of mountain sheep as well as known and potential intermountain corridors throughout their desert range, and to develop conservation strategies on that geographic scale.

Figure 1 is an example of a metapopulation from southeastern California. It is bounded on the north, south, and west by major, fenced interstate highways, and on the east by the Colorado River. Relatively few unfenced, paved roads exist within this metapopulation; thus, with the exception of the Twenty-Nine Palms and Lucerne Valley areas, and an aqueduct partially separating the Coxcomb Mountains from the Granite/Palen and southern Iron mountains to the east and the Turtle Mountains from the unoccupied ranges to the south, there are few physical obstructions to intermountain movements by mountain sheep. Approximately 1,000 mountain sheep permanently inhabit 15 of 31 mountain ranges in this region. Two of the 15 inhabited ranges

Figure 1. Map of a metapopulation of mountain sheep in southeastern California. Stippled mountain ranges currently have resident populations of the approximate size listed. Mountain ranges with N = 0 are extirpated populations; ranges with no N value listed are not known ever to have had resident populations. Arrows indicate documented intermountain movements by mountain sheep.

have been reestablished by translocation (Whipple and Sheephole). Only 8 of the 15 ranges support populations of 50 or more sheep. We have documented movements of mountain sheep between 11 pairs of mountain ranges depicted in Figure 1; the mean distance between those ranges is about 9 km (range = 6–20).

From the standpoint of fragmentation, the population in the Newberry Mountains in the NW corner of Figure 1 is particularly isolated. In fact, as recently as 1982 this population was not known to exist (Weaver 1982). Reestablishing populations along the link between the Rodman and Bullion mountains should be a high priority within this metapopulation. The entire Bullion and Lava Bed Mountains, however, are within the Twenty-Nine Palms Marine Corps Training Center. The Department of Defense is currently pursuing a reintroduction of mountain sheep in the Bullion Mountains in cooperation with the California Department of Fish and Game.

Geographically, the second most notable fragmentation within this metapopulation is the separation of the three populations in the SW corner from the others. This constitutes a much less serious situation than the Newberry Mountains in that the combined population in this area totals about 275 sheep (Fig. 1). Nevertheless, reestablishing a population in the Pinto Mountains would facilitate migration between these three populations and the remainder of the metapopulation.

Within the eastern portion of this metapopulation, reestablishing a population in the Iron Mountains would provide an important connection between the Sheephole/Eagle/Coxcomb/Granite-Palen mountains complex and the occupied ranges to the NE. Given that the former complex contains only about 100 total sheep, reestablishing a population in the Iron Mountains should have priority over such an effort in the Pinto Mountains. The aforementioned disease question, however, will play an important role in the decision to reestablish a permanent population in the Iron Mountains.

Of the ranges not known previously to have had resident mountain sheep populations, the Stepladder Mountains are particularly important as a central "stepping stone" potentially connecting four surrounding populations. The Piute, Little Piute, Ship, and Calumet mountains are only somewhat less central, and three of these are known to have been visited by sheep from adjacent ranges. Similarly, the Lava Bed Mountains have the potential to serve as an important link between the Bullion and Rodman mountains, if populations become established there.

Our discussion has centered around the importance to mountain sheep of unimpeded movement. A similar concern can be extended to other terrestrial species whose primary habitat naturally occurs in disjunct patches but that cross expanses of less desirable habitat between such patches to some extent. Both mule deer (*Odocoileus hemionus eremicus, O. h. crooki*) and mountain lions, where they occur in deserts, probably fit these criteria. Previous discussions of habitat corridors (Simberloff & Cox 1987; Noss 1987) have referred to maintaining or creating corridors of habitats similar to those being connected. The situation considered here differs somewhat in that the disjunct nature of primary habitat patches is natural and the corridor habitat is clearly less desirable to the species involved but is nevertheless used in moving between suitable patches.

Schwartz et al. (1986) concluded, "In general, desert-dwelling mountain sheep populations are sufficiently abundant and juxtaposed, and areas of habitat are still sufficiently large to allow the continued existence of this species throughout much of its historic range." We still have the raw materials; what is needed is a commitment to protect and manage them properly. Only with the recognition that stewardship responsibilities extend beyond areas of "traditional" habitat and what are perceived to be "viable" populations will we assure the long-term stability of desert-dwelling mountain sheep and other vagile species that similarly inhabit naturally fragmented habitat.

Acknowledgments

We thank R. T. Bowyer, A. Y. Cooperrider, D. Ehrenfeld, V. Geist, R. R. Ramey II, O. A. Schwartz, and an anonymous reviewer for critical comments and helpful suggestions. C. Tiernan and K. Quinlan helped prepare the figure, and V. Blankinship translated the abstract into Spanish. This paper was originally presented as an invited paper at the 24th Annual Meeting of the Western Section of the Wildlife Society, during which productive interchange with M. E. Soulé occurred.

Literature Cited

Allendorf, F. W., and R. F. Leary. 1986. Heterozygosity and fitness in natural populations of animals. Pages 57–76 in M. E. Soulé, editor. Conservation biology. Sinauer Associates, Sunderland, Massachusetts.

Bailey, J. A. 1980. Desert bighorn, forage competition, and zoogeography. Wildlife Society Bulletin **8**:208–216.

Berbach, M. W. 1987. The behavior, nutrition, and ecology of a population of reintroduced desert mountain sheep in the Whipple Mountains, San Bernardino County, California. M.S. thesis, California Polytechnic State University, Pomona, California.

Berger, J. 1990. Persistence of different-sized populations: an empirical assessment of recent extinctions in bighorn sheep. Conservation Biology **4**:91–98.

Bleich, V. C. 1986. Early breeding in free-ranging mountain sheep. Southwestern Naturalist **31**:530–531.

Bleich, V. C., L. J. Coombes, and J. H. Davis. 1982*a*. Horizontal wells as a wildlife habitat improvement technique. Wildlife Society Bulletin **10**:324–328.

Bleich, V. C., L. J. Coombes, and G. W. Sudmeier. 1982*b*. Volunteer participation in California wildlife habitat management projects. Transactions of the Desert Bighorn Council **26**:56–58.

Buechner, H. K. 1960. The bighorn sheep in the United States, its past, present, and future. Wildlife Monographs **4**:1–174.

Bureau of Land Management. 1980. Final environmental impact statement and proposed plan. USDI, Bureau of Land Management, Desert District, Riverside, California.

Bureau of Land Management. 1988. Rangewide plan for managing habitat of desert bighorn sheep on public lands. USDI, Bureau of Land Management, Washington, D.C.

Chepko-Sade, B. D., W. M. Shields, J. Berger, et al. 1987. The effects of dispersal and social structure on effective population size. Pages 287–321 in B. D. Chepko-Sade and Z. T. Halpin, editors. Mammalian dispersal patterns. The effects of social structure on population genetics. University of Chicago Press, Chicago, Illinois.

Clark, R. K., D. A. Jessup, M. D. Kock, and R. A. Weaver. 1985. Survey of desert bighorn sheep in California for exposure to selected infectious diseases. Journal of the American Veterinary Medical Association **187**:1175–1179.

Cochran, M. H., and E. L. Smith. 1983. Intermountain movements by a desert bighorn ram in western Arizona. Transactions of the Desert Bighorn Council **27**:1–2.

Cooperrider, A. Y. 1985. The desert bighorn. Pages 473–485 in R. L. DiSilvestro, editor. Audubon wildlife report 1985. National Audubon Society, New York.

DeForge, J. R., C. W. Jenner, A. J. Plechner, and G. W. Sudmeier. 1979. Decline of bighorn sheep (*Ovis canadensis*), the genetic implications. Transactions of the Desert Bighorn Council **23**:63–66.

Dobson, A. P., and R. M. May. 1986. Disease and conservation. Pages 345–365 in M. E. Soulé, editor. Conservation biology. Sinauer Associates, Sunderland, Massachusetts.

Douglas, C. L., and D. M. Leslie, Jr. 1986. Influence of weather and density on lamb survival of desert mountain sheep. Journal of Wildlife Management **50**:153–156.

Elenowitz, A. 1982. Preliminary results of a desert bighorn transplant in the Peloncillo Mountains, New Mexico. Transactions of the Desert Bighorn Council 26:8–11.

Festa-Bianchet, M. 1986. Seasonal dispersion of overlapping mountain sheep ewe groups. Journal of Wildlife Management 50:325–330.

Foreyt, W. J. 1989. Fatal *Pasteurella haemolytica* pneumonia in bighorn sheep after direct contact with clinically normal domestic sheep. American Journal of Veterinary Research 50:341–344.

Foreyt, W. J., and D. A. Jessup. 1982. Fatal pneumonia of bighorn sheep following association with domestic sheep. Journal of Wildlife Diseases 18:163–168.

Frankel, O. H. 1983. The place of management in conservation. Pages 1–14 in C. M. Schonewald-Cox, S. M. Chambers, B. MacBryde, and W. L. Thomas, editors. Genetics and conservation: a reference for managing wild animal and plant populations. Benjamin/Cummings, Menlo Park, California.

Geist, V. 1967. A consequence of togetherness. Natural History 76(8):24, 29–30.

Geist, V. 1971. Mountain sheep, a study in behavior and evolution. University of Chicago Press, Chicago, Illinois.

Geist, V. 1975. On the management of mountain sheep: theoretical considerations. Pages 77–98 in J. B. Trefethen, editor. The wild sheep in modern North America. Winchester Press, New York, N.Y.

Goodson, N. J. 1982. Effects of domestic sheep grazing on bighorn sheep populations: a review. Proceedings of the Biennial Symposium of the Northern Wild Sheep and Goat Council 3:287–313.

Hansen, C. G. 1980. Habitat. Pages 64–79 in G. Monson and L. Sumner, editors. The desert bighorn. University of Arizona Press, Tucson, Arizona.

Harris, L. D. 1984. The fragmented forest: island biogeographic theory and the preservation of biotic diversity. University of Chicago Press, Chicago, Illinois.

Jessup, D. A. 1985. Diseases of domestic livestock which threaten bighorn sheep populations. Transactions of the Desert Bighorn Council 29:29–33.

King, M. M., and G. W. Workman. 1983. Preliminary report on desert bighorn movements on public lands in southeastern Utah. Transactions of the Desert Bighorn Council 27:4–6.

Krausman, P. R., and B. D. Leopold. 1986. The importance of small populations of desert bighorn sheep. Transactions of the North American Wildlife and Natural Resources Conference 51:52–61.

Lande, R. 1988. Genetics and demography in biological conservation. Science 241:1455–1460.

Lehmkuhl, J. F. 1984. Determining size and dispersion of minimum viable populations for land management planning and species conservation. Environmental Management 8:167–176.

McQuivey, R. P. 1978. The desert bighorn sheep of Nevada. Nevada Department of Fish and Game Biological Bulletin 6:1–81.

Monson, G. 1960. Effects of climate on desert bighorn numbers. Transactions of the Desert Bighorn Council 4:12–14.

Noss, R. F. 1987. Corridors in real landscapes: a reply to Simberloff and Cox. Conservation Biology 1:159–164.

Onderka, D. K., and W. D. Wishart. 1984. A major bighorn sheep die-off from pneumonia in southern Alberta. Proceedings of the Biennial Symposium of the Northern Wild Sheep and Goat Council 4:356–363.

Onderka, D. K., and W. D. Wishart. 1988. Experimental contact transmission of *Pasteurella haemolytica* from clinically normal domestic sheep causing pneumonia in Rocky Mountain bighorn sheep. Journal of Wildlife Diseases 24:663–667.

Onderka, D. K., S. A. Rawluk, and W. D. Wishart. 1988. Susceptibility of Rocky Mountain bighorn sheep and domestic sheep to pneumonia induced by bighorn and domestic livestock strains of *Pasteurella haemolytica*. Canadian Journal of Veterinary Research 52:439–444.

Ough, W. D., and J. C. deVos, Jr. 1984. Intermountain travel corridors and their management implications for bighorn sheep. Transactions of the Desert Bighorn Council 28:32–36.

Quinn, J. F.,and A. Hastings. 1987. Extinction in subdivided habitats. Conservation Biology 1:198–208.

Ralls, K. P., H. Harvey, and A. M. Lyles. 1986. Inbreeding in natural populations of birds and mammals. Pages 35–56 in M. E. Soulé, editor. Conservation biology. Sinauer Associates, Sunderland, Massachusetts.

Robinson, R. M., T. L. Hailey, C. W. Livingston, and J. W. Thomas. 1967. Bluetongue in the desert bighorn sheep. Journal of Wildlife Management 31:165–168.

Russo, J. P. 1956. The desert bighorn in Arizona. Arizona Game and Fish Department Wildlife Bulletin 1:1–153.

Sandoval, A. V. 1980. Management of a psoroptic scabies epizootic in bighorn sheep (*Ovis canadensis mexicana*) in New Mexico. Transactions of the Desert Bighorn Council 24:21–28.

Schonewald-Cox, C. M. 1983. Conclusions: guidelines to management: a beginning attempt. Pages 414–445 in C. M. Schonewald-Cox, S. M. Chambers, B. MacBryde, and W. L. Thomas, editors. Genetics and conservation: a reference for managing wild animal and plant populations. Benjamin/Cummings, Menlo Park, California.

Schonewald-Cox, C. M., S. M. Chambers, B. MacBryde, and W. L. Thomas, editors. 1983. Genetics and conservation: a reference for managing wild animal and plant populations. Benjamin/Cummings, Menlo Park, California.

Schwartz, O. A., V. C. Bleich, and S. A. Holl. 1986. Genetics and the conservation of mountain sheep *Ovis canadensis nelsoni*. Biological Conservation 37:179–190.

Seton, E. T. 1929. Lives of game animals. Volume III, Part II. Hoofed Animals. Doubleday Page and Co., New York, N.Y.

Sierra Bighorn Interagency Advisory Group. 1984. Sierra Nevada bighorn sheep recovery and conservation plan. Inyo National Forest, Bishop, California.

Simberloff, D., and J. Cox. 1987. Consequences and costs of conservation corridors. Conservation Biology **1**:63–71.

Soulé, M. E., editor. 1986. Conservation biology: the science of scarcity and diversity. Sinauer Associates, Sunderland, Massachusetts.

Soulé, M. E., and B. A. Wilcox, editors. 1980. Conservation biology. An evolutionary-ecological perspective. Sinauer Associates, Sunderland, Massachusetts.

Stelfox, J. G. 1971. Bighorn sheep in the Canadian Rockies: a history 1800–1970. Canadian Field-Naturalist **85**:101–122.

Weaver, R. A. 1982. Bighorn in California: a plan to determine current status and trends. Administrative Report, California Department of Fish and Game, Sacramento, California.

Wehausen, J. D. 1987. Some probabilities associated with sampling for diseases in bighorn sheep. Transactions of the Desert Bighorn Council **31**:8–10.

Wehausen, J. D. 1988. Cattle impacts on mountain sheep in the Mojave Desert: report II. Unpublished report, California Department of Fish and Game, Bishop, California.

Wehausen, J. D., V. C. Bleich, B. Blong, and T. L. Russi. 1987. Recruitment dynamics in a southern California mountain sheep population. Journal of Wildlife Management **51**:86–98.

Welles, R. E. 1962. What makes a valid observation? Transactions of the Desert Bighorn Council **7**:29–40.

Werner, W. E. 1985. Philosophies of water development for bighorn sheep in southwestern Arizona. Transactions of the Desert Bighorn Council **29**:13–14.

Wilcove, D. S., C. H. McLellan, and A. P. Dobson. 1986. Habitat fragmentation in the temperate zone. Pages 237–256 in M. E. Soulé, editor. Conservation biology. Sinauer Associates, Sunderland, Massachusetts.

Wilcox, B. A., and D. D. Murphy. 1985. Conservation strategy: the effects of fragmentation on extinction. American Naturalist **125**:879–887.

Wilson, L. O., J. Blaisdell, G. Welsh, et al. 1980. Desert bighorn habitat requirements and management recommendations. Transactions of the Desert Bighorn Council **24**:1–7.

Witham, J. H., and E. L. Smith. 1979. Desert bighorn movements in a southwestern Arizona mountain complex. Transactions of the Desert Bighorn Council **23**:20–23.

Comment

Ecological Principles for the Design of Wildlife Corridors

DAVID B. LINDENMAYER
HENRY A. NIX
Centre for Resource and Environmental Studies
The Australian National University
G.P.O. Box 4
Canberra A.C.T. 2601, Australia

Introduction

Networks of wildlife corridors are increasingly being advocated as a key component of strategies for the conservation of biodiversity (Saunders & Hobbs 1991). While they have general support among conservationists, there is a paucity of data on the effectiveness of such corridors for nature conservation. In addition, there are few scientifically-based guiding principles for the evaluation and design of systems of retained areas. Harrison (1992) has drawn attention to these deficiencies and has proposed a theoretical basis for the design of wildlife corridors. While we support this initiative, we have reservations about the generality of the approach outlined by Harrison (1992). He calculates minimum corridor widths for a selected set of terrestrial mammals based on estimates of home range. Our data from studies of forest-dependent arboreal marsupials in southeastern Australia (Lindenmayer 1992*a*, Lindenmayer et al. in press) do not support Harrison's findings and suggest that additional criteria must be considered in developing a theoretical framework for the evaluation and design of wildlife corridors.

Studies of Arboreal Marsupials in Wildlife Corridors

Recent surveys have censured arboreal marsupials in 49 retained linear habitats located within timber production montane ash forests in the Central Highlands of Victoria, southeastern Australia (Lindenmayer 1992*a*; Lindenmayer et al. in press). The primary objectives of the study were (1) to compare the findings from wildlife corridors with those from similar studies of arboreal marsupials in areas of contiguous forest (Lindenmayer et al. 1990, 1991*a*), and (2) to identify corridor attributes that significantly influenced the presence and abundance of arboreal marsupials. The investigation was an extensive one and the study sites varied in width, length, connectivity, habitat quality, and a wide range of other parameters. For example, corridor width varied from 30 m to 264 m, and the total area of sites ranged from 0.8 ha to 14.6 ha (Lindenmayer 1992*a*). Some key data from the study are briefly summarized in Tables 1 and 2.

Findings of Relevance for Corridor Design

The key findings relevant to the evaluation of design of wildlife corridors are outlined below.

(1) Some species of arboreal marsupials, such as the greater glider and the mountain brushtail possum, appeared to survive well in wildlife corridors. Other taxa, such as Leadbeater's possum, sugar glider, and yellow-bellied glider, were rare, although for several species the habitat within many sites was predicted to be suitable (Table 1). Larger species were more commonly recorded in wildlife corridors than were smaller ones (Table 1). These findings appear at first to be counter-intuitive, as larger animals would be expected to be less

Paper submitted August 25, 1992; revised manuscript submitted November 25, 1992.

Table 1. Biological characteristics of various species of arboreal marsupials that were observed in 49 wildlife corridors in the montane ash forests of the Central Highlands of Victoria, southeastern Australia.

Species	*Mean body weight (g)*	*Home range (ha)*	*Diet*	*Social structure*	*No. of corridors observed (max. value = 49)*	*No. of sites where predicted to occur*
Mountain Brushtail Possum	2670	5	Foliage Fungi	Solitary Monagamous	16	22.4 (16.3–28.5)
Greater Glider	1378	1–3	Eucalypt Foliage	Solitary Monagamous	16	13.0 (6.9–19.0)
Common Ringtail Possum	733	1–2	Leaves Fruits Flowers	Colonial Monagamous (?)	2	*
Yellow-Bellied Glider	557	60	Arthropods Insect and Plant Exudates	Colonial Monagamous	2	1.5 (0.9–3.8)
Leadbeater's Possum	145	1–3	Arthropods Insect and Plant Exudates	Colonial Monagamous	1	16.8 (10.9–22.9)
Sugar Glider	128	1–5	Arthropods Insect and Plant Exudates	Colonial Polygamous	3	4.6 (0.7–8.5)
Feathertail Glider	14	Unknown	Arthropods Insect and Plant Exudates	Colonial Polygamous (?)	1	*

*Data on body weight is derived from Lindenmayer et al. (1991*b*). Information on home range, diet, and social structure is based on literature values, in particular Russell (1984) and Davey (1990). Column 7 shows the predicted number of corridors where animals were expected to occur based on models of the habitat requirements of the various species developed in areas of contiguous forest (see Lindenmayer 1992a). The values given are (1) the predicted number of sites summed for each of the 49 corridor sites, and (2) the upper and lower value of the 95% confidence interval about the predicted summed number (in parentheses). The asterisk denotes those species for which there are presently no habitat requirement models.*

commonly observed in small, linear areas of retained forest. However, large species of arboreal marsupials that were most frequently recorded in wildlife corridors were solitary and consumed readily available foods such as leaves (Table 1). Conversely, animals with a colonial social structure that consume widely dispersed food such as insects and plant and animal exudates were rarely encountered. This finding conforms to the theory of central place foraging, which predicts that such species may be disadvantaged in narrow, linear-shaped habitats. The value of this theory for corridor design is further discussed by Recher et al. (1987).

(2) Measures of habitat suitability alone could not explain the presence and abundance of arboreal marsupials in wildlife corridors. In all the models that were developed in the study, the significant explanatory variables included a combination of habitat *and* landscape measures (Table 2).

(3) The "context" of the wildlife corridors in the landscape was found to be important. For example, wildlife corridors that connected gullies to ridges supported more species and a greater abundance of animals than sites confined to a single topographic position such as a midslope. Notably, corridor width was *not* a significant variable in any of the models that were developed (Table 2), despite the range of values recorded from the various sites (30–264 m) (Lindenmayer 1992*a*).

Other aspects of site context are likely to be important. Forest on either side of each of the 49 corridor sites had been recently logged and was aged five years or less. Stands of young secondary regrowth forest do not provide suitable habitat for arboreal marsupials but may, over a period of 40–200 years, be recolonized by such animals (Lindenmayer 1992*b*). Successional changes in the suitability of forest habitat in areas adjacent to a corridor may, in turn, influence the use of the retained area by wildlife. Such changes highlight the potential influence of the status of a surrounding area on the biota within, and thus the landscape context of, a wildlife corridor (Noss 1987).

Conclusions

A number of key findings from studies of arboreal marsupials have important connotations for the design of wildlife corridors. Estimates of habitat suitability, the home range of target taxa, and associated predictions of minimum corridor width alone would appear to be insufficient (Table 1). Indeed, many species of arboreal

Table 2. Significant explanatory variables in generalized linear models of the factors influencing the presence and abundance of arboreal marsupials in retained linear habitats.

Model	*Significant explanatory variables*
Presence/Absence of Greater Glider	No. trees with hollows Aspect of site Order of stream in site
Presence/Absence of Mountain Brushtail Possum	No. trees with hollows Length of site
Presence/Absence of an Animal	No. trees with hollows Dominant tree species Connectivity of site
Number of Species	No. trees with hollows Dominant tree species Connectivity of site

See Lindenmayer 1992a, Lindenmayer et al. in press, for further details of the various models and the methods used to derive the variables. The principal measure of habitat suitability in each of the models is the variable "No. trees with hollows." This reflects the availability of potential nest sites for arboreal marsupials.

marsupials were absent from sites that were wider than their estimated home range (Table 1; Lindenmayer 1992*a*). Given this, the effectiveness of wildlife corridors may be improved if additional design criteria are considered, including (1) site context and connectivity, and (2) the social structure, diet, and foraging patterns of target species.

Although corridor width was not a significant explanatory variable in our study of arboreal marsupials, we do not discount it as a key design feature of wildlife corridors. This parameter has been shown to be important in studies of birds (Stauffer & Best 1980; Arnold & Weeldenburg 1990; Saunders & de Rebeira 1991) and mammals (Dickson & Huntley 1987). However, it is likely that corridor width is not important for all groups of animals.

In summary, general principles for corridor design will need to include a range of design criteria. An understanding of these principles will be required for the development of suitable strategies for the conservation of biodiversity in fragmented and degraded landscapes. Such work may reveal that some species will be poorly conserved by a network of wildlife corridors, as was apparent from our studies of various species of arboreal marsupials. Given this, wildlife corridors alone are insufficient as a strategy for nature conservation.

Perhaps the generation of overarching principles for the evaluation and design of corridors could be assisted by the development of more rigorous definitions that reflect the range of potential values of retained areas for nature conservation. For example, we need to distinguish between those sites that can both support resident animals and facilitate the movement of transient individuals from others that provide only a conduit for the movement of wildlife. Shepherd et al. (1992) recently attempted to do this in planning conservation strategies for forest-dependent fauna in southeastern New South Wales. They labeled relatively wide retained areas as Population Assisting Links (PALs) to emphasize the potential of such areas to contribute to metapopulation viability by being large enough to support both resident and transient animals.

Acknowledgments

Many of the ideas outlined here stem from discussions with Dr. A. Bennett, Dr. K. Viggers, Dr. T. Norton, and Ms. S. May. Ms. S. Kelo kindly typed the tables presented in this paper.

Literature Cited

Arnold, G. W., and J. R. Weeldenburg. 1990. Factors determining the number and species of birds in road verges in the wheatbelt of Western Australia. Biological Conservation **53**:295–315.

Davey, S. M. 1990. The environmental relationships of arboreal marsupials in a eucalypt forest: A basis for Australian forest wildlife management. Ph.D. thesis. The Australian National University, Canberra, Australia.

Dickson, J. G., and J. C. Huntley. 1987. Riparian zones and wildlife in southern forests: The problem and squirrel relationships. Pages 37–39 in Managing Southern Forests for Wildlife and Fish. General Technical Report 50–65. U.S.D.A. Southern Forest and Experiment Station, Atlanta, Georgia.

Harrison, R. L. 1992. Toward a theory of inter-refuge corridor design. Conservation Biology **6**:293–295.

Lindenmayer, D. B. 1992*a*. The distribution and abundance of arboreal marsupials in retained linear strips (wildlife corridors) in timber production forests. Value Adding and Silvicultural Systems Program. Report No. 9. Native Forest Research: September 1992.

Lindenmayer, D. B. 1992*b*. The ecology and habitat requirements of arboreal marsupials in the montane ash forests of the Central Highlands of Victoria. A summary of studies. Value Adding and Silvicultural Systems Program. Report No. 6. Native Forest Research: February 1992.

Lindenmayer, D. B., R. B. Cunningham, M. T. Tanton, A. P. Smith, and H. A. Nix. 1990. The habitat requirements of the mountain brushtail possum and the greater glider in the montane ash-type eucalypt forests of the Central Highlands of Victoria. Australian Wildlife Research **17**:467–478.

Lindenmayer, D. B., R. B. Cunningham, M. T. Tanton, H. A. Nix, and A. P. Smith. 1991*a*. The conservation of arboreal marsupials in the montane ash forests of the Central Highlands of Victoria, south-east Australia. III. Models of the habitat requirements of Leadbeater's possum and the diversity and abundance of arboreal marsupials. Biological Conservation **56**:295–315.

Lindenmayer, D. B., R. B. Cunningham, M. T. Tanton, and H. A. Nix. 1991*b*. Aspects of the use of den trees by arboreal and scansorial marsupials inhabiting montane ash forests in Victoria. Australian Journal of Zoology **39**:57–65.

Lindenmayer, D. B., R. B. Cunningham, and C. F. Donnelly. 1993. The conservation of arboreal marsupials in the montane ash forests of the Central Highlands of Victoria, south-east Australia. IV. The distribution and abundance of arboreal marsupials in retained linear strips (wildlife corridors) in timber production forests. Biological Conservation. In press.

Noss, R. F. 1987. Corridors in real landscapes: A reply to Simberloff and Cox. Conservation Biology **1**:159–164.

Recher, H. F., J. Shields, R. P. Kavanagh, and G. Webb. 1987. Retaining remnant mature forest for nature conservation at Eden, New South Wales: A review of theory and practice. Pages 177–194 in D. A. Saunders, G. W. Arnold, A. A. Burbidge, and A. J. Hopkins, editors. Nature conservation: The role of remnants of vegetation. Surrey Beatty and Sons, Chipping Norton, Australia.

Russell, E. M. 1984. Social behavior and social organisation of marsupials. Mammal Review **14**:101–54.

Saunders, D. A., and C. P. de Rebeira. 1991. Values of corridors to avian populations in a fragmented landscape. Pages 221–240 in D. A. Saunders and R. J. Hobbs, editors. Nature conservation 2: The role of corridors. Surrey Beatty and Sons, Chipping Norton, Australia.

Saunders, D. A., and R. J. Hobbs, editors. 1991. Nature conservation 2: The role of corridors. Surrey Beatty and Sons, Chipping Norton, Australia.

Shepherd, T. G., M. J. Saxon, D. B. Lindenmayer, T. W. Norton, and H. P. Possingham. 1992. A proposed management strategy for the Nalbaugh Special Prescription Area based on guiding ecological principles. South East Forest Series No. 2. Threatened Species Research. N.S.W. National Parks and Wildlife Service, New South Wales, Australia.

Stauffer, D. F., and L. B. Best. 1980. Habitat selection by birds of riparian communities. Journal Wildlife Management **44**:1–15.

The Principle of Nested Subsets and Its Implications for Biological Conservation

BRUCE D. PATTERSON

Division of Mammals
Field Museum of Natural History
Roosevelt Road at Lake Shore Drive
Chicago, Illinois 60605

Abstract: *The nested subset hypothesis states that the species comprising a depauperate insular biota are a proper subset of those in richer biotas, and that an archipelago of such biotas, ranked by species richness, presents a nested series. The pattern characterizes the distributions of mammals in three different archipelagos, and it appears more strongly developed among faunas in the process of* relaxation *(landbridge islands) than in those derived by overwater colonization (oceanic islands). The generality of the nested subset pattern and factors that may produce it are evaluated using distributions of land birds in the New Zealand region. Monte Carlo simulations show that species composition of these insular avifaunas is highly nonrandom, exhibiting significantly nested structure. However, avifaunas on nine oceanic islands lack the structure typical of all islands together or of the 22 landbridge islands. Thus, distinctive patterns of species composition, as well as species number, may distinguish landbridge and oceanic islands.*

Because they are isolated fragments of once-continuous distributions, landbridge islands and their biotas have important implications for long-term biological conservation in preserves. Real or virtual islands that have undergone faunal relaxation support only a fraction of the species expected in equivalent mainland areas. Furthermore, the species inhabiting fragments are not a random collection of those in the source pool, but are rather nested subsets of the species in richer, more intact biotas. The species preserved in such fragments tend to be the most abundant, generalist species that are least in need of special protection.

Paper submitted March 9, 1987; revised manuscript accepted August 27, 1987.

Resumen: *La hipótesis del subgrupo anidado ("nested subset") plantea que las especies que comprenden una biota insular empobrecida representan un subgrupo apropiado de aquellas que se encuentran en biotas más ricas, y que un archipiélago de esas biotas, categorizadas de acuerdo a la riqueza de especies, presenta una serie del tipo "anidado". El patrón caracteriza la distribución de mamíferos en tres archipiélagos diferentes y parece estar desarrollado con más fuerza entre las faunas en proceso de "relaxation" (islas continentalis) que en aquellas derivadas de la colonización sobre agua (islas oceánicas). La mayoría de los patrones de los subgrupos "anidados" y los factores que los podrían producir son evaluados usando la distribución de las aves continentales en la región de Nueva Zelandia. Simulacros en Monte Carlo muestran que las composiciones de las especies de esas avifaunas insulares son altamente no-aleatorias, exibiendo una estructura significativamente anidada. Sin embargo, las avifaunas de nueve islas oceánicas carecen de la estructura típica de todas las islas en conjunto o de cada una de las 22 islas continentalis. Por lo tanto, los patrones distintivos de la composición de las especies, así como su número, pueden caracterizar a las islas continentalis y a las oceánicas. Debido a que son fragmentos aislados de una antigua distribución continua, las islas continentalis y sus biota característica tienen implicaciones importantes para la conservación biológica a largo plazo en las reservas. Las islas reales o virtuales que han sufrido la "relaxation" de su fauna albergan solamente una fracción de las especies supuestas a encontrarse en áreas equivalentes de tierra firme. Además, las especies que habitan en fragmentos no son una colección aleatoria de aquellas en la fuente, sino subgrupos "anidados" de las especies encontradas en biotas más ricas e intactas. Las especies protegidas en tales fragmentos tienden a*

ser las más abundantes y generalistas, las cuales presentan una menor necesidad de protección especial.

Introduction

Studies of island biotas have been important to the development of ecological and evolutionary theory since the time of Darwin and Wallace, and the relevance of such work to isolated biological preserves is clear (cf. Diamond 1975, Wilson & Willis 1975). The seminal insights of MacArthur and Wilson (1967) in *The Theory of Island Biogeography* inaugurated two decades of work on insular distributions, focusing primarily on determinants of species richness. Knowledge of other aspects of island biology, including species composition, interactions, and evolutionary divergence, has developed much more slowly. However, understanding of patterns of species composition and the processes that may affect them is rapidly developing (e.g., Grant & Abbott 1980, Diamond and Gilpin 1982, Colwell & Winkler 1984, Connor & Simberloff 1984, Gilpin & Diamond 1984). A highly structured pattern of distributions on islands termed the nested subset pattern was recently identified in the species composition of mammal faunas in three archipelagos (Patterson & Atmar 1986).

The species composition of biotas in an archipelago can be said to show a nested subset pattern if the species that inhabit a depauperate island collectively comprise a proper (or included) subset of those on richer islands, and if a series of such insular biotas, ordered by species richness, constitute a nested series. This pattern of species composition is based on an underlying nonrandom pattern of species distributions, in which individual species tend to be present in all richer biotas within an archipelago than the most depauperate one in which they are found. Nested and nonnested relationships are shown schematically in Figure 1.

Given the diversity of factors affecting species distributions, natural biotas must rarely, if ever, show the idealized nestedness represented in the left segment of Figure 1. However, a strong and significant pattern of nested subsets was demonstrated for Southwestern mountaintop mammals by Patterson and Atmar (1986), who used computer simulations to populate model archipelagos at random and compared simulated species compositions to those actually observed. The nested pattern of species composition was also found to characterize the composition of terrestrial mammal faunas in two marine archipelagos off the Atlantic and Pacific coasts of North America. However, the faunas of oceanic islands off Baja California did not show the pattern, although all islands and the landbridge islands did so.

Presence or absence of the nested pattern may be related to the historical derivation of biotas. Landbridge islands were directly connected to species-rich mainland areas during glacial episodes of the Pleistocene, and they presumably shared their diverse biotas. Since their disjunction from mainland areas, landbridge islands are thought to have suffered a net loss of species via local extinction (faunal relaxation). Conversely, truly oceanic islands have always lacked dryland connections to potential source pools and are often relatively impoverished, supporting only highly vagile species. Given the

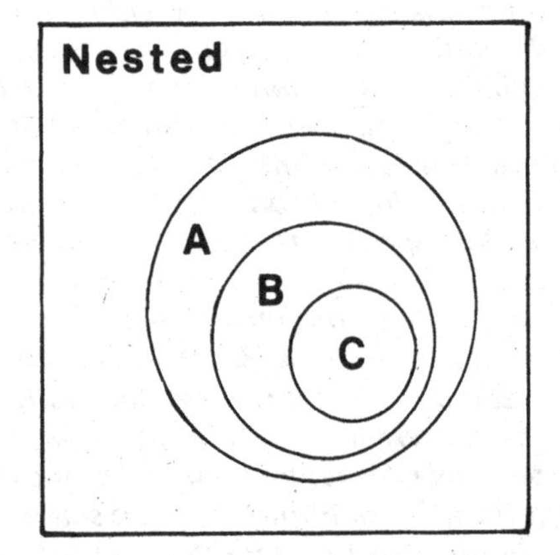

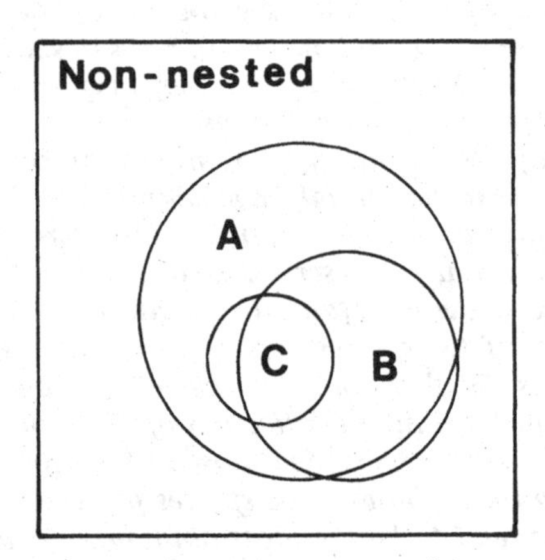

Figure 1. Schematic representation of the nested subset relationship, using Venn diagrams. Two archipelagos are shown, each containing three biotas (circles labeled A, B, *and* C*) of comparable species richness (size). The left archipelago is nested, because the species present (elements contained) in small biotas are also present in richer ones. The right archipelago is nonnested, because island* B *supports species not present in* A, *and* C *supports species that are all present in* A *but not in* B.

results of Patterson and Atmar (1986), it is tempting to speculate that the association between nestedness and biotic derivation via extinction versus colonization is real and general. However, because the oceanic islands off Baja are exceedingly depauperate, each supporting three or fewer species, lack of "nestedness" there may be strictly artifactual. Clearly, additional surveys are needed.

In this paper, a number of questions concerning the nested subset pattern are addressed: Can this distributional pattern be shown for other taxa in other geographic areas? Do landbridge and oceanic islands differ predictably in terms of species composition? Can inferences be drawn regarding the processes that may have produced nested patterns? However produced, do nested patterns of species composition have implications for biological conservation? To investigate these questions, relatively diverse land bird faunas on islands in the New Zealand region are analyzed. Eustatic sea-level drops united many islands to form a "Greater New Zealand" during the Pleistocene, while other islands are remotely situated in deep water and therefore are truly oceanic in a zoogeographic sense. If nested patterns are general and if they are products of relaxation as suggested by the mammal studies, then avifaunas of landbridge islands should exhibit the nested subset pattern, whereas the faunas of oceanic islands should not.

Materials and Methods

Figure 2 depicts the study region, including islands that have adequate avifaunal surveys for analysis. Nine islands are oceanic in a zoogeographic sense by virtue of their remote location and situation in deep water, whereas 22 are landbridge islands that lie on the continental shelf. Distributions of 60 native land and freshwater bird species on these islands were taken directly from Diamond (1984*b*) and are listed in Table 1. Members of three superspecies (the Whitehead-Yellowhead, the Grey warbler-Chatham Island warbler, and the Robin-Chatham Island robin) were combined in Diamond's analysis because of strict complementarity in their geographic distributions and the inherent arbitrariness of the taxonomic rank accorded to such allospecies. The same course is followed here. Alternate treatment of these forms as distinct species would not affect any conclusions or levels of significance in this analysis. Autecological information on the distributions of each of these 57 zoogeographic species is considered in Diamond (1984*b*).

Distributional data were subjected to the analytical procedures described in Patterson and Atmar (1986), treating all islands together, and then landbridge and oceanic islands separately. Islands in each group were first ranked in order of species richness, and species were ranked in order of island occurrences. The resulting presence-absence matrix was then subjected to the following steps: 1) the island of lowest species number supporting species i was identified; 2) all islands of greater species number were examined for the presence of species i; 3) one error for the absence of species i on each island of greater species richness (marked by an asterisk [*] in the presence-absence matrices) was tallied; and 4) the process was repeated for each species in the fauna. The sum of the N_i scores, the N index, tallies the departure of the archipelago as a whole from perfect nestedness, as defined in this study. The N_i values of species for each group of islands (included in Table 1) can be examined individually to evaluate the departure of individual species from the idealized pattern. Because the N_i's were calculated in the context of each group of islands, those for landbridge and oceanic islands do not sum to the N_i for all islands combined.

The significance of a given N index was evaluated by means of Monte Carlo simulations. The simulation programs RANDOM0 and RANDOM1 (presented in HP BASIC in the appendix of Patterson & Atmar 1986) were used to construct 1000 model archipelagos and derive corresponding N indices. Observed values of N were compared to the distribution of simulation scores to evaluate the nonrandomness of observed patterns. Student's t-tests were used to assess the probability that the observed value could have been drawn from the symmetric, approximately normal distributions of simulated values.

Both simulation programs constrained species richness of islands to actual values, but RANDOM1 populated islands with species drawn at random (without replacement) according to actual frequencies of occurrence on islands, whereas RANDOM0 drew species from a uniform probability distribution. Thus, a species found on 10 islands was twice as likely to be selected for an insular fauna during RANDOM1 simulations as a species found on only five islands; the two species had equal probabilities of being drawn during RANDOM0 simulations.

Results

The distribution of 57 species among 31 islands (Tables 2 and 3) is such that there are 21 different classes of species richness (range 2–50 species) and 18 different classes of island occurrences (range 1–23 islands). The distribution of species over all islands tends to show the underlying distributional pattern on which nested species composition is based: species occurrences apparently cluster in the upper-left-hand corner of the ordered distribution matrix, that is, on species-rich islands. This tendency can be exemplified by certain

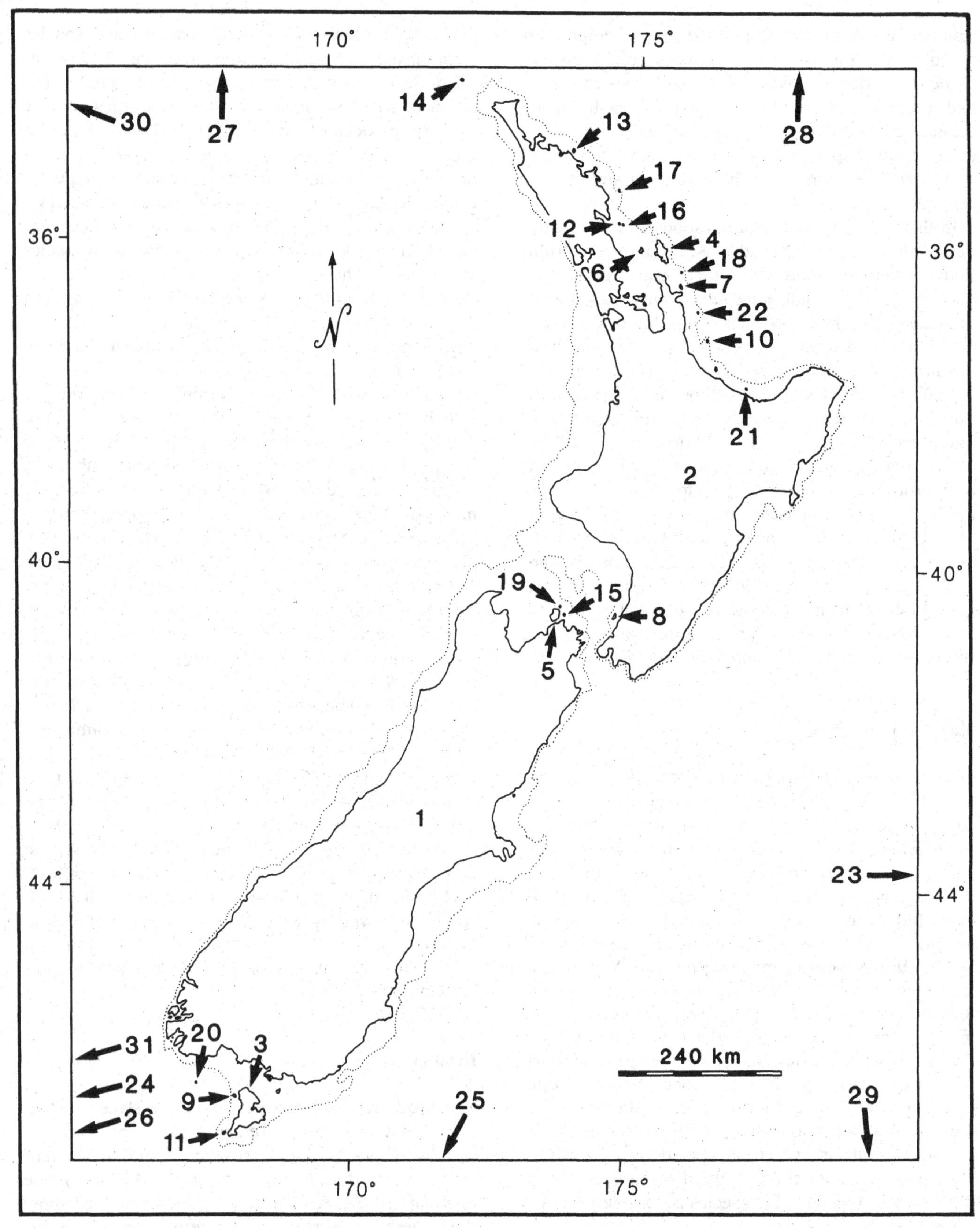

Figure 2. Map of islands in the New Zealand region, showing position of nearshore landbridge islands (1–22) and directions of remote oceanic islands (23–31). Dotted line corresponds to the 100 m bathymetric contour. Islands are: 1, South; 2, North; 3, Stewart; 4, Great Barrier; 5, D'Urville; 6, Little Barrier; 7, Mercury; 8, Kapiti; 9, Codfish; 10, Mayor; 11, Big South Cape; 12, Hen; 13, Cavalli; 14, Great King; 15, Chetwode; 16, Chicken; 17, Poor Knights; 18, Cuvier; 19, Stephens; 20, Solander; 21, Whale; 22, Alderman; 23, Chatham (44°00'S, 175°35'W); 24, Auckland (50°35'S, 166°00'E); 25, Campbell (52°30'S, 169°02'E); 26, Macquarie (54°29'S, 158°58'E); 27, Norfolk (29°05'S, 167°59'E); 28, Raoul (29°15'S, 177°52'W); 29, Antipodes (49°42'S, 178°50'E); 30, Lord Howe (31°28'S, 159°09'E); 31, Snares (48°05'S, 166°34'E).

Table 1. Native land and freshwater bird species in the New Zealand region (*from* Diamond 1984*b*). Letter and number codes of species are those used in Tables 2 and 3. Also tabulated are total insular occurrences and N_i indices (based on distributional matrices) for all islands, landbridge islands, and oceanic islands. N_i values of zero indicate species that are regularly distributed on the most species-rich islands in an archipelago.

			Islands		
Code/Species		No.	Total (N_i)	Landbridge (N_i)	Oceanic (N_i)
A	Kingfisher (*Halcyon sancta*)	23	5	3	2
B	Fantail (*Rhipidura fuliginosa*)	23	5	1	2
C	Silvereye (*Zosterops lateralis*)	23	5	3	1
D	Bellbird (*Anthornis melanura*)	23	3	0	0
E	Red-crowned parakeet (*Cyanoramphus novaezelandiae*)	21	9	3	1
F	Morepork (*Ninox novaeseelandiae*)	19	9	5	2
G	Tui (*Prosthemadera novaeseelandiae*)	19	3	2	1
H	Pigeon (*Hemiphaga novaeseelandiae*)	17	9	6	1
I	Pipit (*Anthus novaeseelandiae*)	17	11	5	3
J	Tomtit (*Petroica macrocephala*)	16	15	5	6
K	Yellow-crowned parakeet (*Cyanoramphus auriceps*)	15	5	4	0
L	Kaka (*Nestor meridionalis*)	14	6	1	1
M	Shining cuckoo (*Chrysococcyx lucidus*)	14	6	1	1
N	Harrier (*Circus approximans*)	13	15	10	0
O	Grey duck (*Anas superciliosa*)	12	18	6	2
P	Fernbird (*Bowdleria punctata*)	12	19	6	7
Q	Robin/Chatham Island robin (*Petroica australis, P. traversi*)	11	3	2	0
R	Banded rail (*Rallus philippensis*)	10	20	9	6
S	Saddleback (*Philesturnus carunculatus*)	10	4	2	—
T	Spotless crake (*Porzana tabuensis*)	9	12	10	2
U	Pukeko (*Porphyrio porphyrio*)	9	11	8	2
V	Long-tailed cuckoo (*Eudynamys taitensis*)	9	11	1	3
W	Grey warbler/Chatham Island warbler (*Gerygone igata, G. albofrontata*)	9	19	16	1
X	Brown teal (*Anas aucklandica*)	8	20	2	3
Y	Falcon (*Falco novaeseelandiae*)	8	8	0	0
Z	Rifleman (*Acanthisitta chloris*)	8	12	9	—
a	Whitehead/yellowhead (*Mohoua albicilla, M. ochrocephala*)	6	5	3	—
b	Kokako (*Callaeas cinerea*)	6	1	0	—
c	Paradise shelduck (*Tadorna variegata*)	4	2	1	0
d	Scaup (*Aythya novaeseelandiae*)	4	9	9	—
e	Weka (*Gallirallus australis*)	4	0	0	—
f	Marsh crake (*Porzana pusilla*)	4	2	0	0
g	Bush wren (*Xenicus longipes*)	4	10	8	—
h	Brown creeper (*Finschia novaeseelandiae*)	4	5	4	—
i	Piopio (*Turnagra capensis*)	4	3	2	—
j	Brown kiwi (*Apteryx australis*)	3	0	0	—
k	Little spotted kiwi (*Apteryx owenii*)	3	1	1	—
l	Shoveler (*Anas rhynchotis*)	3	3	0	0
m	Kakapo (*Strigops habroptilus*)	3	0	0	—
n	Laughing owl (*Sceloglaux albifacies*)	3	0	0	—
o	Stitchbird (*Notiomystis cincta*)	3	5	4	—
p	Crested grebe (*Podiceps cristatus*)	2	0	0	—
q	Dabchick (*Podiceps rufopectus*)	2	0	0	—
r	Grey teal (*Anas gibberifrons*)	2	0	0	—
s	Blue duck (*Hymenolaimus malacorhynchus*)	2	0	0	—
t	Quail (*Coturnix novaeseelandiae*)	2	0	0	—
u	Takahe (*Notornis mantelli*)	2	0	0	—
v	Great spotted kiwi (*Apteryx haastii*)	1	0	0	—
w	Dieffenbach's rail (*Rallus dieffenbachi*)	1	5	—	0
x	Auckland Island rail (*Rallus pectoralis*)	1	15	—	1
y	Chatham Island rail (*Gallirallus modestus*)	1	5	—	0
z	Kea (*Nestor notabilis*)	1	0	0	—
1	Antipodes Island parakeet (*Cyanoramphus unicolor*)	1	26	—	5
2	Orange-fronted parakeet (*Cyanoramphus malherbi*)	1	0	0	—
3	Rock wren (*Xenicus gilviventris*)	1	0	0	—
4	Stephens Island wren (*Xenicus lyalli*)	1	6	5	—
5	Huia (*Heteralocha acutirostris*)	1	1	1	—

Table 2. Distributions of native land and freshwater bird species on landbridge islands in the New Zealand region. Species symbols after Table 1; islands numbered after Diamond (1984*b*). Insular presence indicated by letter; deviations from perfect nestedness, as measured in this study, indicated by asterisks. The overall *N* index equals 158.

Island	*Bird Species*																																																					*Richness*
1	A	B	C	D	F	G	H	E	K	I	J	N	L	M	P	Q	S	R	V	Z	O	T	b	Y	U	a	W	X	d	e	g	h	i	j	k	c	f	m	n	*	p	s	l	r	q	t	u	*	2	v	*	z	3	50
2	A	B	C	D	F	G	H	E	K	I	J	N	L	M	P	Q	S	R	V	Z	O	T	b	Y	U	a	*	X	d	e	g	*	i	j	k	c	f	m	n	o	p	s	l	r	q	t	u	*			5			46
3	A	B	C	D	F	G	H	E	K	I	J	N	L	M	P	Q	S	R	V	Z	O	T	b	Y	U	*	*	X	*	e	g	h	*	j	*	*	f	m	n	*								*						33
5	A	B	C	D	F	G	H	E	K	I	J	N	L	M	P	Q	S	*	V	Z	O	T	b	Y	U	a	*	*	d	e	*	h	i	j	k	c				*								*						31
4	A	B	C	D	F	G	H	E	K	I	J	N	L	M	P	Q	S	R	V	Z	O	T	b	Y	U	a	*	X	*		*	*	*							o								*						28
19	A	B	C	D	F	G	H	*	K	I	J	N	L	M	P	Q	S	R	V	*	*	*	b	Y	*	*	*	*	*		*	*	I							*								4						22
6	A	B	C	D	F	G	H	E	K	*	J	*	L	M	*	Q	S	R	*	Z	*	*			*	a	*	X	*		*	*								o														19
9	A	B	C	D	F	G	H	E	K	*	J	N	L	M	P	*	S	*	V	Z	*	*			*	*	*		*		*	h																						18
8	A	B	C	D	F	G	H	E	K	I	J	*	L	M	*	Q	*	*	V	*	O	*			*	a	*		*		*																							17
12	A	B	C	D	F	G	H	E	K	*	J	*	L	M	*	*	S	*		*	*	*			*		*		*		*																							13
15	A	B	C	D	F	G	H	*	K	I	*	N	L	M	*	Q	*	*		*	*	*			*		*		*		*																							13
11	*	B	C	D	F	G	*	E	K	*	J	*	*	*	P	Q	S	*		*	*	*			*		*		*		g																							12
10	A	B	C	D	F	G	H	*	*	*	*	*	L	M	*			*		*	O	*			*		*		d																									11
7	A	B	C	D	F	*	H	E	*	I	*	N			*			*		*		*			U		*																											10
14	A	B	*	D	F	G	*	E	*	I	*	*			P			R		*		T					*																											10
22	A	B	C	D	*	*	*	E	*	I	*	*			P			*		*		*					W																											8
20	*	B	C	D	*		*		K		J	*								Z							W																											7
17	A	*	*	D	*		*	E		I		N						R				T					*																											7
13	A	B	C	D	*	G	*			I		N															*																											7
16	*	B	*	D	F	G	H				J	*															W																											7
18	A	B	C	D	*		H					*															W																											6
21	A	B			F							N															W																											5

species found on landbridge islands (Table 2). For example, the Bellbird (species D) is found on 21 landbridge islands, which are the 21 most speciose islands; the Kokako (species b) is found on six landbridge islands, the six richest ones; the Marsh crake (species f) on three landbridge islands, the three richest ones; and the Rock wren (species 3) on only the richest landbridge island.

However, this type of distributional regularity, reflected in N_i values of zero, does not hold for a majority of species or apply equally to all groups of islands. Table 1 indicates that, for all islands, only a quarter of the species (14 of 57) exhibit $N_i = 0$; it is noteworthy that none of these species is found on more than four islands, all of which are landbridge islands. By comparison, 19 of the 53 species (36%) found on landbridge islands have landbridge $N_i = 0$, and these species occur on as many as 21 islands. Finally, although fully 10 of the 31 species (32%) found on oceanic islands have oceanic $N_i = 0$, none of these is found on more than two of the nine islands. Clearly, the irregularities of natural distribution patterns demand a statistical approach to determine whether or not the aggregate effects of regular and near-regular distributions can be considered biologically significant. The overall deviation of a distributional matrix from perfect nestedness can be expressed by the N index, a summation across species of N_i values. By this measure, the entire New Zealand archipelago departs from perfect nestedness by 367, amounting to 20.8 percent of the presence-absence matrix.

When island faunas are constructed using RANDOM0 (species richness equals observed; species selected from a uniform probability distribution), the average N index of 1000 model archipelagos is 761 (SD = 22.08), with a minimum value of 694 and a maximum value of 828. The observed value of N therefore lies almost 18

Table 3. Distribution of native land and freshwater bird species on oceanic islands off New Zealand. Conventions are as in Table 2. The overall *N* index equals 53.

Island	*Species*																														*Richness*
23	E	C	O	*	I	G	J	W	U	X	R	H	T	K	B	M	Y	*	L	P	D	Q	*	w	y	l	*	N	*	f	c	26
24	E	C	O	*	I	G	J	*	*	X	*	*	*	K	*	*	Y	*	*	*	D		*				x		*			10
27	E	C	*	A	*	*	*	W	*	*	*	H	*		*	M		F	L	*			*						*			8
28	E	*	O	A	*	G	*		U	*	*		T							*			*						V			7
30	E	C	*	A	*		*	W	U	*	*				B			F		*			*									7
25	*	C	O		I		*			X	*									*												4
29	E			A	I		*				*									*			l									4
26	E		O				*				R									*												3
31							J													P												2

SD from the mean of the simulation values and has a vanishingly small probability of belonging to that distribution of values ($P \ll 0.00001$). However, such nested structure might result from simple differences in the number of island occurrences: a species found on a single island is likely to be found on a speciose one, purely by contingent probability. These differential probabilities of occurrence are incorporated into RANDOM1 models, where species richness of simulated archipelagos equals observed values and species comprising these faunas are selected from a probability distribution weighted by actual occurrence values. Under this scheme, far greater nestedness is evident in the randomized faunas. The average N index of the simulations drops to 591 (SD = 36.3), with a range 455–696. Nevertheless, this value remains more than 6 SD above the actual value, indicating that species composition of New Zealand avifaunas has a nested structure, and that this structure cannot be produced by random subsampling from a source pool under two sets of realistic constraints.

The 22 landbridge islands in the New Zealand archipelago support 53 of the 57 land and freshwater species found in the region (Table 2). Distributions of species among islands are such that there are 17 classes of species richness (5–50 species) and 18 classes of island occurrences (1–21 islands). The N index for the suite of landbridge islands is 158, amounting to 13.6 percent of the distribution matrix. When 1000 model faunas of the landbridge islands are constructed with RANDOM0, the average N index is 453 (SD = 16.3), with a range of 394–504. The actual value is thus 18 SD from the simulated mean and differs significantly from it ($P \ll 0.00001$). When archipelagos are fashioned with RANDOM1, the average N = 364 (SD = 23.6), with extremes in 1000 simulations of 292 and 437. The mean value remains almost 9 SD above the observed value and demonstrates convincingly ($P < 0.00001$) that the nested structure of landbridge island faunas is not simply an artifact of species occurrence frequencies.

Avifaunas on the nine oceanic islands do not exhibit these patterns. A total of 31 species are distributed on oceanic islands (Table 3), showing seven classes of species richness (2–26 species) and 6 classes of island occurrences (1–7 islands). The N index is 53, amounting to 19.0 percent of the distribution matrix. When the faunas of these islands are simulated using RANDOM0, an average N of 60.5 results (SD = 5.74), with a range of 41–77. The extremes of the simulation scores bracket the observed value, and the observed value lies only 1.31 SD below the mean of simulation scores ($P = 0.1$). Results from RANDOM1 simulations of the oceanic islands are even more dramatic: the simulation mean N was 53.3 (SD = 6.36), practically identical to that value actually observed, and obviously not different from it ($P = 0.5$).

Salient features of each set of simulations are tabulated in Table 4.

Discussion

Avifaunas of New Zealand islands exhibit highly nonrandom patterns of distribution. Consistent occurrence of species in rich faunas and their absence in impoverished ones can be seen directly in the distributions of certain species, especially those found on landbridge islands (cf. low N_i indices in Table 1). The simulation results presented in Table 4 show that this regularity characterizes entire avifaunas. The species that comprise a depauperate fauna on New Zealand islands constitute a subset of those in richer faunas, and the suite of insular faunas presents a nested series. However, this nested structure does not typify faunas on truly oceanic islands—species composition of oceanic islands in the New Zealand region is random with respect to nestedness.

Landbridge versus Oceanic Islands

Nested patterns of species composition were also shown for mammal faunas on landbridge islands in three different North American archipelagos but failed to characterize the single group of oceanic islands treated (Patterson & Atmar 1986). This suggests that landbridge and oceanic islands may generally be distinguished by

Table 4. Results of 1000 simulation experiments involving New Zealand islands. Probability values indicate the likelihood that observed N belongs to the distribution of N values from the simulations. (See text for differences between RANDOM0 and RANDOM1.)

	Mean N index	SD	Minimum	Maximum	P
Oceanic and landbridge islands (N_{obs} = 367)					
RANDOM0 simulations	760.99	22.08	694	828	<0.00001
RANDOM1 simulations	591.24	36.34	455	696	<0.00001
Landbridge islands only (N_{obs} = 158)					
RANDOM0 simulations	453.91	16.33	394	504	<0.00001
RANDOM1 simulations	363.73	23.61	292	437	<0.00001
Oceanic islands only (N_{obs} = 53)					
RANDOM0 simulations	60.53	5.74	41	77	≅0.1
RANDOM1 simulations	53.32	6.36	32	71	≅0.5

patterns of species composition as well as species richness (cf. MacArthur & Wilson 1967). Different patterns of species richness for landbridge and oceanic islands are usually attributed to differences in the historical derivation of their biotas (e.g., Lawlor 1986).

Biotas on landbridge islands are thought to have "relaxed" via local extinction from higher species number. Abundant evidence demonstrates that richer, more speciose biotas existed on islands and island-like habitats that were connected to mainland areas by marine regressions or by vegetational displacements during glacial episodes of the Pleistocene (e.g., Hope 1973, Patterson 1984, Heaney 1986). Following areal restriction and increased isolation, such biotas lost more species through local extinction than they gained through recolonization. This greater importance of extinction versus colonization for landbridge islands is reflected in species-area curves with higher slopes than are typical of islands generally (Lawlor 1986) and in lack of correlation between species richness and isolation (Brown 1971).

In contrast, oceanic islands are both currently and historically insular. They characteristically support biotas that are either colonization-limited or in some form of equilibrium between opposing rates of colonization and extinction. Only in exceptional, catastrophic cases, involving volcanism, introductions, or human impact (e.g., Morgan & Woods 1986, Olson & James 1982), do extinction rates on oceanic islands greatly exceed colonization rates. Therefore, species number on oceanic islands often shows a stronger dependence on colonization than extinction, producing characteristically shallow species-area curves and inverse correlations between species richness and measures of isolation. Lawlor (1986) found that species-area slopes for mammals in six oceanic archipelagos averaged half as great as those for mammals on 12 comparable sets of landbridge islands; the two distributions of slopes were nonoverlapping.

It is easy to envision how different historical derivations might also affect the species compositions of landbridge and oceanic faunas. By definition, landbridge islands were once freely colonized by mainland species. Depending on island size, other barriers to dispersal, and the distribution of specific habitats, groups of landbridge islands became populated by a more-or-less common suite of mainland species. Once separated from the mainland, landbridge fragments became supersaturated, containing a continental complement of species in a newly circumscribed area. Such conditions produce increased rates of extinction (MacArthur & Wilson 1967, Brown 1971).

Island biogeographic studies suggest that extinctions from landbridge fragments are often biologically selective. Indeed, the best evidence for selective extinction in nature comes from studies of faunal relaxation. Biogeographic studies involving mammals are particularly compelling because they sometimes include documentation from the Quaternary fossil record as well as trenchant modern patterns (e.g., Hope 1973; Patterson 1984, 1987). Diamond (1984*a*) concluded from a broad literature survey that risk of extinction varies inversely with population size, and so is correlated with a variety of factors, such as island size, body size, trophic position, and habitat specialization, that affect population size (*see also* Terborgh 1974, Karr 1982, Heaney 1984).

Body size, diet, and habitat affinities are generally similar for different insular populations of a species in a given archipelago. In such cases, island size may be the most important determinant of risk of extinction (e.g., Table 3 in Patterson 1984). Correlations between species richness and area and between area and risk of extinction appear to be responsible for the regular pattern of species occurrences in the richer faunas (low N_i). When different species in an archipelago are compared, the influence of biological properties on extinction risk becomes more evident. The relatively small, abundant Bellbird (species D) is found on 23 islands, whereas the large, flightless Brown kiwi (species j) is found on only three islands, even though both species are regularly distributed on the richest landbridge islands. Generally, the most impoverished islands support only the most abundant species while the rarest species are found only in the richest biotas. The fossil record shows that the same vulnerable mammal species disappeared from various smaller islands in the Bass Straits (Hope 1973), and this landbridge archipelago also appears to exhibit a strongly nested pattern of species compositions.

Differential, selective extinction offers a potent explanation both for the number of islands on which different species occur (treated much more fully in Diamond 1984*b*) and for consistent patterns of insular incidence. Local extinctions during relaxation occur independently on different islands in an archipelago. However, if the extinctions are selective, the *sequence* of species loss from each insular biota is roughly comparable, producing the nested pattern. It seems to be the predictable sequence of species loss from landbridge fragments that distinguishes real archipelagos from those produced by either RANDOM0 or RANDOM1.

Complications

It is always difficult to attribute complex patterns, especially those involving an important historical component, to the processes that may have produced them. Numerous factors affect the distribution and occurrence of species on islands, and several important effects are intercorrelated, complicating this simple picture of nested patterns as apparent products of selective species loss.

The geometry of New Zealand islands and their habitats must have had important effects on past and present patterns of colonization (discussed in detail by Diamond 1984*b*; personal communication, 1986). Landbridge islands are distributed along more than 3000 km of coastline off the main islands. While sea-level changes undoubtedly created a unitary "Greater New Zealand" during the Pleistocene (amalgamating islands 1–22 in Fig. 2), it is unlikely that landbridge islands along this entire perimeter were colonized by the same set of species. Species now present on South Island (no. 1) but absent from North Island (no. 2), such as Kea (z) or Brown creeper (h), probably dispersed to nearby southern islands such as Stewart (no. 3), Codfish (no. 9), Big South Cape (no. 11), and Solander (no. 20); their abilities to reach islands at the northern end of the archipelago, such as Cavalli (no. 13) or Great King (no. 14), would have been much more limited. Conversely, North Island species such as Stitchbird (o) and Huia (5) may never have reached islands off the southern tip of South Island. Given nonuniform dispersal of species during the Pleistocene, the regular nested structure evident in Table 2 is especially dramatic.

The current distributions of New Zealand birds have also been profoundly affected by man, with a high proportion of the avifauna being rare, endangered, or extinct. An estimated 45 species became extinct at or soon after the arrival of Europeans (including the moas, 7 waterfowl, 6 rails, 2 hawks, 2 eagles, a snipe, and a crow), and many were lost earlier during colonization by Polynesians. Fully 11 percent of the 318 bird taxa listed in the IUCN Red Data Book are from islands in the New Zealand region (Mills & Williams 1984). Factors thought to underlie these extinctions and endangerments vary according to species and study: habitat destruction, introduced species, direct exploitation by man, and climatic change. Any account of present patterns of species richness and composition in this "wreakage of an avifauna" must acknowledge the severe impacts of Polynesian and European man (cf. Olson & James 1982). It is unknown whether human influences would enhance the selectivity of natural extinctions or tend to mask it, but the selectivity probably varies with type of impact and from species to species and place to place.

But the most important caveats that must accompany this interpretation are that oceanic and landbridge islands typically differ importantly in species richness and geographic isolation, as well as in history. It is thus impossible to attribute biotic differences between them solely to the relative roles of colonization and extinction in their derivation. Landbridge islands generally support richer biotas, and some of the nestedness evident in the biotas of landbridge islands may be a consequence of their greater species richness. For example, Table 2 shows a higher proportion of unexpected absences (asterisks) on the species-poor landbridge islands at the bottom of the table.

When the nestedness of biotas inhabiting the nine most depauperate landbridge islands is examined, the N index is 41 (25% of the matrix). This observed value is significantly lower than the mean of 1000 RANDOM0 simulations ($N = 50.7$; $P < 0.001$), but does not differ significantly from the mean of RANDOM1 simulations ($N = 43.4$; $P > 0.10$). Lack of nestedness among these impoverished landbridge islands, relative to the stringent conditions of RANDOM1, could be due to a variety of biological or methodological factors (e.g., number of *ties* among islands in species richness). However, it does not appear to be a simple consequence of matrix rank. Archipelagos known to support nested subsets of faunas range from seven to 34 islands, with ranges of species richness from 1–13 species to 2–50 species. Those shown to lack nested structure range from nine to 14 islands supporting 1–3 mammal species to 2–26 bird species (Patterson & Atmar 1986).

Geographic isolation also differs considerably between landbridge and oceanic islands. In the New Zealand region, near-shore islands are invariably landbridge islands and remote ones are oceanic (Fig. 2). While faunas on the latter have been certainly derived by over-water colonization, biotas of nearshore landbridge islands are probable products of both local extinctions during faunal relaxation and recurrent colonization. Such considerations have been shown to apply to mammals inhabiting landbridge islands in Penobscot Bay, Maine. Although these islands were colonized during the Pleistocene and suffered faunal relaxation following fragmentation, their current nested species compositions may be affected by modern-day recolonizations (Crowell 1986).

The species that inhabit oceanic islands are perforce good dispersers. Eighty-seven percent (27 of 31) of New Zealand bird species inhabiting oceanic islands also occur on one or more landbridge islands. Such species likely recolonize nearshore landbridge islands from which they have become locally extinct, and such colonizations could influence the significantly nested structure, either enhancing it or partially obscuring it. The remaining 26 species are restricted in New Zealand to landbridge islands and have limited potential for dispersal, owing to either physical or psychological inabilities to cross significant water barriers (Williams 1981, Diamond 1984*b*). Fully 28 percent of the terrestrial bird species are flightless or have very weak powers of flight (Williams 1984).

The fact that oceanic islands never shared an ancestral biota strongly limits their potential for nested subset structure, at least via extinction. Yet nestedness among biotas of oceanic islands because of colonization may be possible for certain corresponding values of vagility and isolation. Differences in dispersal abilities among species could produce nested structure if the islands varied in isolation so as to manifest these differences. Under such conditions, some species could colonize most islands, others reach fewer, and some reach only a handful (*see* Fig. 4 in Patterson & Atmar 1986). However, species of low vagility (e.g., terrestrial mammals and non-crossing bird species) in various island systems and highly vagile species inhabiting very remote islands (Table 3) are demonstrably poor candidates for nestedness (Table 4). Other archipelagos and taxa not yet analyzed may present corresponding degrees of isolation (creating differing opportunities for colonization) and vagility that nested subset structure can be shown to result from colonization. Distributions of birds in the New Hebrides, Bismarcks, and Solomons may represent examples of such taxa-archipelago situations (Diamond, personal communication, 1986).

The extreme geographic isolation of oceanic islands and nonzero colonization rates of landbridge islands loom as the most important factors complicating the hypothesis that nested patterns are produced during biotic relaxation via selective extinctions. To resolve the potential effect of these and other factors on biotic nestedness, additional surveys are needed for archipelagos varying in isolation, species richness, and island number, and covering taxa with a spectrum of vagility.

Biological Conservation

Whatever processes may be involved in its production, the nested subset structure of biotas has important implications for biological conservation. Smaller insular biotas do not represent a random draw of the species in the source pool but instead represent highly nonrandom, nested subsets of those species. Although documented in the biotic structure of all four archipelagos examined, nested subsets appear best defined for landbridge islands.

Although most biological preserves exhibit island-like properties, special, fundamental similarities exist between landbridge islands and nature preserves. Because preserves are progressively made insular by outside habitat disturbances, the derivation of their biotas is closely analogous to those of landbridge islands: a once broadly distributed biota becomes fragmented into a number of isolated pieces, each of which becomes too small to support the entire assemblage (e.g., Diamond 1975, Wilson & Willis 1975). There is now compelling evidence that, even where reserves are relatively large and not fully isolated by inhospitable habitats, extinction rates in fragments greatly exceed colonization rates, generating faunal relaxation (e.g., Newmark 1987).

Fragmented systems appear to be characterized by distinctive patterns of both species richness and composition. First, species-area slopes (z) of fragments tend to be high relative to *normal* island situations, substantially reducing the number of species expected in smaller fragments (Lawlor 1986). Second, as shown here and in Patterson and Atmar (1986), smaller fragments support nested subsets of species found in the source pool. Therefore, small fragments of similar size would tend to support the *same subset* of a richer biota, not different species assemblages. Species richness and composition must figure among the most important considerations in effective conservation strategies.

These apparently general biogeographic patterns constitute stark arguments for allocating the maximum possible area for biological preserves. The ongoing debate over single large or several small (SLOSS) preserves, of equal area, typically focuses on species number (Simberloff & Abele 1982, McIntyre et al. 1984). Because species-area slopes are typically positive but much less than 1.0, a smaller island supports fewer total species than a larger one but supports more species relative to its area. Thus, depending on biotic similarity, two small islands can contain more total species than a larger one. This argument fails to recognize explicitly that studies of fragmented systems, not intact ones, are the proper models for long-term conservation strategies. Because isolated fragments support many fewer species than mainland patches of the same size (heightened z) and because small fragments support nested subsets of the biotas in larger fragments, larger areas provide for more effective conservation over significant periods of time. Only large fragments sustain the rarer species that are often most in need of special protection.

Fragments containing different habitats almost certainly begin faunal relaxation with different species complements. Because the nested subset pattern has been associated with landbridge islands that shared most of their ancestral biotas, its application to situations involving different habitat types must be questioned. Yet New Zealand landbridge islands exhibit significant nestedness, despite probable differences in their species composition at the start of faunal relaxation. The species composition of fragments may actually *converge* over time because of a little-understood relationship between geographic distribution and abundance. Hanski (1982), Bock and Ricklefs (1983), and Brown (1984) have shown that locally abundant species may be geographically widespread and inhabit a higher frequency of sites. Because abundance is an effective safeguard to extinction from demographic causes, and ubiquity is a safeguard against environmental sources of extinction, reserves might be expected to retain widespread, common species and lose local rare ones (Patterson & Atmar 1986).

Acknowledgments

Chief thanks are owed to Wirt Atmar for his continued interest, stimulation, and friendship, and also for running the simulations on the HP 3000 at AICS Research, Inc., in Mesilla Park, New Mexico. J. W. Koeppl and S. M. Lanyon consulted on programming, and L. D. Brady painstakingly proofread the data tables. S. L. Pimm, L. R. Heaney, and especially J. M. Diamond provided a host of useful criticisms and suggestions that significantly improved the manuscript, as did W. Atmar and R. F. Inger. Research supported in part by the Ellen Thorne Smith Bird and Mammal Study Center of Field Museum.

Literature Cited

Bock, C. E., and R. E. Ricklefs. 1983. Range size and local abundance of some North American songbirds: a positive correlation. American Naturalist **122**:295–299.

Brown, J. H. 1971. Mammals on mountaintops: nonequilibrium insular biogeography. American Naturalist **105**:467–478.

Brown, J. H. 1984. On the relationship between abundance and distribution of species. American Naturalist **124**:255–279.

Colwell, R. K., and D. W. Winkler. 1984. A null model for null models in biogeography. Pages 344–359 *in* D. R. Strong, Jr., D. Simberloff, L. G. Abele, A. B. Thistle, editors. Ecological communities: conceptual issues and the evidence. Princeton University Press, Princeton, New Jersey, USA.

Connor, E. F., and D. Simberloff. 1984. Neutral models of species' co-occurrence patterns. Pages 316–331 *in* D. R. Strong, Jr., D. Simberloff, L. G. Abele, A. B. Thistle, editors. Ecological communities: conceptual issues and the evidence. Princeton University Press, Princeton, New Jersey, USA.

Crowell, K. L. 1986. A comparison of relict versus equilibrium models for insular mammals of the Gulf of Maine. Pages 37–64 *in* L. R. Heaney, B. D. Patterson, editors. Island biogeography of mammals. Academic Press and Linnean Society of London, England; Biological Journal of the Linnean Society **28.**

Diamond, J. M. 1975. The island dilemma: lessons of modern biogeographic studies for the design of natural reserves. Biological Conservation **7**:129–146.

Diamond, J. M. 1984*a*. "Normal" extinctions of isolated populations. Pages 191–246 *in* M. H. Nitecki, editor. Extinctions. University of Chicago Press, Chicago, Illinois, USA.

Diamond, J. M. 1984*b*. Distributions of New Zealand birds on real and virtual islands. New Zealand Journal of Ecology **7**:37–55.

Diamond, J. M., and M. E. Gilpin. 1984. Examination of the "null" model of Connor and Simberloff for species co-occurrences on islands. Oecologia **52**:64–74.

Gilpin, M. E., and J. M. Diamond. 1984. Are species co-occurrences on islands non-random, and are null hypotheses useful in community ecology? Pages 297–315 *in* D. R. Strong, Jr., D. Simberloff, L. G. Abele, A. B. Thistle, editors. Ecological communities: conceptual issues and the evidence. Princeton University Press, Princeton, New Jersey, USA.

Grant, P. R., and I. Abbott. 1980. Interspecific competition, island biogeography and null hypotheses. Evolution **34**:332–341.

Hanski, I. 1982. Communities of bumblebees: testing the core-satellite species hypothesis. Annales Zoologici Fennici **19**:65–73.

Heaney, L. R. 1984. Mammalian species richness on islands on the Sunda Shelf, Southeast Asia. Oecologia **61**:11–17.

Heaney, L. R. 1986. Biogeography of mammals in SE Asia: estimates of rates of colonization, extinction and speciation. Pages 127–165 *in* L. R. Heaney, B. D. Patterson, editors. Island biogeography of mammals. Academic Press and Linnean Society of London, England; Biological Journal of the Linnean Society **28.**

Hope, J. H. 1973. Mammals of the Bass Strait Island. Proceedings of the Royal Society of Victoria **85**:163–195.

Karr, J. R. 1982. Avian extinction on Barro Colorado Island, Panama: a reassessment. American Naturalist **119**:220–239.

Lawlor, T. E. 1986. Comparative biogeography of mammals on islands. Pages 99–125 *in* L. R. Heaney, B. D. Patterson, editors. Island biogeography of mammals. Academic Press and Linnean Society of London, England; Biological Journal of the Linnean Society **28.**

MacArthur, R. H., and E. O. Wilson. 1967. The theory of island biogeography. Princeton University Press, Monographs in Population Biology **1**, Princeton, New Jersey, USA.

McIntyre, M. E., E. C. Young, and C. M. King, editors. 1984. Biological reserve design in mainland New Zealand. Journal of the Royal Society of New Zealand **14**:1–45.

Mills, J., and G. Williams. 1984. The status of endangered New Zealand birds. Pages 1107–1120 *in* M. Archer, G. Clayton, editors. Vertebrate zoogeography and evolution in Australasia. Hesperian Press, Carlisle, Western Australia.

Morgan, G. S., and C. A. Woods. 1986. Extinction and the zoogeography of West Indian land mammals. Pages 167–203 *in* L. R. Heaney, B. D. Patterson, editors. Island biogeography of mammals. Academic Press and Linnean Society of London, England; Biological Journal of the Linnean Society **28.**

Newmark, W. D. 1987. A land-bridge island perspective on mammalian extinctions in western North American parks. Nature **325**:430–432.

Olson, S. L., and H. F. James. 1982. Fossil birds from the Hawaiian Islands: evidence for wholesale extinction by man before western contact. Science **217**:633–635.

Patterson, B. D. 1984. Mammalian extinction and biogeography in the southern Rocky Mountains. Pages 247–293 *in* M. H. Nitecki, editor. Extinctions. University of Chicago Press, Chicago, Illinois, USA.

Patterson, B. D. 1987. Local extinctions as a critical factor in the historical zoogeography of boreal mammals in the Southwest. Presented at the Annual Meeting of the American Society of Mammalogists. Albuquerque, New Mexico, June 1987.

Patterson, B. D., and W. Atmar. 1986. Nested subsets and the structure of insular mammalian faunas and archipelagos. Pages 65–82 *in* L. R. Heaney, B. D. Patterson, editors. Island biogeography of mammals. Academic Press and Linnean Society of London, England; Biological Journal of the Linnean Society **28.**

Simberloff, D., and L. G. Abele. 1982. Refuge design and island biogeographic theory: effects of fragmentation. American Naturalist **120:**41–50.

Terborgh, J. 1974. Preservation of natural diversity: the problem of extinction prone species. BioScience **24:**715–722.

Williams, G. R. 1981. Aspects of avian island biogeography in New Zealand. Journal of Biogeography **8:**439–456.

Williams, G. R. 1984. In the flightless tradition: New Zealand. Pages 1103–1105 *in* M. Archer, G. Clayton, editors. Vertebrate zoogeography and evolution in Australasia. Hesperian Press, Carlisle, Western Australia.

Wilson, E. O., and E. O. Willis. 1975. Applied biogeography. Pages 522–534 *in* M. L. Cody, J. L. Diamond, editors. Ecology and Evolution of Communities. Belknap Press of Harvard University Press, Cambridge, Massachusetts, USA.

Nested Subsets and the Distribution of Birds on Isolated Woodlots

JOHN G. BLAKE

Department of Biology
University of Missouri — St. Louis
8001 Natural Bridge Road
St. Louis, Missouri 63121, U.S.A.

Abstract: *Distribution of bird species among isolated habitat patches (e.g., woodlots in an agricultural landscape) often appears to be nonrandom; species present in small, species-poor patches also are found in larger patches that support more species. Bird communities form "nested subsets" (after Patterson & Atmar 1986) if all species found in small faunas also are found in more species-rich assemblages. Occurrence of a nested subset pattern implies an underlying, nonrandom pattern of species distributions. I used computer simulations to analyze the degree of nestedness exhibited by bird communities in east-central Illinois. Results demonstrated that the distribution of bird species breeding in isolated woodlots (1.8 to 600 ha) differed significantly from that predicted by a random distribution model; species assemblages were more "nested" than expected by chance. Most species present in small, species-poor woodlots also were found in larger, species-rich woodlots. As groups, species requiring forest interior habitat for breeding and species wintering in the tropics showed highly nested distributions. In contrast, short-distance migrants and species breeding in forest edge habitat showed more variable distribution patterns; species recorded on smaller woodlots were not always recorded on larger, more species-rich woodlots. Apparent absences from larger woodlots may have reflected real distribution patterns or insufficient sampling of edge habitats. These results support previous conclusions that small habitat patches are insufficient for preservation of many species.*

Resumen: *La distribución de las especies de aves entre lotes de hábitat aislados (por ejemplo, lotes boscosos en un campo agricola) parece con frecuencia no ser al azar; las especies presentes en lotes pequeños, pobres en especies, también se han encontrado en lotes más grandes que alojan a más especies. Las especies de aves forman "subgrupos reproductivos" (Patterson & Altmar, 1986), si todas las especies que se encuentran en faunas pequeñas también son encontradas en conjuntos más ricos en especies. La presencia de un patrón de subgrupo reproductivo implica un subyacente patrón de distribución no al azar de las especies. Se utilizaron simulacros computarizados para analizar el grado de subgrupos que se presentaban en las comunidades de aves en el este-centro de Illinois. Los resultados demuestran que la distribución de las especies de aves que anidan en lotes boscosos aislados (1.8 a 600 hectareas) difieren significativamente de los que son predecidos por un modelo de distribución al azar; las agrupaciones de las especies fueron más "subconjuntos" de lo que se esperaba por casualidad. La mayoria de las especies presetes en lotes boscososs pequeños y pobres en especies también fueron encontrados en lotes más grandes y ricos en especies. Como grupos, las especies que requieren de hábitat del interior del bosque para anidar y las especies que invernan en los trópicos mostraron altas distribuciones de subconjuntos. En contraste, los migratorios de cortas distancias y las especies que anidan en los límites del bosque mostraren patrones de distribución más variables; las especies registradas en lotes boscosos pequeños no siempre se registraron en lotes más grandes y más ricos en especies. Las aparentes ausencias en los lotes grandes pueden reflejar los patrones de distribución reales o un muestreo insuficiente en los ecotonos. Estos resultados apoyan previas conclusiones en el sentido de que pequeños lotes de hábitat son insuficientes para la preservación de muchas especies.*

Paper submitted November 6, 1989; revised manuscript accepted April 25, 1990.

Introduction

Fragmentation of habitats (e.g., forest) into smaller, isolated patches (e.g., woodlots) affects species distribution patterns (e.g., Willis 1979; Burgess & Sharpe 1981; Lynch & Whigham 1984; Wilcove, McLellan, & Dobson 1986; Blake & Karr 1987). The observation that number of species found in a given patch of habitat is related to the area of that patch (e.g., MacArthur & Wilson 1967; many subsequent) has led many to argue that large reserves are required to preserve many species (e.g., May 1975; Diamond 1976; Whitcomb et al. 1976; Blake & Karr 1984; Willis 1984) and has engendered substantial debate.

Much argument over reserve design and size has focused on number of species likely to be preserved (Simberloff & Abele 1976 and subsequent replies [SLOSS debate]). What species are present and likely to be preserved often is more important, however, than actual number of species (Diamond 1976; Blake & Karr 1984). Studies on avifaunas of isolated woodlots in eastern North America have shown that the distribution of species among woodlots of different area follows a distinct pattern. Most species present on small, isolated woodlots also occur on larger, more species-rich forest patches (e.g., Galli, Leck, & Forman 1976; Whitcomb et al. 1981; Ambuel & Temple 1983; Freemark & Merriam 1986; Blake & Karr 1987).

Such a structured (i.e., nonrandom) distribution pattern has been termed a "nested subset pattern" by Patterson and Atmar (1986; Patterson 1987). A truly nested pattern implies that if a series of faunas is ordered by species richness, smaller faunas will be subsumed within each successively larger fauna (see Patterson [1987], Fig. 1). A nested subset pattern has important implications for conservation. If, as a general rule, all species in smaller communities also occur in larger assemblages, then a series of small tracts is likely to be inhabited by the same set of species. The species most susceptible to fragmentation or area effects will be absent from small tracts.

Patterson and Atmar (1986) used computer simulations to demonstrate that nonmigratory mammal faunas of mountaintops and land-bridge islands formed highly nested subsets; most species found in smaller faunas also were represented in larger faunas. A nested structure also was shown for bird faunas of land-bridge islands off the coast of New Zealand (Patterson 1987). The nested (nonrandom) structure of mountaintop and land-bridge island faunas was attributed largely to faunal relaxation (i.e., selective extinctions following isolation, not balanced by recolonization). Bird and mammal faunas of oceanic islands, by contrast, did not form nested subsets (Patterson & Atmar 1986; Patterson 1987), suggesting that different processes (e.g., colonization ability) were important in determining the distribution of species among islands.

Bird faunas of forest fragments and other habitat patches in the temperate zone differ in many respects from nonmigratory faunas characteristic of mountaintops and true islands; woodlots are "recolonized" every spring by migrants that comprise most of the breeding species. Also, resident species often wander widely during the nonbreeding season and are not confined to a single woodlot (Howe 1984; Blake 1987). Although habitat intervening between woodlots (agricultural fields) is not suitable for long term support of forest species, it can be used temporarily. Thus, number of species present on woodlots is determined by such factors as resource availability, species interactions (competition, predation), and habitat structure.

Here I follow procedures developed by Patterson and Atmar (1986) in an analysis of bird communities found in isolated woodlots in east-central Illinois. Larger woodlots in east-central Illinois support more species overall, more species that breed in forest interior habitat, and more species that winter in the tropics than smaller woodlots (Blake & Karr 1987). In contrast, short-distance migrants (species wintering north of the tropics) and species breeding primarily in forest edge habitats were found in small as well as large woodlots (Blake & Karr 1984, 1987). Thus, the extent to which faunas exhibit a nested distribution pattern is probably influenced by migratory strategy and preferred breeding habitat of the component species. By comparing observed distribution patterns to those based on computer simulations, I am able to evaluate the degree to which faunas depart from a random distribution among woodlots.

Methods

Bird Community Composition

I sampled bird communities in 12 isolated woodlots (1.8 to 600 ha) in east-central Illinois in both 1979 and 1980. Land in east-central Illinois is devoted primarily to agriculture (Iverson et al. 1989); little forest cover exists (Table 1). Woodlots are typically well separated from one another, forming isolated patches in wide expanses of corn and soybeans. Clearing for agriculture often extends right to woodlot borders (personal observation), leaving little "edge" habitat. Thus, the transition from forest to field usually is abrupt.

All woodlots included in this study possessed a mature canopy and well-developed understory, did not show evidence of recent disturbance, and were isolated from other woodlots by farmland. Observation points (two or more per woodlot) were about 150 m apart. Number of census points increased with woodlot area

Table 1. Amount of forest cover (in thousands of hectares) and percentage of county covered by forest for counties where woodlots included in this study were located; area (ha) of woodlots in each county are given. Data are from Iverson et al. (1989:15–16).

		1978		1982	
County	*Woodlots (ha)*	*Forest*	*Percentage*	*Forest*	*Percentage*
Champaign	1.8, 24, 24, 28	1.5	0.6	2.6	1.0
Macon	65	4.8	3.2	6.3	4.2
McLean	65, 118	2.8	0.9	3.8	1.3
Piatt	2.3, 5.1, 6.5, 16.2, 600	2.4	2.1	2.9	2.6

(see Tables 3, 4) and were distributed to sample the range of habitats present. Each point was censused at least three times during the breeding season; censuses lasted 15 min per point. Breeding was assumed on the basis of active nests, fledged young, obvious territorial behavior, or presence within a woodlot throughout the summer. Birds flying over the canopy, wide-ranging species (e.g., birds of prey, crows), and nocturnal species were not included. I categorized species by migratory status and preferred breeding habitat, using data from the literature (Bohlen 1978; Kendeigh 1982) and personal observations.

Analyses

I followed procedures described by Patterson and Atmar (1986; see also Patterson 1987), and the following is a synopsis from their paper. I used data from point counts to create presence/absence matrices for species in the 12 woodlots; each year was treated separately. Separate presence/absence matrices were constructed for all species combined and for groups based on migratory status and preferred breeding habitat. The assemblage of species found in one woodlot was considered "nested" if all species present in that woodlot also were found in all woodlots supporting a greater total number of species. Departures from perfect nestedness ($N = 0$) were calculated by counting all absences of a species from assemblages in more species-rich woodlots (partial nestedness score [N_{io}] observed for each species; $i = 1$ to total number of species for the year). For example, if a species occurred in a woodlot supporting 5 species, was present in woodlots supporting 10, 15, and 25 species, but was absent from a woodlot with 20 species, the species would have an $N_{io} = 1$ (i.e., one absence from a more species-rich assemblage). Partial nestedness scores were summed over all species to give an index of observed nestedness (N_o) for the entire set of species. A low total N_o indicates that most species present in smaller, species-poor woodlots also were present in larger woodlots (i.e., a more nested distribution). Large deviations from nestedness indicate that species distribution were more variable (i.e., smaller faunas were not always subsets of larger faunas).

The extent to which woodlot bird communities departed from a perfectly nested pattern was assessed via Monte Carlo simulations, using the program provided by Patterson and Atmar (1986:81–82), modified for use on an AT&T PC6300. By comparing observed distribution patterns to those generated by computer simulation, I was able to assess the probability that the observed distribution (i.e., degree of nestedness) occurred simply by chance. A distribution significantly different than random would suggest that composition of woodlot bird communities was influenced by biological processes (e.g., species interactions, habitat suitability).

Patterson and Atmar (1986) presented two procedures for selecting species to form random communities. The first (RANDOM0) based species selections on a uniform probability distribution. I used only their second program (RANDOM1) to create assemblages by selecting species from a probability distribution weighted by the actual number of woodlots each species occurred in. Such a weighted distribution provides a more realistic constraint on the construction of random assemblages than does a uniform probability distribution (Patterson & Atmar 1986). The program was run 1,000 times to create random woodlot assemblages. Species were grouped by migratory status and habitat preference; separate analyses were conducted for each group. Separate partial nestedness scores were determined for each species. Observed (N_{io}) values were compared to the mean of simulated values (N_{is}) for each group of species by t-tests (Sokal & Rohlf 1981).

Results

Forty-six (1979) or 49 (1980) species were recorded as breeding residents (Table 2). Number of species per woodlot ranged from 9 on the smallest (1.8 ha) to 40 on the largest (600 ha) (Tables 3, 4). Examination of species presence/absence matrices revealed that bird assemblages were not perfectly nested; species present on one woodlot usually, but not always, were present on all woodlots with more total species. Most species (48% in 1979; 53% in 1980) had partial nestedness scores of 0, indicating no absences from woodlots with more species. An additional 24% (1979) or 20% (1980) had only one absence ($N_{io} = 1$) from more species-rich woodlots (Tables 2–4).

Table 2. List of bird species from woodlots in east-central Illinois, number of woodlots the species occurred in (Woods), and partial nestedness scores (N_i) for observed (N_{io}) and simulated (N_{is}) data.

	1979			*1980*					
Species	*Woods*	N_{io}	N_{is}	*Woods*	N_{io}	N_{is}	*MIG*[a]	*HAB*[b]	*Species Code*
Red-headed Woodpecker *Melanerpes erythrocephalus*	12	0	1.61	11	1	1.75	PR	IE	A
Great Crested Flycatcher *Myiarchus crinitus*	12	0	1.78	12	0	1.73	LDM	IE	B
Blue Jay *Cyanocitta cristata*	12	0	1.69	12	0	1.65	PR	IE	C
American Robin *Turdus migratorius*	12	0	1.35	12	0	1.52	SDM	E	D
Northern Cardinal *Cardinalis cardinalis*	12	0	1.49	12	0	1.56	PR	IE	E
Indigo Bunting *Passerina cyanea*	12	0	1.60	12	0	1.59	LDM	E	F
Red-eyed Vireo *Vireo olivaceus*	11	1	1.89	10	0	2.12	LDM	IE	G
Common Grackle *Quiscalus quiscula*	11	1	1.66	11	1	1.58	SDM	E	H
European Starling *Sturnus vulgaris*	9	3	2.33	12	0	1.56	PR	E	I
House Sparrow *Passer domesticus*				4	8	3.49	PR	E	J
Downy Woodpecker *Picoides pubescens*	11	0	1.29	12	0	1.57	PR	IE	K
Northern Flicker *Colaptes auratus*	11	0	1.87	11	0	1.89	SDM	IE	L
Eastern Wood-Pewee *Contopus virens*	11	0	1.61	12	0	1.39	LDM	IE	M
Yellow-billed Cuckoo *Coccyzus americanus*	11	0	1.61	10	0	1.92	LDM	IE	N
Mourning Dove *Zenaida macroura*	4	7	3.24	7	1	2.70	SDM	E	O
Gray Catbird *Dumetella carolinensis*	7	4	2.96	6	5	2.85	SDM	IE	P
Kentucky Warbler *Oporornis formosus*	7	4	2.96	3	0	3.98	LDM	I	Q
House Wren *Troglodytes aedon*	8	1	2.44	8	3	2.31	SDM	E	R
Brown-headed Cowbird *Molothrus ater*	10	0	1.73	11	0	1.75	SDM	E	S
American Goldfinch *Carduelis tristis*				10	1	1.96	SDM	E	T
Brown Thrasher *Toxostoma rufum*	5	4	2.84	4	6	3.57	SDM	E	U
Wood Thrush *Hylocichla mustelina*	10	0	1.85	8	0	2.09	LDM	IE	V
Rufous-sided Towhee *Pipilo erythrophthalmus*	6	3	2.84	8	2	2.31	SDM	IE	W
Red-bellied Woodpecker *Melanerpes carolinus*	9	0	2.11	10	0	1.77	PR	IE	X
Scarlet Tanager *Piranga rubra*	7	2	2.92	7	0	2.83	LDM	I	Y
Acadian Flycatcher *Empidonax virescens*	6	1	3.17	5	1	3.14	LDM	I	Z
Red-winged Blackbird *Agelaius phoeniceus*	4	3	3.39	5	2	3.28	SDM	E	a
Yellow-breasted Chat *Icteria virens*	1	6	3.54	2	4	3.87	LDM	E	b
Chipping Sparrow *Spizella passerina*	2	5	4.02	1	2	3.80	SDM	E	c
Common Yellowthroat *Geothlypis trichas*	5	1	3.48	5	6	3.42	SDM	IE	d

Table 2. Continued.

Species	1979 Woods	1979 N_{io}	1979 N_{is}	1980 Woods	1980 N_{io}	1980 N_{is}	MIG[a]	HAB[b]	Species Code
Tufted Titmouse *Parus bicolor*	6	0	2.58	6	0	2.85	PR	IE	e
White-breasted Nuthatch *Sitta carolinensis*	5	0	3.06	5	0	3.35	PR	I	f
Black-capped Chickadee *Parus atricapillus*	5	0	3.68	5	0	3.04	PR	IE	g
Blue-gray Gnatcatcher *Polioptila caerulea*	5	0	3.99	2	0	3.49	LDM	IE	h
Veery *Catharus fuscescens*	5	0	3.82	2	1	4.04	LDM	I	i
Hairy Woodpecker *Picoides villosus*	4	1	3.24	7	1	2.67	PR	I	j
Northern Oriole *Icterus galbula*	3	2	3.70	2	6	3.70	LDM	E	k
Northern Parula *Parula americana*	3	2	3.80	1	0	3.20	LDM	IE	l
Rose-breasted Grosbeak *Pheucticus ludovicianus*	3	1	3.93	5	0	3.21	LDM	IE	m
White-eyed Vireo *Vireo griseus*	1	2	3.78	1	1	3.35	LDM	IE	n
Yellow-throated Warbler *Dendroica dominica*	2	1	3.68	1	0	3.72	LDM	I	o
American Redstart *Setophaga ruticilla*	2	1	3.99				LDM	I	p
Yellow-throated Vireo *Vireo flavifrons*	2	1	3.82	4	5	3.13	LDM	IE	q
Cerulean Warbler *Dendroica cerulea*	1	1	3.24	1	0	3.17	LDM	I	r
Ovenbird *Seiurus aurocapillus*	1	0	3.64	5	2	3.07	LDM	I	s
Ruby-throated Hummingbird *Archilochus colubris*	1	0	3.39	1	0	2.90	LDM	E	t
Hooded Warbler *Wilsonia citrina*	1	0	2.73				LDM	I	u
Brown Creeper *Certhia americana*	1	0	3.16	1	0	3.01	PR	I	v
Cedar Waxwing *Bombycilla cedrorum*				1	4	3.61	PR	E	w
Field Sparrow *Spizella pusilla*				2	1	3.89	SDM	E	x
Song Sparrow *Melospiza melodia*				1	1	2.86	SDM	E	y

[a] *Migratory status: LDM = long-distance migrant; SDM = short-distance migrant; PR = permanent resident.*
[b] *Breeding habitat: I = forest interior; E = forest edge; IE = forest interior and edge.*

Most species that were recorded in both 1979 and 1980 had the same partial nestedness scores in each year (48%) or had scores that differed by only 1 between years (25%). Thus, occurrence of species in woodlots showed a similar pattern between years with respect to presence or absence in woodlots with different numbers of species. Only five species (11% of total; Mourning Dove, Yellow-throated Vireo, Kentucky Warbler, Common Yellowthroat, and Northern Oriole) had partial nestedness scores that differed between 1979 and 1980 by at least 4 (Table 2), indicating substantial variation in distribution pattern between years for those species.

Four of the above five species often are associated with edge habitats and may have been overlooked in some counts. The Mourning Dove showed the largest change in N_{io} scores between years (N_{io} = 7 in 1979 and 1 in 1980). Only the Kentucky Warbler was associated primarily with forest interior habitat. In 1979, a single pair was recorded in a small (2.3 ha) woodlot that supported 15 species, but no Kentucky Warblers were found in four of ten woodlots with more species (Table

Table 3. Distribution of breeding birds on woodlots in east-central Illinois during 1979. Presence of a species in a woodlot is indicated by letter code (see Table 2); absences from woodlots with more species (i.e., deviations from perfect nestedness) are indicated by asterisks. Area (in hectares), total species, and number of sample points are given for each woodlot.

Woodlot			
Area	*Sp*	*Pts*	*Bird Species*
600	39	14	A B C D E F G H I K L M N * P Q R S * V * X Y Z a * c * e f g h i j k l m * o p q * s t u v
65	34	8	A B C D E F G H I K L M N * P Q R S U V W X Y Z a * * d e f g h i j k l * * * * * r
118	32	6	A B C D E F G H I K L M N O * Q R S * V W X Y Z * * * d e f g h i j * * m p q
28	32	6	A B C D E F * H I K L M N O P Q R S U V W X Y * a * * d e f g h i * * * m n o
65	28	8	A B C D E F G H I K L M N * * Q R S * V * X Y Z * * * * e f g h i j k l
24	24	5	A B C D E F G H I K L M N * P * R S U V W X Y Z * * * d e
24	22	5	A B C D E F G H I K L M N * P * * S U V * X * Z a b c
6.5	19	3	A B C D E F G H * K L M N * P Q R S * V W X *
16.2	17	4	A B C D E F G H * K L M N O * * S V X Y
5.1	17	3	A B C D E F G H * K L M N * * * R S U V W
2.3	15	2	A B C D E F G * I K L M N O P Q
1.8	9	2	A B C D E F G H I

3). In contrast, the smallest woodlot the warbler was recorded in 1980 was 65 ha (34 species).

The species with the largest N_{io} score (8) was the House Sparrow in 1980. (House Sparrows were not recorded during 1979.) Distribution of this species depends largely on presence of farm buildings. Thus, it may be absent from many woodlots not located near farms.

Despite the variation between years in distribution of some species, the overall pattern was the same in both years. Most species that occurred on species-poor woodlots also occurred on woodlots with more species (Tables 3, 4). Consequently, total nestedness scores (N_o) comprised 16–17% of the total presence/absence matrix (Table 5) (i.e., absences accounted for about 16% of all possible occurrences of a perfectly nested distribution). Observed departures from nestedness were significantly less than expected based on results of 1,000 simulations (Table 5); thus, species were distributed in a nonrandom fashion. Observed total nestedness scores for 1979 and 1980 were about six standard deviations less than the means for simulated assemblages when all species were considered together.

Long-distance migrants and species that breed preferentially in forest interior habitat also formed highly nested distributions (Table 5). In 1979, the observed N_o value for forest interior species was only marginally ($P < 0.06$) less than that based on computer simulations. The higher value in 1979 ($N_o = 7$) compared to 1980 ($N_o = 2$) primarily resulted from the distribution of the Yellow-throated Warbler ($N_{io} = 3$ in 1979 and 0 in 1980).

Means of observed partial nestedness scores were largest (i.e., more absences from species-rich relative to species-poor woodlots) for short-distance migrants and edge-inhabiting species (Table 6); observed values (N_{io}) were not significantly different from those predicted by a random distribution model (N_{is} values). In contrast, observed values were significantly lower than simulated values for all other groups, except permanent residents in 1980. The higher observed value for permanent residents in 1980 (mean $N_{io} = 1.08$) relative to 1979 (mean = 0.36) was due to the high partial nestedness score of the House Sparrow (see above). With the House Sparrow omitted, the mean observed N_{io} for permanent residents in 1980 was close to that for 1979 and

Table 4. Distribution of breeding birds on woodlots in east-central Illinois during 1980. Presence of a species in a woodlot is indicated by letter code (see Table 2); absences from woodlots with more species (i.e., deviations from perfect nestedness) are indicated by asterisks. Area (in hectares), total species, and number of sample points are given for each woodlot.

Woodlot			
Area	*Sp*	*Pts*	*Bird Species*
600	40	18	A B C D E F H I * K M L P R S T * G N * W X O q V j * Y a s e Z * f g m * x Q * i h * * o l r t v
65	34	10	A B C D E F H I J K M L P R S T * G N * W X * * V j * Y a s e Z * f g m * * Q * * h n y
118	33	8	A B C D E F H I * K M L * R S T * G N * W X O q V j k Y * s e Z * f g m * Q c i
28	33	6	A B C D E F H I J K M L P R S T d G N U W X O * V j * Y a * e * b f g m * x
65	30	10	A B C D E F H I * K M L * R S T d G N * W X O * V j * Y a * e Z * f g m w
24	29	6	A B C D E F H I J K M L P R S T d G N U W X O * V * * Y * s e Z b
24	26	6	A B C D E F H I * K M L P R S T d G N U * X O * V j * Y a s
16.2	19	4	A B C D E F H I * K M L * * S * * G N * * X q V j k
6.5	19	3	A B C D E F H I * K M L * * S T * G N * W X O q
5.6	17	3	* B C D E F H I * K M L * * S T * G N U W X
2.3	15	2	A B C D E F * I * K M L P R S T d
1.8	11	2	A B C D E F H I J K M

Table 5. Departure from perfect nestedness (N index and percentage of the presence/absence matrix) based on actual distribution pattern of species (Sp. = number of species) among woodlots (Woods = number of woodlots occupied by at least one species) and on 1,000 simulations (mean, SD, min, max). Results of t-tests comparing the observed index (N_o) to that based on simulations (N_t) are given. Separate analyses were conducted for each group and year.

Year	Sp.	Woods	Observed N_o Index (and %)	Simulation Results					
				Mean N_t	SD	Min.	Max	t =	P <
All species									
1979	46	12	59 (17%)	127.4	11.17	91	152	6.09	0.001
1980	49	12	65 (16%)	133.1	11.57	105	159	5.86	0.001
Long-distance migrants									
1979	23	12	24 (17%)	61.2	9.00	38	84	4.11	0.001
1980	21	12	17 (14%)	55.3	8.36	34	79	4.56	0.001
Forest-interior breeders									
1979	12	10	7 (14%)	14.1	3.77	3	22	1.88	0.06
1980	10	8	2 (5%)	9.3	2.61	2	16	2.80	0.01

significantly ($P < 0.001$) less than the mean for simulated data.

Discussion

A general pattern has emerged from studies of bird communities of isolated woodlots: small patches typically are dominated by generalist species, species able to breed and forage in a variety of habitats. With increasing size of woodlots, species with more specialized resource requirements increase in importance (Martin 1981; Whitcomb et al. 1981; Ambuel & Temple 1983; Blake & Karr 1987). Studies that have compared observed distribution patterns to those based on random distribution models report similar conclusions (Blake 1983; Blake & Karr 1984; Moller 1987). Similarity of results from studies conducted in different regions provides strong evidence that observed patterns represent real consequences of forest fragmentation.

Results of this study further indicate that occurrence of bird species in isolated woodlots in east-central Illinois is distinctly nonrandom. Instead, species distribution patterns follow a regular, predictable sequence: most species that occurred in woodlots supporting few species, which typically are smaller in area than species-rich woodlots (Blake & Karr 1987), also occurred in woodlots with a greater total number of species. These results thus extend the concept of nested subsets from nonmigratory faunas of mountaintops and true islands (Patterson & Atmar 1986; Patterson 1987) to primarily migratory faunas of mainland habitat islands. A nested pattern implies that biological processes, rather than chance, underlie observed distributions.

Bird communities in temperate regions are dominated by migrants that winter in the tropics. Breeding areas are "recolonized" every spring and individuals depart every fall. Successful colonization has two necessary components: first, the species must be able to locate the particular habitat patch; second, it must be able to breed there. Differential colonization ability, at least

Table 6. Mean partial nestedness (N_i) values for observed and simulated data, calculated from data in Table 2. Results of t-tests comparing means of observed and simulated data are shown. Sp. = number of species in each group.

Group	Sp.	Observed		Simulated		t =	P <
		Mean	SD	Mean	SD		
1979							
Long-distance migrant	23	1.09	1.47	3.06	0.88	5.51	0.001
Short-distance migrant	12	2.42	2.27	2.65	0.84	0.33	0.90
Permanent resident	11	0.36	0.92	2.42	0.77	5.69	0.001
Forest interior	12	0.92	1.16	3.30	0.39	6.74	0.001
Interior and edge	21	0.71	1.15	2.58	0.96	5.72	0.001
Forest edge	13	2.46	2.44	2.71	0.91	0.35	0.90
1980							
Long-distance migrant	21	0.95	1.80	2.94	0.81	4.62	0.001
Short-distance migrant	15	2.07	2.05	2.64	0.81	1.00	0.40
Permanent resident	13	1.08	2.36	2.45	0.82	1.98	0.06
Forest interior	10	0.50	0.71	3.30	0.47	10.40	0.001
Interior and edge	21	0.95	1.91	2.39	0.73	3.23	0.01
Forest edge	18	2.22	2.44	2.77	0.91	0.90	0.50

with respect to locating or reaching a given woodlot, is not likely to produce a nested pattern. In east-central Illinois, even the smallest woodlot will probably be visited each spring by most returning migrants; over 80 species were recorded from a 1.8 ha woodlot (unpublished data). Further, previous analyses (Blake & Karr 1987; unpublished data) indicated that species richness of woodlots was not related to several measures of isolation (e.g., distance to nearest woodlot >2 ha, nearest woodlot >24 ha, nearest river with riparian vegetation). Instead, total species richness was correlated primarily with woodlot area.

Species are likely to differ with respect to breeding success on different-sized woodlots. Fragmentation of forests may reduce breeding success through increased nest predation (Wilcove 1985; Martin 1988), nest parasitism (Brittingham & Temple 1983; Temple & Cary 1988; Robinson 1990), food limitation (Blake 1983; Martin 1987), microhabitat loss (Karr 1982), or other effects. If a species cannot maintain a successful breeding population in a given area, continued presence of that species will depend on dispersal to that area of young produced elsewhere ("rescue effect" of Brown & Kodric-Brown 1977; also Leck, Murray, & Swinebroad 1988). Periodic occurrence of species such as Kentucky Warbler in small woodlots may reflect dispersal by young birds. Lack of persistence in such sites may be due, in part, to poor breeding success. Recent evidence suggests, in fact, that the current breeding success of most neotropical migrants breeding in central Illinois is probably too low to maintain populations without immigration from elsewhere (Robinson 1990).

As forests become increasingly fragmented, fewer young may be produced and the probability of dispersal to other woodlots may decrease. Thus, as forest fragments become more isolated, "faunal relaxation" may occur on woodlots as it occurs on land-bridge islands. Over time, it is possible that breeding populations of some species will disappear from woodlots in east-central Illinois. A similar situation may occur in other regions dominated by fragmented habitats. A nested species distribution pattern thus may develop because of differential ability among species to breed successfully in different-sized patches, rather than differential ability to reach a given patch.

Variation in habitat quality also may influence species distribution patterns, as shown by the Yellow-throated Warbler in this study. This species occurred on a 28 ha woodlot with four other forest interior species but was not recorded from woodlots of 65 to 118 ha with six or seven forest interior species in 1979. Occurrence in the smaller woodlot was probably due to the presence of a large river along the border of the smaller woodlot. The Yellow-throated Warbler is associated primarily with bottomland forest, a habitat not present in the 65 and 118 ha woodlots.

Whatever the specific mechanism producing a nested pattern, such a pattern has important implications for conservation. As many have suggested, small, species-poor reserves are less likely to support a full complement of species, no matter what the total area. Some species, typically those most in need of protection, will occur only in larger patches.

Acknowledgments

This paper has benefited from the comments of B. A. Loiselle, G. J. Niemi, B. D. Patterson, and two anonymous reviewers. Original field work for this study was supported by the U.S. Fish and Wildlife Service, Contract No. 14-16-0009-023, to J. R. Karr while the author was at the University of Illinois.

Literature Cited

Ambuel, B., and S. A. Temple. 1983. Area-dependent changes in the bird communities and vegetation of southern Wisconsin forests. Ecology **64**:1057–1068.

Blake, J. G. 1983. Trophic structure of bird communities in forest patches in east-central Illinois. Wilson Bulletin **95**:416–430.

Blake, J. G. 1987. Species-area relationships of winter residents in isolated woodlots. Wilson Bulletin **99**:243–252.

Blake, J. G., and J. R. Karr. 1984. Species composition of bird communities and the conservation benefit of large versus small forests. Biological Conservation **30**:173–187.

Blake, J. G., and J. R. Karr. 1987. Breeding birds of isolated woodlots: area and habitat relationships. Ecology **68**:1724–1734.

Bohlen, H. D. 1978. An annotated check-list of the birds of Illinois. Illinois State Museum Popular Science Series Number 9.

Brittingham, M. C., and S. A. Temple. 1983. Have cowbirds caused forest songbirds to decline? BioScience **33**:31–35.

Brown, J. H., and A. Kodric-Brown. 1977. Turnover rates in insular biogeography: effect of immigration on extinction. Ecology **58**:445–449.

Burgess, R. L., and D. M. Sharpe, editors. 1981. Forest island dynamics in man-dominated landscapes. Springer-Verlag, New York.

Diamond, J. M. 1976. Island biogeography and conservation: strategy and limitation. Science **193**:1027–1029.

Freemark, K. E., and H. G. Merriam. 1986. Importance of area and habitat heterogeneity to bird assemblages in temperate forest fragments. Biological Conservation **36**:115–141.

Galli, A. E., C. F. Leck, and R. T. Forman. 1976. Avian distribution patterns in forest islands of different sizes in central New Jersey. Auk **93**:356–365.

Howe, R. W. 1984. Local dynamics of bird assemblages in small forest habitat patches in Australia and North America. Ecology **65**:1585–1601.

Iverson, L. R., R. L. Oliver, D. P. Tucker, P. G. Risser, C. D. Burnett, and R. G. Rayburn. 1989. Forest resources of Illinois: an atlas and analysis of spatial and temporal trends. Illinois Natural History Survey Special Publication 11.

Karr, J. R. 1982. Avian extinction on Barrow Colorado Island, Panama: a reassessment. American Naturalist **119**:220–239.

Kendeigh, S. C. 1982. Bird populations in east-central Illinois: fluctuations, variations, and development over a half-century. Illinois Biological Monographs, Number 52.

Leck, C. F., B. G. Murray, Jr., and J. Swinebroad. 1988. Long-term changes in the breeding bird populations of a New Jersey forest. Biological Conservation **46**:145–157.

Lynch, J. F., and D. F. Whigham. 1984. Effects of forest fragmentation on breeding bird communities in Maryland, USA. Biological Conservation **28**:287–324.

MacArthur, R. H., and E. O. Wilson. 1967. The theory of island biogeography. Princeton University Press, Princeton, New Jersey.

Martin, T. E. 1981. Limitation in small habitat islands: chance or competition? Auk **98**:715–733.

Martin, T. E. 1987. Food as a limit on breeding birds: a life-history perspective. Annual Review of Ecology and Systematics **18**:453–487.

Martin, T. E. 1988. Habitat and area effects on forest bird assemblages: is nest predation an influence? Ecology **69**:74–84.

May, R. M. 1975. Island biogeography and the design of nature preserves. Nature **254**:177–178.

Moller, A. P. 1987. Breeding birds in habitat patches: random distribution of species and individuals? Journal of Biogeography **14**:225–236.

Patterson, B. D. 1987. The principle of nested subsets and its implications for biological conservation. Conservation Biology **1**:323–334.

Patterson, B. D., and W. Atmar. 1986. Nested subsets and the structure of insular mammalian faunas and archipelagoes. Biological Journal of the Linnean Society **28**:65–82.

Robinson, S. K. 1990. Effects of forest fragmentation on nesting songbirds. Illinois Natural History Survey Reports, No. 296.

Simberloff, D. S., and L. G. Abele. 1976. Island biogeography theory and conservation practice. Science **191**:285–286.

Sokal, R. R., and F. J. Rohlf. 1981. Biometry: the principles and practice of statistics in biological research. 2nd ed. W. H. Freeman, San Francisco, California.

Temple, S. A., and J. R. Cary. 1988. Modeling dynamics of habitat-interior bird populations in fragmented landscapes. Conservation Biology **2**:340–347.

Whitcomb, R. F., J. F. Lynch, P. A. Opler, and C. S. Robbins. 1976. Island biogeography and conservation: strategy and limitations. Science **193**:1030–1032.

Whitcomb, R. F., C. S. Robbins, J. F. Lynch, B. L. Whitcomb, M. K. Klimkiewicz, and D. Bystrak. 1981. Effects of forest fragmentation on avifauna of the eastern deciduous forest. Pages 123–205 in R. L. Burgess and D. M. Sharpe, editors. Forest island dynamics in man-dominated landscapes. Springer-Verlag, New York.

Wilcove, D. S. 1985. Nest predation in forest tracts and the decline of migratory songbirds. Ecology **66**:1211–1214.

Wilcove, D. S., C. H. McLellan, and A. P. Dobson. 1986. Habitat fragmentation in the temperate zone. Pages 237–256 in M. E. Soulé, editor. Conservation biology: the science of scarcity and diversity. Sinauer Associates, Sunderland, Massachusetts.

Willis, E. O. 1979. The composition of avian communities in remanescent woodlots in southern Brazil. Papeis Avulsos de Zoologia (Sao Paulo) **33**:1–25.

Willis, E. O. 1984. Conservation, subdivision of reserves, and the anti-dismemberment hypothesis. Oikos **42**:396–398.

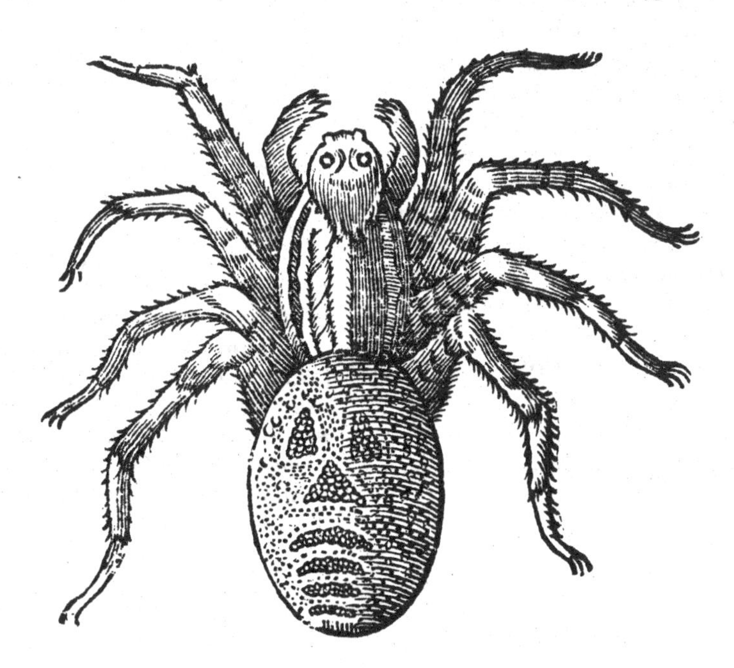

Review

The Effect of Edge on Avian Nest Success: How Strong Is the Evidence?

PETER W. C. PATON
Department of Fisheries and Wildlife
and Utah Cooperative Fish and Wildlife Research Unit
Utah State University
Logan, UT 84322, U.S.A.

Abstract: *Wildlife biologists historically considered the edge between adjacent habitat types highly productive and beneficial to wildlife. A current dogma is that edges adversely affect a wide range of avian species by increasing depredation and parasitism rates of nests. I critically evaluated existing empirical evidence to test whether there was a gradation in nest success as a function of distance from an edge. Researchers investigating this question have been inconsistent in their experimental designs, making generalizations about edge-effect patterns difficult. The majority of studies I examined found nest success varied near edges, with both depredation rates (10 of 14 artificial nest studies, and 4 of 7 natural nest studies) and parasitism rates (3 of 5 studies) increasing near edges. In addition, there was a positive relationship between nest success and patch size (8 of 8 studies). The most conclusive studies suggest that edge effects usually occur within 50 m of an edge, whereas studies proposing that increased depredation rates extend farther than 50 m from an edge are less convincing. Prior research has probably focused on distances too far from an edge to detect threshold values, and future research should emphasize smaller scales: 100–200 m from an edge at 20–25 m increments. Researchers often use relatively arbitrary habitat characteristics to define an edge. Therefore, I propose that only openings in the forest canopy with a diameter three times or more the height of the adjacent trees should be included in edge analyses. This review suggests that fragmentation of eastern North American temperate forests could lead to increased nest predation and parasitism, and there is need to determine if similar processes occur in other forested regions of North America.*

Paper submitted August 3, 1992; revised manuscript accepted July 18, 1993.

El efecto de borde sobre el éxito en la nidificación: ¿Cuan firme es la evidencia?

Resumen: *Históricamente, los biólogos de fauna silvestre consideraron a los bordes entre tipos de habitats adyacentes como altamente productivos y beneficiosos para la fauna silvestre. Un dogma actual sostiene que los bordes afectan adversamente a un amplio rango de especies de aves al incrementar las tasas de predación y parasitismo de los nidos. Yo evalué criticamente la evidencia empírica existente para establecer si existía un gradiente en el éxito de los nidos en función de la distancia a un borde. Los investigadores que analizaron esta cuestión han sido inconsistentes en el diseño de sus experimentos, haciendo difícil las generalizaciones acerca de los patrones del efecto de bordes. La mayoría de los estudios que examiné establecieron que el éxito de los nidos varió cerca de los bordes, observandose un incremento tanto en las tasas de predación (10 de 14 estudios de nidos artificiales y 4 de 7 estudios de nidos naturales) como de parasitismo (3 de 5 estudios) cerca de los bordes. Además existió una relación positiva entre el éxito de los nidos y el tamaño de los parches (8 de los 8 estudios). Los estudios más decisivos sugieren que los efectos de borde ocurren usualmente a menos de 50 m del borde, mientras que los estudios que proponían que los incrementos en las tasas de predación se extendían más allá de 50 m del borde fueron poco convincentes. Investigaciones anteriores se concentraron, posiblemente, en distancias muy alejadas del borde como para detectar valores de umbral y la investigación futura debería enfatizar menores escalas; 100–200 m del borde con incrementos de 20–25 m. Los investigadores usan frecuentemente características del hábitat relativamente arbitrarias para definir un borde. Por consiguiente propongo que sólo se incluyan en los análisis de borde aberturas en el canopeo forestal que posean un diámetro su-*

perior a 3 veces la altura de los árboles adjacentes. Esta revisión sugiere que la fragmentación de las forestas del este de Norte América podría conducir a un incremento en la predación y parasitismo de nidos, y que es necesario determinar si procesos similares ocurren en otras regiones forestales de Norte América.

Introduction

Many biologists consider the edge between adjacent habitat types a positive feature of the landscape for wildlife (see Kremsater & Bunnell 1992). This belief is based in part on Leopold's (1933) "law of interspersion," which postulated that increases in the amount of edge habitat resulted in higher population densities (Guthery & Bingham 1992). Leopold's edge-effects hypothesis led many wildlife biologists to create edges to enhance habitat for game species (for example, Yoakum et al. 1980: 402). Within the past decade, however, wildlife biologists have pointed out that Leopold's hypothesis pertained only to certain species and have challenged the idea that edges benefit most wildlife (see Reese & Ratti 1988; Yahner 1988).

Much of the recent interest in edge is due to avian population declines in fragmented landscapes, as exemplified by neotropical migrants in eastern North America (Askins et al. 1990; Hagan & Johnston 1992). In a recent summary of breeding season mortality factors for 32 species of neotropical migrants, Martin (1992) identified nest predation as the most important cause of nest failure. Therefore, determining what factors influence nest predation is imperative if biologists hope to successfully manage many avian populations.

One of the most frequently cited explanations for population declines in fragmented landscapes is higher nest depredation rates near edges, hereafter referred to as edge effects. Gates and Gysel (1978) developed the "ecological trap" hypothesis, which postulated that nest predation rates were density-dependent, with greater nest densities and a concomitant increase in depredation rates near edges. Subsequent studies have refuted this hypothesis, however, based on data with artificial nests showing that nest success did not decline near edges (Angelstam 1986; Ratti & Reese 1988).

Because the data are equivocal, I was interested in the circumstances when edge effects appear to be an important biological phenomenon. My purpose is to (1) summarize existing literature that quantified the relationship between nest loss (either by predation or parasitism) and distance from an edge (or patch size), (2) re-analyze these data where possible, (3) summarize the conditions when predation and parasitism rates change near edges (for example, degree of contrast between adjoining habitats, forested versus unforested habitats, type of predator), and, finally, (4) discuss what conclusions I believe realistically can be drawn from existing empirical evidence.

Methods

I have divided this review into two sections. The first section, on distance as an independent variable, deals with studies that used artificial or natural nests to quantify the relationship between nest success and distance from an edge. I attempted to include in this review all studies that have directly measured nest success as a function of distance from an edge. The second section, on patch size as an independent variable, analyzes studies that indirectly investigated edge effects by quantifying the relationship between nest success and patch size. There is a large body of literature focusing on the relationship between patch size and avian species richness (for example, Lovejoy et al. 1986; Hagan & Johnston 1992), but there are fewer published papers that have quantified the relationship between patch size and nest success.

Study-site habitat types ranged from tropical forests in Central America (one study), coniferous forests in western North America (two studies), deciduous forests in eastern North America (five studies), and a mixture of habitat types in Europe (six studies; Table 1). Because nests were placed across a broad spectrum of different habitat types, they were exposed to a variety of potential predators (Table 1). Five of the studies (Yahner & Wright 1985; Angelstam 1986; Andrén & Angelstam 1988; Møller 1989; Yahner 1991) found that corvids were the primary predators at their study sites, whereas red squirrels (*Tamiasciurus hudsonicus*) were in Alberta (Boag et al. 1984), skunks (*Mephitis mephitis*) were in Maine (Vickery et al. 1992), and snakes were in grasslands of Illinois (Best 1978). The other studies did not identify a principal predator.

I tested whether nest success was independent of either distance from an edge or of patch size using a likelihood-ratio chi square (G) test (Agresti 1990). Many of the studies presented here originally used a likelihood-ratio test to analyze their data (Yahner & Wright 1985; Møller 1988, 1989; Berg et al. 1992). Each nest was considered an independent observation (although see below for a discussion of Gates & Gysel [1978] and Chasko & Gates [1982]). I considered p values <0.1 significant, as I wanted to be statistically conservative to minimize Type II errors (concluding there was no evidence for edge effects, when in fact they existed) and was not as concerned with Type I errors (concluding edge effects exist when they do not).

Most artificial nests studies reviewed here used either real chicken or quail (*Coturnix coturnix*) eggs, with

Table 1. Summary of primary predators and methods used in nest predation studies.

Reference	*Location*	*Primary Predators*[a]	*Type of Nest*[b]	*Type of Eggs (n)*[c]	*Exposure Days*
Artificial Nests					
Andrén & Angelstam 1988	Sweden	C[1] RF PM	OG	C (2)	8
Angelstam 1986	Sweden	C[1] RF PM	OG	C (2)	14
Avery et al. 1989	Scotland	ST RF C[2]	OG	C (1)	8
Berg et al. 1992	Sweden	ND	OG	Q (2)	28
Boag et al. 1989	Alberta	RS DM C[3]	OG	Q (5–7)	22
Burger 1988	Missouri	C[4] M[1]	W-CG	Q (1–3)	7
Carlson 1989	Missouri	ND	CG	Q (1)	7
Gibbs 1991	Costa Rica	M[2] B SN	W-OG/OE	Q (1)	7
Møller 1989	Denmark	C[5]	OG/CG	P (3)	7
Ratti and Reese 1988	Idaho	C[6] CH RS	W-OG/OE	Q (2)	12
Wilcove et al. 1986	Maryland	CH C[7] RF	W-OG	Q (3)	7–25
Small & Hunter 1988	Maine	RF RS SS	W-OG	C (3)	8
Storch 1991	W. Germany	M C[8]	OG	C (3)	35
Yahner & Wright 1985	Pennsylvania	C[4] R RF	OG	C (3)	6
Natural Nests					
Best 1978	Illinois	SN			
Chasko & Gates 1982	Maryland	CH RF C[7]			
Gates & Gysel 1978	Michigan	C[7] CH M B			
Yahner 1991	Pennsylvania	C[4]			
Vickery et al. 1991	Maine	SS			

[a] *B = birds, C = corvids* ([1]Corvus corax, C. cornix, C. monedula, Pica pica; [2]C. cornix; [3]Perisoreus canadensis; [4]C. brachyrhynchos; [5]C. cornix, P. pica, Garrulus glandarius; [6]Cyanocitta stelleri, P. c., C. corax, C. b.; [7]Cyanocitta cristata, C. b.; [8]C. corax, C. corone, G. g., Nucifraga caryocatactes), *CH = chimpmunk* (Tamais *spp.*), *DM = deer mouse* (Peromyscus maniculatus), *M = mammals* ([1]Didelphis virginiana, Procyon lotor, Mephitis mephitis, Sciurus carolinensis; [2]Nasua narica, Mustela frenata, Eira barbara, Potos flavus, Sciurus granatensis, Tayassu tajacu, [3]*PM, RF, RS,* Meles meles, Mustela nivalis), *ND = no data, PM = Pine Marten* (Martes martes), *R = red fox* (Vulpes vulpes), *RS = red squirrel* (Tamiasciurus hudsonicus), *SN = snakes, SS = striped skunk* (Mephitis mephitis), *ST = stoat* (Mustela erminea).
[b] *CG = covered, on ground, OE = open, elevated (1–3 m high), OG = open, on ground, W = in wicker basket.*
[c] *C = chicken, P = plasticine, Q = quail* (Coturnix coturnix).

clutch sizes ranging from one to seven eggs (Table 1). The only exception was Møller (1989), who used plasticine eggs to determine the species of corvid predators preying on nests based on bill markings on the artificial eggs. The majority of artificial nest studies exposed eggs to predators for 6–8 days (8 of 14; 57%) and used open, ground nests. This type of nest generally experiences the highest depredation rates (Ratti & Reese 1988; Gibbs 1991; Martin 1992; but see Yahner et al. 1989). For example, Møller (1989) observed that open nests were depredated at a higher rate than partially covered nests, 64% versus 36% respectively.

Distance as an Independent Variable

Predation Rates of Artificial Nests

I re-analyzed data from 14 studies, subdivided into 33 "treatments," that used artificial bird nests to investigate the relationship between distance from an edge and nest success (Table 2). Table 2 emphasizes the inconsistencies in the experimental designs of these studies. The large variance in the distance increments among the 14 studies makes it extremely difficult to ascertain anything conclusive about edge-effect patterns. Only one study, Ratti and Reese (1988), spaced nests at equidistant intervals to adequately determine where threshold values might occur (Table 2). Some studies did not place any artificial nests near the edge (≤50 m; Avery et al. 1986; Berg et al. 1992), others spaced nests relatively far apart (>300 m; Wilcove et al. 1986; Angelstam 1988; Gibbs 1991), several studied nests 1000 m or more from an edge (Angelstam 1986; Wilcove et al. 1986; Avery et al. 1989), and others placed nests only 50 m or less from an edge (Boag et al. 1984; Yahner & Wright 1985).

Notwithstanding this tremendous variation in experimental designs, a majority of the studies (10 of 14; 71%) had at least one treatment that found significant variation in nest success near an edge (Table 2). I next tested whether treatments conformed to Gates and Gysel's (1978) hypothesis that nest success decreased near edges and found most studies supported their hypothesis. Based on 19 treatments that found differential nest success near an edge, 15 had poorer nest success near an edge ($X^2 = 3.5$, d.f. $= 1$, $p = 0.062$; Table 2). One interesting exception was Storch (1991), who found that artificial nest success decreased farther from the edge and that Capercaillie (*Tetrao urogallus*) selected nest sites near the edge (<25 m), suggesting that the increased cover near the edge provided refuge from nest predators.

Based on my review of these studies, at least four generalizations concerning landscape patterns can be made. First, although more studies are needed with nests located nearer to edges, these data suggest that nest predation rates are greatest at less than 50 m from

Table 2. Studies investigating depredation rates of avian nests (% nest loss) as a function of distance from an edge.*

Distance from edge (m)													Type of	
0	25	50	75	100	200	300	500	1000	1500	>1500	P[a]	N[b]	edge[c]	References
Artificial Nests														
50		61		54		61					0.836	92	FA-CF	Andrén & Angelstam 1988
50		33		13			8				0.004	88	CF-FA	Andrén & Angelstam 1988
		67			59			70	60	61	0.885	150	CF-FA	Angelstam 1986
				44			37	33	30		<0.001	1512	MO-CF	Avery et al. 1989
		61				58					0.790	96	BO-WO	Berg et al. 1992
		38				7					0.002	64	BO-WO	Berg et al. 1992
		58				64					0.723	40	BO-WO	Berg et al. 1992
		35				61					0.015[f]	84	BO-WO	Berg et al. 1992
		20				9					0.057	88	BO-WO	Berg et al. 1992
		15				10					0.604	68	BO-WO	Berg et al. 1992
58	70										0.232	90	DF-TR	Boag et al. 1984
	65	37									<0.001	324	DF-FI	Burger 1988:30
		30	11								<0.001	270	PR-WO	Burger 1988:49
95					34	23	24				<0.001	400	DF-FI	Carlson 1989
		72		64	76	72					0.611	200	DF-FI	Carlson 1989
60							24				<0.001	147	TF-SG	Gibbs 1991[d]
10							24				0.073[f]	120	TF-FA	Gibbs 1991[d]
23							10				0.036	147	TF-SG	Gibbs 1991[e]
10							10				1.000	120	TF-FA	Gibbs 1991[e]
77	85				28						<0.001	180	FI-CF	Møller 1989[g]
40	40				30						0.417	180	FI-CF	Møller 1989[h]
77	75			55							0.019	180	CF-FI	Møller 1989[g]
40	47			25							0.039	180	CF-FI	Møller 1989[h]
	40	57	47	63	30	53					0.113	180	CF-FI	Ratti and Reese 1988[i]
17	44	64	77	71	65	67					<0.001[f]	191	CF-FI	Ratti and Reese 1988[j]
	13	10	3	0	7	0					0.057	180	FI-CF	Ratti and Reese 1988[i]
7	24	13	10	20	17	7					0.337	208	FI-CF	Ratti and Reese 1988[j]
											>0.100	508	CF-UK	Small and Hunter 1988[k]
45		68		75							0.003[f]	135	CF-CC	Storch 1991
29						20	15	15			0.612	84	DF-FI	Wilcove et al. 1986[l]
95							45				<0.001	40	DF-FI	Wilcove et al. 1986[m]
100							60				<0.001	40	DF-FI	Wilcove et al. 1986[n]
55[k]		55[k]									>0.360	384	DF-CC	Yahner and Wright 1985
Natural Nests														
90		82	96	83	100						0.442	135	PR-WO	Best 1978[o]
51	37	30	31								0.116	209	DF-CC	Chasko and Gates 1982[p]
33		41	12								0.004	136	DF-CC	Chasko and Gates 1982[p]
	27	20		11							<0.001	562	DF-FI	Gates and Gysel 1978[p]

Table 2. Continued.

Distance from edge (m)													Type of	
0	25	50	75	100	200	300	500	1000	1500	>1500	P[a]	N[b]	edge[c]	References
											<0.100	350	PR-WO	Johnson & Temple 1990[k]
		82		42	30						<0.001	276	DF-FI	Temple and Cary 1988
60	50	41									0.26	95	DF-CC	Yahner 1991
	33		67	62	38	88	75	50	50		0.40	60	GR-DF	Vickery et al. 1992

* P *values are based on re-analysis for this summary and may differ from original published values.*
[a] *Likelihood ratios (G^2) testing whether nest fate was independent of distance from edge.*
[b] *Total number of nests.*
[c] *Nests were located in boldface habitat: CF = conifer forest, CC = clearcut, DF = Deciduous forest, FI = Field, FA = Farmland, GR = grassland, MO = moorland, OG = old-growth forest, PR = prairie, TF = tropical forest, TR = Trail, UK = unknown, WO = woodland.*
[d] *Ground nest.*
[e] *Elevated nests.*
[f] *Nest loss was greater farther from edge.*
[g] *Open ground nests.*
[h] *Covered ground nests.*
[i] *Abrupt edge.*
[j] *Feathered edge.*
[k] *Exact predation rates not given; only* P *values were presented.*
[l] *7-d experiment.*
[m] *14-d experiment.*
[n] *25-d experiment.*
[o] *Based on my analysis of Best 1978, Fig. 4.*
[p] *% loss calculated for each egg, not for each nest.*

an edge, based on three lines of evidence. Only within 25 m of an edge was there more than 80% nest loss (Table 2). Studies that did not have nests located less than 50 m and more than 50 m from an edge often did not find evidence for edge effects (Boag et al. 1984; Yahner & Wright 1985; three of six treatments by Berg et al. 1992). Finally, the only two studies that had nests spaced relatively close together found that edge effects occurred less than 50 m from an edge (Burger 1988; Ratti & Reese 1988).

Other studies that found support for edge effects were more difficult to interpret. For example, Wilcove et al. (1986:251) inferred a linear decline in depredation rates up to 600 m from an edge and suggested edge effects could extend from 300 to 600 m into the forest. Yet my re-analysis of his data found that only nests exposed for 14 and 25 days at 0 and 600 m from an edge showed statistically significant evidence for edge effects (Table 2). There is no reason to believe this was a linear relationship, however, and a threshold value could occur anywhere from 0 to 600 m from the edge. A similar dilemma occurs with data gathered by Møller (1989), Gibbs (1991), and Berg et al. (1992), as their nests were distributed 100 m or more apart.

A second generalization is that significant edge effects were as likely to occur in unforested habitats (7 of 13 treatments) as in forested habitats (13 of 20 treatments) ($G = 0.41, p > 0.5$; Table 2). Third, depredation rates in northern conifer forests appear to be greater in forested habitats than in adjacent unforested habitat (Andrén & Angelstam 1988; Ratti & Reese 1988), although work by Møller (1989) did not support this idea. This may be because corvids are among the primary nest predators in boreal forests, and unforested habitat lacks perches used by many avian predators for nest searching (Ratti & Reese 1988).

Current biotic studies are inconclusive concerning the influence of the type of edge on edge effects. Ratti and Reese (1988) studied plots with an abrupt edge and a "feathered edge" (that is, a shelterwood timber removal). They found predation rates were greater on either side of a feathered edge compared to an abrupt edge, although predation rates within the feathered edge were relatively low. In contrast, Yahner et al. (1989) found no variation in nest success based on edge contrast.

Although my re-analysis of 10 of the 14 studies found statistical support for edge effects, at least one study (Avery et al. 1986) probably had no biological significance. This was due to minimal variation in nest success (30–44%) coupled with a large sample size (Table 2). Also, once Avery et al. (1986) factored out vegetational differences, they found no evidence to support edge effects. I found two significant relationships in the data of Ratti and Reese (1988), where they reported none. One discrepancy was due to the conservative p value I used, and the other was their feathered-edge forest plot, where I included an edge datum (17%) in my computations and they excluded it.

It is my opinion that three of four studies that found no evidence for edge effects were dubious tests of any edge effects hypothesis. Two studies (Boag et al. 1984;

Yahner & Wright 1985) did not place nests far enough away from an edge; that is, their artificial nests were potentially located within a relatively homogeneous band of high predation pressure, and predation rates might have declined beyond the area they sampled. Boag et al. (1984) compared predation rates only for nests less than 15 m to those greater than 20 m from a trail. Yahner and Wright (1985) placed nests at 0 m and 50 m from the edge of 1-ha forest fragments, in stands they concluded were composed entirely of edge habitat.

Another type of problem was the relatively arbitrary definition many researchers used to define an edge. Angelstam (1986) placed nests 20–40 m from roadsides at 500 m intervals yet assumed that nests were located as far as 1.5 km from the forest-farmland edge. I believe Angelstam's (1986) data might have been more properly classified as less than 50 m from an edge, because nests were placed near a road. This is especially true because corvids often follow roads to look for potential prey, and 75% of the nests in this study were taken by corvids (Angelstam 1986:367). The fourth study that found no evidence for edge effects, Small and Hunter (1988), did not present their data in a format conducive to re-analysis.

Ratti and Reese (1988) had a rigorous experimental design with one potential problem. They had two study plots, 2.4 ha and 6.2 ha, which were both small enough to be within the confines of a corvid territory. A different experimental design, which used a larger number of study plots and increased the distance between artificial nests, would reduce the probability that a minimal number of predators were responsible for all depredations (see Laurance & Yensen [1991] for further discussion).

Predation Rates of Natural Nests

Four of the seven studies (57%) that examined natural nest success had at least one significant treatment supporting the existence of edge effects (Table 2). In contrast to artificial nest experiments, natural nests were usually not studied farther than 100 m from an edge, with the exception of the studies by Temple and Cary (1988) and Vickery et al. (1992). Nevertheless, edge effects were still found at this relatively small scale, again suggesting that edge effects generally occur close to edges.

Johnson and Temple (1990) had two distance categories, less than 45 m or 45 m or more, and they found that nests near an edge had higher depredation rates. Their data presentation precluded further analyses, however, because nest success was based on exposure days rather than apparent nest success. Best (1978: Figure 4) did not directly analyze edge effects, but I estimated edge effects based on his sketch of the location of nests. Predation pressure was very high at his shrub-grassland study area, with 76% of all nests lost to snakes (Best 1978). Failed nests were evenly distributed throughout his study area. Vickery et al. (1992) did not find any evidence for edge effects in grasslands where striped skunks were the principal predator. In fact, nest success was greatest at less than 50 m from the woodland edge, although this relationship was not statistically significant.

Gates and Gysel (1978) and Chasko and Gates (1982) found a positive relationship between nest hatching success and distance from the edge. However, data from both studies should be interpreted with caution. They presented predation rate information for each egg rather than for individual nests (J. E. Gates, personnel communication), making re-analysis of their data impossible. For example, Gates and Gysel (1978:876) calculated fledging success based on 562 eggs rather than 194 nests. Individual nests were the independent sampling unit, because once one egg was taken from a nest, the probability increases that all eggs in the nest will be preyed on.

Parasitism Rates of Natural Nests

Three of five studies concluded that parasitism rates of Brown-headed Cowbirds (*Molothrus ater*) declined away from edges (Table 3). Gates and Gysel's (1978) data suggested a similar relationship, but their trend was not significant.

Johnson and Temple (1990) found all five potential host species at their tallgrass prairie site were more likely to be parasitized by cowbirds when nests were less than 45 m from a wooded edge and in small fragments of prairie (<32 ha). Best (1978:18) studied nesting Field Sparrows (*Spizella pusilla*) in the shrub-grasslands of central Illinois and found that parasitized nests averaged 13.4 m from an adjacent woodland, whereas successful nests averaged 31.5 m from the edge. Brittingham and Temple (1983) worked in a 1000-ha deciduous forest in Wisconsin and found that cowbirds were much more likely to parasitize nests in open habitat (>40% open, forest canopy within 200 m of nests) compared to contiguous forested habitat (<20% open).

Further evidence suggesting that habitat fragmentation results in increased cowbird nest parasitism is provided by Robinson (1992), who worked in a fragmented landscape in Illinois and found that 76% of neotropical migrant nests were parasitized. In contrast, Sherry and Holmes (1992) have not documented any cowbird parasitism in over 20 years of research at their field site in an unfragmented forest in New Hampshire.

Patch Size as an Independent Variable

Predation of Artificial and Natural Nests

All eight studies I reviewed demonstrated a positive relationship between patch size and nest success, with the

Table 3. Summary of studies investigating cowbird parasitism rates (% nests parasitized) of natural avian nests as a function of distance from an edge.

Distance from edge (m)												Habitat	
0	25	50	75	100	200	300	500	1000	>1000	P[a]	N[b]	type[c]	References
		65		46	36	18				0.004	105	DF	Brittingham and Temple 1983[d]
	17	5	0	0	0					0.013	171	PR	Best 1978[e]
		25	14							0.015	350	PR	Johnson and Temple 1990
7	0	19	6							0.216	62	DF	Chasko and Gates 1982
	15	8	14							0.445	40	DF	Chasko and Gates 1982
12	10	3	6							0.395	164	DF	Gates and Gysel 1978

[a] *Likelihood ratio (G^2) testing whether parasitism rates were independent of distance from an edge.*
[b] *Total number of nests.*
[c] *DF = deciduous forest, PR = prairie.*
[d] *Edge was considered any gap in the forest canopy ≥0.2 ha (not 0.02 ha in the paper, S. Temple, personal communication).*
[e] *Based on my analysis of Best 1978, Figure 4.*

exception of one treatment of Møller (1988) (Table 4). As with studies that investigated distance as an independent variable, there was a great deal of variation in the size spectrum and habitat types of patches, making generalizations difficult (Table 4). Burger (1988) tested for edge and area effects simultaneously and found that predation rates were more closely associated with distance from an edge. Much of the variation in depredation rates was probably due to variation in predator abundances associated with individual habitat patches (Table 1). For example, Wilcove (1985) found higher predation rates in urban forest fragments than in rural fragments, and Møller (1988) showed that predation rates increased regardless of patch size when a breeding pair of magpies (*Pica pica*) was present. For the most part, only patches less than 10 ha in size had relatively high depredation rates (≥50% nest predation rates; Table 4). If one assumes that edge effects occur 50 m or less from an edge,

Table 4. Summary of studies investigating depredation rates of avian nests (% nest loss) as a function of island size.

Island size (ha)												*Forest*	
<1	*<5*	*10*	*25*	*50*	*100*	*200*	*300*	*1000*	*>1000*	*P[a]*	*N[b]*	*type[c]*	*References*
Artificial nests													
83	46	54	59	35			37			<0.001	324	DF	Burger 1988:28
		39	19	9	15		30			<0.001	270	PR	Burger 1988:50
		46				25				<0.001	250	TR	Gibbs 1991[d]
		33				13				<0.001	250	TR	Gibbs 1991[e]
										<0.100	350	PR	Johnson & Temple 1990[f]
83 78 72 44 55 44										0.040	108	DF	Møller 1988[g]
61 72 50 39 11 17										<0.001	108	DF	Møller 1988[h]
		31			19	7				<0.001	2684	DF	Telleria & Santos 1992
			33	66	16 11 12	27	9	3		<0.001	506	DF	Small & Hunter 1988
	71[i]48[j]					48	18		2	<0.001	360	DF	Wilcove 1985[k]
68	19	9								<0.001	420	DF	Yahner and Scott 1988
Natural nests													
100 60 50 44 32 31										0.031	89	DF	Møller 1988[g]
50 36 37 25 26 20										0.650	80	DF	Møller 1988[h]

[a] *Likelihood ratio (G^2) testing whether nest fate was independent of island size.*
[b] *Total number of nests.*
[c] *TR = tropical forest, DF = deciduous forest, CF = conifer forest.*
[d] *Ground nests.*
[e] *Elevated nests.*
[f] *Unable to re-analyze, only* p *value presented.*
[g] *Magpies absent.*
[h] *Magpies present.*
[i] *Mean for suburban woodlots.*
[j] *Mean for rural woodlots.*
[k] *Not all values for this study are included; see original reference.*

then an 10-ha patch would maintain 4.7 ha of core habitat with reduced predation pressure, whereas assuming 100-m-wide edge effects results in a core area of only 1.4 ha.

Discussion

Our current understanding of the biotic and abiotic effects of edges is still in its infancy (Laurance & Yensen 1991; Saunders et al. 1991; Bierregaard et al. 1992; Kremsater & Bunnell 1992). Much of the research on the potential changes in depredation rates near edges is based on work with artificial nests. Research with artificial nests should be interpreted with caution, however, because there is evidence to suggest that avian predators using visual cues are much more likely to prey on artificial nests, whereas mammalian predators using olfactory cues are primarily attracted to natural nests (Willebrand & Marcström 1988). Therefore, there may be a poor correlation between the predation rates of artificial nests and natural nests (see Reitsma et al. 1990), and studies using artificial nests may be biased towards predators using visual cues. In fact, when researchers were able to identify the principal predator, many studies I reviewed found that avian predators were responsible for the majority of artificial nest losses. However, artificial nests may provide an estimate of relative predation rates.

There are wide discrepancies in the vegetation characteristics researchers use to classify edges; therefore, there is a need to formalize the criteria used to include patches in edge analyses. For example, Boag et al. (1984) determined that man-made and game trails were wide enough to form edges in lodgepole pine (*Pinus contorta*) forests, whereas Angelstam et al. (1986) ignored edges created by roads in boreal forests and assumed edges were the nearest farmland patch. Burger (1988) used any woody vegetation to define edges at her prairie study site. Brittingham and Temple (1983) used 0.2-ha (45-m radius) openings in the forest canopy to define the source of edge habitat, as this patch size was the smallest they could detect from their aerial photos. This latter definition is purely a function of the scale of aerial photos, however, rather than a meaningful biological scale. Finally, many researchers neglect to provide readers with an accurate assessment of the landscape where the nest studies took place.

Using the forest canopy height to classify edges would be one way to eliminate ornithologists' past arbitrary criteria, because all studies I reviewed focused on the juxtaposition of forested (tropical, deciduous, coniferous) and unforested habitat (moorlands, fields, farmlands, clearcuts, prairies, grasslands). The silviculture literature suggests that the microclimatic conditions at the center of an opening with a diameter 2–3 times the height of the surrounding trees are similar to those of larger openings (Smith 1986:206). Therefore, researchers could include in edge analyses any opening in the forest canopy with a diameter three or more times the adjacent tree height, while smaller openings would be excluded. However, certain species, such as cowbirds, might perceive smaller openings as edges (Brittingham & Temple 1983). These same criteria could be used for unforested habitats (for example, prairies) with small woodlands, by ignoring woodlands with a diameter less than three times the mean tree height. Another approach would be for researchers to clearly state the biological reasons why a particular habitat feature is an edge to the organisms under investigation.

In addition, all the studies in this review focused on anthropogenic changes in the landscape, so called "induced" edges, even though there are also natural, long-lived, "inherent" edges (Yahner 1988). The biological distinctions between inherent and induced edges are relatively unexplored and need further research.

My review of the existing literature suggests that nest success declines near edges, but on a smaller scale than some authors have suggested (for example, Wilcove et al. 1986). The current evidence, although equivocal, suggests that predation and parasitism rates are often significantly greater within 50 m of an edge, and studies suggesting that edge effects occur beyond this potential threshold value are less convincing. This conclusion should be interpreted with caution, however, due to the lack of consistency in the experimental designs of these studies.

This leads to another important point. Research on abiotic and vegetational processes suggests that edge effects generally occur less than 50 m into a forest stand (Ranney et al. 1981; Kapos 1989; Laurance & Yensen 1991). Therefore, given this empirical evidence and the fact that my analyses suggest nest edge effects usually occur within 50 m of an edge, ornithologists should probably focus future edge research on a much smaller scale to quantify threshold values for depredation rates of artificial and natural nests. More data are needed within 100–200 m of the ecotone between forest and unforested habitat, and at smaller distance increments (about 20–25 m). Laurance and Yensen (1991) also provide some useful insights into study designs.

Two aspects of edge effects that have been relatively unexplored are the relationship between predator densities and nest success and how nest predators (or brood parasites) search for nests. Andrén (1992) found marked variation in the species of corvid predators using different habitat types in Sweden. Møller (1989) observed that Hooded Crows (*Corvus cornix*) wandered across habitat types, magpies concentrated their activities along the edge, and jays (*Garrulus glandarius*) were woodland specialists, indicating that knowledge of the autecology of nest predators in a particular habitat is vital to understanding potential depredation problems. Norman and Robertson (1975) reported that female cowbirds often remain motionless for hours searching for potential hosts nests; therefore, perches that provide a good view of surrounding habitat may be a critical habitat feature to female cowbirds. This is possibly why cowbirds usually only parasitized nests within 13 m of the woods-grassland ecotone (Best 1978). Normand and Robertson (1975) also reported that cowbirds actively search edge habitat trying to flush incubating birds.

Perches with a clear view of the surrounding habitat are probably also very important to nest predators (Ratti & Reese 1988; Møller 1989). Avery et al. (1989) found no biological evidence for edge effects when placing nests 100 m or more from the woodlands-ecotone. However, if avian predators use the woods for perch sites to locate nests in adjacent unforested habitat, 100 m might have been too far to find hidden artificial nests. The same argument holds true for Berg et al. (1992), and nests possibly should have been placed within 50 m of the woods to determine if edge effects occurred on a smaller scale.

Clearly strong evidence exists that avian nest success declines near edges. But more data are needed concerning potential threshold values for edge effects in a variety of landscape patterns and habitat types (Laurance & Yensen 1991). In North America, I found no studies that investigated edge effects in the forests of the southeastern or northwestern United States, even though these two areas are among the largest timber producers on the continent. Finally, future research should focus on realistic landscape scales with solid experimental designs to quantify the spatial extent of nest depredation and parasitism.

Acknowledgments

I am grateful to Susan Durham for advice on the appropriate statistical methods to use on these type of data. Thomas C. Edwards, Jr., John A. Bissonette, Susan Durham, John Faaborg, Kimberly A. Sullivan, David S. Wilcove, and one anonymous reviewer provided comments on various drafts of this paper.

Literature Cited

Agresti, A. 1990. Categorical data analysis. Wiley and Sons, New York.

Andrén, H. 1992. Corvid density and nest predation in relation to forest fragmentation: A landscape perspective. Ecology **73**:794–804.

Andrén, H., and P. Angelstam. 1988. Elevated predation rates as an edge effect in habitat islands: Experimental evidence. Ecology **69**:544–547.

Angelstam, P. 1986. Predation on ground-nesting birds' nests in relation to predator densities and habitat edge. Oikos **47**:365–373.

Askins, R. A., J. F. Lynch, and R. Greenburg, 1990. Population declines in migratory birds in eastern North American. Current Ornithology **7**:1–57.

Avery, M. I., F. L. R. Winder, and V. M. Egan. 1989. Predation on artificial nests adjacent to forestry plantations in northern Scotland. Oikos **55**:321–323.

Berg, Å., S. G. Nilson, and U. Boström. 1992. Predation on artificial wader nests on large and small bogs along a north-south gradient. Ornis Scandinavica **23**:13–16.

Best, L. B. 1978. Field sparrow reproductive success and nesting ecology. Auk **95**:9–22.

Bierregaard, R. O., Jr., T. E. Lovejoy, V. Kapos, A. A. dos Santos, R. W. Hutchings. 1992. The biological dynamics of tropical rainforest fragments. Bioscience **42**:859–866.

Boag, D. A., S. G. Reebs, and M. A. Schroeder. 1984. Egg loss among spruce grouse inhabiting lodgepole pine forests. Canadian Journal of Zoology **62**:1034–1037.

Brittingham, M. C., and S. A. Temple. 1983. Have cowbirds caused forest songbirds to decline? Bioscience **33**:31–35.

Burger, L. D. 1988. Relations between forest and prairie fragmentation and depredation of artificial nests in Missouri. M.S. thesis. University of Missouri, Columbia, Missouri.

Carlson, K. 1989. Studies of artificial nest predation rates in Missouri. Unpublished summer research fellowship report. University of Missouri, Columbia, Missouri.

Chasko, G. G., and J. E. Gates. 1982. Avian habitat suitability along a transmission-line corridor in an oak-hickory forest region. Wildlife Monograph **82**:1–41.

Gates, J. E., and L. W. Gysel. 1978. Avian nest dispersion and fledging success in field-forest ecotones. Ecology **59**:871–883.

Gibbs, J. P. 1991. Avian nest predation in tropical wet forest: An experimental study. Oikos **60**:155–161.

Guthery, F. S., and R. L. Bingham. 1992. On Leopold's principle of edge. Wildlife Society Bulletin **20**:340–344.

Hagan, J. M., and D. W. Johnston. 1992. Ecology and conservation of neotropical migrant landbirds. Smithsonian Institution Press, Washington, D.C.

Johnson, R. G., and S. A. Temple. 1990. Nest predation and brood parasitism of tallgrass prairie birds. Journal of Wildlife Management **54**:106–111.

Kapos, V. 1989. Effects of isolation on the water status of forest patches in the Brazilian Amazon. Journal of Tropical Ecology **5**:173–185.

Kremsater, L. L., and F. L. Bunnell. 1992. Testing responses to forest edges: The example of black-tailed deer. Canadian Journal of Zoology **70**:2426–2435.

Laurance, W. F., and E. Yensen. 1991. Predicting the impacts of edge effects in fragmented habitats. Biological Conservation **55**:77–92.

Leopold, A. 1933. Game management. Charles Scribner's Sons, New York.

Lovejoy, T. E., R. O. Bierregaard, Jr., A. B. Rylands, J. R. Malcolm, C. E. Quintela, L. H. Harper, K. S. Brown, Jr., A. H. Powell, G. V. N. Powell, H. O. R. Schubart, and M. B. Hays. 1986. Edge and other effects of isolation on Amazon forest fragments. Pages 257–285 in M. E. Soulé, editor. Conservation biology. Sinauer Associates, Sunderland, Massachusetts.

Martin, T. E. 1988. Habitat and area effects on forest bird assemblages: Is nest predation an influence? Ecology **69**:74–84.

Martin, T. E. 1992. Breeding productivity considerations: What are the appropriate habitat features for management? Pages 455–473 in J. M. Hagan and D. W. Johnston, editors. Ecology and conservation of neotropical migrant landbirds. Smithsonian Institution Press, Washington, D.C.

Møller, A. P. 1988. Nest predation and nest site choice in passerine birds in habitat patches of different size: A study of magpies and blackbirds. Oikos **53**:215–221.

Møller, A. P. 1989. Nest site selection across field-woodland ecotones: The effect of nest predation. Oikos **56**:240–246.

Norman, R. F., and R. J. Robertson. 1975. Nest-searching behavior in the Brown-Headed Cowbird. Auk **92**:610–611.

Ranney, J. W., M. C. Bruner, and J. B. Levenson. 1981. The importance of edge in the structure and dynamics of forest islands. Pages 67–95 in R. L. Burgess and D. M. Sharpe, editors. Forest island dynamics in man-dominated landscapes. Springer-Verlag, New York.

Ratti, J. T., and K. P. Reese. 1988. Preliminary test of the ecological trap hypothesis. Journal of Wildlife Management **52**:484–491.

Reese, K. P., and J. T. Ratti. 1988. Edge effect: A concept under scrutiny. Pages 127-136 in 53rd Transactions of the North American Wildlife Resources Conference.

Reitsma, L. R., R. T. Holmes, and T. W. Sherry. 1990. Effects of removal of red squirrels, *Tamiasciurus hudsonicus*, and eastern chipmunks, *Tamias striatus*, on nest predation in a northern hardwood forest: An artificial nest experiment. Oikos **57**:375–380.

Robinson, S. K. 1992. Population dynamics of breeding neotropical migrants in a fragmented Illinois landscape. Pages 408–418 in J. M. Hagan and D. W. Johnston, editors. Ecology and conservation of neotropical migrant landbirds. Smithsonian Institution Press, Washington, D.C.

Sunders, D. A., R. J. Hobbs, and C. R. Margules. 1991. Biological consequences of ecosystem fragmentation: A review. Conservation Biology **5**:18–32.

Sherry, T. W., and R. T. Holmes. 1992. Population fluctuations in a long-distance neotropical migrant: Demographic evidence for the importance of breeding season events in the American Redstart. Pages 431–442 in J. M. Hagan and D. W. Johnston, editors. Ecology and conservation of neotropical migrant landbirds. Smithsonian Institution Press, Washington, D.C.

Small, M. F., and M. L. Hunter. 1988. Forest fragmentation and avian nest predation in forested landscapes. Oecologia **76**:62–64.

Smith, D. M. 1986. The practice of silviculture. Wiley & Sons, New York.

Storch, I. 1990. Habitat fragmentation, nest site selection, and nest predation risk in Capercaillie. Ornis Scandinavica **22**:213–217.

Temple, S. A., and J. R. Cary. 1988. Modeling dynamics of habitat-interior bird populations in fragmented landscapes. Conservation Biology **2**:340–347.

Vickery, P. D., M. L. Hunter, and J. V. Wells. 1992. Evidence of incidental nest predation and its effects on nests of threatened grassland birds. Oikos **63**:281–288.

Wilcove, D. S. 1985. Nest predation in forest tracts and the decline of migratory songbirds. Ecology **66**:1211–1214.

Wilcove, D. S., C. H. McLellan, and A. P. Dobson. 1986. Habitat fragmentation in the temperate zone. Pages 237–256 in M. E. Soulé, editor. Conservation biology. Sinauer Associates, Sunderland, Massachusetts.

Willebrand, T., and V. Marcström. 1988. On the danger of using dummy nests to study predation. Auk **105**:378–379.

Yahner, R. H. 1988. Changes in wildlife communities near edges. Conservation Biology **2**:333–339.

Yahner, R. H. 1991. Avian nesting ecology in small even-aged aspen stands. Journal of Wildlife Management **55**:155–159.

Yahner, R. H., and D. P. Scott. 1988. Effects of forest fragmentation on depredation of artificial nests. Journal of Wildlife Management **52**:158–161.

Yahner, R. H., and A. L. Wright. 1985. Depredation of artificial ground nests: Effects of edge and plot age. Journal of Wildlife Management **49**:508–513.

Yahner, R. H., T. E. Morrell, and E. S. Rachael. 1989. Effects of edge contrast on depredation of artificial avian nests. Journal of Wildlife Management **53**:1135–1138.

Yoakum, J., W. P. Dasmann, H. R. Sanderson, C. M. Nixon, and H. S. Crawford. 1980. Habitat improvement techniques. Pages 329–404 in S. D. Schemnitz, editor. Wildlife management techniques manual. The Wildlife Society, Washington, D.C.

The Effects of Fencelines on the Reproductive Success of Loggerhead Shrikes

REUVEN YOSEF*

Department of Zoology
The Ohio State University
Columbus, OH 43210, U.S.A. and
Archbold Biological Station
Lake Placid, FL 33852, U.S.A.

Introduction

The Loggerhead Shrike is important as an indicator species of environmental quality because of its exclusively predatory feeding habits and close association with agricultural areas. This shrike has been of special concern to conservationists for over two decades because it has undergone a steady decline in numbers since the turn of the century (Graber et al. 1973). Many factors have been implicated in the decline (see Anderson & Duzan 1978; Blumton 1989).

Numerous recommendations have been made for the enhancement of roadside grass strips and hedgerows for use by nesting birds (see Warner et al. 1987). Changes in agricultural practices and the subsequent loss of grasslands and hedgerows (Vance 1976) have lead to greater breeding densities in species that use fencerows as nest sites. Studies have found that these densities may be 10–30 times greater in fencerows than in more natural breeding habitat (Basore et al. 1986). To date, most studies have concentrated on the impact on game species of this human-induced breeding concentration in fencelines (see Warner et al. 1987). Lately, the use of such linear habitats has been documented for nongame species too, especially passerines (Basore et al. 1986).

One of the few passerine species so far studied in the context of linear habitats is the Loggerhead Shrike. Luukkonen (1987) found that shrikes nested closer to roadsides than expected, that the productivity of roadside pairs was half that of pairs breeding in other habitats, and that most losses were attributed to nest predation. DeGeus (1990) thought that linear habitats attracted birds to areas where heavy predation limited production to levels below those needed for replacement.

Fencelines on a working cattle ranch should be analogous to roadside hedgerows in that many "hedgerow-fenceline-nesting" species and their predators are concentrated at great densities. Also, cattlemen often prefer monoculture pastures, and thus trees and bushes are found almost exclusively along fencelines. Where nests of breeding species are concentrated in a linear fashion, it should be most profitable for a predator to search along such corridors. Such behavior by predators would result in a decrease in fitness levels of corridor-nesting prey species. I postulated that species nesting in fencelines are affected in a manner similar to those in other linear habitats (Santos & Telleria 1992). I tested the prediction that Loggerhead Shrike nests in trees and shrubs away from fencelines would suffer fewer losses from predation than would those in fenceline substrates.

Study Area and Methods

To assess the effects of linear habitats on Loggerhead Shrike reproductive success, I studied 27 breeding pairs

Paper submitted November 9, 1992; revised manuscript accepted April 5, 1993.

Present address: International Ornithological Center, P.O. Box 774, Eilat 88000, Israel.

during the 1992 breeding season (mid-February to mid-June) at the MacArthur Agro-ecology Research Center (MAERC) of the Archbold Biological Station, Florida. MAERC is a 4200-ha working cattle ranch with extensive Bahia-grass pastures. Barbed-wire fences bound the pastures. Through the years, cabbage palm (*Sabal palmetto*), live oak (*Quercus virginianus*), wax myrtle (*Myrcus cerifera*), and a few other species of trees and bushes have colonized these fencelines, creating a series of linear habitats that are used by shrikes for nesting. Numerous other bird species (such as Mockingbird, *Mimus polyglottos,* Northern Cardinal, *Cardinalis cardinalis,* Red-winged Blackbird, *Agelaius phoeniceus,* Boat-tailed Grackle, *Quiscalus major,* Rufous-sided Towhee, *Pipilo erythrophthalmus*) nested in fenceline vegetation during my study.

During the 1990 and 1991 breeding seasons, we trapped and banded shrikes for individual recognition with aluminum and color bands from the Fish & Wildlife Service. Color-marked shrike pairs were observed intensively during Spring 1992 in order to locate nests. Only those nests discovered within two days of the first egg being laid were included in the analysis, a procedure which greatly reduced uncertainty about the number of days of exposure to predation. Because predators were observed to forage up to 10 m away from fencelines, I classified nests as in a fencerow if they were situated within 10 m of a fenceline, and in pasture if situated beyond that distance. Snelling (1968) found that human visits to nests attracted predators. Therefore, I visited all nests at equal frequencies (every third day) to ascertain the stage of the reproductive cycle. Data were pooled for all pairs that nested in each of the two habitat types, so that probability of survival from the day an egg was laid to fledging could be calculated using Mayfield's exposure-day method (1961, 1975).

I searched the vicinity of depredated nests for signs of the predator. In all cases, I ascribed predation to mammals if the nest structure was damaged and to avian or reptilian predators if the nest was intact and only the contents were missing. In addition to examining habitat type as a factor influencing reproductive success, I checked for an effect of distance from fenceline on reproductive success, and on the proportion of eggs laid that was later lost to predation. I used each nesting attempt as an independent sampling unit and compared various measures of reproductive effort in an attempt to isolate incompetent pairs (at hiding or defending the nest) within a habitat and between habitats, to verify that no pair biased my analyses. Unless otherwise stated, data are presented as means ± SD. All percentages were arc-sine transformed prior to analysis. I employed simple regression and Chi-square tests for comparison of data from the two nesting habitats, and chose $p = 0.05$ as the minimum acceptable level of significance.

Results

During the 1992 breeding season, the first nest was initiated on February 19, and the nesting season ended on June 18 when the last successful brood fledged. The 27 Loggerhead Shrike pairs laid 64 clutches (four pairs laid more than one replacement clutch), of which 22 (34%) were renestings after a previous clutch had been lost to either predation or inclement weather, and 15 (23%) were second clutches after the first brood had successfully fledged. The overall nesting success was 51%. Predation accounted for 30 (86%) of the 36 nests destroyed. Mammals destroyed 21 (60%) nests, and avian or reptilian predators nine (26%) nests. Clutch size averaged 3.8 ± 0.8 eggs (range 2–5). A total of 245 eggs was laid, of which 113 (46%) hatched and 90 (37%) fledged (an average of 3.33 young/pair/season).

Sixteen pairs nested in fenceline vegetation. Nests were placed in cabbage palm, live oak, wax myrtle, blackberry (*Rubus betulifolius*), or southern elderberry (*Sambucus canadensis*). Of the 16 first clutches, nine were depredated and two were lost to inclement weather. Replacement clutches were laid within 10 days of loss. Thirteen of the 17 replacement clutches laid in fencelines were depredated. Five (31%) pairs successfully fledged two broods (Table 1). The average incubation period for successful nests was 15.1 ± 1.2 days, and the average nestling period for successful nests was 15.8 ± 0.8 days.

Of the 16 first clutches, 17 replacement and six second clutches (total 39 clutches) placed in fenceline vegetation, 25 failed during an aggregate of 412 nest-days during incubation, a daily mortality rate of 0.061 nests per day. Predation accounted for 22 nest failures, and three nests were destroyed by inclement weather. The probability of a clutch surviving from the day the first egg was laid until hatching was 39.0 ± 0.5%.

Nest mortality during the nestling period was 1.0%

Table 1. Reproductive success of Loggerhead Shrikes nesting in fencelines and pastures in south-central Florida in 1992.

Parameter	*Fenceline*	*Pasture*	*Combined*
Clutch size	3.6 ± 0.9 (39)	4.0 ± 0.8 (25)	3.8 ± 0.9
Incubation Period (days)	9.7 ± 4.6 (39)	11.9 ± 3.2 (25)	10.6 ± 4.2
Nestling Period (days)	5.0 ± 7.1 (39)	9.7 ± 7.6 (25)	6.8 ± 7.6
Nestlings/Nest	1.3 ± 1.8 (39)	2.6 ± 2.0 (25)	1.8 ± 2.0
Fledglings/Nest	0.9 ± 1.5 (39)	2.1 ± 1.9 (25)	1.4 ± 1.7
Fledglings/Pair	2.2 ± 2.6 (16)	4.8 ± 3.1 (11)	3.3 ± 3.1
Hatching Success	35.0 (53)	61.2 (63)	46.1
Fledging Success	26.0 (37)	38.8 (40)	31.4
% Nests that Fledged >1 Young	33.3 (13)	60.0 (15)	42.2
% Pairs that Fledged >1 Young	56.3 (9)	81.8 (9)	66.7

Data are presented as means ± SD, and sample sizes are in parentheses.

per nest-day, and the survival rate to fledging of young from a nest that had successfully hatched young was 92.0 ± 0.4%. The overall nest survival for a nest placed in a fencerow was 36%, and the daily survival rate was 94.0 ± 5.3%. The probability of an egg surviving to hatch was 13.0 ± 0.2%, and of a nestling fledging, 67.0 ± 1.8%. The overall production rate was 8.0 ± 0.5%.

Eleven pairs nested in trees or bushes that were dispersed in pastures. Nests were placed in blackberry bushes, cabbage palms, or southern elderberry. Of the 11 first clutches, four were depredated and two were lost to inclement weather. In all six cases, replacement clutches were laid within seven days. Two of five replacement clutches were depredated. Seven (64%) pairs successfully fledged two broods during the nesting season (Table 1). The average incubation and nestling periods for successful nests were 14.7 ± 0.5 and 15.6 ± 0.5 days, respectively.

Of the 25 nests placed in vegetation away from fencelines, seven failed during incubation. Daily mortality rate was 3% (318 nest-days). Predation accounted for the failure of eight nests, and two others were destroyed by inclement weather. The overall probability of a nest surviving to the hatching stage was 64.0 ± 22.6% (Table 2).

Nest mortality during the nestling period was 1.0% per nest-day, and the survival rate of nestlings was 85.0%. The cumulative nest survival for a nest placed away from fencerows was 54.0 ± 10.2%. The overall probability of an egg surviving the incubation period was 64.0 ± 30.1%, and of a hatched nestling fledging, 73.0 ± 28.8%.

The overall production rate for shrike nests in combined habitats was 40.0 ± 16.7%. The proportion of offspring preyed on was 0.62 in fencelines and 0.30 in pastures. Pasture eggs had a five-fold greater chance of survival than did fenceline eggs (40% versus 8%, Table 2). The probability of a nest surviving through the nesting cycle was greater for pasture nests (54% compared to 36%), probably because the probability of an egg surviving to the hatching stage in pastures was almost double that in fencelines (64% compared to 33%). Reproductive success increased significantly with distance from fencelines ($r^2 = 0.094$, 63 DF, $p = 0.0136$). Fledging success, once the nestling stage had been reached, was not very different between the two nesting habitats (86% compared to 73%). However, both the number of young fledged and the overall production were significantly greater for pasture nests (Chi square test, $p = 0.0001$).

In order to examine whether birds nesting in fencelines were of lower quality, various measures of reproductive effort were compared. No statistically significant differences were found between pairs in the two habitats in clutch size ($z = -0.8$, $p = 0.4236$), incubation period ($z = -0.9$, $p = 0.367$), nestling period ($z = -0.7$, $p = 0.5002$), or number of unhatched eggs ($z = -1$, $p = 0.3173$).

Table 2. Mayfield estimates of probabilities of survival per day for Loggerhead Shrike nests in fencelines and pastures in south-central Florida in 1992.

Parameter	Fenceline	Pasture
Daily Nest Survival (incubation)	0.94	0.97
Daily Nest Survival (nestling)	0.99	0.99
Cumulative Nest Survival	0.36	0.54
Daily Egg Survival	0.93	0.97
$P_{Egg\ Surviving\ Incubation}$*	0.39	0.62
$P_{Survival\ During\ Hatching}$	0.94	0.90
Daily Nestling Survival	0.98	0.99
% Nest Survival	0.67	0.73
Production*	0.08	0.42

Probabilities were calculated from 39 nesting attempts by 16 pairs in fencelines and on 25 attempts by 11 pairs in pastures. Production is the probability that eggs at the start of incubation will produce fledgings. Asterisks denote statistically significant differences.

Discussion

These results support the hypothesis that Loggerhead Shrikes nesting away from linear habitats suffer fewer nest losses and therefore attain greater reproductive success than shrikes nesting in fencelines. The probability of nest survival for the pairs that nested out in pastures in south-central Florida was similar to data for natural habitats in 10 other studies (Table 3). However, the probability of nest survival for the pairs nesting in fencelines (36%) was the lowest documented. Also, a significant relationship existed between nest survivability and distance from the fenceline. These results indicate that pairs nesting in fencelines suffered greater predation than those in pastures.

An alternative explanation to the increased-predation

Table 3. Percent nest survival for Loggerhead Shrikes in North America.

State	$P_{nest\ survival}$	Source
Alabama	43	Siegel 1980
Colorado	66	Porter et al. 1975
Florida (Pastures)	54	Present study
(Fencelines)	36	
(Total)	51	
Illinois	80	Graber et al. 1973
Illinois	72	Anderson and Duzan 1978
Indiana	57	Burton 1990
Minnesota	62	Brooks and Temple 1990
Missouri	69	Kridelbaugh 1982
New York	50	Novak 1989
Oklahoma	46	Tyler 1992
South Carolina	65	Gawlik 1988
South Carolina	75	Gawlik and Bildstein 1990
Virginia	62	Luukkonen 1987
Virginia	55	Blumton 1989

hypothesis for nests in fencelines is that the birds there were of low quality and had been excluded from better habitats by higher quality conspecifics. However, a comparison of nesting parameters (clutch size, incubation period, nestling period, unhatched eggs/nest; Table 1) showed no statistically significant differences between the pairs nesting in the two habitat patch types.

Goransson et al. (1975) found that nest predation was positively related to nest density and suggested that predator learning was one likely explanation for higher predation rates on more densely packed nests. Martin (1988) found that predators increased their searching intensity when they encountered an increasing frequency of nests containing eggs. These findings are reinforced by the fact that "systematic" predators use linear habitats as travel corridors and are thus able to find nests easily when they are at high densities (Crabtree et al. 1989). Although I have no quantitative documentation, I observed that most potential mammalian predators (such as raccoon, *Procyon lotor,* opossum, *Didelphis marsupialis,* bobcat, *Lynx rufus*), reptilian predators (such as indigo snake, *Drymarchon corais,* yellow rat snake, *Elaphe obsoleta,* corn snake, *E. guttata*), and avian predators (such as Audubon's Crested Caracara, *Polyborus plancus cheriway,* Barred Owl, *Strix varia*) of shrike eggs and nestlings in my study area walked or flew along fencelines, and they appeared to avoid crossing open pastures. Thus, even if not intentionally concentrating their search for nests in fencelines, they increased their chances of accidentally finding nests there. Crabtree et al. (1989) found that striped skunks (*Mephitis mephitis*) specifically searched for nests in linear strips. Audubon's Crested Caracaras perch on fencelines and intently follow the activities of parent shrikes to tending nests, apparently in an attempt to find the nest (James N. Layne, personal communication).

Although the combined reproductive success (51%) of Loggerhead Shrikes in this study is at the lower end of the results from other studies (Table 3), it is representative for the species. However, the nest success of fenceline pairs compared with that of pasture pairs (33% versus 60%) has important implications for conservation and management efforts.

In Florida, rapid conversion of natural habitats to pasture, and more recently to citrus groves, has created evolutionarily instantaneous changes in habitat structure and vegetation. Conversion to pasture increases linear habitats (fencelines and hedgerows) and reduces natural vegetation, forcing most tree- and shrub-nesting passerines to the only available habitat, the remaining fencelines and hedgerows. Such fencelines are the major travel corridors available to predator species (Crabtree et al. 1989), thus increasing the predation frequency on the nesting species.

In conclusion, this study illustrates that hedgerows and fencelines are not necessarily optimal or even adequate nesting habitat, contrary to some earlier assumptions. If we wish to preserve biodiversity in general, and Loggerhead Shrikes in particular, it is important to distinguish habitats that are "sinks" and to delineate the factors that make them so.

Acknowledgments

I gratefully acknowledge financial support from the Frank M. Chapman Award of the American Museum of Natural History, and research grants from the Animal Behavior Society and the American Ornithologist's Union. Comments of T. C. Grubb, Jr., K. Tarvin, T. A. Bookhout, A. S. Gaunt, J. Bart, E. G. Bolen, and an anonymous reviewer improved earlier drafts of this manuscript. This is contribution 14 of the MacArthur Agroecology Research Center of the Archbold Biological Station.

Literature Cited

Anderson, W. C., and R. E. Duzan. 1978. DDE residues and eggshell thinning in Loggerhead Shrikes. Wilson Bulletin. **90**:215–220.

Basore, N. S., L. B. Best, and J. B. Wooley, Jr. 1986. Bird nesting in Iowa no-tillage and tilled crop-land. Journal of Wildlife Management **50**:19–28.

Blumton, A. K. 1989. Factors affecting Loggerhead Shrike mortality in Virginia. M. S. thesis. Virginia Polytechnic Institute and State University, Blacksburg, Virginia.

Brooks, B. L., and S. A. Temple. 1990. Dynamics of Loggerhead Shrike population in Minnesota. Wilson Bulletin **102**:441–450.

Burton, K. M. 1990. An investigation of population status and breeding biology of the Loggerhead Shrike in Indiana. M.S. thesis. Indiana University, Bloomington, Indiana.

Crabtree, R. L., L. S. Broome, and M. L. Wolfe. 1989. Effects of habitat characteristics on Gadwall nest predation and nest-site selection. Journal of Wildlife Management **53**:129–137.

DeGeus, D. W. 1990. Productivity and habitat preferences of Loggerhead Shrikes inhabiting roadsides in a midwestern agroenvironment. M.S. thesis, Iowa State University, Ames, Iowa.

Gawlik, D. E. 1988. Reproductive success and nesting habitat of Loggerhead Shrikes and relative abundance, habitat use, and perch use of Loggerhead Shrikes and American Kestrels in South Carolina. M.S. thesis. Winthrop College, Rock Hill, South Carolina.

Gawlik, D. E., and K. L. Bildstein. 1990. Reproductive success and nesting habitat of Loggerhead Shrikes and in north-central South Carolina. Wilson Bulletin **102**:37–48.

Goransson, G., J. Karlsson, S. G. Nilsson, and S. Ulfstrand. 1975. Predation on birds' nests in relation to antipredator aggression and nest density: An experimental study. Oikos **26**:117–120.

Graber, R. R., J. W. Graber, and E. L. Kirk. 1973. Illinois birds: Laniidae. Biology Notes 83, Illinois Natural History Survey, Urbana, Illinois.

Kridelbaugh, A. L. 1982. An ecological study of Loggerhead Shrikes in central Missouri, M.S. thesis. University of Missouri, Columbia, Missouri.

Luukkonen, D. R. 1987. Status and breeding ecology of the Loggerhead Shrike in Virginia. M.S. thesis. Virginia Polytechnic Institute, Blacksburg, Virginia.

Martin, T. E. 1988. On the advantage of being different: Nest predation and the coexistence of bird species. Proceedings of the Academy of Sciences **85**:2196–2199.

Mayfield, H. 1961. Nesting success calculated from exposure. Wilson Bulletin **73**:255–261.

Mayfield, H. 1975. Suggestions for calculating nesting success. Wilson Bulletin **87**:456–466.

Novak, P. G. 1989. Breeding ecology and status of the Loggerhead Shrike in New York State. M.S. thesis. Cornell University, Ithaca, New York.

Porter, D. K., M. A. Strong, J. B. Giezentanner, and R. A. Ryder. 1975. Nest ecology, productivity and growth of the Loggerhead Shrike on the shortgrass praire. Southwest Naturalist **19**:429–436.

Santos, T., and J. L. Telleria. 1992. Edge effects on nest predation in Mediterranean fragmented forests. Biological Conservation **60**:1–5.

Siegel, M. S. 1980. The nesting ecology and dynamics of the Loggerhead Shrike in the blackbelt of Alabama. M.S. thesis. University of Alabama, Tuscaloosa, Alabama.

Snelling, J. C. 1968. Overlap in feeding habits of Red-winged Blackbirds and Common Grackles nesting in a cattail marsh. Auk **85**:560–585.

Tyler, J. D. 1992. Nesting ecology of the Loggerhead Shrike in southwestern Oklahoma. Wilson Bulletin **104**:95–104.

Vance, D. R. 1976. Changes in land use and wildlife populations in southeastern Illinois. Wildlife Society Bulletin **4**:11–15.

Warner, R. E., G. B. Joselyn, and S. L. Eetter. 1987. Factors affecting roadside nesting by pheasants in Illinois. Wildlife Society Bulletin **15**:221–228.

Forests Too Deer: Edge Effects in Northern Wisconsin

WILLIAM S. ALVERSON*
Botany Department
Birge Hall
University of Wisconsin
Madison, Wisconsin 53706, U.S.A.

DONALD M. WALLER†
Botany Department
Birge Hall
University of Wisconsin
Madison, WI 53706, U.S.A.

STEPHEN L. SOLHEIM
Botany Department
Birge Hall
University of Wisconsin
Madison, WI 53706, U.S.A.

Abstract: *Browsing by white-tailed deer (*Odocoileus virginianus*) can profoundly affect the abundance and population structure of several woody and herbaceous plant species. Enclosure studies and population surveys reveal that past and current deer densities as low as 4 deer/km² may prevent regeneration of the once common woody species, Canada yew (*Taxus canadensis*), eastern hemlock (*Tsuga canadensis*), and white cedar (*Thuja occidentalis*), as well as several herbaceous species. Prior to European settlement, forests in northern Wisconsin contained relatively sparse deer populations (<4/km²), but extensive timber cutting in the late nineteenth century boosted deer populations. Continued habitat fragmentation resulting from scattered timber harvests and the creation of "wildlife openings" to improve deer forage maintain these high densities throughout much of the Northeast.*

Because deer wander widely, the effects of high deer densities penetrate deeply into remaining stands of old and mature forest, greatly modifying their composition. Thus, abundant early successional and "edge" habitat, and the high deer densities they engender, represent significant external threats

* *Requests for reprints should be addressed to this author.*
† *This author presented this paper at the first annual meeting of the Society for Conservation Biology, Bozeman, Montana, June 25, 1987.*

Resumen: *El pastoreo y ramoneo del venado de cola blanca (*Odocoileus virginianus*) en la zona norteña del estado de Wisconsin, puede afectar profundamente a la abundancia y estructura de población de varias especies de plantas herbáceas y leñosas. Antes de la Colonia, los bosques norteños de Wisconsin eran habitat de poblaciones relativamente pequeñas de venados (menos de 4 por Km²) pero las extensas talas de estos bosques, a finales del siglo 19, dió paso a un incremento poblacional de estas especies. Estudios recientes de unidades de exclusión y catastros poblacionales revelan que, en efecto, una densidad de solo 4 ejemplos por Km² logra impedir la regeneración de las especies de* Taxus canadensis, Tsuga candensis *y* Thuja occidentalis, *muy conocidas otrora, como tambien varias especies herbáceas. La fragmentación de habitats, que ha resultado de la tala dispersa de arboles y la creación, de "claros silvestres" con propósitos de incrementar el forraje para los venados, ha aumentado las poblaciones de estas especies a lo largo de la región del nordeste.*

Debido a que un gran número de venados vagan ampliamente en la región, los efectos sobre los espacios remanentes de bosque maduro son considerables modificandose significativamente su composición. De esta manera, incipientes y extensos ecotonos, ademas de las grandes densidades de poblaciones de venados que estos albergan, representan una amenaza externa sobre estas comunidades de bosque.

to these plant communities. We hypothesize that establishing large (200–400 km²) continuous areas of maturing forest, especially in conjunction with increased hunting, could reduce local deer densities and so provide a simple and inexpensive method for retaining species sensitive to the deleterious effects of browsing.

Nos permitimos suponer que el establecimiento de grandes y continuas áreas de bosques en desarrollo (200–400 Kms), conjuntamente con el incremento en la actividad de caza, podría reducir localmente la densidad de población de los venados y de esta forma disponerse de un método simple y poco costoso que permita conservar estas expecies sujetas al deterioro por efectos del pastoreo y ramoneo.

> [M]obile animals greatly affect plant life, so that a small virgin forest may *appear* to be natural when actually it has been profoundly affected by forces applied to animals, waters, or climate at points far distant. (Thus the deer populations determined by laws passed at Lansing, by hunters . . . , and by lumbering operations . . . have apparently exterminated the ground hemlock [yew] from the "virgin" forest of Mountain Lake.)
>
> A. Leopold (1938)

Ever since the pioneering work of Leopold (1936) on habitat manipulation, wildlife biologists have strived to boost populations of game species by creating clearings and other areas of sharp transition between two or more types of plant community. Indeed, the traditional meaning of the term "edge effect" was the local increase in the diversity and abundance of animal species found along the boundary between two habitat types (Leopold 1936; Swift 1946; Dahlberg & Guettinger 1956; Yahner 1988). In seeking to explain this phenomenon, a number of other edge effects have been noted, including microclimatic changes in temperature, light, and humidity; altered tree species composition due to increased colonization by shade-intolerant and exotic plants; invasions by insects; and increased parasitism, predation, and competition by "weedy" birds and mammals (Ranney, Bruner, & Levenson 1981; Matthiae & Stearns 1981; Guntenspergen 1983; Brittingham & Temple 1983; Wilcove 1985; Janzen 1983, 1986; Wilcove, McLellan, & Dobson 1986; Yahner & Scott 1988).

Because the apparently beneficial effects of habitat tend to be local (on the order of a few hundred meters at most), wildlife managers often try to establish small clearings throughout a forested area. These efforts, the simultaneous creation of edge via other ongoing human disturbances, and controls on hunting have resulted in abundant populations of game and other edge-loving species. As reflected in the leading quote, however, some naturalists recognize a darker side to edges. Within conservation biology, the term "edge effects" is now usually used to refer to increased predation and parasitism of vulnerable animals in the vicinity of edges (e.g., Temple & Cary 1988). We would like to extend this connotation to include the deleterious effects of herbivores on sensitive plant species within stands of mature forest. While younger forest can undoubtedly buffer older forests against many microclimatic and biological edge effects (as assumed for western U.S. forests by Harris 1984), such a matrix might also threaten diversity by facilitating the invasion of successional plants and animals capable of interfering with species restricted to older communities (Janzen 1983, 1986).

Here, we review evidence that herbivory can profoundly alter plant community composition in the Northeast. We became concerned that such efforts could be widespread after learning of Hough's (1965) 20-year field study in the Allegheny National Forest in Pennsylvania, where he found the understory of a large (1650 ha) tract of virgin hemlock-hardwood forest to be severely damaged by deer browsing. Because herbivorous mammals wander widely and can invade even areas of ostensibly unfavorable habitat, such edge effects penetrate much farther than those previously reported for the region. This raises the important policy question of how plant species diversity is to be retained in areas subject to regional management for high deer densities. For concreteness and relevance, we primarily discuss interactions between white-tailed deer (*Odocoileus virginianus*) and various plants in northern Wisconsin (scientific names in Table 1). Policy decisions are now pending regarding this issue in two National Forests located there (Task Force 1986; U.S. Forest Service 1986).

The Context: Northern Wisconsin

Land survey records and detailed analyses of remnant forest stands allow a reasonable reconstruction of presettlement forest conditions (Curtis 1959; Finley 1976). Upland mesic forest habitats were predominantly old-growth (200–300 years old), with only 17–25% of their area occupied by successional communities (Canham & Loucks 1984). This relationship has now been reversed, with small patches of old and mature growth occupying less than 5% of the forest within a matrix of younger successional communities. The pre-Columbian forest contained about 64% "hardwood" by area and consisted mostly of the hemlock-hardwood community type. Intact, mature examples of this forest type are now relegated to token occurrence, primarily in existing or proposed "Research Natural Areas" 12 to 260 ha in size.

Because white pine, once a major component of northern Wisconsin forests, was preferred by early loggers, its abundance was drastically reduced. Hemlocks were then cut preferentially to service the tanning trade. Finally, with increasing demand for hardwood

Table 1. Common and scientific names of organisms mentioned in the text.

Aspen	*Populus tremuloides* Michx.
Black ash	*Fraxinus nigra* Marsh.
Blunt-leaved orchid	*Habenaria obtusata* (Pursh) Richards.
Buckthorn	*Rhamnus cathartica* L.
Flowering dogwood	*Cornus florida* L.
Hemlock, eastern hemlock	*Tsuga canadensis* (L.) Carr.
Honeysuckle	*Lonicera tatarica* L., *L. morrowi* Gray, *and their hybrid, L.* X *bella* Zabel
Indian cucumber-root	*Medeola virginiana* L.
Large-flowered trillium	*Trillium grandiflorum* (Michx.) Salisb.
Leatherwood	*Dirca palustris* L.
Purple fringed orchid	*Habenaria psycodes* (L.) Spreng.
Redbud	*Cercis canadensis* L.
Showy lady's-slipper orchid	*Cypripedium reginae* Walt.
Sugar maple	*Acer saccharum* Marsh.
Tall northern bog orchid	*Habenaria hyperborea* (L.) R. Br.
Yellow birch	*Betula alleghaniensis* Britton
Yew, Canada yew, ground hemlock	*Taxus canadensis* Marsh.
Yew-tree	*Taxus baccata* L.
White cedar	*Thuja occidentalis* L.
White oak	*Quercus alba* L.
White pine	*Pinus strobus* L.
Wood sorrel	*Oxalis acetosella* L.
Yellow lady's-slipper orchid	*Cypripedium calceolus* L.
Brainworm	*Parelaphostrongylus tenuis*
Canadian lynx	*Lynx canadensis* Kerr
Deer, white-tailed deer	*Odocoileus virginianus* (Zimmerman)
Deer tick	*Ixodes*
Elk, American elk	*Cervus elaphus* Linnaeus
Moose	*Alces alces* (Linneaus)
Mountain lion	*Felis concolor* Linnaeus
Timber wolf	*Canis lupus* Linnaeus
Wolverine	*Gulo gulo* (Linnaeus)
Woodland caribou	*Rangifer tarandus* (Linnaeus)

during and after World War I, most of the area was clearcut. Wisconsin's two National Forests, the Chequamegon and Nicolet, were created in the 1920s and 30s and now occupy 3,420 and 2,650 km^2, respectively (Fig. 1).

Aspen is the preeminent early successional tree species of the region. Its wind-dispersed seeds, clonal propagation, and fast growth allow it to quickly occupy large areas. Partly because it freely root-sprouts following fire or cutting, it has increased from about 1% on these National Forests presettlement to about 26% now (U.S. Forest Service 1986*c*). Other disturbance and edge-adapted species were originally rather scarce and limited to tree-fall gaps, riparian habitats, and areas of forest recently blown down or burned. They are now quite

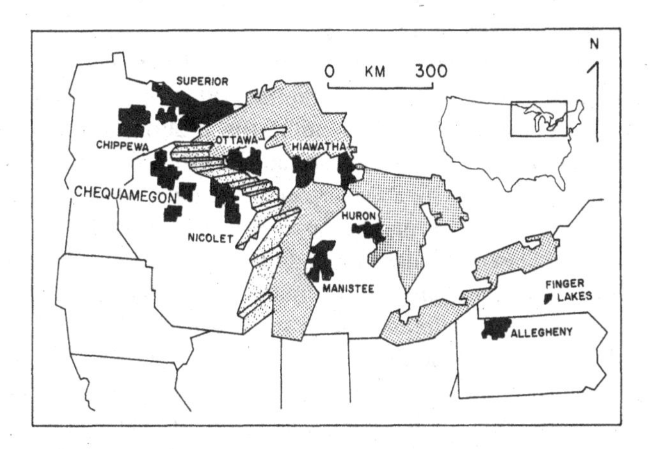

Figure 1. The U.S. National forests of the Great Lakes region. Stippled areas represent the Great Lakes and solid lines the boundaries between states. The Chequamegon and Nicolet National Forests of northern Wisconsin are 3,420 and 2,650 km^2, respectively. Source: Modified from U.S.D.A. Forest Service map.

common throughout these forests. Many weedy, exotic plant species have colonized heavily disturbed habitats, but the less disturbed habitats have not yet been seriously invaded (unlike the situation in southern Wisconsin, where buckthorn and honeysuckle have invaded even "intact" forests [Barnes & Cottam, 1974]).

Not surprisingly, major changes in Wisconsin's fauna have also occurred during the last century. Moose, elk, and woodland caribou, as well as predators like the wolverine, have all been extirpated. Forest disturbance does not fully explain why these species were lost, but is certainly a contributing factor (Jackson 1961; Gates, Clarke, & Harris 1983). Timber wolf and Canadian lynx continue to occupy sections of both forests but are quite scarce, largely due to human activity. Although it was assumed to be extirpated, there have been several recent sightings of the mountain lion (Lewis & Craven 1987).

Population Densities of Deer

Severe winters and wide expanses of virgin timber lacking undergrowth originally produced marginal habitat for the white-tailed deer in the northern Great Lakes region (Swift 1946; Schorger 1953; Dahlberg & Guettinger 1956, Blouch 1984; but see Habeck & Curtis 1959). As stated by the U.S. Forest Service, "Species associated with aspen and other early successional stages were present but in low numbers. Early settler's notes indicate few deer and other game animals in northern Wisconsin" (1986*c*, p. D5).

Deer populations in northern Wisconsin were originally less than 4/km^2 of range, and probably as low as

2/km^2 over large areas of range (total surface area minus the area of lakes, rivers, urban areas, and large farms). Populations began to rise in the mid-1850s and peaked in the Forest in the 1930s and 1940s at about 14/km^2 due to extensive favorable habitat and protective hunting laws (Swift 1946; Dahlberg & Guettinger 1956; McCaffery 1986). During the last 25 years, densities in northern Wisconsin have ranged from 5 to 12/km^2 (McCaffery 1986) and are now estimated at 2 to 9 deer/km^2 in the northern units of Wisconsin's National Forests (F. Haberland, personal communication, data for 1985 and 1986). The stated goals of the Forest Service for deer production are much higher, calling for sufficient habitat to support 31,952 deer in the Chequamegon, or 9.3 deer/km^2 (U.S. Forest Service 1986*a*, p. B4). Goals of the Wisconsin Department of Natural Resources are for 4 to 8 deer/km^2 overwintering in the same area (F. Haberland, personal communication).

Habitat management for deer in the Upper Great Lakes region entails establishing winter range, young aspen growth, oaks for acorns, and openings in the forest to supply grasses and other pasturage (McCaffery 1984, 1986). In keeping with these traditional management practices, the management plans for the national forests intersperse small (<20 ha) timber cuts and openings designed to boost deer and other game populations throughout both Forests (U.S. Forest Service 1986). Collectively, these comprise at least 14% of the Forests' areas and should result in a uniform, abundant distribution of deer, a goal explicitly embraced by the Forest Service (Task Force 1986).

Effects of Deer Browsing on Woody Plants

The damage deer do to crop and natural vegetation has been extensively studied, and depends on deer density. High deer populations slow the regeneration of several commercial species, causing significant economic losses (Graham 1954; Marquis 1981; Redding 1987). However, as deer densities in northern Wisconsin have declined to levels below that which threatens commercial forestry or deer range per se, concern for deer damage to vegetation has virtually disappeared (McCaffery 1986).

Partial lists of preferred deer foods in Wisconsin all agree that Canada yew, eastern hemlock, and white cedar are highly preferred by deer during winter months (DeBoer 1947; Swift 1948; Cottam & Curtis 1956; Dahlberg & Guettinger 1956; Beals, Cottam, & Vogl 1960). Yew is severely damaged by deer because it is both sought out and does not recover well after browsing. For some time, there has been little or no reproduction of yew in most of the region (Stearns 1951; Curtis 1959) with many populations now lost from known sites of prior occurrence. Surviving populations exist on rocky outcrops that are inaccessible to deer, or as scattered individuals browsed nearly to ground level. The only other area in Wisconsin where yew populations are known to be extensive and healthy is in the tribal lands of the Menominee Reservation (Waller et al., in prep.) where year-round hunting limits deer densities (see below). On the nearby Apostle Islands in Lake Superior, Beals, Cottam, & Vogl (1960) found yew common on islands with few or no deer, yet yew were lacking in mainland forests. Leuthold (1980) has shown a similar decline of the related yew-tree in Switzerland and predicts extirpation of the species unless active protection occurs. Besides direct browsing losses, Canada yew suffers indirectly via a novel mechanism that has only recently been recognized: browsing skews floral sex ratios which, in turn, limit the availability of pollen to the point where it becomes limiting and reproduction is impaired (Allison 1987).

Like yew, eastern hemlock and white cedar are quite sensitive to deer browsing. Although these trees are capable of growing tall enough to escape browsing, deer can severely impair reproduction by preventing seedling and sapling recruitment, particularly since slow-growing seedlings and saplings of these species are vulnerable to browsing for decades (Hough 1965; Rogers 1978). Browsing is particularly conspicuous within winter deer yards in Wisconsin where hemlock and white cedar are reproducing poorly or not at all (Task Force 1986).

Deer enclosure studies carried out in Wisconsin, through other parts of the Northeast, and elsewhere show dramatic differences in survival and reproduction of hemlock, yew, white cedar, and other species within fenced enclosures compared to exposed individuals outside the protected areas (Graham 1954; Dahlberg & Guettinger 1956; Stoeckeler, Strothman, & Krefting 1957; Marquis 1974; Blewett 1976; Kroll, Goodrum, & Behrman 1986; Tilghman, in press; Fig. 2). These observations all support the results of Goff (1967), Anderson & Loucks (1979), and Waller et al. (in prep.), in which hemlock only exhibits a healthy population structure within Wisconsin on certain islands, in the Menominee Reservation, and within deer enclosures. Similar results hold for white cedar (Blewett 1976 and references therein), with striking differences in stem height and density within and outside enclosures.

The enclosure illustrated contains a population of hemlock with hundreds of individuals representing all seedling and sapling age classes, yet the surrounding area outside the enclosure contains only a few individuals, which either show signs of recent browsing or are shorter than the winter snow cover. Nearby, a showcase grove of old-growth hemlocks is virtually devoid of recent hemlock reproduction, despite falling within a deer management unit with a reported population density of only 2 deer/km^2. Its understory is composed mainly of

Figure 2. Deer enclosure at Fould's Creek, Chequamegon National Forest, Wisconsin, viewed from the top of the 4 m high fence, which bisects the photograph. Vigorous growth of hemlock can be seen within the forty-year-old enclosure (left side of photograph). Source: W. S. Alverson, February, 1988.

stunted, gnarled sugar maple seedlings and saplings bearing the distinctive mark of heavy browsing by deer (cf. Fig. 1 in Switzenberg, Nelson, & Jenkins 1985; Fig. 1 in Stoeckeler, Strothman, & Krefting 1957).

Some researchers question whether deer browsing alone has caused the conspicuous changes in hemlock reproduction in the upper Midwest (Webb, King, & Patric 1956; Tierson, Patric, & Behrend 1966). Stearns (1951) suggested that changes in climate or catastrophic storms allowed greater hemlock regeneration during certain periods in the past (reviewed by Eckstein 1980). This seems unlikely, however, both from the enclosure studies cited above and because cycles of hemlock reproduction are asynchronous between noncontiguous stands and appear to be governed by internal stand dynamics (Hett & Loucks 1976).

Even more definitively, Frelich & Lorimer (1985) documented changes in size-class distributions and extensive browse damage to young hemlock that appear directly attributable to deer browsing in the Porcupine Mountains of Michigan's Upper Peninsula. A forest near the Lake Superior shore with an estimated deer density of 10/km^2 (winter) suffered almost complete annihilation in some size classes, while inland sites with lower winter deer densities (2/km^2) exhibited unimpaired reproduction. They rejected the hypothesis that climate was responsible for the differences in hemlock reproduction by demonstrating that herbivory was the causal factor. A model they constructed predicts eventual exclusion of hemlock by hardwood species in the coastal sites within 200 years if deer densities remain high.

In a study designed specifically to test whether high deer densities prevent reproduction in these species, we compared hemlock's population structure within the Menominee Reservation to its structure within the adjacent Nicolet National Forest (Waller et al., in prep.). Because the Menominee Reservation allows hunting year-round, deer densities are lower than in surrounding areas (ca. 1–2/km^2; Morehouse & Becker 1966; O. Rongstad, personal communication). While almost half of the stands within the Menominee showed substantial hemlock reproduction, less than 6% of those in the Forest did. The Nicolet stands exhibit drastically reduced seedling abundance, especially relative to the number of adult trees (Fig. 3).

Other Species Affected by High Deer Densities

Deer can affect the composition of entire communities and not just individual woody plant species. For example, in Great Smoky Mountains National Park, areas subject to intensive deer browsing close to openings lost

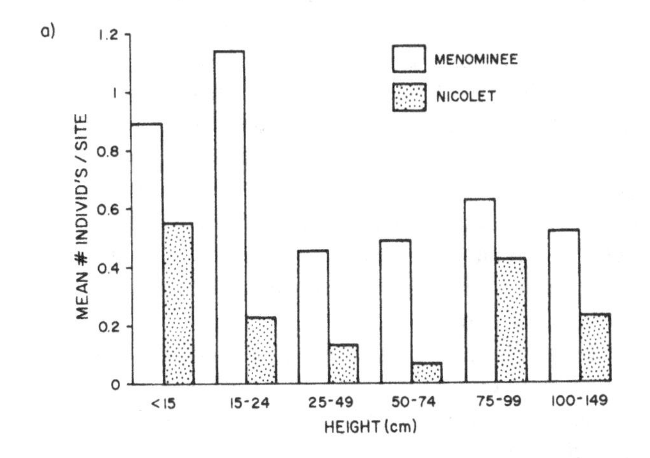

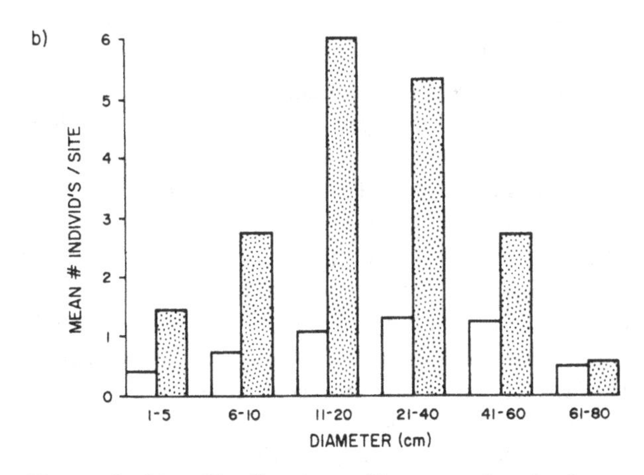

Figure 3. Size distributions of eastern hemlock seedlings (a) and adult trees (b) in the Menominee Reservation and the adjacent Nicolet National Forest. Deer densities are much lower in the Menominee Reservation due to year-round hunting. Source: Data from Waller, Judziewicz, Alverson, & Solheim (in prep.).

more than a quarter of their total species richness compared to control areas (Bratton 1979). White oak, redbud, and flowering dogwood all appeared to be significantly affected by deer browsing. Not just plant species are affected, either. Moose are thought to be excluded from many areas by infection with brainworm, a parasite carried by deer. Decreases in deer abundance could enhance the chances for immigrant moose now drifting into northern Wisconsin to become successfully reestablished, as recently occurred in New York (Hicks & Stumvoll 1985). Lyme disease, carried primarily by deer ticks, also appears to be increasing in many areas in response to increased deer abundance, but no effects on animal communities are yet evident.

Many herbaceous species are also favored by deer, including the showy and yellow lady's slipper orchids, the blunt-leaf orchid, the tall northern bog orchid, and the purple fringed orchid. Many of these could be experiencing reduced reproductive success and/or local extirpation due to intensified deer herbivory (Cottam & Curtis 1956; L. Lipsey, personal communication; personal observation). Enclosure studies in the Allegheny National Forest of northwestern Pennsylvania demonstrated that deer populations of 4/km^2 caused significant reductions in the abundance of Indian cucumber-root and large-flowered trillium, both of which also occur in northern Wisconsin (Tilghman, in press). Like yew, herbs are highly susceptible to herbivory by deer because they never outgrow the zone of accessibility (approximately 2 m).

Changes in canopy composition could also result in changes in community composition in other forest strata. For example, observed and predicted losses of mature hemlock and other northern hardwood canopies in the Forest could cause the eventual loss of shrubs and herbs like leatherwood and wood sorrel that tend to be restricted to these habitats (Stearns 1951; personal observation). As evergreen conifers lose dominance to deciduous species like sugar maple and black ash, light regimes change drastically and understory species can be expected to respond. Furthermore, as remaining habitats of older hemlock and white cedar become smaller and more isolated, further gradual but inexorable losses of the many species restricted to these habitats are expected via "relaxation" (MacArthur & Wilson 1967). How quickly this occurs obviously depends on species-specific characteristics, but analogous losses of herb species have been documented for isolated forests in southern Wisconsin (Hoehne 1981).

Recommended Deer Densities

Assuming that deer browsing has caused the observed reductions in reproduction of hemlock, yew, and white cedar, it becomes important to determine that density of deer below which successful regeneration is possible. While few studies directly addressing this question exist, existing work allows us to infer which densities are clearly incompatible with successful reproduction. For example, studies of deer carrying capacity routinely suggest that densities of 8 deer/km^2 are compatible with good range management. However, studies of carrying capacity are normally undertaken out of primary concern for healthy populations of deer or commercial timber species, especially their ability to fulfill their auxiliary role as "cover" or "browse" for deer. Such studies provide little or no assurance that the impact of deer is uniformly benign. For example, Tubbs, Jacobs, & Cutler (1983) combine data for hemlock with numerous other species composing the "northern hardwood types," and obscure the problematic relationship with deer: "The northern hardwood types can support relatively high populations of deer without serious injury; damage will be minimal if management practices favor dense reproduction and vigorous shoot growth (Jacobs 1969)" (p. 122). Yet Jacobs considers only the ability of sugar maple to survive under such deer densities, not hemlock, yellow birch, or yew, all important components of the northern hardwood forest. Furthermore, it is sugar maple that replaces hemlock in this region as the latter is browsed (Anderson & Loucks 1979; Frelich & Lorimer 1985).

Densities of 8/km^2 appear far too high if maintaining the diversity of all plants and animals is the management objective, as reviewed above. Instead, deer densities approximating presettlement conditions for substantial periods of time appear necessary to ensure the survival and healthy reproduction of hemlock, yew, and other sensitive plant species. Existing meager data suggest this density to be less than 4 deer/km^2, and possibly as low as 1–2 deer/km^2. Precise figures cannot yet be stated because of the lack of thorough, species-specific studies in our region. A wildlife biologist currently studying deer movements in the area suggests that young hemlock occur in areas where deer densities approach 2/km^2 (O. Rongstad, personal communication).

Discussion

Browsing by elevated populations of white-tailed deer appears to constitute a major edge effect in the forests of northern Wisconsin and perhaps other parts of the Northeast. Deer affect forest composition through direct, well-documented negative effects on several woody plant species and through direct and indirect damage to many herbaceous species. Failure to acknowledge these ecological interactions and plans to maintain dense populations of deer (8–9.3/km^2) by state and federal land stewards work directly against the preservation of these components of natural diversity.

What steps could be taken to protect the viability of species sensitive to browsing at high deer densities? Habitats suitable for deer-sensitive species could be created in at least three ways: enclosures, increased hunting, or habitat management to reduce deer densities. Other means of controlling deer density exist (such as deer repellents and birth control), but these are unlikely to be viable solutions (Redding 1987; Marquis 1987).

Enclosures are now being used in northwestern Pennsylvania for regenerating commercial forest stands (Marquis 1987), but are extremely expensive to construct and maintain (Kochel & Brenneman 1987; U.S. Forest Service 1986c, p. F63). Alternatively, commercial seedling caps can be used to protect individual seedlings, also at great cost. Both enclosures and seedling protectors appear best suited to regenerate small local stands of a single target species such as hemlock or yew. However, such a solution, unless extended to a complete set of other sensitive plant species (many of which must still be unknown), offers no general relief. The same problem applies to the use of silvicultural techniques aimed at regenerating single species (Eckstein 1980; Johnson & Booker 1983; Tubbs, Jacobs, & Cutler 1983; Wendel et al. 1983; Marquis 1987). Such methods for hemlock require soil scarification, removal of litter, fencing and/or partial canopy removal while reducing the area's attractiveness to deer. These intrusive management techniques are prohibitively expensive on a large scale and could still cause or permit damage to other species sharing the habitat. It would also be an obvious mistake to assume that protecting hemlock (or any other particular species) somehow protects the overall diversity of the hemlock-hardwood forest community. At present, there exists neither the knowledge nor the will to create active programs of species-specific management for all deer-sensitive species in these communities.

Increased hunting pressure can also decrease deer populations locally (Morehouse & Becker 1966; Creed et al. 1984). While some uncertainty exists as to the relative importance of hunting versus deer behavior and habitat quality in determining deer population levels (McCaffery 1986), few doubt that increased hunting pressure would reduce browsing, especially if coupled with habitat alteration. Whether hunting alone could reduce deer densities to 2/km^2 is unclear, however, particularly since most hunters prefer to hunt in areas of known high deer density.

Species sensitive to high deer densities could also be protected by habitat management if vegetation capable of supporting only reduced deer densities could be established. This would involve running conventional game management practices in reverse. Instead of increasing edge habitat and young browse, large blocks of forest would be allowed to mature naturally to the point where they become inferior deer habitat (Fig. 4). Such areas would have to be large and continuous enough to create core areas relatively free of the edge effects produced by deer and could be created by redistributing management activities in public (and perhaps private) forests. For example, efforts to harvest timber and improve deer habitat could be confined to 80% of each National Forest, leaving 20% in one or two large, contiguous blocks that would eventually become, through natural succession, habitats unfavorable to dense deer populations. This, in fact, represents the actual recommendation made in a formal appeal process involving Wisconsin's National Forests (Task Force 1986). The proposed biotic reserves were termed "Diversity Maintenance Areas."

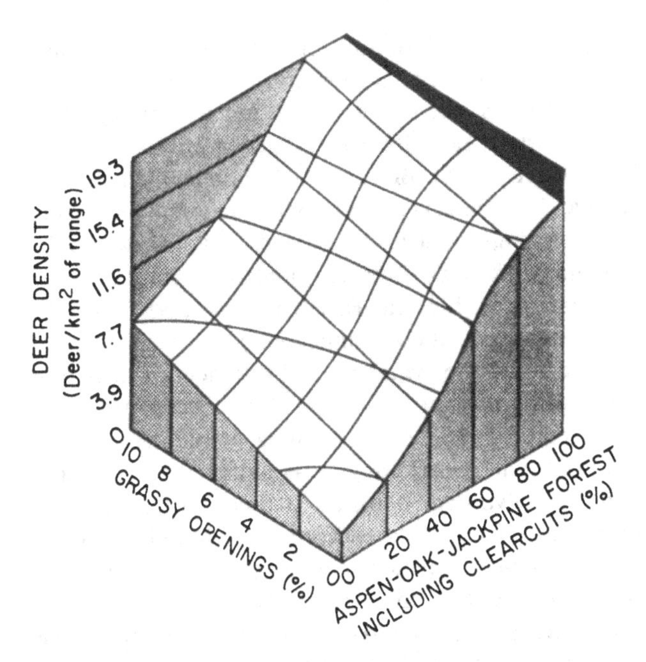

Figure 4. Theoretical model of the influence of habitat on the population densities of white-tailed deer in northern Wisconsin. Units of deer density, originally given in deer/mi^2, have been converted to deer/km^2; density figures are comparable to those given in the text. One purpose of Diversity Maintenance Areas would be to create large blocks of habitat corresponding to the lowermost surface of the slope. Source: Modified with permission from McCaffery (1986).

The crucial question concerning this final alternative lies with size: How large must a block of old forest be to effectively reduce deer densities? The literature on deer movements is extensive but insufficient by itself to resolve the size issue (e.g., Tierson et al. 1985). Winter ranges of individual deer tend to be less than 480 ha in Minnesota (Rongstad & Tester 1969), with summer home ranges somewhat larger. Winter to summer range movements for adult deer averaged 5.6 km in a Wisconsin study, with 90% of the deer moving 12 km or less (Dahlbert & Guettinger, 1956). An eight-year study in

Michigan's Upper Peninsula found that the mean annual dispersal distance between winter yards and the following November kill site was 13.8 km for hunter-killed deer; yearly mean distances ranged from 10.9 to 20.2 km (Verme 1973). Bratton (1979) concluded that intensive deer impacts did not extend beyond 1 km away from openings in the Cades Cove region of Great Smoky Mountains National Park. Fortunately, a study on deer movements within the Forest is now underway (O. Rongstad, personal communication).

If we apply the average travel distance for deer of 8 km used by the Forest Service in its management plan (Task Force 1986), blocks of unfavorable habitat with radii of 8 km, constituting areas of ca. 200 km^2, might serve to halve deer densities at their center points. If unfavorable habitat like old growth is found to reduce travel distances, smaller areas might suffice. Prudence would dictate the conservative course of first designating larger areas, then reducing them if penetrations were found to be of shorter range. If the mean dispersal distance were reduced to 2 km in a circular block of unfavorable habitat, the block would still need to have a radius of 7 km to have a 1:1 ratio of edge to interior habitat. Historical patterns of movement and other particular features of deer behavior also clearly influence how such habitat blocks would function (Tierson et al. 1985). Such information is limited, making it difficult to predict *a priori* exactly how large the blocks of unfavorable habitat need to be to protect sensitive species.

Mature or old-growth forest blocks of this scale are much larger than any existing old-growth areas in northern Wisconsin. The two congressionally designated wilderness areas in the Chequamegon National Forest are 1,710 and 2,660 ha in size, comparable in size to Hough's (1965) 1,650 ha severely damaged study area in Pennsylvania. These wilderness areas are imbedded in a matrix of young forest containing extensive deer habitat. As more forest lands in northern Wisconsin mature, deer densities should decline slowly (McCaffery 1986). However, the National Forest plans call for increased timber harvests, which all convert nearly 50% of their area into new successional habitats during the next 50 years. This makes it unlikely that the wilderness areas will experience consistently low herbivory by deer over the foreseeable future, even when the wilderness areas themselves become old. In fact, the scattered, shifting pattern of timber harvests and the creation of additional "wildlife openings" proposed by the Forest Service instead promotes a Forest-wide homogenization of habitat via deer edge effects.

Conclusions

Our understanding of edge effects is still in its infancy. Edge effects on the scale of several km resemble those already suggested for other forests (Janzen 1983, 1986), but remain politically controversial due to their management implications. Studies are now under way that should eventually allow us to tailor the size and shape of reserves specifically to retain a full complement of plant and animal species. Such areas, if they prove unnecessarily large, can always be reduced in size at a later time, but cannot be expanded without losing decades of forest growth.

Maintaining the proposed 200 to 400 km^2 reserves of contiguous habitat within the National Forests to retain species sensitive to deer browsing or otherwise dependent on forest interior habitats would be simple and inexpensive (Task Force 1986). Such areas would also be freely available for a wide variety of other uses, including hunting (intensified for deer, if possible), fishing, snowmobiling, skiing, hiking, camping, and small-scale wood removal. They would, however, exclude commercial-scale timber harvests, "wildlife openings," and new road construction, all of which create large amounts of edge habitat. Such management would probably not reduce the total deer populations of the area, but would alter the spatial distribution of deer, allowing local reductions in deer abundance and consequent survival of sensitive species. Encouragingly, the staff of the Chequamegon National Forest concluded that such areas could be created without losing jobs or sacrificing timber production or other outputs of their Forest (J. Wolter, personal communication 1986). Disappointingly, the regional office of the U.S. Forest Service, perhaps concerned about the precedent such areas would establish, reversed this decision without any formal scientific or economic review. Currently these issues are being considered by the chief of the U.S. Forest Service.

Acknowledgments

We thank O. Rongstad, C. Lorimer, L. Tyrrell, J. Kotar, E. Judziewicz, K. McCaffery, F. Haberland, and D. Wilcove for their comments on drafts of this paper or for making unpublished data available. The E.K. and O.N. Allen Fund of the University of Wisconsin Herbarium provided funds for publication.

Literature Cited

Allison, T. D. 1987. The reproductive biology of Canada yew (*Taxus canadensis*) and its modification by herbivory: Implications for wind pollination. Ph.D. dissertation, University of Minnesota, Minneapolis.

Anderson, R. C., and O. L. Loucks. 1979. White-tail deer (*Odocoileus virginianus*) influence on the structure and composition of *Tsuga canadensis* forests. Journal of Applied Ecology **16**:855–861.

Barnes, W. J., and G. Cottam. 1974. Some autecological studies of the *Lonicera* x *bella* complex. Ecology **55**:40–50.

Beals, E. W., G. Cottam, and R. J. Vogl. 1960. Influence of deer on the vegetation of the Apostle Islands, Wisconsin. Journal of Wildlife Management **24(1)**:68–80.

Blewett, T. J. 1976. Structure and dynamics of the McDougal Springs lowland forest. M.S. thesis, University of Wisconsin, Madison.

Blouch, R. I. 1984. [Deer in the] Northern Great Lakes States and Ontario forests. Pages 391–410 *in* L. K. Halls, editor. The White-tailed Deer: Ecology and Management. Stackpole Books, Harrisburg, Pennsylvania.

Bratton, S. P. 1979. Impacts of white-tailed deer on the vegetation of Cades Cove, Great Smoky Mountains National Park. Proceedings of the Annual Conference of the Southeastern Association of Fish & Wildlife Agencies **33**:305–312.

Brittingham, M. C., and S. A. Temple. 1983. Have cowbirds caused forest songbirds to decline? Bioscience **33**:31–35.

Canham, C. D., and O. L. Loucks. 1984. Catastrophic windthrow of the presettlement forests of Wisconsin. Ecology **65(3)**:803–809.

Cottam, G., and J. T. Curtis. 1956. The effect on deer of the botanical composition of deer yards and a method for measuring it. Unpublished interim report on project 15-933 to Wisconsin Conservation Department of Natural Resources. Botany Department, University of Wisconsin, Madison.

Creed, W. A., F. Haberland, B. E. Kohn, and K. R. McCaffery. 1984. Harvest management: The Wisconsin experience. Pages 243–260 *in* L. K. Halls, editor. *The White-tailed Deer: Ecology and Management. Stackpole Books, Harrisburg, Pennsylvania.*

Curtis, J. T. 1959. The Vegetation of Wisconsin. University of Wisconsin Press, Madison. 657 p.

Dahlberg, B. L., and R. C. Guettinger. 1956. *The White-tailed Deer in Wisconsin.* Wisconsin Conservation Department, Madison.

DeBoer, S. G. 1947. The deer damage to forest reproduction survey. Wisconsin Conservation Bulletin **12(10)**:1–23.

Eckstein, R. G. 1980. Eastern hemlock (*Tsuga canadensis*) in north-central Wisconsin. Research Report No. 104. Wisconsin Department of Natural Resources, Rhinelander.

Finley, R. W. 1976. Original vegetation cover in Wisconsin. North Central Forest Experiment Station, U.S.D.A. Forest Service, St. Paul, Minnesota.

Frelich, L. E., and C. G. Lorimer. 1985. Current and predicted long-term effects of deer browsing in hemlock forests in Michigan, U.S.A. Biological Conservation **34**:99–120.

Gates, D. M., C. H. D. Clarke, and J. T. Harris. 1983. Wildlife in a changing environment. Pages 52–80 *in* S. L. Flader, editor. The Great Lakes Forest: An Environmental and Social History. University of Minnesota Press, Minneapolis.

Goff, F. G. 1967. Upland vegetation. Pages 60–90 *in* C. J. Milfred, G. W. Olson, and F. D. Hole, editors. Soil Resources and Forest Ecology of Menominee County, Wisconsin. University of Wisconsin Geology and Natural History Survey Bulletin No. 85, Soil Series No. 60. University of Wisconsin Extension, Madison.

Graham, S. A. 1954. Changes in northern Michigan forests from browsing by deer. Transactions of the Nineteenth North American Wildlife Conference **19**:526–533.

Guntenspergen, G. 1983. The minimum size for nature preserves: Evidence from southern Wisconsin forests. Natural Areas Journal **3(4)**:38–46.

Habeck, J. R., and J. T. Curtis. 1959. Forest cover and deer population densities in early northern Wisconsin. Transactions Wisconsin Academy Sciences, Arts and Letters **48**:49–56.

Harris, L. D. 1984. The Fragmented Forest: Island Biogeography Theory and the Preservation of Biotic Diversity. University of Chicago Press.

Hett, J. M., and O. L. Loucks. 1976. Age structure models of balsam fir and eastern hemlock. Journal of Ecology **64(3)**:1029–1044.

Hicks, A., and R. Stumvoll. 1985. An old friend returns. The Conservationist (New York) **39(4)**:9–13.

Hoehne, L. M. 1981. The groundlayer vegetation of forest islands in an urban-suburban matrix. Pages 41–54 *in* R. L. Burgess and D. M. Sharpe, editors. Forest Island Dynamics in Man-dominated Landscapes. Springer-Verlag, New York.

Hough, A. F. 1965. A twenty-year record of understory vegetational change in a virgin Pennsylvania forest. Ecology **46(3)**:370–373.

Jackson, H. T. 1961. Mammals of Wisconsin. University of Wisconsin Press, Madison.

Jacobs, R. D. 1969. Growth and development of deer-browsed sugar maple seedlings. Journal of Forestry **67**:870–874.

Janzen, D. H. 1983. No park in an island: Increased interference from outside as park size decreases. Oikos **41**:402–410.

Janzen, D. H. 1986. The eternal external threat. Pages 286–303 *in* M. E. Soulé, editor. Conservation Biology. Sinauer Associates, Sunderland, Massachusetts.

Johnson, W. F., and R. G. Booker. 1983. Northern white cedar. Pages 105–108 *in* R. M. Burns, editor. Silvicultural Systems for the Major Forest Types of the United States. U.S.D.A. Forest Service Agriculture Handbook No. 445. U.S. Department of Agriculture, Washington, D.C.

Kochel, J., and R. Brenneman. 1987. Use and effectiveness of electric fencing in protecting clearcuts from deer browsing. Pages 118–123 *in* Proceedings of the Symposium on Deer, Forestry and Agriculture: Interactions and Strategies for Management. Allegheny Society of American Foresters, Warren, Pennsylvania.

Kroll, J. C., W. D. Goodrum, and P. J. Behrman. 1986. Twenty-seven years of over-browsing: Implications to white-tailed deer management on wilderness areas. Pages 294–303 *in* D. L. Kulhavy and R. N. Conner, editors. Wilderness and Natural Areas in the Eastern United States: A Management Challenge. Center for Applied Sciences, School of Forestry, Stephen F. Austin University, Nacogdoches, Texas.

Leopold, A. 1936. Game Management. Charles Scribner's Sons, New York.

Leopold, A. 1938. Private report to the Huron Mountain Club. 18 p.

Leuthold, C. von. 1980. The ecological and phytosociological situation of the yew-tree in Switzerland. Veröffentlichungen des geobotanischen Institutes ETH, Stiftung Rübel, Zurich: 67:1–217.

Lewis, T. L., and S. R. Craven. 1987. Mountain lions in Wisconsin? Maybe. Wisconsin Natural Resources. **11(1)**:21–25.

MacArthur, R. H., and E. O. Wilson. 1967. The Theory of Island Biogeography. Princeton University Press, Princeton, New Jersey.

Marquis, D. A. 1974. The impact of deer browsing on Allegheny hardwood regeneration. Research Paper NE-308. U.S.D.A. Forest Service, Northeastern Forest Experiment Station, Upper Darby, Pennsylvania.

Marquis, D. A. 1981. Effect of deer browsing on timber production in Allegheny hardwood forests in northwestern Pennsylvania. Research Paper NE-475. U.S.D.A. Forest Service, Northeastern Forest Experiment Station, Broomall, Pennsylvania.

Marquis, D. A. 1987. Silvicultural techniques for circumventing deer damage. Pages 125–136 *in* Proceedings of the Symposium on Deer, Forestry and Agriculture: Interactions and Strategies for Management. Allegheny Society of American Foresters, Warren, Pennsylvania.

Matthiae, P. E., and F. Stearns. 1981. Mammals in forest islands in southeastern Wisconsin. Pages 55–66 *in* R. L. Burgess and D. M. Sharpe, editors. Forest Island Dynamics in Man-dominated Landscapes. Springer-Verlag, New York.

McCaffery, K. 1984. Fat deer laugh at winter. Wisconsin Natural Resources **8(6)**:17–19.

McCaffery, K. 1986. On deer carrying capacity in northern Wisconsin. Pages 54–69 *in* R. J. Regan and S. R. Darling, compilers. Transactions of the Twenty-second Northeast Deer Technical Committee. Vermont Fish and Wildlife Department, Waterbury, Vermont.

Morehouse, M., and R. J. Becker. 1966. Menominee County deer. Wisconsin Conservation Bulletin **31(4)**:20–21.

Ranney, J. W., M. C. Bruner, and J. B. Levenson. 1981. The importance of edge in the structure and dynamics of forest islands. Pages 67–96 *in* R. L. Burgess and D. M. Sharpe, editors. Forest Island Dynamics in Man-dominated Landscapes. Springer-Verlag, New York.

Redding, J. C. 1987. Impact of deer on forest vegetation and timber production in northern Pennsylvania. Pages 23–32 *in* Proceedings of the Symposium on Deer, Forestry and Agriculture: Interactions and Strategies for Management. Allegheny Society of American Foresters, Warren, Pennsylvania.

Rogers, R. S. 1978. Forests dominated by hemlock (*Tsuga canadensis*): Distribution as related to site and postsettlement history. Canadian Journal of Botany **56**:843–854.

Rongstad, O. J., and J. R. Tester. 1969. Movements and habitat use of white-tailed deer in Minnesota. Journal of Wildlife Management **33(2)**:366–379.

Schorger, A. W. 1953. The white-tailed deer in early Wisconsin. Transactions of the Wisconsin Academy Sciences, Arts and Letters **42**:197–247.

Stearns, F. W. 1951. The composition of the sugar maple–hemlock–yellow birch association in northern Wisconsin. Ecology **32**:245–265.

Stoeckeler, J. H., R. O. Strothman, and L. W. Krefting. 1957. Effect of deer browsing on reproduction in the northern hardwood-hemlock type in northeastern Wisconsin. Journal of Wildlife Management **21**:75–80.

Swift, E. 1946. A History of Wisconsin Deer. Bulletin No. 323, Wisconsin Conservation Department, Madison.

Swift, E. 1948. Wisconsin Deer Damage to Forest Reproduction Survey: Final Report. Publication 347, Wisconsin Conservation Department, Madison.

Switzenberg, D. F., T. C. Nelson, and B. C. Jenkins. 1955. Effect of deer browsing on quality of hardwood timber in northern Wisconsin. Forest Science **1**:61–67.

Task Force. 1986. Statement of Reasons in Appeal of the U.S.D.A. Forest Service Record of Decision of August 11, 1986 Approving the Land and Resource Management Plan and Final Environmental Impact Statement for the Chequamegon National Forest. Sierra Club and Wisconsin Forest Conservation Task Force, Madison, Wisconsin. (Contains several U.S.D.A. Forest Service documents obtained under the Freedom of Information Act. Copies available at cost from Walter Kuhlman, of Boardman, Suhr, Curry, and Field. P.O. Box 927, Madison, WI 53701-0927.)

Temple, S. A., and J. R. Carey. 1988. Modelling dynamics of habitat interior bird populations in fragmented landscapes. Conservation Biology **2** (this issue).

Tierson, W. C., E. F. Patric, and D. F. Behrend. 1966. Influence of white-tailed deer on the logged northern hardwood forest. Journal of Forestry **64**:801–805.

Tierson, W. C., G. F. Mattfeld, R. W. Sage, Jr., and D. F. Behrend. 1985. Seasonal movements and home ranges of white-tailed deer in the Adirondacks. Journal of Wildlife Management **49(3)**:760–769.

Tilghman, N. G. In Press. Impacts of several densities of white-tailed deer on regeneration of forests in northwestern Pennsylvania. Journal of Wildlife Management.

Tubbs, C. H., R. D. Jacobs, and D. Cutler. 1983. Northern hardwoods. Pages 121–127 *in* R. M. Burns, editor. Silvicultural Systems for the Major Forest Types of the United States. U.S.D.A. Forest Service Agriculture Handbook No. 445, U.S. Department of Agriculture, Washington, D.C.

U.S. Forest Service. 1986. *a, b, c, d.* Chequamegon National Forest Management Plan and associated documents: *a.* Land and Resource Management Plan; *b.* Final Environmental Impact Statement; *c.* Appendix to Final Environmental Impact Statement; *d.* Record of Decision. Chequamegon National Forest Headquarters, U.S.D.A. Forest Service. Park Falls, Wisconsin.

Verme, L. J. 1973. Movements of white-tailed deer in upper Michigan. Journal of Wildlife Management **37(4)**:545–552.

Webb, W. L., R. T. King, and E. F. Patric. 1956. Effect of white-tailed deer on a mature northern hardwood forest. Journal of Forestry **54**:391–398.

Wendel. G. W., L. Della-Bianca, J. Russell, and K. F. Lancaster. 1983. Eastern white pine including eastern hemlock. Pages 131–134 *in* R. M. Burns, editor. Silvicultural Systems for the Major Forest Types of the United States. U.S.D.A. Forest Service Agriculture Handbook No. 445. U.S. Department of Agriculture, Washington, D.C.

Wilcove, D. S. 1985. Nest predation in forest tracts and the decline of migratory songbirds. Ecology **66**:1211–1214.

Wilcove, D. S., C. H. McLellan, and A. P. Dobson. 1986. Habitat fragmentation in the temerate zone. Pages 237–256 *in* M. E. Soulé, editor. Conservation Biology. Sinauer Associates, Sunderland, Massachusetts.

Yahner, R. H. 1988. Changes in wildlife communities near edges. Conservation Biology **2** (this issue).

Yahner, R. H., and D. P. Scott. 1988. Effects of forest fragmentation on depredation of artificial nests. Journal of Wildlife Management **52**:158–161.

Eastern Hemlock Regeneration and Deer Browsing in the Northern Great Lakes Region: A Re-examination and Model Simulation

DAVID J. MLADENOFF

Natural Resources Research Institute
University of Minnesota, Duluth
5013 Miller Trunk Highway
Duluth, MN 55811, U.S.A.

FOREST STEARNS

Department of Biological Sciences
University of Wisconsin–Milwaukee
P.O. Box 413
Milwaukee, WI 53201, U.S.A.

Abstract: *White-tailed deer (*Odocoileus virginianus*) populations are currently extremely high in the upper Great Lakes region, and browsing may have severe negative effects on many forest species, including the eastern hemlock (*Tsuga canadensis*), a former forest dominant. We suggest that this problem must be examined in an expanded spatial and temporal context. It may be incorrect to generalize from limited evidence on one scale of observation (stand level) to explain a regional, long-term failure of tree species regeneration. We reviewed hemlock life-history characteristics, its long-term behavior on a regional scale, and evidence from recent studies in the context of long-term forest ecosystem processes. We then conducted 400-year computer simulations within a neutral model framework (see Caswell 1976), relative to browsing, to test whether other factors might adequately explain the lack of hemlock regeneration in the northern lake states. The results suggest that browsing is not the critical step blocking hemlock forest re-establishment on a regional scale. We suggest that the interaction of climate, disturbance, hemlock life history, ecosystem processes, and historical land use produce positive feedbacks that prevent hemlock regeneration usually before deer browsing can occur. There is need*

Paper submitted April 29, 1992; revised manuscript accepted March 12, 1993.

Regeneración del abeto del este y ramoneo por ciervos en la región norteña de los Grandes Lagos: Una reexaminación y un modelo de simulación

Resumen: *Las poblaciones actuales de ciervos* Odocoileus virginianus *están muy arriba en la región superior de los Grandes Lagos, y el ramoneo puede tener severos efectos negativos para muchas especies forestales incluyendo el abeto del este (*Tsuga canadensis*), un exdominador del bosque. Sin embargo sugerimos que este problema debe ser examinado en un contexto espacial y temporal expandido. Sería incorrecto generalizar evidencias limitadas a una escala de observación (nivel de stand) para explicar el fracaso de regeneración a largo plazo de una especie de árbol. Hemos revisado las características de la historia de vida del abeto, comportamiento a largo plazo a una escala regional, y evidencia a partir de estudios recientes en el contexto de procesos a largo plazo de ecosistemas forestales. Luego simulamos 300 años usando el contexto del modelo neutral (sensu Caswell 1976), en relación al ramoneo, para testear si otros factores podrían explicar adecuadamente la falta de regeneración de abetos en el norte de los Estados de los Lagos. Los resultados sugieren que el ramoneo no es el paso crítico que liga el reestablecimiento del bosque de abeto a una escala regional. Sugerimos que la interacción del clima,*

to modify current single-species forest management and silviculture, and to manage forests at a landscape scale in a way that maintains regional biodiversity and sustainable forest ecosystems. These modifications should include reduced deer levels and maintenance of larger patches of mature ecosystems. But we caution against broadly invoking the negative effects of any single factor such as browsing as a basis for management changes without a thorough assessment of all factors.

perturbaciones, historia de vida del abeto, procesos ecosistémicos, y uso histórico de la tierra produce retroalimentaciones positivas que previenen la regeneración del abeto usualmente antes que el ramoneo por ciervos pueda llegar a ocurrir. Es necesario modificar el manejo y la silvicultura monoespecíficos, y manejar los bosques a una escala del paisaje de forma tal que mantenga la biodiversidad regional y los ecosistemas forestales sostensibles. Estas modificaciones deberían incluir niveles reducidos de ciervos y el mantenimiento de patches más grandes de ecosistemas maduros. Sin embargo, prevenimos contra la invocación amplia de los efectos negativos de cualquier factor en particular, tal como el ramoneo, como la base de los cambios de manejo sin una exhaustiva evaluación de todos los factores.

Introduction

Critical factors influencing the early stages of plant establishment (especially for long-lived trees) are sometimes overlooked in vegetation studies that focus on the influence of a single detrimental factor such as herbivory. The complexity of ecosystems implies that many factors are involved in determining the patterns of forest succession and community development. Therefore, an analysis based on a multiple hypothesis approach (Chamberlin 1892) appears more appropriate than restriction to a single factor.

Recently, Alverson et al. (1988) suggested that heavy browsing resulting from high white-tailed deer (*Odocoileus virginianus*) populations is the primary factor preventing the regeneration of eastern hemlock (*Tsuga canadensis*) in northern Wisconsin and neighboring areas. Hemlock is of particular interest because it has been reduced from its former position as a regional dominant in the northern lake states to where it now occupies only 0.5% of the landscape (Eckstein 1980). We believe that the question of hemlock regeneration requires a critical examination in an expanded spatial and temporal context, especially since hemlock utilizes several modes of establishment and occurs in several different situations. In attempting to explain the current greatly reduced role of hemlock, we will consider the life-history characteristics of hemlock and the long-term behavior of the species on a regional scale and in the light of long-term forest ecosystem processes. Since "... early stages of growth are especially critical for the establishment and survival of hemlock" (Olson et al. 1959), a conclusion which Coffman (1978) and others have further demonstrated, we will document the importance of site conditions, developmental biology, and past establishment patterns to indicate the vital role of these factors in hemlock regeneration.

With the background developed in this review, we will examine current evidence on the role of white-tailed deer in the failure of hemlock regeneration (Alverson et al. 1988). We will develop a neutral model framework (see Caswell 1976) in relation to browsing effects. In this context we propose that a variety of intrinsic ecosystem processes and hemlock life-history characteristics interacting within climatic constraints explain the failure of hemlock recovery both regionally, and to a considerable extent locally, without invoking the effects of browsing. We will then examine the results of this neutral model using computer simulations of forest succession.

Clearly, herbivores can have a major impact on vegetation. In north temperate regions of eastern North America, white-tailed deer population densities above 3–5 animals/km^2 can seriously affect the reproduction or survival of several woody species. The occurrence and effects of browsing are well documented (see Swift 1948; Stoeckeler et al. 1957; Jordan & Sharpe 1967; Anderson & Loucks 1979; Frelich & Lorimer 1985), but consideration of all evidence concerning hemlock reproduction suggests that many factors are involved in preventing its recovery as a regional dominant. Examination of the evidence suggests that Alverson et al. (1988), among others, have erred in part by mixing scales of observation (Allen & Starr 1982) in invoking site-specific phenomena as explanations for regional conditions.

There is no direct evidence that deer populations in some manner reduced to 2/km^2 (or eliminated completely) will result in the widespread regeneration of hemlock forests (Alverson et al. 1988). Certainly, in a limited number of favorable sites and with favorable stochastic events, elimination of browsing will permit more regeneration. However, hemlock—which in presettlement time was a major component of several forest types (Godman & Lancaster 1990)—is most unlikely ever again to reach its earlier extent as a major codominant in stands with white pine (*Pinus strobus*) or yellow birch (*Betula alleghienensis*), or as a common associate in northern hardwood stands with species such as sugar maple (*Acer saccharum*) and basswood (*Tilia americana*) and in other forest types.

We wish to state clearly that we share the view of many observers that recent deer densities have been unusually high and that the associated browsing intensity may pose threats to plant and animal species more characteristic of late-successional and old-growth ecosystems. We also agree that there is a need to manage forests at the landscape scale, with larger patches of mature and old-growth forest, to restore and maintain a diverse array of functioning ecosystems and biodiversity (Stearns 1990; Mladenoff & Pastor 1993; Mladenoff et al. 1993). These steps alone, however, will not restore hemlock to its presettlement dominance.

Discussion

Hemlock Life-History Characteristics

SEED AND SEEDLING DEVELOPMENT AND SITE REQUIREMENTS

Hemlock produces small seed, and seeds less abundantly than most of its associated tree species (Olsen et al. 1959; U.S. Department of Agriculture 1974; Godman & Mattson 1976).

Seed shedding begins in September and continues into the winter. Without disturbance, the small seed is often buried under fresh leaf litter. The seed is dormant when shed, preventing germination before winter. After cold stratification for at least 60 days, the seed will germinate rapidly when the temperature rises in the spring (Olson et al. 1959). In one study, hemlock seed was found to germinate at a rate greater than 60% in the spring following seedfall (Mladenoff 1990), but reduced viability (25%) is common (Godman & Lancaster 1990). Hemlock seed germinates most successfully in a moist, relatively warm site (7° to 18° C), and the seedlings are highly susceptible to desiccation (Olson et al. 1959; Coffman 1978; Godman & Lancaster 1990). After three months, the root may be less than 6 to 10 cm in length, while the short shoot bears only a few needles. Seedlings from northern areas may be limited to as few as 3 to 9 needles in the first year. When buds form, growth ceases (Olson et al. 1959). Seedlings of this size are easily buried in the autumn by leaf fall from sugar maple and other deciduous species.

The slow growth of the young seedling and its susceptibility to desiccation are critical factors in establishment. Olson et al. (1959) reported that drying stratified seed only once delayed germination, and as germination proceeded drying for 2 to 6 hours resulted in serious injury or death for the young seedling. Coffman (1978) has shown that repeated drying slows the rate of germination and that fewer seeds germinate. The roots of hemlock seedlings are soon desiccated. They cannot penetrate hardwood litter since the surface layers of leaves dry out rapidly, nor can seedlings survive on conifer litter (Barnes 1991).

Seedling establishment may occur in three different situations, in each of which the nature of the substrate influences the critical factor, that of a consistent source of available moisture for the seedling. On upland sites, ground fires may consume the litter, exposing bare mineral soil. Erosion produces a similar result on steep slopes. Where partially shaded, such sites remain moist, and young seedlings can become established (Jordan & Sharpe 1967; Barnes 1991).

Loamy sands or sandy and silty loams appear to be the most suitable substrates (F. Goff in Milfred et al. 1967; Rogers 1978). Some shade is beneficial because radiation increases water loss and the stems of young seedlings are susceptible to sunscald. Opportunity for the young roots to enter moist mineral soil accounts for the successful establishment of hemlock seedlings on north-facing road cuts and on steep slopes where mineral soil is exposed. Similarly, the exposed mineral soil of a tip-up mound may provide a suitable substrate.

While hemlock seedlings are often seen growing on a recently fallen log or fresh stump, there is little chance that these seedlings will survive to maturity because drought will usually eliminate them before the wood has decomposed sufficiently to retain moisture. Once the wood is lying on the ground and has reached an advanced state of decomposition, however, roots of hemlock seedlings can penetrate the rotten wood. The seedling roots may be confined within the decayed wood until the seedling is several feet tall. Roots growing in rotting wood are extensively branched and may form numerous mycorrhizae (Stearns 1951). The moss covering found on logs and stumps may help to retain moisture and is itself an indication of moist conditions. Hemlock is most readily established in sites that remain moist continuously, such as acid hardwood-conifer swamps, valley bottoms, and mineral soils adjacent to lakes and wetlands (Westfield 1933; Rogers 1978; Davis et al. 1986; Barnes 1991). In those sites, adequate moisture, decaying logs (often a result of windthrow of canopy trees), and other suitable seedbeds are the factors contributing to survival.

Once established, hemlock seedlings grow slowly for the first 10 or 15 years; with adequate light and limited competition, later growth is more rapid. Hemlock is highly tolerant of shade, but saplings, small trees, and even mature trees respond quickly to opening of the canopy with increased growth.

Hemlock is shallowly rooted, a characteristic making it especially susceptible to drought (Secrest et al. 1940) as well as fire and windthrow. During the pine logging period, more than five billion board feet of hemlock were destroyed by fire in Wisconsin alone (Roth 1898). However, ground fires can produce excellent seedbeds for hemlock regeneration. A number of authors (see Graham 1941; Maissurow 1941; F. Goff in Milfred et al. 1967; Eckstein 1980; Simpson et al. 1990; Barnes 1991) have noted the importance of fire in promoting hemlock seedling establishment. As F. Goff states in Milfred et al. (1967):

> Hemlock may become established in very high densities following surface fire beneath an intact overstory. In many places along the Shawano-Menominee County (Reservation) line where fire burned into the Menominee pine-hemlock stands from adjacent cut-over areas, a dense hemlock "fire-fringe" may be observed. Trees 1–6 inches dbh often exceed 1000 per acre in such areas.

Others have pointed out the role of wind in creating substrates for seedling establishment.

The coincidence of the essential components—an adequate seed crop, several wet years, and disturbance adequate to produce a seedbed—occurs infrequently, resulting in episodic establishment of hemlock reproduction in all save continuously moist sites. This episodic pattern can be traced well back into presettlement time and has been noted by several workers (see Lloyd 1900; Hough & Forbes 1943; Stearns 1947; Hett & Loucks 1976). Such coincidence of climatic conditions with disturbance is now much less frequent than it used to be (Hix & Barnes 1984). In the younger stands that predominate today, there is also far less well-decayed large woody debris to serve as a favorable substrate than in the few remaining old-growth stands (Tyrrell 1991).

HERBIVORY

Woody twigs provide a major source of winter nourishment for several herbivores, including the snowshoe hare (*Lepus americanus*) and the white-tailed deer. Browsing is primarily a winter activity in which twigs and buds are consumed along with attached conifer needles. In summer, white-tailed deer range widely, consuming leaves of deciduous trees, forbs, and ground vegetation (Dahlberg & Guettinger 1956; Hine & Nehls 1980). In the case of eastern hemlock, a preferred browse species, deer can destroy young trees by removing buds and needles (Swift 1948; Stoeckeler et al. 1957). Where deer are abundant, stems that are visible above the snow line will be heavily browsed. These heavily browsed stems have lost terminal buds and most photosynthetic tissue and rarely survive to produce a sapling. However, if in fact browsing alone is responsible for the failure of hemlock regeneration, one must assume that there are abundant hemlock seedlings large enough to be browsed.

Great Lakes Forests and Long-Term Ecosystem Dynamics

HOLOCENE AND SETTLEMENT HISTORY

Forest ecosystems as they existed at the time of European settlement differed from the forests of today not only because of the human activities that have occurred since then, but also as a result of species interactions with the biotic and abiotic conditions that formed the presettlement forest. It is doubtful, with a changed and constantly changing climate and altered ecosystem interactions, that removing any one factor, such as deer browsing, would alone allow these extensive forests to return. Viewed in the context of forest development and change through the Holocene, the presettlement forests do not represent a benchmark, but rather a snapshot in time (Stearns 1990).

Palynological evidence (Webb 1974; Bernabo & Webb 1977; Swain 1978; Davis 1981, 1987) provides the broad temporal context for understanding the presettlement forest and the relative absence today of a once widespread species, such as hemlock. After the glacial ice retreated, tree species migrated into this region at different rates and times. Hemlock successfully invaded the northern portion of the western Great Lakes only recently, reaching its current limit in northern Wisconsin about 1800 years BP (before present) and its presettlement abundance only in the past 1000–2000 years (Davis 1981; Davis et al. 1986). The hemlock–northern hardwoods community existed as a dominant assemblage for approximately 2000 years, only several generations for a long-lived species such as hemlock (Davis 1981). Hemlock was also codominant with white pine (*Pinus strobus*) in many stands (Roth 1898) and was often present in patches that were associated with and integrated into northern hardwoods (see Potzger 1946; Stearns 1951; Goff in Milfred et al. 1967).

It is instructive to consider these long-term changes in a regional context of ecosystem interactions and hemlock life-history characteristics. The Holocene hemlock invasion occurred in a landscape that had been dominated by other conifers, spruce, and pine, for 6000–7000 years (10,000–3500 years BP). In the millennia since retreat of the glacial ice, soil and forest-floor development were controlled largely by conifer leaf litter. Conifer litter exerts a profound effect on soil factors by depressing the availability of essential nutrients, particularly nitrogen. By reducing competition from more nutrient-demanding hardwoods, this effect favored the less demanding conifers and produced a positive feedback within the system (Baxter in Milfred et al. 1967; Pastor et al. 1984). Such soil conditions would have favored the expansion of hemlock in these forests once it was no longer limited by dispersal (Davis et al. 1986). Although no causal relationship can be clearly documented, tree ages indicate that the extensive hemlock stands existing at the time of settlement were established during the geologically brief Little Ice Age (Lamb 1966; Bryson & Murray 1977; Swain 1978; Frelich & Lorimer 1991), that lasted from the late 1300s to the mid-1800s.

RECENT CONTROLLING FACTORS

In an ecosystem context, the current landscape of the northern lake states is a successional forest mosaic caused by widespread human activity at a time (1850 to present) when the climate was perhaps becoming less favorable for hemlock (Fig. 1). During the post-settlement period, hemlock was logged heavily, both for lumber and bark (Roth 1898; Corrigan 1976; Karamanski 1989). The heavy cutting, frequent destructive fires, and discrimination against hemlock as a low-value species resulted in eliminating the seed source over extensive areas of its original range. These region-wide disturbances contributed to the following changes: (1) the destruction of the conifer litter–controlled forest floor that had favored the nutrient cycling requirements of conifers; (2) the elimination of large-diameter woody debris that provided establishment sites for seedlings; (3) regional elimination of seed sources by logging, fire, and forest management; and (4) creation of pure deciduous successional seres that, with fire suppression, caused positive feedbacks that (*a*) maintain deciduous litter seedbeds that further favor hardwood establishment while discouraging hemlock, and (*b*) further alter soil properties and nutrient cycling through deciduous litter input to favor deciduous species (Pastor & Mladenoff 1992).

This explanation is consistent with several lines of recent evidence. Hemlock and hardwood patches and even small single-tree gaps have different nitrogen mineralization dynamics (Mladenoff 1987). Also, Frelich et al. (1993) have shown that adjacent old-growth hemlock and hardwood stands, rather than undergoing reciprocal replacement, tend to have maintained their separate identities since the arrival of hemlock 2000–3000 years BP. An adequate seed source and appropriate site conditions are both regionally sparse, however,

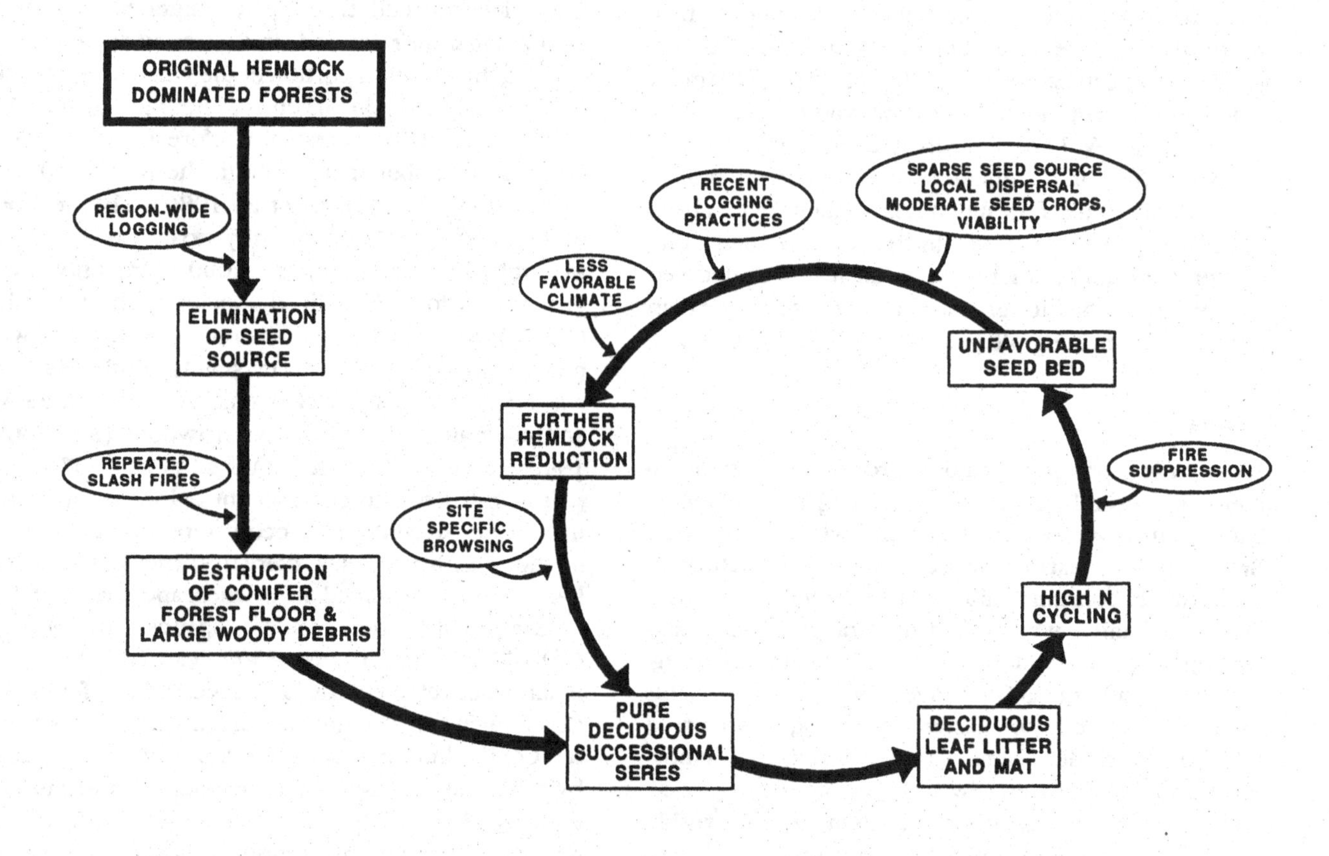

Figure 1. Dynamics of ecosystem processes and life-history characteristics influencing regional hemlock loss and recruitment failure in the western Great Lakes region.

and the probability of their co-occurrence on a landscape with suitable climatic conditions is extremely small (Fig. 1).

In this long-term ecosystem context, large deer populations can provide an additional positive feedback on certain sites where hemlock reproduction does occur and deer browse in winter. Deer contribute as a feedback in the system because of the same factor that limits regional seed availability. Hemlock has been reduced on the regional landscape to scattered remnant stands. Many of these few remaining stands are on low, wet-mesic sites where hemlock may reproduce more consistently than on well-drained uplands. These protected sites with high-density conifer cover and browse are favored as winter deer yards, for thermal cover and food (Eckstein 1980). This fact, coupled with the large deer populations of today, clearly places stress on remnant hemlock stands. But this is not evidence that deer browsing serves as the control for hemlock regeneration across the regional landscape (Fig. 1).

Evidence from Recent Studies

Exclosures that fence out deer or other large herbivores have often been used to demonstrate browsing impact, as indeed they do. But exclosures provide a limited regional view. Most of the handful of exclosures established in the 1940s and 1950s in the lake states were built in areas with large winter deer populations. Stands of existing hemlock were chosen where reproduction was also observed or suspected (Stoeckeler et al. 1957; Anderson & Loucks 1979). It is important to note that because exclosure locations were usually chosen where dramatic results were expected, these exclosures hardly constitute an adequate or unbiased sample of regional forest conditions.

As a specific example, the Foulds Creek exclosure (pictured by Alverson et al. 1988) was established in 1946 in a large deer yard where deer concentrations have reached as high as 230/km^2. The site was also particularly favorable for hemlock reproduction, with a mixed hardwood/conifer cover, windthrown fir, and thin litter (C. Wiita, Wisconsin Department of Natural Resources, personal communication).

The work by Frelich and Lorimer (1985) is cited by Alverson et al. (1988) as definitive in illustrating the effects of deer browsing, rather than climate or other factors, in limiting hemlock reproduction. Frelich and Lorimer (1985) described the effect of deer on hemlock in a Lake Superior shoreline stand within the Porcupine Mountains of upper Michigan. The lakeshore stands are occupied as deer yards in winter, and hemlock reproduction there is heavily browsed. Frelich and Lorimer (1985) logically conclude that deer would eventually prevent maintenance of the lakeshore hemlock forest. They found that other park stands farther from the lake within the Carp River valley have lower winter deer populations than lakeshore stands and sustain successful hemlock reproduction and recruitment.

There are many upland areas within the park, however, more distant from the lake influence than is the Carp River valley, that contain abundant mature hemlock. These stands have still lower browsing evidence than the valley bottom stands, but they usually lack hemlock seedlings (Mladenoff 1990). The results of Hett and Loucks (1976) indicating asynchronous hemlock reproduction events between noncontiguous stands also suggest the importance of site-specific conditions and infrequent, episodic events in hemlock reproduction.

Alverson et al. (1988) present data from a study (Fig. 2) comparing hemlock seedling abundance and adult trees from sites stated to have two different deer pop-

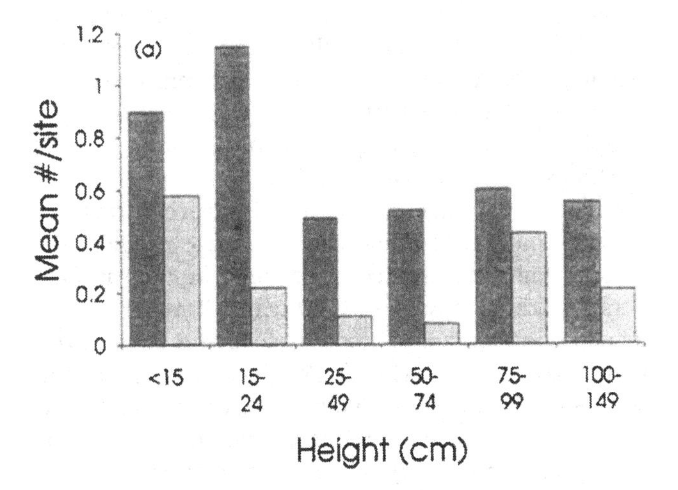

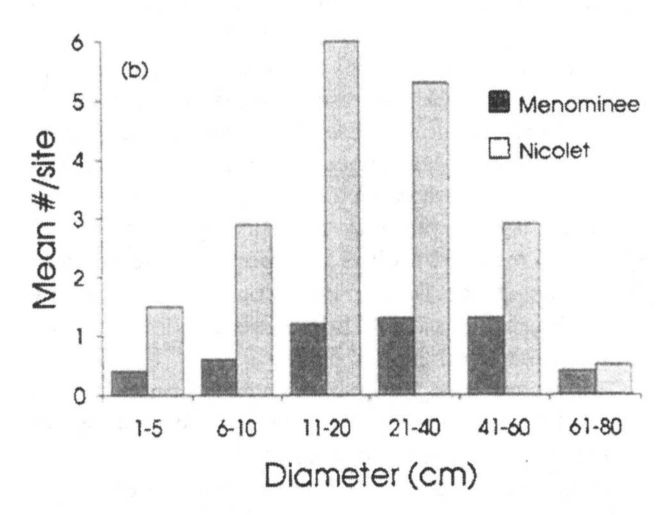

Figure 2. Size distributions of eastern hemlock seedlings (a) and adult trees (b) in the Menominee Reservation and the adjacent Nicolet National Forest. Deer densities are much lower in the Menominee Reservation due to year-round hunting. Caption and graphs (redrawn) from Alverson et al. 1988.

ulation and browsing levels. Stands on tribal lands of the Menominee Reservation, with lower deer densities because of year-round hunting, are compared to Nicolet National Forest stands. Alverson et al. (1988) suggest that the data demonstrate that, despite mature hemlock trees being more abundant in the Nicolet stands (Fig. 2*b*), hemlock seedlings are less abundant in the Nicolet stands than in the Menominee sample, apparently as a result of greater deer browsing on the national forest (Fig. 2*a*).

But these data do not permit an unambiguous conclusion. The different deer population levels in the Menominee and Nicolet stands are not documented. The tree-size class graphs presented are difficult to interpret without area units as labeled ("Mean # individuals/site"). Methods, such as stand selection, location, conditions or histories, number of sites, site size, stand composition, and sampling methods are not provided. Error bars or statistical tests of the differences in various size classes are also not included on the graphs. For seedlings, the means for each size class compare differences equalling less than one individual seedling per sample, with a total range of approximately 0.1–1.2 seedlings over all classes (Fig. 2).

Given such data, alternative interpretations are equally possible. It would be informative to know the ages of stems represented by seedling, sapling, and understory tree-size classes (1–20 cm in diameter), and it may not be accurate to graph these classes subjectively under "adult trees" (Fig. 2). These are likely all understory size classes, in fact indicating greater hemlock recruitment over time in the Nicolet than on the Menominee, except in the very youngest size classes. Given the mean number of stems shown, hemlock seedlings in the smallest height class (<15 cm) and those above 75 cm do not appear to be significantly different between the Menominee and Nicolet stands. This may suggest that although the middle sizes are browsed, as much reproduction and recruitment has taken place in the Nicolet stands as in the Menominee. Over the lifetime of the canopy trees, which can exceed 400 years (Godman & Lancaster 1990), hemlock may be reproducing adequately to maintain itself as an important component in the Nicolet stands. Data on the amount of other canopy or seedling species would be necessary to interpret the meaning of these hemlock data.

The Menominee Reservation stands contain much successful reproduction, but several major differences besides deer numbers are involved. For example, stand histories (including fire), other species present, soils, and the relative abundance of organic substrates such as downed logs must be considered. Again, it is not our main purpose to assert that deer do not negatively affect hemlock reproduction, but rather to suggest that hemlock reproduction is limited across the northern lake states by a variety of causes. The Nicolet-Menominee comparison of the importance of browsing appears weak as a general case.

Modeling Results

MODEL DESCRIPTION

We used a neutral model framework (see Caswell 1976; Pastor et al. 1987) to examine the hypothesis that hemlock life-history characteristics, ecosystem interactions, and climate adequately explain the regional lack of hemlock. To examine this hypothesis and explore the stand dynamics of hemlock and its canopy species associates, we conducted model runs using LINKAGES, a well-validated forest ecosystem/succession model (Pastor & Post 1986). LINKAGES is a member of the widely used JABOWA/FORET family of forest "gap models" (Botkin et al. 1972; Shugart 1984; Botkin 1993). Gap models simulate forest succession through the establishment, growth, and death of trees in plots that correspond to the size of single canopy trees. Species life-history variables are parameterized based on empirical, published silvics data (see Burns & Honkala 1990). LINKAGES includes ecosystem processes such as litter decomposition and nitrogen mineralization as well as demographics. Tree species respond to soil moisture, nitrogen availability, and light levels, as constrained by site factors and climate—in this case without large disturbances. This makes the model particularly appropriate for examining our hypothesis that these ecosystem factors, the life-history responses of hemlock, historical changes, and a lack of large disturbances are of prime importance in understanding the past and current behavior of hemlock on the landscape.

OLD-GROWTH HEMLOCK SIMULATION

We conducted model runs with succession beginning from composite stand data from field samples taken from stands dominated by old-growth hemlock in Vilas county, in north central Wisconsin (Figure 3*a*) (Mladenoff et al. 1993). These stands are located on typical hemlock-hardwood sites, on loamy soils of morainal landforms (Curtis 1959; Attig 1985). The data represent old-growth stands dominated by hemlock (>60% canopy basal area) (Figure 4), which commonly occurred across northern Wisconsin and upper Michigan (Curtis 1959).

The model was run under current climate conditions (1950–1980 temperature and precipitation), not as a prediction but to isolate other important variables. Climate varies randomly in the simulation within the standard deviations of the 1950–1980 period. We ran the model with the assumption of no browsing impact, to maximize possible hemlock success under current climate and the modal site conditions simulated. Simula-

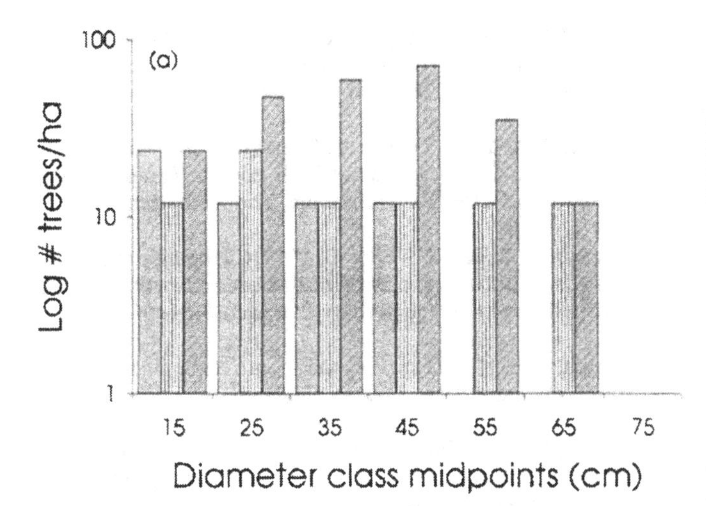

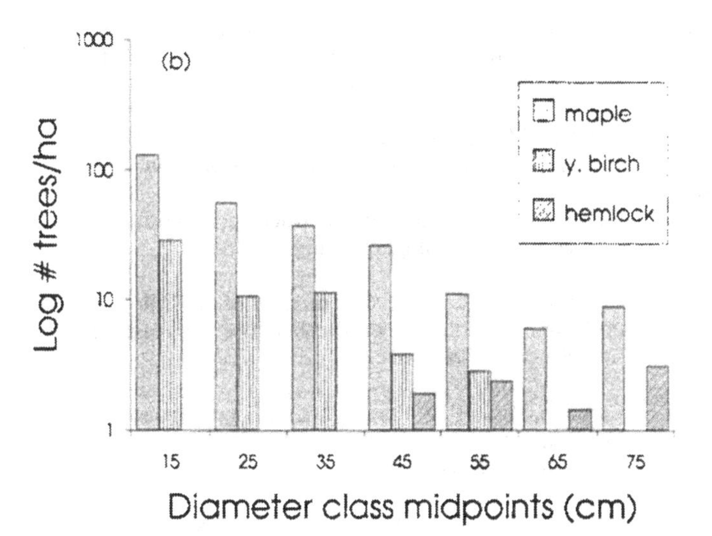

Figure 3. Original (current) stand composition data (log density of major canopy species by size class) (a) used to begin model simulations, and (b) final stand composition data produced from 400-year simulation of old-growth hemlock on loamy uplands.

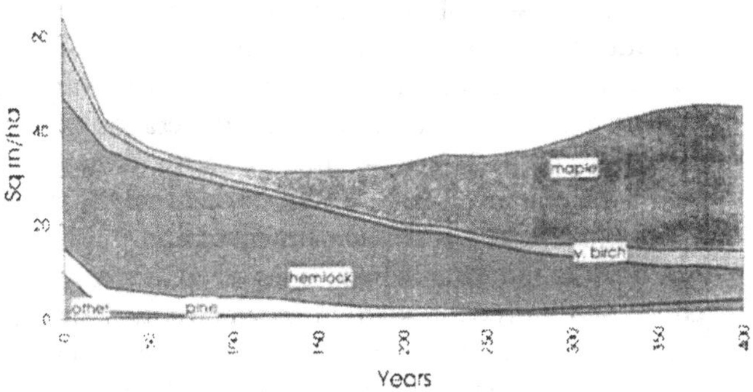

Figure 4. Model output of basal area change during 400-year simulation of old-growth hemlock on loamy uplands.

tions assume that seed of all species is available, with successful establishment an outcome of appropriate seedbed availability, species interactions, and site factors (Pastor & Post 1986). The current hemlock-dominated stand was simulated through 400 years of change, with gaps resulting from the death of individual trees, a condition typical of these stands over most of the regional landscape (Figure 3*b*).

For the first 100–125 years, the basal area of hemlock and the codominants yellow birch and white pine decline as the large canopy trees die (Figure 4). Sugar maple is initially present in small amounts in the canopy and in smaller size classes, but survival and recruitment are low, probably limited by the hemlock stand environment and low nitrogen availability. With the death of the large canopy trees, sugar maple gradually begins to take advantage of canopy gaps and higher nitrogen availability as deciduous species transform the stand and forest floor. This trend accelerates, leading to a stand largely dominated by sugar maple (Figures 3*b* and 4). For the first 100–200 years, lack of hemlock recruitment could be ascribed to effects of deer over the past 75 years. Even without browsing, however, hemlock continues to decline, without significant recruitment throughout the entire simulation. The low-level persistence of hemlock at the 300-year point in the simulation results from a few long-lived individuals. Extending the simulation to 500 years (not shown) indicated the same trend continuing, with greater sugar-maple dominance and less hemlock. The model simulation is therefore consistent with our hypothesis. Specifically, the model output suggests that other factors—site, climate, and disturbance—are limiting hemlock reproduction on typical upland sites before deer browsing can come into play.

COARSE WOODY DEBRIS SUBSTRATE SIMULATION

We conducted a second model run to examine the availability of downed log substrates as seedbeds for hemlock in young and old-growth stands. The lower level of well-decayed, coarse woody debris in younger stands (<250 years old) compared to older (>300 years old) old-growth stands (Tyrrell 1991) is an outcome of regional forest change (Figure 1) not addressed by Alverson et al. (1988). In stands lacking disturbance that would create favorable soil seedbeds, large, partially decayed logs and stumps have been found to be favorable establishment sites (see Stearns 1951; Goder 1961; Godman & Lancaster 1990). This large woody debris would need to be present at levels and sizes and in decay classes at least approaching the presettlement condition to promote recovery of hemlock.

This second simulation compares the amounts of woody debris present in young stands that now characterize the region to that in old-growth stands. The model was run to simulate recovery and stand regener-

factor, and less well documented conditions—such as climatic events and wildfire in combination with historical changes including past logging—are clearly the primary influences limiting extensive regeneration of hemlock. Deer reduction alone, under current climatic and disturbance regimes, will not produce hemlock forests approximating the presettlement condition.

A broader question at issue is the need for those of us concerned with such matters as species diversity and the implications of human activity and climate change to approach these problems accepting the probability that more than one causal factor will be involved.

There is a real need to modify traditional single-species management and silvicultural practices to restore and maintain a more balanced array of regional biodiversity in the forest landscape. These goals should include deer herd reduction and habitat modifications to favor late-successional ecosystems and species favoring larger intact patches of forest. We believe that justification for such restoration and management goals must be based on a thorough analysis of all causal factors. The case of eastern hemlock regeneration is a useful illustration of why these issues must be carefully assessed.

Acknowledgments

We thank Phyllis Barnidge for drafting Figure 1. The model simulations in LINKAGES were programmed and run by Joel Boeder, who also produced the output figures. We appreciate comments on the manuscript and assistance we received with LINKAGES from John Pastor. We wish to thank Ronald Eckstein, Kieth McCaffery, and Lucy Tyrrell for their suggestions on drafts of this paper and for providing unpublished data. We are also grateful to Burton Barnes, Hal Salwasser, and two anonymous reviewers for helping us focus our presentation. This is contribution number 111 of the Center for Water and the Environment, Natural Resources Institute, University of Minnesota, Duluth.

Literature Cited

Allen, T. F. H., and T. B. Starr. 1982. Hierarchy: Perspectives for ecological complexity. University of Chicago Press, Chicago, Illinois.

Alverson, W. S., D. M. Waller, and S. L. Solheim. 1988. Forests too deer: Edge effects in northern Wisconsin. Conservation Biology 2:348–358.

Anderson, R. C., and O. L. Loucks. 1979. Whitetail deer (*Odocoileus virginianus*) influence on structure and composition of *Tsuga canadensis* forests. Journal of Applied Ecology 16:855–861.

Attig, J. W. 1985. Pleistocene geology of Vilas County, Wisconsin. Information Circular 50. Wisconsin Geological and Natural History Survey, Madison, Wisconsin.

Barnes, B. V. 1991. Deciduous forests of North America. Pages 219–344 in E. Röhrig and B. Ulrich, editors. Temperate deciduous forests. Elsevier, Amsterdam, The Netherlands.

Bernabo, J. C., and T. Webb, III. 1977. Changing patterns in the Holocene pollen record of northeastern North America: A mapped summary. Quaternary Research 8:64–96.

Botkin, D. B. 1993. Forest dynamics: An ecological model. Oxford University Press, Oxford, England.

Botkin, D. B., J. F. Janak, and J. R. Wallis. 1972. Some ecological consequences of a computer model of forest growth. Journal of Ecology 60:849–872.

Burns, R. M., and B. H. Honkala, coordinators. 1990. Silvics of North America. Agriculture Handbook 654. Forest Service, U.S. Department of Agriculture, Washington, D.C.

Bryson, R. A., and T. J. Murray. 1977. Climates of hunger. University of Wisconsin Press, Madison, Wisconsin.

Caswell, H. 1976. Community structure: A neutral model analysis. Ecological Monographs 46:327–354.

Chamberlin, T. C. 1892. The method of multiple working hypotheses. Republished by the Institute for Humane Studies, Inc., Stanford, California.

Coffman, M. S. 1978. Eastern hemlock germination influenced by light, germination media, and moisture content. The Michigan Botanist 17:99–103.

Corrigan, G. A. 1976. Calked boots and cant hooks. Privately published. 2nd edition, 1986, Northwood Inc., Ashland, Wisconsin.

Curtis, J. T. 1959. The vegetation of Wisconsin. University of Wisconsin Press, Madison, Wisconsin.

Dahlberg, B. L., and R. C. Guettinger. 1956. The white-tailed deer in Wisconsin. Technical Wildlife Bulletin 14. Wisconsin Conservation Department, Madison, Wisconsin.

Davis, M. B. 1981. Quaternary history and the stability of forest communities. Pages 132–153 in D. C. West, H. H. Shugart, and D. B. Botkin, editors. Forest succession. Springer-Verlag, New York.

Davis, M. B. 1987. Invasions of forest communities during the Holocene: Beech and hemlock in the Great Lakes region. Pages 373–393 in A. J. Gray, M. J. Crawley, and P. J. Edwards, editors. Colonization, succession and stability. Blackwell, Oxford, England.

Davis, M. B., K. D. Woods, S. L. Webb, and R. P. Futyma. 1986. Dispersal versus climate: Expansion of *Fagus* and *Tsuga* into the Upper Great Lakes region. Vegetatio 67:93–103.

Eckstein, R. G. 1980. Eastern hemlock (*Tsuga canadensis*) in north central Wisconsin. Research Report 104. Wisconsin Department of Natural Resources, Madison, Wisconsin.

ation following clearcutting, on the same site type as simulated above. The simulation produced a mixed aspen-birch–dominated stand that gradually succeeded to sugar maple after about 75 years, typical of much of the region today (Figure 5*a*). Model output includes all dead wood in the stand, including both snags and downed logs, of which the major proportion is downed logs (Figure 5*b*). The model indicates woody debris present in old-growth stands (250 years old) of approximately 30.0 Tons/ha. It requires 80–100 years for this level to be reached in new stands (Figure 5*b*), ages that are now being reached by only the oldest second-growth stands. Much of the landscape is covered by stands of 50–80 years of age. The oscillations through time reflect the death of tree cohorts, since the stand began as even-aged. Thus the first peak in woody debris (80–120-year-old stands) is primarily aspen and birch, which are relatively small in size and decay rapidly. These oscillations in woody debris may be less apparent in stands where partial canopy disturbances more

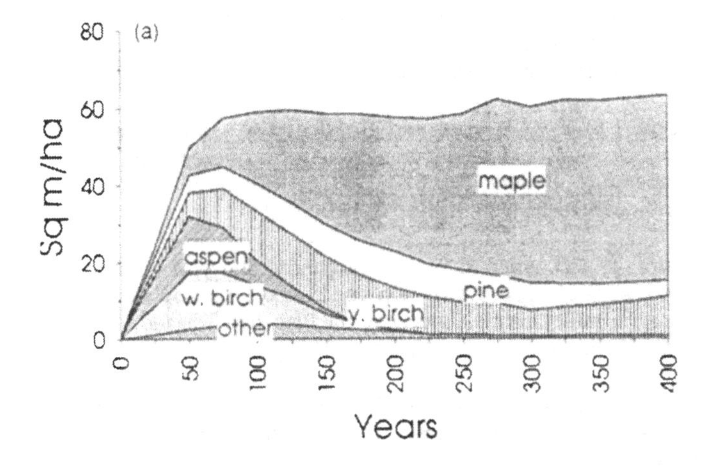

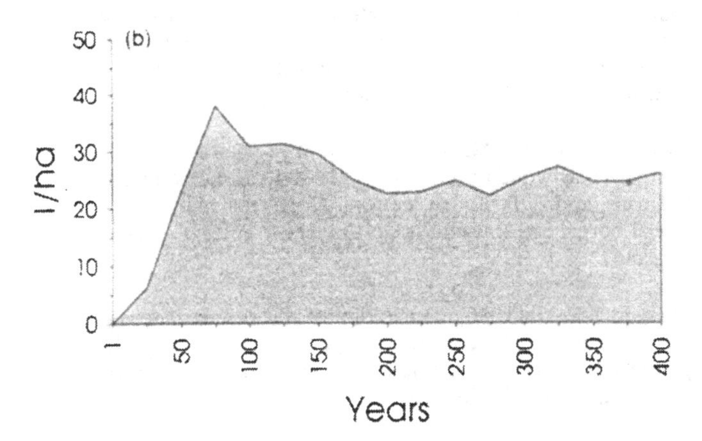

Figure 5. Model output of simulated stand recovery on loamy uplands following clearcutting on the same site as simulated in Figures 3 and 4, typical of the region today. Graphs of (a) tree species basal area and (b) dead wood biomass during 400-year simulation beginning from bare ground.

quickly cause an uneven-aged condition to develop. Aside from the quantity of woody debris, small size is a qualitative difference in the downed wood between young and old growth that affects its suitability as a seedbed. In the simulation it required 200–250 years for large trees (>25 cm diameter at breast height) to grow and begin to die in significant numbers, which as logs would persist long enough on the forest floor to provide a stable starting substrate for seedlings. Field data from 25 stands, with hemlock logs only reaching 720 m^2/ha in stands more than 250 years (Tyrrell 1991), appear to validate the results of the model. Also, the larger and more persistent logs of conifer species—hemlock, white pine, and white cedar (*Thuja occidentalis*)—rarely or never occur in these modeled young stands (Figure 5*a*). Total dead wood can vary greatly between stands and on different sites (Tritton & Siccama 1990; Tyrrell 1991), but these simulated values represent the relative difference of old and young stands on similar sites. Based on model simulations, there is both a great quantitative and qualitative reduction in these organic establishment sites in stands that reflect current regional conditions, a reduction clearly evident on the ground (Tyrrell 1991).

Conclusions

Alverson et al. (1988) and others have stated that white-tailed deer are controlling the regeneration of eastern hemlock. They have generalized a causative explanation for regional conditions from stand-level phenomena at selected locations by mixing the scales of observation and causal effects. Hemlock reproduction is clearly influenced by a number of variables, including climate, disturbance, regional land-use history, and life-history characteristics. A combination of either a moist organic substrate or ground layer disturbance with sustained favorable weather conditions is essential for seed germination and seedling establishment. This combination of suitable seedbed and climatic conditions now rarely occurs. Even if these conditions were to occur, past forest use (management) practices have resulted in elimination of the seed source from most of the area where the species was a regional dominant but now occupies only about 0.5% of the landscape. Before white-tailed deer can have an impact, hemlock seeds must be produced and germinated and seedlings grown tall enough to be visible above the snow line so that they are available for deer to browse.

Our point is not to suggest that deer do not have a serious, site-specific impact on hemlock regeneration and on other species, or that they have not had an influence in the recent past. Instead the effect of deer must be kept in perspective as only one of several pertinent factors. The basic biology of hemlock is a critical

Frelich, L. E., and C. G. Lorimer. 1985. Current and predicted long-term effects of deer browsing on hemlock forests in Michigan, USA. Biological Conservation **34**:99–120.

Frelich, L. E., and C. G. Lorimer. 1991. Natural disturbance regimes in hemlock-hardwood forests of the upper Great Lakes region. Ecological Monographs **61**:145–164.

Frelich, L. E., R. R. Calcote, M. B. Davis, and J. Pastor. 1993. Patch formation and maintenance in an old-growth hemlock-hardwood forest. Ecology **74**:513–527.

Goder, H. A. 1961. Hemlock reproduction and survival on its border in Wisconsin. Transactions of the Wisconsin Academy of Science, Arts, and Letters **50**:175–182.

Godman, R. M., and K. Lancaster. 1990. *Tsuga canadensis* (L.) Carr. eastern hemlock. Pages 604–612 in R. M. Burns and B. H. Honkala, coordinators. Silvics of North America. Agriculture Handbook 654. Forest Service, U.S. Department of Agriculture, Washington, D.C.

Godman, R. M., and G. A. Mattson. 1976. Seed crops and regeneration problems of 19 species in northeastern Wisconsin. Research Paper NC-123. North Central Forest Experiment Station, Forest Service, U.S. Department of Agriculture, St. Paul, Minnesota.

Graham, S. A. 1941. Climax forests of the Upper Peninsula of Michigan. Ecology **22**:355–362.

Hett, J. M., and O. L. Loucks. 1976. Age structure models of balsam fir and eastern hemlock. Journal of Ecology **64**:1029–1044.

Hine, R. L., and S. Nehls, editors. 1980. White-tailed deer population management in the North Central States. Proceedings of 1979 symposium. North Central Section, The Wildlife Society, Lafayette, Indiana.

Hix, D. M., and B. V. Barnes. 1984. Effects of clear-cutting on the vegetation and soil of an eastern hemlock dominated ecosystem, western Upper Michigan. Canadian Journal of Forestry Research **14**:914–923.

Hough, A. F., and R. D. Forbes. 1943. The ecology and silvics of forests in the high plateaus of Pennsylvania. Ecological Monographs **13**:299–320.

Jordan, J. S., and W. M. Sharpe. 1967. Seeding and planting hemlock for ruffed grouse cover. Research Paper NE-83. Northeast Forest Experiment Station, Forest Service, U.S. Department of Agriculture, Washington, D.C.

Karamanski, T. J. 1989. Deep woods frontier: A history of logging in northern Michigan. Wayne State University Press, Detroit, Michigan.

Lamb, H. H. 1966. The changing climate. Methuen, London, England.

Lloyd, F. E. 1900. Seed and seedlings of the hemlock, *Tsuga canadensis*. Journal of the New York Botanical Garden **1**:97–100.

Maissurow, D. K. 1941. The role of fire in the perpetuation of virgin forests of northern Wisconsin. Journal of Forestry **39**:201–207.

Milfred, C. J., G. W. Olson, and F. D. Hole, with chapters by F. P. Baxter and F. G. Goff; W. A. Creed and F. Stearns. 1967. Soil resources and forest ecology of Menominee County, Wisconsin. Bulletin 85, Soil Series 60. Wisconsin Geological and Natural History Survey, Madison, Wisconsin.

Mladenoff, D. J. 1987. Dynamics of nitrogen mineralization and nitrification in hemlock and hardwood treefall gaps. Ecology **68**:1171–1180.

Mladenoff, D. J. 1990. The relationship of the soil seed bank and understory vegetation in old-growth northern hardwood-hemlock treefall gaps. Canadian Journal Botany **68**:2714–2721.

Mladenoff, D. J., and J. Pastor. 1993. Sustainable forest ecosystems in the northern hardwood and conifer region: Concepts and management. In G. A. Aplet, J. T. Olson, N. Johnson, and V. A. Sample, editors. Defining sustainable forestry. Island Press, Washington, D.C. In press.

Mladenoff, D. J., M. A. White, J. Pastor, and T. R. Crow. 1993. Comparing spatial pattern in unaltered old-growth and disturbed forest landscapes. Ecological Applications **3**:293–305.

Olson, J. S., F. W. Stearns, and H. Nienstaedt. 1959. Eastern hemlock seeds and seedlings: Response to photoperiod and temperature. Bulletin No. 620. Connecticut Agricultural Experiment Station, New Haven, Connecticut.

Pastor, J., J. D. Aber, C. A. McClaugherty, and J. M. Melillo. 1984. Aboveground production and N and P cycling along a nitrogen mineralization gradient on Blackhawk Island, Wisconsin. Ecology **65**:256–268.

Pastor, J., and W. M. Post. 1986. Influence of climate, soil moisture and succession on forest carbon and nitrogen cycles. Biogeochemistry **2**:3–27.

Pastor, J., R. H. Gardner, V. H. Dale, and W. M. Post. 1987. Successional changes in nitrogen availability as a potential factor contributing to spruce declines in boreal North America. Canadian Journal of Forestry Research **17**:1394–1400.

Pastor, J., and D. J. Mladenoff. 1992. The southern boreal—northern hardwood border. Pages 216–240 in H. H. Shugart, R. Leemans, and G. B. Bonan, editors. A systems analysis of the global boreal forest. Cambridge University Press, Cambridge, England.

Potzger, J. E. 1946. Phytosociology of the primeval forest in central northern Wisconsin and upper Michigan and a brief post-glacial history of the Lake Forest formation. Ecological Monographs **16**:213–234.

Rogers, R. S. 1978. Forests dominated by hemlock (*Tsuga canadensis*): Distribution as related to site and postsettlement history. Canadian Journal of Botany **56**:843–854.

Roth, F. 1898. On the forestry conditions of northern Wisconsin. Bulletin 1. Wisconsin Geological and Natural History Survey, Madison, Wisconsin.

Secrest, H. C., H. J. MacAloney, and R. C. Lorenz. 1940. Causes of decadence of hemlock at the Menominee Indian Reservation, Wisconsin. Journal of Forestry **39**:3–12.

Shugart, H. H. 1984. A theory of forest dynamics: An investigation of the ecological implications of several computer models of forest succession. Springer-Verlag, New York.

Simpson, T. B., P. E. Stewart, and B. V. Barnes. 1990. Landscape ecosystems and cover types of the reserve area and adjacent lands of the Huron Mountain Club. Occasional paper No. 4. Huron Mountain Wildlife Foundation, Big Bay, Michigan.

Stearns, F. 1947. The composition of the sugar maple–hemlock–yellow birch association in northern Wisconsin. Ph.D. dissertation. University of Wisconsin–Madison, Madison, Wisconsin.

Stearns, F. 1951. The composition of the sugar maple–hemlock–yellow birch association in northern Wisconsin. Ecology **32**:245–265.

Stearns, F. 1990. Forest history and management in the northern midwest. Pages 107–122 in J. M. Sweeney, editor. Management of dynamic ecosystems. North Central Section, The Wildlife Society, West Lafayette, Indiana.

Stoeckeler, J. H., R. O. Strothmann, and L. W. Krefting. 1957. Effect of deer browsing on reproduction in the northern hardwood-hemlock type in northeastern Wisconsin. Journal of Wildlife Management **21**:75–80.

Swain, A. M. 1978. Environmental changes during the past 2000 years in North-Central Wisconsin: Analysis of pollen, charcoal, and seeds from varved lake sediments. Quaternary Research **10**:55–68.

Swift, E. F. 1948. Wisconsin deer damage to forest reproduction survey. Publication 347. Wisconsin Conservation Department, Madison, Wisconsin.

Tritton, L., and T. G. Siccama. 1990. What proportion of standing trees in forests of the northeast are dead? Bulletin of the Torrey Botanical Club **117**:163–166.

Tyrrell, L. 1991. Patterns of coarse woody debris in old-growth hemlock-hardwood forests of northern Wisconsin and western upper Michigan. Ph.D. dissertation. University of Wisconsin–Madison, Madison, Wisconsin.

U.S. Department of Agriculture Forest Service. 1974. Seeds of woody plants in the United States. Agriculture Handbook No. 450. Washington, D.C.

Webb, T., III. 1974. A vegetational history from northern Wisconsin: Evidence from modern and fossil pollen. American Midland Naturalist **92**:12–34.

Westveld, R. H. 1933. The relation of certain soil characteristics to forest growth and composition in the northern hardwood forest of northern Michigan. Technical Bulletin 135. Michigan Agricultural Experiment Station, E. Lansing, Michigan.

Avian Survival Rates and the Extinction Process on Barro Colorado Island, Panama

JAMES R. KARR

Department of Biology
Virginia Polytechnic Institute and State University
Blacksburg, VA 24061, U.S.A.

Abstract: *Selective extinction following isolation of habitat patches may be due to biogeographical (e.g., island size or isolation) and ecological (species natural histories, interspecific interactions) factors, or their interactions. Among the demographic and life history attributes commonly associated with high extinction probability are small populations, large size of individuals, and population variability. Long-term capture-recapture data from forest habitat in central Panama permit an examination of the association between mainland survival rates and extinction on a nearby land-bridge island. Species of birds that no longer occur on Barro Colorado Island (BCI), Panama, have, on average, lower survival rates on the adjacent mainland than species that have persisted on BCI. Moreover, of the species that no longer occur on BCI, those with lower mainland survival rates generally disappeared earlier from the island. My analysis provides little evidence of a relationship between extinction and population size. Recolonization of BCI from the adjacent mainland by the forest undergrowth species studied here is unlikely. Reduced reproductive success on BCI combined with naturally low adult survival rates seems to be responsible for these BCI extinctions. High nest predation and/or altered landscape dynamics are probable agents in the low reproductive success. The methods used here could be employed in other circumstances to identify fragmentation-sensitive species.*

Resumen: *La extinción selectiva después del aislamiento de fragmentos de hábitat puede ser debida a factores biogeográficos (por ejemplo, el tamaño de la isla o el aislamiento) y ecológicos (biología de las especies, interacciones interespecificas) o a la interacción de ambos. Dentro de los atributos demográficos y biológicos comúnmente asociados con la alta probabilidad de extinción se encuentran las poblaciones pequeñas, los individuos de gran tamaño y la variabilidad dentro de las poblaciones. Los datos obtenidos durante largos peréodos de tiempo en el hábitat de boaque del centro de Panamá, permito examinar la asociación entre los indices de sobrevivencia en tierra firme y la extinción en una isla cercana, conectada con tierra firme. Las especies de aves que ya no se encuentran en la isla de Barro Colorado (IBC) en Panamá tienen, en promedio, indices da sobrevivencia más bajos en la tierra firme adjacente que las especies que han-persistido en la IBC. Además, de las especies que ya no se encuentran en la IBC, aquéllas que téienen bajos indices da sobrevivencia en tierra firme generalmente desaparecieron primero en las islas. De cualquier modo mi análisis ofrece poca evidencia de que exista una relación entre la extinción y el tamaño de la población. La recolonización de la IBC por las especies del sotobosque en la tierra firme adjacente estudiadas aquí es poco probable. El reducido éxito reproductivo en la IBC combinado con los naturalmente bajos indices de sobrevivencia parecen ser los responsables de estas extinciones. La alta predación de nidos y/o la dinámica alterada del paisaje son los probables agentes en el bajo éxito reproductivo. Los métodos utilizados aqui podrían ser usados en otras circunstancias para identificar a las especies sensibles a la fragmentación.*

Paper submitted January 25, 1989; revised manuscript accepted January 15, 1990.

Permanent loss of species (extinction) has devastating consequences on maintenance of ecological systems, on preservation of the genetic diversity needed to maintain the biological integrity of the earth's life-support system, and on availability of medical and agricultural products. Yet rates of extinction and endangerment have increased at an alarming rate as populations are isolated through reduction and fragmentation of their natural habitats. Regrettably, knowledge of the biological attributes that make species susceptible to extinction is at best fragmentary.

Barro Colorado Island (BCI), Panama, a land-bridge relief isolated in 1914 by construction of the Panama Canal is an ideal site for theoretical and empirical exploration of the number and identity of bird species that disappeared from BCI since isolation Willis 1974, Willis and Eisenman 1979; Wilson and Willis 1975; Terborgh 1974. These researchers documented extinctions of about 60 species from BCI since its isolation; over two-thirds disappeared as areas of second-growth forest were replaced by mature forest. According to these authors, about 15 to 18 of the species lost were classed as forest birds. Using a somewhat different approach, Karr (1982*a*, *b*) argued that 50 to 60 species of forest birds present on the mainland adjacent to BCI are not present on BCI. Nearly all of these might reasonably be expected on BCI, so the actual number of extinct forest species is probably much higher than the 15 to 18 estimated earlier.

Extinction of birds on BCI has been attributed to ecological truncation (Wilson & Willis 1975), differential loss of bird species of undergrowth (Wilson & Willis 1975; Karr 1982*b*) and of wet and foothill forest (Karr 1982*a*), and loss of species with variable populations (Karr 1982*b*). Efforts to reintroduce species extirpated from BCI (Morton 1985; Morton 1987) and studies using artificial nests (Loiselle & Hoppes 1983; K. Sieving, in preparation) have identified predation as an additional cause of extinction. Neither abundances on the mainland nor dominant food resources were good predictors of species likely to be missing from BCI (Karr 1982*a*, *b*).

Here I present estimates of survival rates and population sizes, derived from Jolly-Seber capture-recapture models for open populations, for birds of tropical forest undergrowth. I test two predictions derived from the general hypothesis that probability of extinction on a land-bridge island is inversely related to annual survival rates of the same species on the nearby mainland. Prediction 1: Species with low survival rates on the mainland near BCI are more likely to be missing from BCI than species with higher survival rates. Prediction 2: Survival rates for species still present on BCI do not differ between mainland and island study areas. My results provide the first evidence for a relationship between mainland survival rate and extinction on a nearby land-bridge island.

Study Areas and Methods

I conducted a capture-recapture study of undergrowth birds in moist lowland forests of central Panama. Birds were captured, banded with uniquely numbered aluminum bands, and released in two areas: a mainland site (Karr & Freemark 1983) in Parque Nacional Soberanía 8 km east of BCI, and an island site on Shannon Trail on Barro Colorado Island (Karr 1982*a*, *b*). Mainland data were collected on a 90-ha study area from 1977–1988 (Karr et al. 1990) and island data were collected on a 2-ha area from 1980–1988 (Karr 1982*a*, *b*). Comparisons of data between the two sites (Prediction 2) are based on sampling efforts similar in both timing of sampling and sample area.

I estimated demographic parameters with Jolly-Seber stochastic models for open populations (Brownie et al. 1986) with four capture-recapture models (1) standard Jolly-Seber model (Jolly 1965, 1982; Seber 1965; Brownie & Robson 1983) with time-specific capture and survival probabilities; (2) model B with constant survival probability; (3) model D with constant survival and capture probability; and (4) model 2 with time-specific variation in capture and survival probabilities. In addition, model 2 allows different survival probabilities for newly banded versus previously banded individuals during each time interval. Because these models include parameters for both survival and capture probabilities, they are superior to estimates that equate survival and recapture rates (Nichols & Pollock 1983; Pollock et al. 1990). Finally, the complement of estimates of survival rate includes both mortality and permanent emigration. Emigration followed by territory establishment seems rare at both mainland and island sites (Karr et al. 1990).

Decisions about which model was most appropriate for a particular species were based on two types of tests: goodness-of-fit tests assessing the fit of a model to a species, and between-model tests (Brownie et al. 1986; Pollock et al. 1985). Three estimated parameters are used in this analysis (1) annual survival rate — the mean estimated annual survival rate for individuals in the study population; (2) capture probability — the estimated probability that an animal alive at time i is captured in the ith sample; and (3) population size — the mean estimated population size over the study period.

Mainland data for 20 sample periods from 1977 to 1986 were used to estimate parameters for 25 species of tropical forest passerines (Karr et al. 1990). In addition, I netted during an 8-year period at the island plot. An equivalent data set (same time periods and area size;

similar sample efforts) was extracted from mainland data (1980–1988) to provide paired samples from the mainland (data from Limbo Hunt Club) and island (Shannon 4 on BCI). Each data set contained 10 sample periods and involved about 5,000 net hours of sampling between 1980 and 1988.

Results

Eight of the 25 mainland species for which estimates were obtained no longer occur on BCI. Mean annual survival rates (Fig. 1) for the 8 species missing from BCI are lower ($\bar{x} \pm se$; 0.50 ± 0.05) than those for the 17 species still present on BCI (0.59 ± 0.02; pooled t-test statistic = −2.07, p = 0.025, one-tailed test, d.f. = 23; Levene's test for equality of variances $F_{1,23}$ = 2.59). Survival of the "missing" species is 15% below the survival of "present" species. Given these survival rates, a birth rate (m_x) of 1.6949 is necessary to maintain a stable population in the present species (l_x m_x = 0.59 × 1.6949 = 1.00), whereas populations of missing species with a birth rate of 15% lower would decline by 50% every four years (assuming non–age-structured populations). Thus, although the absolute difference may appear small, its demographic consequence is considerable. Further, survival rates seem to indicate not only which species are at greatest risk of extinction, but also the sequence of their loss for species known to have occurred on BCI (Table 1). (Virtually nothing is known about renesting frequency in these species but 23 of 25 have two-egg clutches and two have two- or three-egg clutches [Karr et al. 1990].)

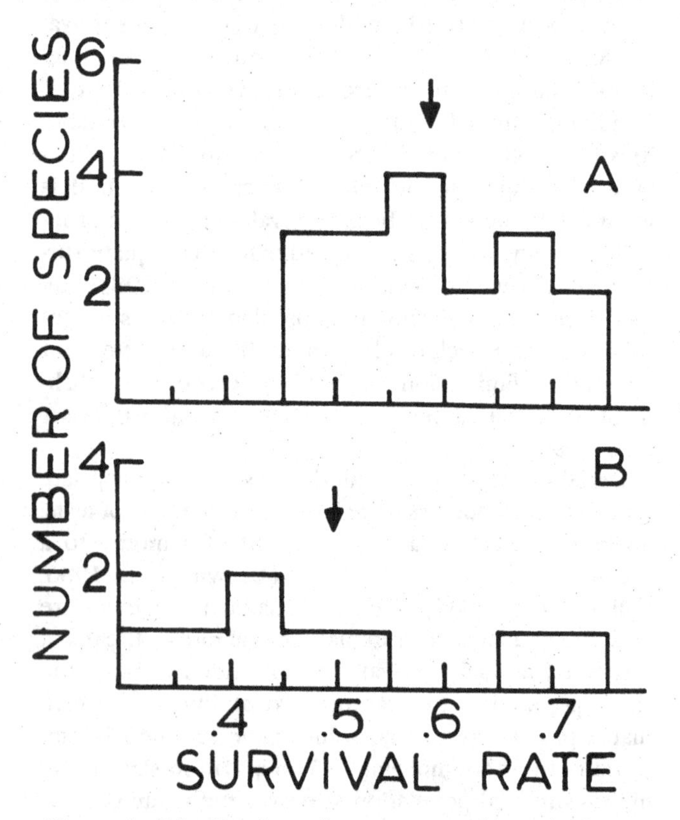

Figure 1. Distribution of mean annual survival rates (1977–1986) for 25 species of passerine birds in Parque Nacional Soberanía, Panama, in two groups (A) Still present on BCI; (B) No longer present on BCI. Arrows indicate means.

Two species now extinct on BCI (*Phaenostictus mcleannani* and *Automolus ochrolaemus*) had high survival rates on the mainland. The only common life history attribute for these species is that both are the largest species in their foraging guilds, an attribute often associated with a high extinction probability. *Phaenostictus* is an obligate ant-follower and *Automolus* gleans insects from dead leaves suspended in the forest undergrowth. The population estimate of 53 for *Phaenostictus* is well above average; the estimate of 13 for *Automolous* is near the low (8) for my study species.

As in an earlier analysis (Karr 1982*b*), I found little evidence that species rare at the mainland site were more likely to have gone extinct on BCI than abundant species (Fig. 2). The estimated mean population size at the mainland site for the missing (23 ± 6) and present (47 ± 14) species were not statistically different (pooled t-test statistic = 1.30, p = 0.103, one-tailed test using square root transformation, d.f. = 23; Levene's test for equality of variances $F_{1,23}$ = 0.56). The difference between these means is in the predicted direction, but over half the difference is due to the high population of *Pipra mentalis*. Excluding that species from the analysis reduces the estimated mean population size for the species still present on BCI to 35 ± 8. Insufficient data were available for a two-way analysis (survival rate and population size) of extinction pattern.

The demographies of birds on and off BCI at and shortly after isolation will never be known. However, my data allow comparisons of demographies for one

Table 1. Annual survival rate ($\bar{x} \pm se$) for the period 1977–1986 at Parque Nacional Soberanía, Panama, and last record from Barro Colorado Island (BCI) for eight species missing from BCI.

Species	*Annual survival rate*	*Last BCI record*
Microbates cinereiventris	0.33 ± 0.09	Never recorded
Myiobius sulphureipygius	0.40 ± 0.06	1950's
Formicarius analis	0.42 ± 0.07	1951
Myrmornis torquata	0.45 ± 0.08	Never recorded
Cyphorhinus phaeocephalus	0.46 ± 0.05	1961
Pipra coronata	0.51	Never recorded
Automolus ochrolaemus	0.66	1966
Phaenostictus mcleannani	0.73 ± 0.05	1977

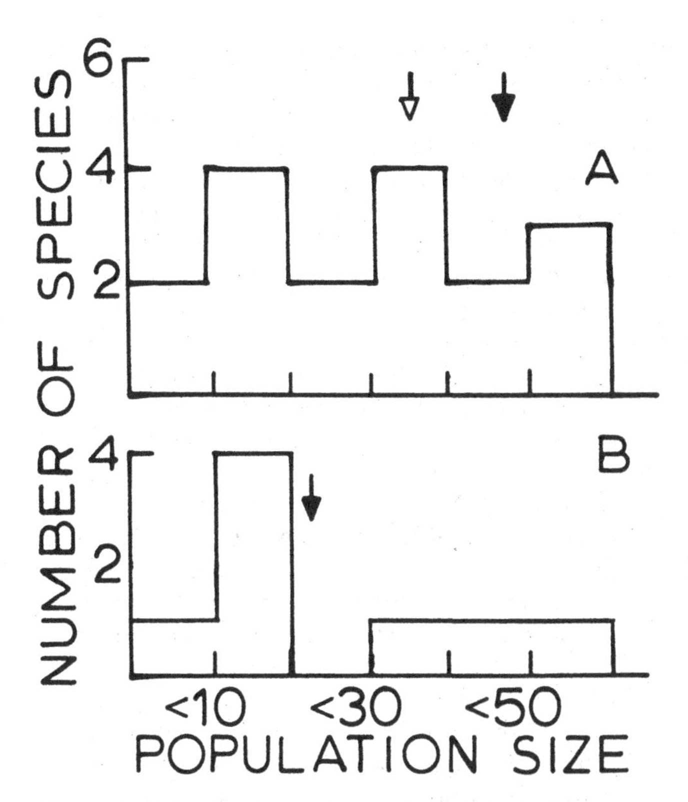

Figure 2. Distribution of mean population sizes (1977–1986) for 25 species of passerine birds in Parque Nacional Soberanía, Panama, in two groups (A) Still present on BCI; (B) No longer present on BCI. Arrows indicate means. Open arrow means without Pipra mentalis. *All species with population size greater than 50 in right cell.*

species present at both sites from 1980 to 1988. For *Pipra mentalis,* survival probability, capture probability, and population size do not differ between the mainland and island study sites (Table 2). A second analysis grouping all captures of species present at both island and mainland sites also shows no differences in demographic parameters between mainland and island populations (Table 3). These limited data suggest that the demographies of species still present on BCI have not been significantly altered by their isolation on that island or by the loss of a number of species with which they co-occur on the mainland.

Table 2. Population parameter estimates and associated test statistics for *Pipra mentalis* at Barro Colorado Island (BCI) and Limbo Hunt Club (LHC), Parque Nacional Soberanía, Panama, 1980–1988.

Demographic parameter	*BCI*[a] $\bar{x}$*(se)*	*LHC*[a] $\bar{x}$*(se)*	*Test statistic* z^b	*p*
Survival probability	0.62(0.07)	0.65(0.06)	−0.35	0.73
Capture probability	0.17(0.03)	0.16(0.03)	0.12	0.90
Population size	73.0(42.2)	117.3(55.8)	−0.63	0.53
Model[c]	D	D		
Goodness of fit[d]	x^2_{11} = 5.8	x^2_{19} = 20.7		
P	0.15	0.36		

[a] Number of captures:individuals in analysis—BCI (132:110), LHC (186:140).
[b] z statistic computed as in Brownie et al. (1985).
[c] Model D involves constant survival and capture probability.
[d] Goodness of fit of data set to Model identified above.

Table 3. Population parameter estimates and associated test statistics for all data combined for species present at both Barro Colorado Island (BCI) and Limbo Hunt Club (LHC), Parque Nacional Soberanía, Panama, 1980–1988.

Demographic parameter	*BCI*[a] $\bar{x}$*(se)*	*LHC*[a] $\bar{x}$*(se)*	*Test statistic* z^b	*p*
Survival probability[c]				
New captures	0.18(0.06)	0.13(0.02)	0.86	0.39
Old captures	0.67(0.25)	0.62(0.48)	0.09	0.93
Capture probability[d]	0.37(0.04)	0.34(0.04)	0.73	0.47
Model	2	2		
Goodness of fit[e]	x^2_{12} = 8.2	x^2_{16} = 7.9		
P	0.77	0.64		

[a] Number of captures:individuals in analysis—BCI (499:351); LHC (528:366).
[b] z statistic computed as in Brownie et al. (1985).
[c] Model 2 provides survival rate estimates for both newly banded (new) and previously banded (old) birds.
[d] Applies only to marked birds.
[e] Goodness of fit of data set to model noted above.

Discussion

Extinction of species on land-bridge islands occurs when a population is no longer able to sustain itself. (Recolonization from the nearby mainland is unlikely in the forest undergrowth species considered here [Willis 1974; Terborgh 1974; Karr 1982*a, b*].) Causes of extinction vary by species and circumstance; they may represent systematic (deterministic) pressures or stochastic perturbations (Shaffer 1981). Stochastic sources of extinction include natural catastrophes and demographic, genetic, and environmental accidents. The causes of some bird extinctions on BCI seem clear. About two-thirds of the extinctions documented by Willis and Eisenmann (1979) occurred as second-growth habitat was converted through succession to forest; that is, habitat requirements of the species are no longer met on the island. Even some forest species may not find the mosaic of forest environments that allows them to survive the full range of environmental extremes (e.g., the local landscape is not adequate to provide refuges during climate extremes; Karr 1982*b*).

These and other factors that may contribute to extinction may not be simply tied to deterministic or stochastic processes. Stochastic elements may play a role through their effect on the timing of extinctions, such as rate of plant succession (environmental stochasticity) or the frequency of climate extremes (natural catastrophes). Further, other aspects of the biota may combine systematic and stochastic elements. Predator abundances may have been altered on BCI (as proposed by many with experience on BCI) or microclimatic conditions may be altered, an especially common event at the edge of forest fragments (Temple & Cary 1988). Because BCI is a mostly well-drained mountaintop without a permanent stream, the regional decline in rainfall (about 8 mm/year) experienced this century in Central Panama (Rand & Rand 1982) may be exacerbating the microclimate problem. Finally, stochastic fluctuations in life-history parameters may also be responsible for extinction (Lande 1987).

The loss of species with low survival rates on BCI may be due to reduced survival of post-fledging individuals or to reduced reproductive success (higher egg and/or nestling mortality). Altered adult survival rate might result from direct predation on adults at nests or be an indirect result of higher rates of nest loss. Increased egg production for replacement clutches in insects (Talamy & Denno 1982) and in vertebrates (Tinkle 1969; Martin 1987; Nur 1989) decreases adult survival rates, presumably because of trade-offs between energy used for maintenance and for reproduction (Slagsvold 1982).

The differential extinction of species with lower survival rates raises an additional theoretical problem. Assuming that species with low adult survival rates are r-selected, conventional wisdom would expect them to be less susceptible to extinction because of high reproductive rates. They must invest more energy in reproduction through multiple broods and/or greater effort in parental care, which are often associated with higher adult mortality rates. Their high rate of extinction on BCI calls this expectation into question.

Although a definitive judgment is not possible with existing data, two patterns observed in this study suggest that the loss of species with low survival rates is due to reduced reproductive success. First, species still present on BCI have similar survival rates and population sizes on and off the island, suggesting that the island may not have altered adult survival rates in extinct species. Survival rates of Blue Tits (*Parus caeruleus*) did not differ between mainland and island sites (Blondel et al., in press) despite substantial differences in other demographic attributes. Second, the strong association between low survival rate on the mainland and extinction on BCI suggests a predisposition to extinction.

The mechanisms and processes responsible for the apparent reduced reproductive success cannot be clearly defined at this time, but three alternatives come to mind. Perhaps species with low survival rates persist on the mainland in a patchy landscape. Both temporal and spatial variability in forest in central Panama is of consequence to birds (Karr et al. 1982; Karr & Freemark 1983). Twenty-seven of 30 species analyzed exhibited significant spatial and/or temporal variability in activity along vegetation and/or microclimatic gradients on the mainland (Karr & Freemark 1983). Each species seemed to seek habitat optima in the context of current environmental conditions on diurnal, seasonal, and between-year time scales. Many species may require a complex mosaic of habitats among which they move on various spatial and temporal scales (permanent streams with moist sheltered valleys to serve as refuges in dry periods are lacking on BCI; Karr 1982*a*, *b*). The complex landscape may be critical to the population in the same sense that during each year migratory species depend on availability of breeding and wintering areas as well as stopover sites between those areas. Species in the tropical forest landscape may be less dependent on the full mosaic of habitats each year than species with annual migrations. But if the full range of habitats is required only every 5 or 10 years, the evolutionary consequence of habitat truncation will be the same: truncation of the regional species list. The tendency for species that have disappeared from BCI to have variable populations in a small area of contiguous forest on the mainland is in concert with this scenario, if, as suggested earlier (Karr 1982*b*), this population variability is due to movement among habitats. Inadequate representation of certain patch types on BCI would lead to increased extinction probability for species that require movement of individuals among patches in a more complex mainland landscape.

Alternatively, one might invoke the concept of a metapopulation similar to the metapopulation of Levins (1969, 1970) but with landscape units that produce excess offspring (sources — Pulliam 1988) mixed with other units (sinks) that do not. Some patches may even "wink" on and off as sources with natural environmental variation. In this situation, species with low survival rates may face an additional problem on an island where limited "source" areas prevent maintenance of their population, a situation similar to the extinction thresholds (minimum proportion of suitable habitat necessary for population persistence) of Lande (1987).

Finally, BCI may experience high predation rates (Willis 1974; Terborgh 1974; Karr 1982*a*). Several years ago, Loiselle and Hoppes (1983) demonstrated that predation rates on artificial nests containing Coturnix Quail eggs were much higher on BCI than on the nearby mainland. Recently, a more comprehensive study by K. Sieving (manuscript) confirmed that predation rates on artificial nests are higher on BCI than the adjacent mainland. In her study, nests were constructed and placed to mimic the nests of specific bird species, some still present on BCI and others disappeared from the island. Most importantly, she found that predation rates on artificial nests increased disproportionately on BCI for nests of extinct relative to extant species. The combination of low survival rates and differential increases in the rates of nest predation may be fatal to resident populations which, because of isolation on an island, do not experience immigration.

Whichever of these scenarios (or some hybrid alternative) proves to be true, extinctions are a consequence of a complex interaction of factors such as low survival, high predation, and aspects of landscape dynamics.

Both theoretical and empirical work (Hotker 1988; Lande 1988; Mertz 1971; Smith 1988) demonstrates the importance of adult survival rates to the viability of bird populations. The association between annual adult survival rate on the mainland and extinction on BCI adds a new dimension to study of the ecology of extinction. Connections between demography (specifically population size and rate of change) and environmental effects, including environmental catastrophes, have been explored on theoretical (MacArthur & Wilson 1967; Leigh 1975, 1981; Goodman 1987) and empirical grounds (Pimm et al. 1988). Pimm et al. (1988) noted the influence of demographic attributes other than population size (e.g., body size, migratory status, and population variability) in determining extinction probability. They further speculated that these other factors may increase in importance over longer time scales. My results confirm their speculation about the importance of additional demographic attributes in the extinction process, adding mean annual survival rate to the list. Finally, I note that the methods used here could be developed for other species, habitats, or locations to identify species threatened by habitat fragmentation before land management options are exercised.

Acknowledgments

This work was supported by Earthwatch, the National Science Foundation (DEB #82-06672), the National Geographic Society, and the American Philosophical Society. RE.NA.RE. kindly granted permits for research in Parque Nacional Soberanía. J. Brawn, J. Cranford, M. Dionne, C. Kellner, T. Martin, J. Nichols, S. Pimm, and K. Sieving commented constructively on an earlier draft and J. Nichols and J. Hines contributed advice and assistance in analysis of capture-recapture data.

Literature Cited

Blondel, J., R. Pradel, and J.-D. Lebreton. Survival rates in two populations of Blue Tit (*Parus caeruleus*), a test of theory on population biology on an island. (In press.)

Brownie, C., J. F. Hines, and J. D. Nichols. 1986. Constant-parameter capture-recapture models. Biometrics **42**:561–574.

Brownie, C., and D. S. Robson. 1983. Estimation of time-specific survival rates from tag-resighting samples: a generalization of the Jolly-Seber model. Biometrics **39**:437–453.

Diamond, J. M. 1972. Biogeographic kinetics: estimation of relaxation times for avifaunas of southwest Pacific islands. Proceedings of the National Academy of Science, USA. **69**:3199–3203.

Goodman, D. 1987. The demography of chance extinctions. Pages 11–34 in M. E. Soulé, editor. Viable populations. Cambridge University Press, Cambridge, England.

Hotker, H. 1988. Lifetime reproductive output of male and female Meadow Pitpits *Anthus pratensis*. Journal of Animal Ecology **57**:109–117.

Jolly, G. M. 1965. Explicit estimates from capture-recapture data with both death and immigration-stochastic model. Biometrics **52**:225–247.

Jolly, G. M. 1982. Mark-recapture models with parameters constant in time. Biometrics **38**:301–321.

Karr, J. R. 1982*a*. Population variability and extinction in a tropical land-bridge island. Ecology **63**:1975–1978.

Karr, J. R. 1982*b*. Avian extinction on Barro Colorado Island, Panama: a reassessment. American Naturalist **119**:220–239.

Karr, J. R., and K. E. Freemark. 1983. Habitat selection and environmental gradients: dynamics in the "stable" tropics. Ecology **64**:1481–1494.

Karr, J. R., J. D. Nichols, M. K. Klimkiewicz, and J. D. Brawn. 1990. Survival rates of tropical and temperate forest birds: will the dogma survive? American Naturalist. **136**:277–291.

Karr, J. R., D. W. Schemske, and N. V. L. Brokaw. 1982. Temporal variation in the understory bird community of a tropical forest. Pages 441–453 in E. G. Leigh, A. S. Rand, and D. M. Windsor, editors. The ecology of a tropical forest; seasonal rhythms and long-term changes. Smithsonian Institution Press, Washington, D.C.

Lande, R. 1987. Extinction thresholds in demographic models of territorial populations. American Naturalist **130**:624–635.

Lande, R. 1988. Demographic models of the Northern Spotted Owl (*Strix occidentalis caurina*). Oecologia **75**:601–607.

Leigh, E. G. 1975. Population fluctuations, community stability and environmental variability. Pages 51–73 in M. L. Cody and J. M. Diamond, editors. Ecology and evolution of communities. Harvard University Press, Cambridge, Massachusetts.

Leigh, E. G. 1981. The average lifetime of a population in a varying environment. Journal of Theoretical Biology **90**:213–239.

Levins, R. 1969. Some demographic and genetic consequences of environmental heterogeneity for biological control. Bulletin of the Entomological Society of America **15**:237–240.

Levins, R. 1970. Extinction. Lectures in Mathematical and Life Sciences **2**:75–107.

Loiselle, B. A., and W. G. Hoppes. 1983. Nest predation in insular and mainland lowland forest in Panama. Condor **85**:93–95.

MacArthur, R. H., and E. O. Wilson. 1967. The theory of island biogeography. Princeton University Press, Princeton, New Jersey.

Martin, T. E. 1987. Food as a limit on breeding birds: a life-history perspective. Annual Review of Ecology and Systematics **18:**453–487.

Mertz, D. B. 1971. The mathematical demography of the California Condor population. American Naturalist **150:**437–453.

Morton, E. S. 1985. The realities of reintroducing species to the wild. Pages 147–148 in R. J. Hoage, editor. Animal extinctions. Smithsonian Institution Press, Washington, D.C.

Morton, E. S. 1987. Reintroduction as a method of studying bird behavior and ecology. Pages 165–172 in W. R. Jordan III, M. E. Gilpin, and J. D. Aber, editors. Restoration ecology: a synthetic approach to ecological research. Cambridge University Press, New York.

Nichols, J. D., and K. H. Pollock. 1983. Estimation methodology in contemporary small mammal capture-recapture studies. Journal of Mammology **64:**253–260.

Nur, N. 1989. The cost of reproduction in birds: an examination of the evidence. Ardea **76:**155–168.

Pimm, S. L., H. L. Jones, and J. Diamond. 1988. On the risk of extinction. American Naturalist **132:**757–785.

Pollock, K. H., J. E. Hines, and J. D. Nichols. 1985. Goodness-of-fit tests for open capture-recapture models. Biometrics **41:**399–410.

Pollock, K. H., J. D. Nichols, C. Brownie, and J. E. Hines. 1990. Statistical inference for capture-recapture experiments. Wildlife Monographs No. 107.

Pulliam, H. R. 1988. Sources, sinks, and population regulation. American Naturalist **132:**652–661.

Rand, A. S., and W. M. Rand. 1982. Variation in rainfall on Barro Colorado Island. Pages 47–59 in E. G. Leigh, Jr., A. S. Rand, and D. M. Windsor, editors. The ecology of a tropical forest: seasonal rhythms and long-term changes. Smithsonian Institution Press, Washington, D.C.

Seber, G. A. F. 1965. A note on multiple-recapture census. Biometrika **52:**249–259.

Shaffer, M. 1981. Minimum population sizes for species conservation. BioScience **31:**131–134.

Slagsvold, T. 1982. Clutch size variation in passerine birds: the nest predation hypothesis. Oecologia **54:**159–169.

Smith, J. M. N. 1988. Determinants of lifetime reproductive success in the Song Sparrow. Pages 154–172 in T. H. Clutton-Brock, editor. Reproductive success: studies of individual variation in contrasting breeding systems. University of Chicago Press, Chicago, Illinois.

Talamy, D. W., and R. F. Denno. 1982. Life history trade-offs in *Gargaphia solani* (Hemiptera: Tingidae): the cost of reproduction. Ecology **63:**616–620.

Temple, S. A., and J. R. Cary. 1988. Modeling dynamics of habitat — interior bird populations in fragmented landscapes. Conservation Biology **2:**340–347.

Terborgh, J. W. 1974. Preservation of natural diversity: the problem of extinction-prone species. BioScience **24:**715–722.

Tinkle, D. W. 1969. The concept of reproductive effort and its relation to the evolution of life histories in lizards. American Naturalist **103:**501–516.

Williamson, M. 1981. Island populations. Oxford University Press, Oxford, England.

Willis, E. O. 1974. Populations and local extinctions of birds on Barro Colorado Island, Panama. Ecological Monographs **44:**153–169.

Willis, E. O., and E. Eisenmann. 1979. A revised list of birds on Barro Colorado Island, Panama. Smithsonian Contributions to Zoology **291:**1–31.

Wilson, E. O., and E. O. Willis. 1975. Applied biogeography. Pages 522–534 in M. L. Cody and J. M. Diamond, editors. Ecology and evolution of communities. Harvard University Press, Cambridge, Massachusetts.

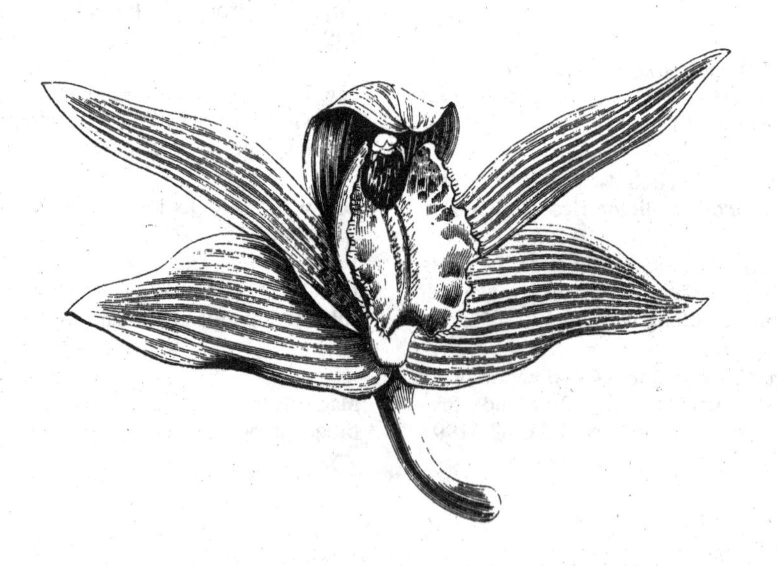

Forest Fragmentation and Bird Extinctions: San Antonio Eighty Years Later

GUSTAVO H. KATTAN*

Instituto Vallecaucano de Investigaciones Científicas
Museo de Ciencias Naturales
Cali, Colombia

HUMBERTO ALVAREZ-LÓPEZ
MANUEL GIRALDO

Departamento de Biología
Universidad del Valle
Cali, Colombia

Abstract: *We report on the extent of bird extinctions at San Antonio, a fragmented cloud forest site in the western Andes of Colombia, for which surveys dating back to 1911 and 1959 are available. In 1911, 128 forest bird species were present in San Antonio. Twenty-four species had disappeared by 1959, and since then 16 more species have gone locally extinct, for a total of 40 species or 31% of the original avifauna. We analyzed patterns of extinction in relation to geographic distribution and foraging guilds. We found that in this montane assemblage, being at the limits of the altitudinal distribution was the main correlate of extinction; 37% of the extinct species were at the upper limit of their altitudinal distribution. We also found that the most vulnerable guilds were the understory insectivores and the large canopy frugivores. Our study illustrates the extent of bird extinctions that are currently undocumented in the highly fragmented forests of the northern Andes, where the absence of baseline information on the fauna of unaltered forests is a limiting factor for the development of conservation and management plans. We stress the need to establish data bases and long-term monitoring projects for the Andean fauna.*

Fragmentación forestal y extinción de pájaros: San Antonio 80 años más tarde

Resumen: *En este artículo presentamos los resultados de una evaluación de los efectos de la fragmentación sobre la riqueza de la avifauna en un bosque andino. San Antonio es un cerro en la Cordillera Occidental de Colombia, que forma parte de un sistema de parches de bosque relictuales, con un área total de unas 700 ha, a elevaciones entre 1800 y 2100 m. La avifauna de San Antonio fue inventariada en 1911, 1959 y 1985. Durante 1990 realizamos un nuevo inventario, para compararlo con los anteriores. A principios de siglo la avifauna del bosque consistía de 128 especies. Para 1959 habían desaparecido 24 especies, y desde entonces al presente desaparecieron 16 más, para un total de 40 especies extinguidas localmente, es decir, el 31% de la avifauna del bosque original. Analizamos patrones de extinción en relación con distribución geográfica y gremios tróficos. Encontramos que para esta comunidad de montaña, las aves más vulnerables fueron las que se encontraban en su límite altitudinal de distribución; el 37% de las especies extinguidas se encontraban en su límite altitudinal superior. En cuanto a gremios tróficos, los más vulnerables fueron los insectívoros de sotobosque y los frugívoros grandes del dosel. Nuestros resultados indican que la fragmentación de bosques en un gradiente altitudinal puede ocasionar extinciones masivas, pues aísla poblaciones pequeñas de especies en sus límites altitudinales, y porque interrumpe las rutas de migración altitudinal. Destacamos la limitación que supone para este tipo de estudio la carencia de información sobre la composición de la avifauna (y de la biota en general) de selvas andinas inalteradas. Es obvia la necesidad de aprovechar de manera inmediata toda oportunidad de crear y fortalecer bases de datos sobre estos aspectos, así como la iniciación de proyectos de monitoreo a largo plazo.*

* *Present address: NYZS The Wildlife Conservation Society, Apartado 1845, Cali, Colombia.*

Paper submitted March 31, 1992; revised manuscript accepted September 15, 1992.

Introduction

Modification or removal of large areas of natural vegetation often results in a mosaic of isolated habitat fragments scattered across the landscape. Habitat fragmentation, especially in tropical forests, produces changes in the physical environment and extinction of species due to factors such as area reduction and isolation of the fragments (Lovejoy et al. 1986; Saunders et al. 1991). Area and isolation are important in determining the number of species that a habitat island will maintain when equilibrium between immigration and extinction is reached (MacArthur & Wilson 1967). Immigration, however, depends on the presence of a source area for propagules, and for isolated fragments extinction may be the dominant process (Pickett & Thompson 1978). The study of fragmentation, therefore, has become synonymous with the study of extinction processes (Simberloff 1988).

A number of studies have shown that bird extinctions occur following fragmentation of tropical forests. Extinctions have been documented using two different approaches. One compares bird-species diversity in fragments of different sizes. These studies have found that small fragments are less diverse than large fragments or continuous forest (Willis 1979; Newmark 1991). The species composition of small fragments is usually a nested subset of the assemblage found in larger fragments, which suggests that species go extinct in a nonrandom, predictable pattern (Willis 1979; Blake 1991; Newmark 1991). A second, more direct approach compares the composition of pre- and post-fragmentation assemblages at a given site (Willis 1974; Leck 1979; Bierregaard & Lovejoy 1989). While this approach clearly has the advantage of showing that extinctions actually occurred, it requires the existence of prefragmentation surveys, which are rarely available in the neotropics. This method was used experimentally in the Brazilian Amazon, where a variety of organisms were surveyed, and then the forest was fragmented in a design that left fragments of different sizes and degrees of isolation. Post-isolation surveys have shown high rates of extinction in small fragments (Lovejoy et al. 1986; Bierregaard & Lovejoy 1986, 1989).

Besides documenting extinctions, studies of fragmentation have attempted to identify patterns of vulnerability, to identify which species or groups of species are more susceptible to extinction (Willis 1979; Terborgh & Winter 1980; Karr 1990; Newmark 1991). A common approach is to consider foraging guilds. For example, large canopy frugivores and large understory insectivores have emerged as two extinction-prone neotropical guilds (Willis 1974, 1979; Terborgh 1986; Bierregaard & Lovejoy 1989; Kattan 1992). High vulnerability has also been associated with a variety of life-history and ecological traits, such as large body size, small population size, habitat specialization, and low survival rates (Terborgh & Winter 1980; Karr 1990; Newmark 1991; Kattan 1992). Understanding patterns of vulnerability not only has theoretical interest, but also has important implications for the design and management of reserves.

All studies of fragmentation in the Neotropics have been carried out in lowland rain forests. San Antonio is a fragmented cloud forest site in the western Andes of Colombia, for which several bird surveys, dating as far back as 1911, are available. These historic, pre-isolation surveys provided a rare opportunity to document extinctions and to analyze patterns of extinction in order to identify vulnerable species or guilds. We examined patterns of extinction in relation to foraging guilds; for canopy frugivores and understory insectivores, we tested whether there was any relationship between frequency of extinction and body size. Because this locality is at the geographic or altitudinal limits of distribution for many species, we also tested for any relationship between frequency of extinction and altitudinal/geographic distribution.

Study Area and Methods

The Cerro de San Antonio (also called Cerro de La Horqueta) is a montane summit (2100 m elevation) in the Cordillera Occidental, 15 km west of Cali, on the road to Buenaventura, Colombia. The cloud forests of San Antonio and its surrounding area, the "El Dieciocho" site (km 18 of the Cali-Buenaventura road) and forests on the ridge along the road to Pavas (see Fig. 1), were fragmented during the first half of the century, but the remaining fragments have not changed much since the 1960s. At present, this region is an archipelago of isolated forest fragments in a matrix of small farms and suburban houses. The nearest "mainland" is the 150,000-ha Farallones de Cali National Park, about 10 km to the south. The total area of the fragments we surveyed is about 700 ha, at an elevation of 1800–2100 m (Fig. 1). A detailed description of the San Antonio forest is given by Kattan et al. (1984).

The avifauna of the San Antonio region was surveyed in 1911 by F. M. Chapman of the American Museum of Natural History (Chapman 1917), who collected birds at San Antonio and along the road to Pavas. In 1959, A. H. Miller conducted a one-year study of the avifauna of San Antonio–El Dieciocho (Miller 1963). More recently, a one-year study was conducted at Finca Zíngara, about 4 km north of El Dieciocho (Giraldo 1985; see Fig. 1). These studies provide a data base to which we compared a survey we made during 1989–1990. While it is difficult to know precisely the area of forest surveyed by the older studies, we estimate that we covered a much larger area. Thus our data probably provide a

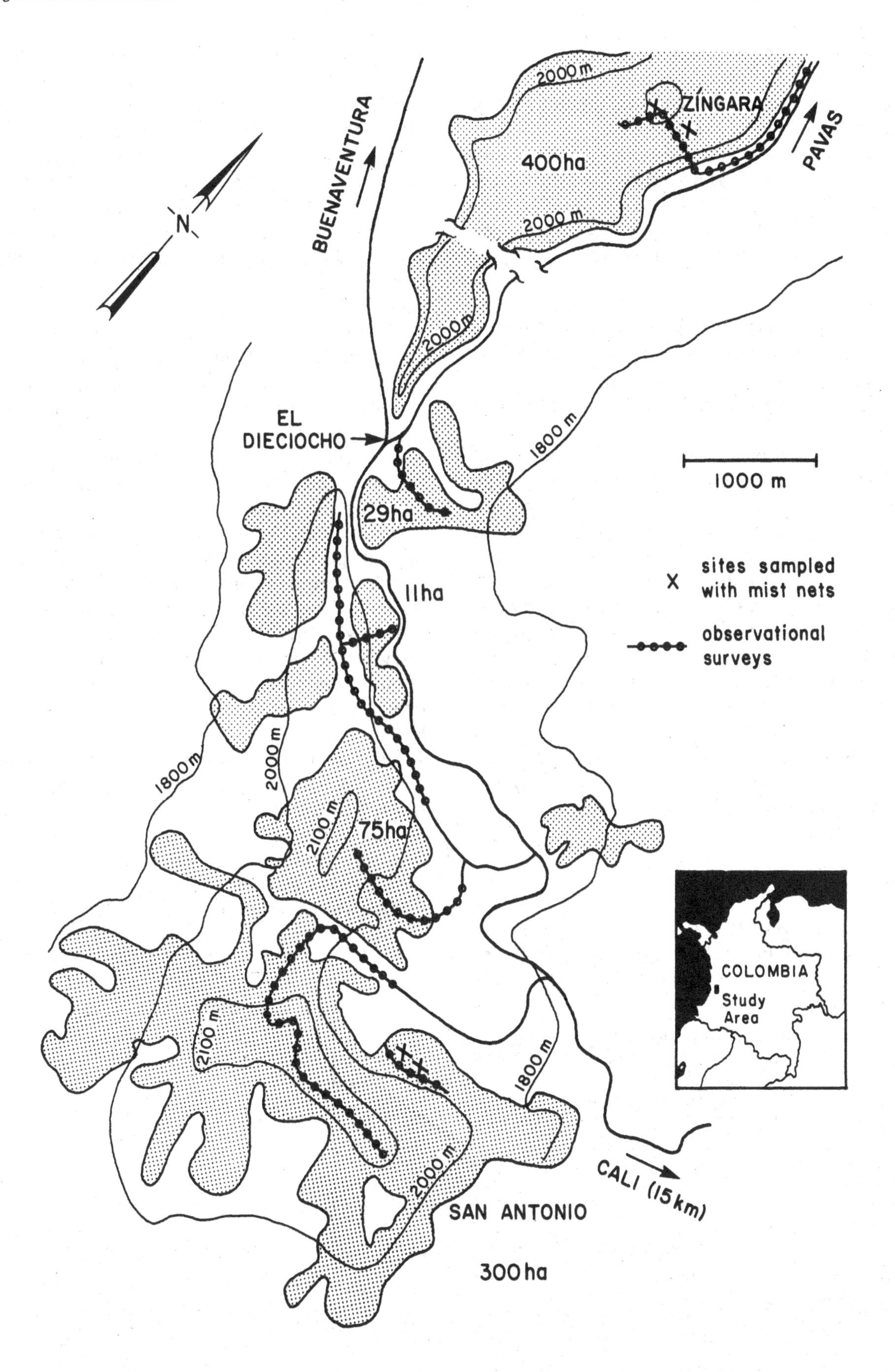

Figure 1. Map of the San Antonio–El Dieciocho region, on the western range of the Colombian Andes, showing the study sites.

minimum estimate of the number of extinctions that have occurred. In this paper we analyze extinctions at the regional level and do not consider the distribution of birds in the individual fragments.

We surveyed the avifauna of San Antonio, Finca Zíngara, and several small fragments between them on the eastern slope of the Cordillera (Fig. 1). For brevity, here we will refer to the whole area as San Antonio. Between October 1989 and February 1991 we made seven surveys using mist nets, for a total of 1360 net-hours. In each survey, 12 nets were placed in different microhabitats (forest edge, interior, and gaps) and opened for two days (24 h). We also made 16 observational surveys, for a total of 233 person-hours of observation. The paths followed in these surveys are shown in Figure 1. In addition, we made one three-day survey with 12 mist nets and several observational surveys at Reserva Hato Viejo in the Farallones de Cali National Park, the closest large tract of unbroken forest, which presumably supports a bird fauna similar to the original San Antonio avifauna.

We included in the analysis only forest-dwelling birds—birds that require to some extent the presence of a closed-canopied forest—even though some species may also use second-growth habitats. We excluded species of open habitats, as well as accidentals. To analyze patterns of extinction, we assigned species to foraging guilds, defined according to Willis (1979), with some modifications (see Table 1). We also examined the relationship between geographic and altitudinal distribution and local extinction. Distributional and body-size data were obtained from Hilty and Brown (1986).

Results

In 1911 F. Chapman (1917) collected a total of 110 forest bird species in the San Antonio region. Miller (1963) recorded 18 additional species that Chapman failed to collect. Assuming these 18 species were present in San Antonio in Chapman's time, which is reasonable given their distributions, the original forest bird fauna of San Antonio was composed of 128 species. In 1959, 24 of the species recorded by Chapman had disappeared (Miller 1963), and since then 16 more species have gone locally extinct, for a total of 40 species or 31.2% of the original avifauna (Fig. 2). Presently, 92 forest bird species persist in San Antonio (including four species recorded by us that neither Chapman nor Miller reported).

Because previous studies have found that large body size is a correlate of vulnerability, particularly for canopy frugivores and understory insectivores, we first inspected the relationship between body size and local extinction in these two guilds. Among canopy frugivores, large species were significantly more likely to become extinct than were smaller species. Extinct species were significantly larger than those still present (Mann-Whitney U, $p = 0.0005$). The size distribution of present species was strongly skewed to the right; note that there are no extant species larger than 45 cm (Fig. 3). In contrast, the distribution of extinct species was not skewed, and there were extinctions in most size categories. Understory insectivores, on the other hand, did not exhibit a significant difference in size between extinct and extant species (Mann-Whitney U, $p = 0.15$; Fig. 4).

Based on the above results, we divided canopy frugivores, but not understory insectivores, into large and small. We defined large canopy frugivores as those species larger than 25 cm total length (mostly nonpasserines). Patterns of extinction by foraging guilds were then analyzed by comparing the distribution of extinct species with an expected distribution, which was gen-

Table 1. Guild distribution of species extinct at San Antonio.

Guild	*No. of Extinct Species* obs.	*No. of Extinct Species* exp.	*Total No. Species*	*% Extinct*
Large Canopy Frugivores	12	5.6	18	66.7
Small Canopy Frugivores	5	6.9	22	22.7
Ground Frugivores	1	1.9	6	16.7
Understory Frugivores	4	4.7	15	26.7
Carnivores	1	0.9	3	33.3
Nectarivores	1	4.1	13	7.7
Trunk Insectivores	3	4.1	13	23.1
Understory Insectivores	9	6.2	20	45.0
Canopy Insectivores	3	2.5	8	37.5
Edge Insectivores	1	3.1	10	10.0

Note: For the chi square test, some categories were pooled to obtain expected values greater than 5.

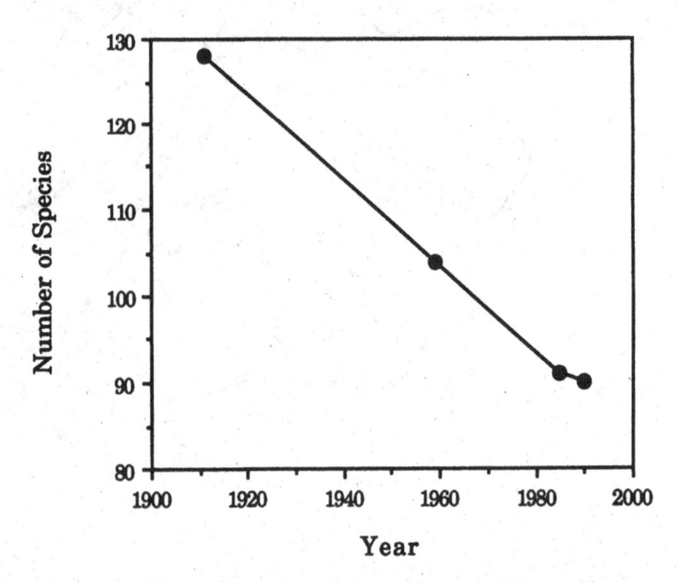

Figure 2. Loss of species at San Antonio through the years 1911–1990.

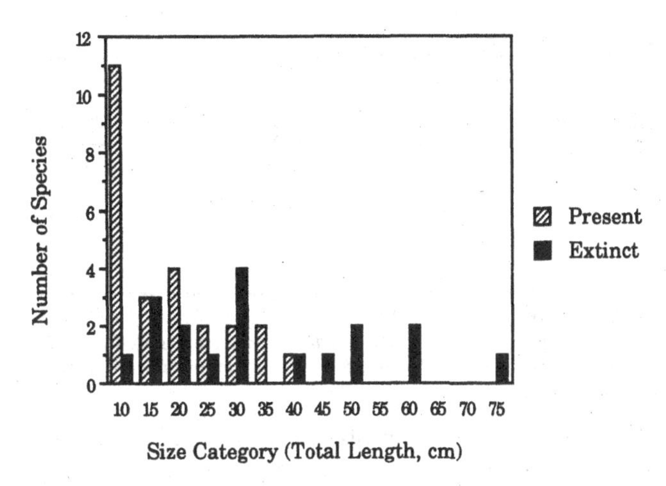

Figure 3. Frequency distribution of sizes of present and extinct species of canopy frugivores.

erated assuming that the proportion of extinct species in each guild would be the same as the 31% overall extinction rate. The difference between the observed and expected distributions was marginally significant ($\chi^2 = 12.34$, df = 6, $p = 0.056$), which suggests a tendency for some guilds to be more susceptible to extinction than others. As expected, the more vulnerable guilds were the large canopy frugivores and the understory insectivores (Table 1).

We next inspected the relationship between geographic and altitudinal distribution and local extinction (Table 2). Species at the limits of their distribution exhibited high rates of extinction. Seven of the species had a distributional range at mid-elevations (500–2000 m) on the humid Pacific slope of the Western Cordillera. These species had marginal populations on the eastern slope, near low montane passes, and showed high rates of extinction. Similarly, species at their altitudinal limits showed high rates of extinction. Nineteen of 29 species

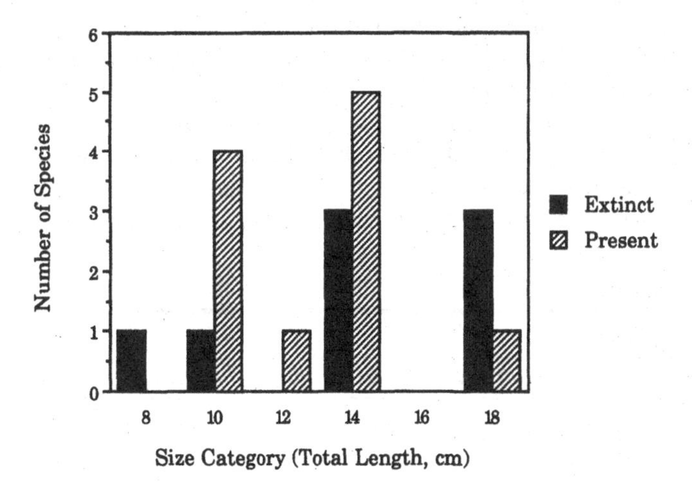

Figure 4. Frequency distribution of sizes of present and extinct species of understory insectivores.

for which San Antonio was either the upper or lower limit of their altitudinal distribution are locally extinct (Table 2). A patchy distribution (associated with species that are known only from isolated and localized populations) also appears to be a factor correlated with a high extinction probability, although this sample is too small to be conclusive. In contrast, species for which San Antonio is well within the limits of their altitudinal and geographic range exhibited low rates of extinction. A two-way *G*-test of independence indicated that the probability of extinction depends on geographic distribution; species at the distributional limits (geographic or altitudinal) are more prone to extinction than species within their distributional range ($G = 12.51$, df = 1, $p < 0.01$).

For the 13 extinct species for which San Antonio is well within their distributional range, we identified some possible causes of extinction (Table 3). Hunting, capture for the pet trade, and predation by domestic dogs may have been important proximate factors in the extinction of some species (Lehmann 1957; personal observation), although fragmentation made them vulnerable by decreasing their populations. Toucan Barbets (*Semnornis ramphastinus*) from San Antonio, for example, were heavily traded during the 1950s (Lehmann 1957) and are presently extinct. Deterioration or absence of certain microhabitats may be the reason for the absence of four species. Many microhabitats, such as streams, may not be represented in forest remnants because of a sampling effect, or because they deteriorate due to a natural process of dessication following fragmentation (Lovejoy et al. 1986). In addition, in San Antonio many streams have been totally or partially piped to supply water for houses, which has made this microhabitat unavailable for wildlife. Several birds that are always found in association with streams, such as *Rupicola peruviana* and *Habia cristata*, are locally extinct. Finally, there are four canopy frugivores that are known to require large home ranges (such as the *Amazona mercenaria*), and the decrease in habitat may have caused their local extinction.

Discussion

Our results indicate that 40 species of birds have gone locally extinct in the San Antonio region in the western Andes of Colombia, probably as a result of forest fragmentation. There are records that show that these species were present at this site at the beginning of the century (Chapman 1917), and most of them are present at the Farallones de Cali National Park, a large tract of forest that harbors an avifauna similar to San Antonio (personal observation). Twenty-four of the species were extinct by 1959 (Miller 1963), when the region was

Table 2. Relationship between geographic and altitudinal distribution and local extinction.

	*Guild**										
Distribution	*LCFR*	*SCFR*	*GRFR*	*USFR*	*CAR*	*NEC*	*TRIN*	*USIN*	*CAIN*	*EDIN*	*TOTAL*
Upper Limit	3	1		1		1	2	4	2	1	15/25
Lower Limit	2	1						1			4/4
Pacific Distribution		1					1	2			4/7
Patchy Distribution		1			1						2/2
Center of Distribution	7	2	1	1				2			13/83

** LCFR = large canopy frugivores; SCFR = small canopy frugivores; GRFR = ground frugivores; USFR = understory frugivores; CAR = carnivores; NEC = nectarivores; TRIN = trunk and twig insectivores; USIN = understory insectivores; CAIN = canopy insectivores; EDIN = edge insectivores. Total = number of extinct species as proportion of total number of species in each distribution category in original avifauna.*

already transformed into a mosaic of forest fragments, small farms, and suburban houses. Although there has been almost no further deforestation in the last 20 years, extinctions have continued to occur and presumably will continue to occur in the near future. Some of the species that are still present have very small populations and are highly vulnerable. For example, two species of quetzal (*Pharomacrus antisianus* and *P. auriceps*) existed in San Antonio, and both are present in sympatry in the Farallones de Cali National Park (personal observation). One of the species (*P. auriceps*) is locally extinct in San Antonio, and we estimate that only five pairs of the other species remain in approximately 700 ha of forest fragments. Quetzals are involved in altitudinal migrations (Powell et al. 1991; personal observation), and, with San Antonio almost completely isolated from forests at other elevations, their probability of survival is low.

A major pattern that emerged from this study is that species at the limits of their altitudinal or geographic distribution are particularly prone to extinction. This pattern has been suggested previously by Terborgh and Winter (1980), and it affects species in all trophic guilds (Table 2). In contrast, species that are well within their geographic and altitudinal range exhibit relatively low vulnerability. Most of the species within their distributional range that are extinct are frugivores.

Species at their distributional limits may be vulnerable for a variety of reasons. First, they may be at their physiological or ecological limits, as is probably the case with the species from the Pacific slope, which occur at very low population densities in the drier forests of the eastern slope. Second, fragmentation prevents altitudinal movements and thus disrupts migration and recolonization. At San Antonio, isolation is almost complete from lower elevations on the eastern slope. Many of the extinct birds are species from the slopes of the Cauca Valley, which are largely deforested, and some species may be regionally extinct (such as *Ramphastos swainsonii* and *Psarocolius angustifrons*; Lehmann 1957; Alvarez-López et al. 1991). Migration routes to higher elevations in the Farallones de Cali are also disrupted. Some connection exists with the lowland forests of the Pacific slope, however, and migrations on this slope may still occur. For instance, the hummingbird *Eutoxeres aquila*, a lowland species, was captured seasonally at San Antonio.

Our results partially support the hypothesis that large canopy frugivores are extinction-prone. Large canopy frugivores may be vulnerable because they depend on spatially and temporally patchy resources (Willis 1979). During periods of low fruit availability, species may depend on a few pivotal or keystone resources or may have to migrate to track resources (Howe 1984; Terborgh 1986; Loiselle & Blake 1991; Levey & Stiles 1992). Habitat fragmentation would easily disrupt these seasonal movements and fragile dependence on keystone resources. In contrast to large frugivores, small canopy frugivores tend to be habitat generalists and depend on small, carbohydrate-rich fruits abundant in gaps and edges (Levey 1988). For example, a large proportion of the small frugivores at San Antonio are tanagers, mainly of the genus *Tangara* (Emberizidae, Thraupinae). These partially insectivorous birds use a variety of habitats, including forest canopy, edges, gaps, and second-growth woodlands, making them less reliant on pri-

Table 3. Possible causes of extinction for species within limits of distribution.

	*Guild**										
Cause	*LCFR*	*SCFR*	*GRFR*	*USFR*	*CAR*	*NEC*	*TRIN*	*USIN*	*CAIN*	*EDIN*	*TOTAL*
Area Effect	3	1									4
Microhabitat Loss	1	1						2			4
Pet Trade				1							1
Hunting	3										3
Dog Predation			1								1

** Guilds as in Table 2.*

mary forest (Kattan, Alvarez, and Giraldo, unpublished data).

Although we found that understory insectivores tend to be susceptible to extinction, we did not find a relationship between susceptibility and body size within the guild. Vulnerability of large understory insectivores is sometimes related to specialization on army ants. Because these ants are among the first organisms to disappear in small fragments (Bierregaard & Lovejoy 1989), most of the large insectivores that go extinct are professional ant-followers (Willis 1974; Bierregaard & Lovejoy 1989). Army ants, however, are rare or absent in Andean highlands (Hilty 1974) and have never occurred at our site. Hence their absence cannot account for the extinction of understory insectivores. Other factors, such as altitudinal distribution and absence of certain microhabitats, are apparently more important in the extinction of these birds at San Antonio.

Thus, in addition to food habits, other factors such as body size, habitat specialization, population size, and taxonomic affiliation may interact to affect the vulnerability of species to extinction (Terborgh & Winter 1980; Kattan 1992). For example, the toucans (Ramphastidae) were identified by Terborgh and Winter (1980) as vulnerable. Of the four species originally present at San Antonio, two small species persist (*Aulacorhynchus prasinus* and *A. haematopygus*) and two large species (*Andigena nigrirostris* and *Ramphastos swainsonii*) are extinct. Body size may have determined the vulnerability of the two extinct species, but both are also at the limits of their altitudinal distribution, and it is difficult to separate the two factors.

Body size and initial abundance may have been the factors determining the local extinction of one of the two large parrots of San Antonio. Both species are similar in total length, but *Aratinga wagleri* is a long-tailed species, while *Amazona mercenaria* is a short-tailed, heavy-bodied species. *Aratinga* used to be abundant on the Cauca Valley and the eastern slope of the western Andes, where flocks of up to 300 individuals were recorded (Lehmann 1957); at least two flocks of 10–20 individuals persist in the San Antonio region. *Amazona*, in contrast, is a rarer species that forms small flocks (8–10 individuals) throughout its range (Ridgely 1981) and is presently extinct at San Antonio.

The extent of bird extinctions reported in this study, and the pattern of vulnerability associated with altitudinal limits, are especially noteworthy in the face of the extensive deforestation occurring in the Colombian Andes. Fragmentation of forests on an altitudinal gradient has the potential to produce massive extinctions because it isolates small populations of species at their altitudinal limits, and because it disrupts migration routes. The Andean region of Colombia is heavily populated and extensively deforested. Although it represents only 15% of the country's surface, it contributes 39 of 70 species of threatened birds (Hilty 1985; Orejuela 1985; Collar & Andrew 1988). Some of the birds extinct at San Antonio are listed in the International Council for Bird Preservation's list of threatened birds, while others that still persist are vulnerable at both the local and country levels (Hilty 1985; Collar & Andrew 1988; Alvarez-López et al. 1991). San Antonio and the Farallones de Cali National Park are at the limits of one of the areas of highest endemism in the continent, and should be considered of the highest priority in conservation programs (Terborgh & Winter 1983; Saavedra & Freese 1986).

This study illustrates the extent of bird extinctions that are currently undocumented in the Andes of Colombia and other countries, where the absence of baseline information on the fauna of unaltered forests is a limiting factor for the evaluation of wildlife status and the development of conservation and management plans. The few existing studies on Andean bird communities in Colombia have all been done in fragmented forests (see Orejuela et al. 1979; Orejuela et al. 1982; Giraldo 1985; Johnels & Cuadros 1986; Cuadros 1988); there is no documentation on what a pristine bird community would be like. We would like to stress the urgent need for establishing data bases on the Andean fauna and initiating long-term monitoring projects in as many places as possible.

Acknowledgments

We are especially grateful to Sr. Luis Felipe Carvajal for allowing us to work on his property, and to Carolina Murcia for logistical support. We thank Carla Restrepo, Peter Feinsinger, and Douglas Levey for making valuable suggestions that improved the manuscript, and F. Putz for suggesting a better way to present data. The study was partially supported by the Fundación para la Promoción de la Investigación y la Tecnología, Banco de La República, Bogotá, Colombia.

Literature Cited

Alvarez-López, H., G. H. Kattan, and M. Giraldo. 1991. Estado del conocimiento y la conservación de la avifauna del Departamento del Valle del Cauca. Pages 335–354 in Memorias del Primer Simposio de Fauna del Valle del Cauca. Instituto Vallecaucano de Investigaciones Científicas, Cali, Colombia.

Bierregaard, R. O., and T. E. Lovejoy. 1986. Birds in Amazonian forest fragments: Effects of insularization. Pages 1564–1579 in Acta XIX Congressus Internationalis Ornithologici, Ottawa, Canada.

Bierregaard, R. O., and T. E. Lovejoy. 1989. Effects of forest fragmentation on Amazonian understory bird communities. Acta Amazonica 19:215–241.

Blake, J. G. 1991. Nested subsets and the distribution of birds in isolated woodlots. Conservation Biology **5**:58–66.

Chapman, F. M. 1917. The distribution of bird life in Colombia: A contribution to a biological survey of South America. Bulletin of the American Museum of Natural History **36**:1–729.

Collar, N. J., and P. Andrew. 1988. Birds to watch: The International Council for Bird Preservation world checklist of threatened birds. ICBP Technical Publication No. 8. Smithsonian Institution Press, Washington, D.C.

Cuadros, T. 1988. Aspectos ecológicos de la comunidad de aves en un bosque nativo en la Cordillera Central en Antioquia (Colombia). Hornero **13**:8–20.

Giraldo, M. 1985. Estructura y composición de la comunidad aviaria en un bosque montano húmedo de la Cordillera Occidental. Tesis, Departamento de Biología, Universidad del Valle, Cali, Colombia.

Hilty, S. L. 1974. Notes on birds at swarms of army ants in the highlands of Colombia. Wilson Bulletin **86**:479–481.

Hilty, S. L. 1985. Distributional changes in the Colombia avifauna: A preliminary blue list. Pages 1000–1012 in P. A. Buckley et al., editors. Neotropical ornithology. Ornithological Monograph 36. American Ornithologists' Union, Washington, D.C.

Hilty, S. L., and W. L. Brown. 1986. A guide to the birds of Colombia. Princeton University Press, Princeton, New Jersey.

Howe, H. F. 1984. Implications of seed dispersal by animals for tropical reserve management. Biological Conservation **30**:261–281.

Johnels, A., and T. Cuadros. 1986. Species composition and abundance of bird fauna in a disturbed forest in the Central Andes of Colombia. Hornero **12**:235–241.

Karr, J. R. 1990. Avian survival rates and the extinction process on Barro Colorado Island, Panama. Conservation Biology **4**:391–397.

Kattan, G. H. 1992. Rarity and vulnerability: The birds of the Cordillera Central of Colombia. Conservation Biology **6**:64–70.

Kattan, G., C. Restrepo, and M. Giraldo. 1984. Estructura de un bosque de niebla en la Cordillera Occidental, Valle del Cauca, Colombia. Cespedesia **13**:23–43.

Leck, C. F. 1979. Avian extinctions in an isolated tropical wet-forest preserve, Ecuador. Auk **96**:343–352.

Lehmann, F. C. 1957. Contribuciones al estudio de la fauna de Colombia XII. Novedades Colombianas **1(3)**:101–156.

Levey, D. J. 1988. Tropical wet forest treefall gaps and distributions of understory birds and plants. Ecology **69**:1076–1089.

Levey, D. J., and F. G. Stiles. 1992. Evolutionary precursors of long-distance migration: Resource availability and movement patterns in neotropical landbirds. American Naturalist **140**:447–476.

Loiselle, B. A., and J. G. Blake. 1991. Temporal variation in birds and fruits along an elevational gradient in Costa Rica. Ecology **72**:180–193.

Lovejoy, T. E., R. O. Bierregaard, A. B. Rylands, J. R. Malcolm, C. E. Quintela, L. H. Harper, K. S. Brown, A. H. Powell, G. V. N. Powell, H. O. R. Schubart, and M. B. Hays. 1986. Edge and other effects of isolation on Amazon forest fragments. Pages 257–285 in M. E. Soulé, editor. Conservation biology: The science of scarcity and diversity. Sinauer Associates, Sunderland, Massachusetts.

McArthur, R. H., and E. O. Wilson. 1967. The theory of island biogrography. Princeton University Press, Princeton, New Jersey.

Miller, A. H. 1963. Seasonal activity and ecology of the avifauna of an equatorial cloud forest. University of California Publications in Zoology **66**:1–74.

Newmark, W. D. 1991. Tropical forest fragmentation and the local extinction of understory birds in the eastern Usambara Mountains, Tanzania. Conservation Biology **5**:67–78.

Orejuela, J. E. 1985. Tropical forest birds of Colombia: A survey of problems and a plan for their conservation. Pages 95–115 in A. W. Diamond and T. E. Lovejoy, editors. Conservation of tropical forest birds. ICBP Technical Publication No. 4. Paston Press, Norwich, Connecticut.

Orejuela, J. E., R. J. Raitt, and H. Alvarez-López. 1979. Relaciones ecológicas de las aves en la Reserva Forestal de Yotoco, Valle del Cauca. Cespediesia **8**:7–28.

Orejuela, J. E., G. Cantillo, J. E. Morales, and H. Romero. 1982. Estudio de la comunidad aviaria en una pequeña isla de hábitat de bosque premontano húmedo cerca a Argelia, Valle, Colombia. Cespedesia **11**:103–120.

Pickett, S. T. A., and J. N. Thompson. 1978. Patch dynamics and the design of nature reserves. Biological Conservation **13**:27–37.

Powell, G., R. Bjork, and M. L. Avila Hernández. 1991. Estudio de la migración del quetzal, *Pharomacrus mocinno*. Abstracts, IV Congreso de Ornitología Neotropical, Quito, Ecuador.

Ridgely, R. S. 1981. The current distribution and status of mainland neotropical parrots. Pages 233–384 in R. F. Pasquier, editor. Conservation of New World parrots. ICBP Technical Publication No. 1. Smithsonian Institution Press, Washington, D.C.

Saavedra, C. J., and C. Freese. 1986. Prioridades biológicas de conservación en los Andes tropicales. Parques **11(2–3)**:8–11.

Saunders, D. A., R. J. Hobbs, and C. R. Margules. 1991. Biological consequences of ecosystem fragmentation: A review. Conservation Biology 5:18–32.

Simberloff, D. 1988. The contribution of population and community biology to conservation science. Annual Review of Ecology and Systematics 19:473–511.

Terborgh, J. 1986. Keystone plant resources in the tropical forest. Pages 330–344 in M. E. Soulé, editor. Conservation biology: The science of scarcity and diversity. Sinauer Associates, Sunderland, Massachusetts.

Terborgh, J., and B. Winter. 1980. Some causes of extinction. Pages 119–133 in M. E. Soulé and B. A. Wilcox, editors. Conservation biology, an evolutionary-ecological approach. Sinauer Associates, Sunderland, Massachusetts.

Terborgh, J., and B. Winter. 1983. A method for siting parks and reserves, with special reference to Colombia and Ecuador. Biological Conservation 27:45–58.

Willis, E. O. 1974. Populations and local extinctions of birds on Barro Colorado Island, Panama. Ecological Monographs 44:153–169.

Willis, E. O. 1979. The composition of avian communities in remanescent woodlots in southern Brazil. Papeis Avulsos Zoologia 33:1–25.

Geographic Range Fragmentation and Abundance in Neotropical Migratory Birds

BRIAN A. MAURER
Department of Zoology
Brigham Young University
Provo, UT 84602, U.S.A.

S. GREG HEYWOOD
Department of Zoology
Brigham Young University
Provo, UT 84602, U.S.A.

Abstract: *Populations of neotropical migrant landbirds have experienced significant declines in recent years. We investigated potential consequences of these declines by examining the relationship between abundance and fragmentation of geographic ranges of species on the North American breeding grounds. We estimated areographic fragmentation using the box dimension of a species' geographic range and demographic fragmentation using the fractal dimension of the semivariance function calculated from samples of population abundance across species' geographic ranges. We found a negative relationship between average abundance and demographic fragmentation for neotropical migrants, but not for residents. We also showed that demographic fragmentation and areographic fragmentation are inversely related for residents, but not for neotropical migrants. These results imply that neotropical migrants may be more sensitive to extinction than are residents.*

Rango geográfico, fragmentación y abundancia en pájaros migratorios neotropicales

Resumen: *Las poblaciones de pájaros terrestres migratorios neotropicales han experimentado una declinación significativa en años recientes. Nosotros investigamos las consecuencias potenciales de esta declinación examinando la relación entre abundancia y fragmentación del rango geográfico de especies criadas en Norte América. Nosotros estimamos la fragmentación areográfica utilizando el método de "box dimension" del rango geográfico de especies y la fragmentación demográfica utilizando dimensión fractal de la función de semivarianza calculada a partir de las muestras de abundancia de población a través del rango geográfico de especies. Nosotros encontramos que existe una relación negativa entre la abundancia promedio y la fragmentación demográfica para los migrantes neotropicales, pero no para los residentes. Nosotros también demostramos que la fragmentación demográfica y la fragmentación areográfica estan inversamente relacionadas para los residentes, pero no para los migrantes neotropicales. Estos resultados implican que los migrantes neotropicales pueden ser más susceptibles a la extinción que los residentes.*

Introduction

Recent concern over the decline of neotropical migratory birds has focused on problems encountered by these species in both their breeding habitats in North America and their wintering habitats throughout Central and South America (Askins et al. 1990; Finch 1991). On the breeding grounds, problems seem to result from fragmentation of once relatively continuous deciduous forests, the primary breeding habitat of most of these species (Robbins 1979; Whitcomb et al. 1981; Robbins et al. 1989*a*; Finch 1991). On the wintering grounds, North American migrants face habitat loss due to trop-

Paper submitted November 15, 1991; revised manuscript accepted July 17, 1992.

ical deforestation and increased levels of potentially harmful pesticides (Robbins et al. 1989*b;* Terborgh 1989; Finch 1991).

Evidence from studies in forest fragments in North America suggests that as breeding habitat becomes fragmented, individuals have more difficulty in successfully reproducing (Gates & Gysel 1978; Wilcove 1985; Temple & Cary 1988) and, consequently, populations are thought to be less stable and more likely to go locally extinct (Whitcomb et al. 1981; Lynch & Whigham 1984, Temple & Cary 1988). Theoretical arguments, however, have not resolved whether metapopulation persistence is enhanced or decreased by dividing the population into small fragments (Wright & Hubbell 1983; Goodman 1987; Quinn & Hastings 1987, 1988; Gilpin 1988; Hastings 1991). The result of the decline of metapopulations across a landscape should be the loss of the species from the landscape. But most species occur across a wide range of landscapes from one region to the next within their geographic range. A species will not be in danger of extinction unless it declines throughout its geographic range, with individual metapopulations declining simultaneously. Therefore, in order to assess the threat to a species of its decline within local and regional populations, data are necessary from extensive samples collected across the entire geographic range of a species. The consequences of breeding habitat fragmentation for neotropical migrants ultimately are expressed on the geographical scale. Many of these species breed in different kinds of habitats in different portions of their geographic range (Collins 1983*a*, 1983*b;* Shy 1984). Studies that restrict attention to relatively localized collections of forest fragments of similar habitat may be of limited use when extrapolated to the entire geographic range of a species. The basic assumption upon which the methods developed in this paper are based is that as metapopulations of a species decline, the distribution of abundance across a species' geographic range becomes more patchy and, consequently, the entire geographic range becomes more fragmented.

The goal of this paper is to introduce a method of quantifying fragmentation of a species' geographic range and to use this measure to evaluate the relationship between fragmentation and abundance for several taxa that contain a large proportion of species that migrate to the neotropics during the nonbreeding season. By documenting spatial patterns of population abundance and how they are related to the commonness of a species, it should be possible to obtain some understanding of the processes underlying population change at continental scales. In a more practical vein, such analyses should be useful in identifying species at risk of major population decline across their geographic ranges.

It is important to point out that there are at least two potential meanings to the general concept of fragmentation. The first meaning is related to the pattern of presence and absence of the species of interest. If one takes the set of all locations on a continent where the species is present and compares it to the set of locations where the species is absent, then the outline or boundary between the two sets represents a shape on a two-dimensional plane, like the coastline of an island archipelago. The amount of fragmentation of that shape we refer to as the *areographic fragmentation* of a species' geographic range.

The second meaning is related to the spatial distribution of population density across a species' geographic range. A species may occupy a relatively continuous set of habitats across its geographic range but may experience unequal productivity in different habitats (Pulliam 1988; Maurer & Brown 1989). If such is the case, then population density, or some measure of productivity, may be unevenly distributed across space. Picture a mountain range, where the height of the peaks represents population abundance or productivity. A species with steep density peaks and precipitous density valleys could be considered to have a more fragmented geographic range than one where density changes more smoothly from place to place. This kind of fragmentation we call *demographic fragmentation.* It is not necessary that demographic and areographic fragmentation be the same, since they reflect different biological processes. Areographic fragmentation corresponds to metapopulation models that emphasize patterns of presence or absence (see Hanski 1991; Villard et al. 1992), while demographic fragmentation is related to models that represent population dynamics within and among patches (see Pulliam 1988; Hastings 1991).

Methods

Spatial Distribution of Population Densities

Data on the spatial distribution of population abundances of 86 species from the passerine taxa Tyrannidae (19 spp.), Vireonidae (9 spp.), Parulinae (43 spp.), and Icterinae (15 spp.) were obtained from the North American Breeding Bird Survey (BBS) conducted annually by the U.S. Fish and Wildlife Service and the Canadian Wildlife Service. The BBS is a series of annual counts that have been conducted by volunteer ornithologists for over 25 years. Each June, over 2000 censuses are conducted. A census consists of a transect 40 km in length with 50 counting stations located 0.8 km apart. All birds located within a specified radius around each station are recorded. Abundances are expressed in counts per 40-km route. Robbins et al. (1986) describe the field methods used in the BBS in more detail. Each species was classified as either a neotropical migrant or a temperate migrant/resident based on the classification given in Finch (1991) and supplemented by the current list of migratory status of species maintained by the Neotropical Migratory Bird Conservation Program (S. A. Gauthreaux, personal communication). In this paper we use the terms "temperate migrants" and "residents" interchangeably to refer to species that do not travel long distances to the neotropics during the nonbreeding season.

We averaged the abundance of each species over all years in which it was detected on each census transect. Estimates of the spatial pattern of population abundance for each species were calculated using a geostatistical technique called kriging (Ripley 1981; Burrough 1986; Robertson 1987). Kriging estimates spatial patterns generated by a process with weighted interpolations of the pattern between grid blocks for which samples are available. Formally, this is done by calculating the semivariance function, $\gamma(h)$, for pairs of sample points located h units apart as

$$\gamma(h) = \frac{\sum_{i=1}^{N(h)} (x_i - x_{i+h})^2}{2N(h)},$$

where x_i and x_{i+h} are the abundances of a species at the ith pair of sampled points h units apart, and $N(h)$ are the number of pairs of points h units apart. A parametric function is then fit to the estimated semivariance and used to generate weights to interpolate between sample points (Burrough 1986).

A Measure of Areographic Fragmentation

Areographic fragmentation is related to the degree to which the border of a species' geographic range is convoluted or jagged, and to how many "islands" are separated from the main body of the geographic range. Mandelbrot (1983) has discussed a number of techniques that can be used to measure such patterns. We used the "box dimension" as a measure of areographic fragmentation.

The box dimension belongs to a family of measurements called "fractal" dimensions. Fractal geometry is an extension of geometry that provides mathematical models for natural shapes. The only shapes considered in classical geometry have smooth boundaries and integer dimensions: a square is two-dimensional and a cube is three-dimensional. In fractal geometry, shapes can have jagged boundaries and non-integer dimensions. These non-integer dimensions provide a way to describe the complexity of a jagged boundary. The relative amount of jaggedness of a boundary is often related to processes that generate the pattern. A number of useful applications of fractal dimensions have been found for ecological problems (Palmer 1988, 1992; Sugihara & May 1990; Milne 1991, 1992).

The box dimension is a fractal measure based on the rate at which the number of boxes that cover a pattern increase with the decreasing size of the box (Maurer 1994). It is calculated by choosing a range of box sizes and, for each size, counting the minimum number of boxes needed to cover the geographic range of a species (Barnsley 1988). In this case, we used the boundary of a species' geographic range as the part of continent for which the kriging procedure estimated abundance as less than or equal to zero. The logarithm of the number of boxes of size $n(h)$ is regressed against the logarithm of $1/h$. The slope of this regression is the box dimension. If D_B is the box dimension, then the linear regression equation used to estimate the box dimension is

$$\log n(h) = \beta_0 + D_B \log (1/h).$$

The box dimension ranges between 1 and 2. Values of the box dimension close to 1 indicate a relatively high amount of perimeter compared with area of geographic range (Maurer 1994). This would be the case if a species had a highly convoluted geographic range boundary. Thus, areographic fragmentation increases with decreasing box dimension.

There are no theoretical considerations that might give a priori expectations for the form of the relationship between abundance and areographic fragmentation. On the one hand, species that are common and widespread may be able to penetrate deeply into unfavorable habitat near the edge of the geographic range, so the edge might be relatively jagged. In this case, one would expect a positive relationship between areographic fragmentation and abundance. On the other hand, species that are rare and narrowly distributed might be limited by many factors at the edge of their geographic ranges, so the boundary would be highly fragmented. In this case, there would be a negative relationship between areographic fragmentation and abundance. It is not clear which, if either, of these alternatives is to be expected.

A Measure of Demographic Fragmentation

Burrough (1983) suggested that for many geological phenomena, such as variation in soil properties, a number of factors contribute to the overall spatial pattern. Since these factors often operate on different spatial scales, the pattern may be exceedingly complex. The statistical patterns generated by such processes can be modeled by a fractional Brownian process, a random spatial process constrained by the spatial scales of the factors responsible for the process. Such a process assumes that the smoothness of the spatial pattern is related to the spatial "persistence" or autocorrelation of the biological or geophysical processes generating the pattern. For population abundance, if resource quality varies relatively smoothly from one place to the next with respect to the species' needs, then the population processes will change more slowly across space than would be the case for a species for which resource quality varies drastically across the landscape (see also the discussion in Sugihara & May 1990). This persistence of the spatial processes generating the pattern can be measured by the fractal dimension of the semivariance function (Palmer 1988). Patterns generated by a process in which the conditions that determine it change slowly across the landscape have lower fractal dimensions than processes where those conditions change rapidly. Burrough (1983) showed that the fractal dimension of the pattern is related to the semivariance function by the slope of the linear regression of log $\gamma(h)$ against log h. If slope is denoted by b, then the fractal dimension of the pattern (D_S) is given by

$$D_S = 2 - \frac{b}{2}.$$

The fractal dimension of the semivariance function varies between 1 and 2. When it is 2, there is complete spatial independence in the distribution of abundances across the geographic range; when it is close to 1, there is a high degree of spatial dependence (Palmer 1988). Generally, when the logarithm of the semivariance function is plotted against the logarithm of the lag (h), then a nearly straight line is obtained. However, there are some situations in which this plot is nonlinear. Such nonlinearities often suggest different processes operating at different scales (Palmer 1988). Our approach in this paper was to consider only the linear estimate of the fractal dimension as a first approximation. This resulted in some estimates of the fractal dimension of the semivariance slightly exceeding 2. For species with such estimates, we assumed that the high value of the estimate reflected a high degree of spatial dependence even if other processes might also contribute to the observed pattern.

Central regions of the species' geographic range will have the highest density and will represent those locations where conditions are most favorable to the species. As conditions deteriorate with respect to the ecological needs of the species, densities will decrease (Brown 1984; James et al. 1984; Maurer & Brown 1989). Brown (1984) postulated that species that are common and abundant across a landscape will have the tendency to use a wide variety of resources and therefore will be found in a larger number of habitat types than a species with a more restricted distribution. Although Brown (1984) argued that most of the time species should have only a single density peak across their geographic range, his insight can be generalized to patterns of geographic range fragmentation. A common species that can use a wide variety of resources will be found in a large number of habitats across its geographic range and consequently will generally have densities that change relatively smoothly from place to place. Hence, its geographic range will have a low degree of demographic fragmentation and a relatively small value for the fractal dimension associated with the semivariance function calculated from its densities. A rare species will have a relatively narrow set of useful resources, so its density will change more drastically from one location to the next than would be the case for a common species. The density peaks and valleys across its geographic range will be steeper, so a rare species will have a high degree of demographic fragmentation. Consequently, a rare species will have a relatively high fractal dimension associated with its semivariance function. This prediction was tested by plotting the fractal dimension of the semivariance function against the logarithm of average population abundance for each of the 87 species used in this study.

Results

We hypothesized that rare species should generally have higher degrees of demographic fragmentation than common species. Across all taxa, the relationship between the fractal dimension calculated from the semivariance function and the logarithm of abundance showed no relationship (Fig. 1). This was primarily due to the fact that a number of species of blackbirds (Icterinae) had relatively high abundances coupled with a fractal dimension of around 1.95. Analysis of covariance indicated that the relationship between demographic fragmentation and abundance was significantly different among the four taxa (Table 1). Two of the three taxa that contained neotropical migrants had significantly negative relationships between fractal dimensions and abundance. The third of these taxa, Vireonidae, had a nonsignificant relationship. When the relationship between demographic fragmentation and abundance was estimated for neotropical migrants and residents separately, a significant negative relationship was obtained for neotropical migrants but not for species that do not migrate to the neotropics (Fig. 1).

Surprisingly, there were no significant relationships between areographic fragmentation and abundance for either neotropical migrants or residents (Fig. 2). The reason for this lack of relationship is not clear, but it indicates that the box dimension is not directly sensitive to the abundance of a species across its geographic range. We found no significant correlations between areographic fragmentation and abundance for any of the four families considered.

The relationship between demographic fragmenta-

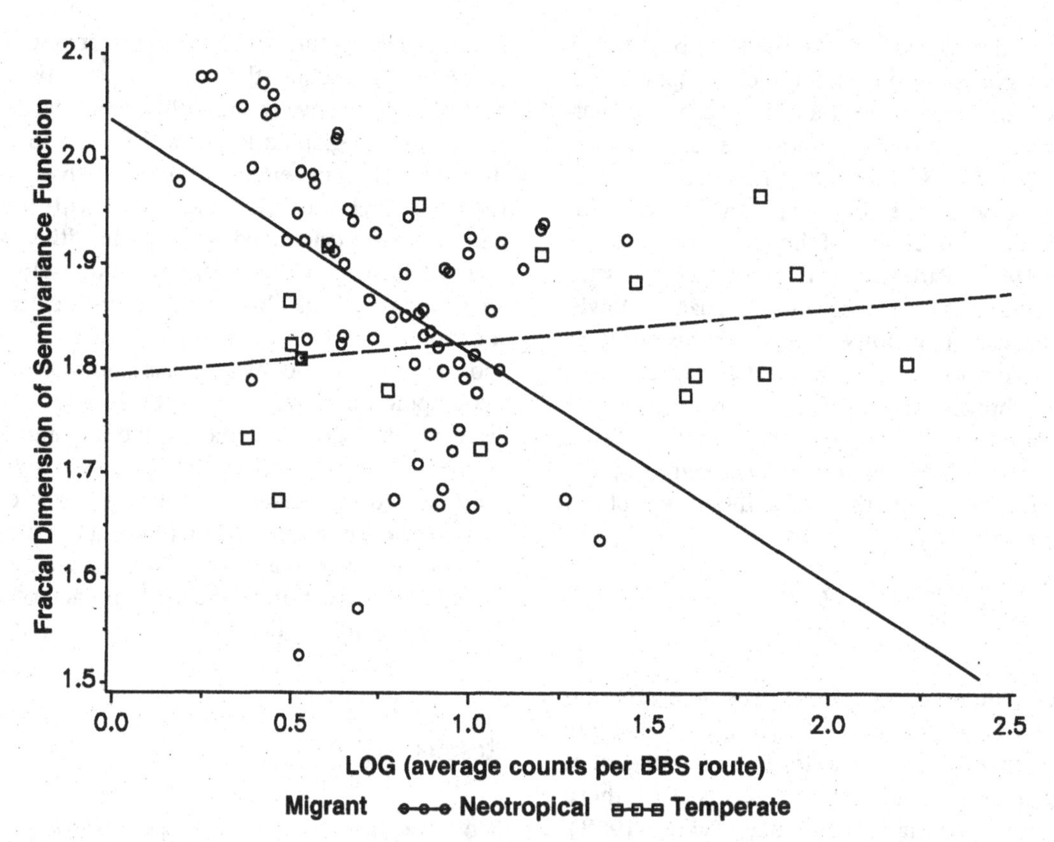

Figure 1. Relationship between the logarithm of average abundance and degree of demographic fragmentation measured by the fractal dimension of the semivariance function for neotropical migrants and residents during the breeding season. Note that demographic fragmentation declines with increasing abundance for neotropical migrants but not for residents. Lines were estimated by least squares regression.

tion and areographic fragmentation was different for neotropical migrants and residents. While there was no relationship between the two measures of fragmentation for neotropical migrants, there was a significant positive relationship between the box dimension (D_B) and the semivariance dimension (D_S) for residents (Fig. 3). Thus, for resident species only, higher degrees of areographic fragmentation were associated with *lower* degrees of demographic fragmentation. Interestingly, species with the highest levels of areographic fragmentation tended to be colonial, such as the Boat-Tailed Grackle (*Quiscalus major*) and the Tricolored Blackbird (*Agelaius tricolor*).

Discussion

Our data indicate that there are fundamental differences in how populations of neotropical migrant and resident bird species fill the North American continent. Neotropical migrant species that are abundant also tend to have relatively gradual increases and decreases in abundance across the continent, but there is no such trend for resident species. Residents that have highly convoluted geographic range boundaries tend to show relatively gradual changes in abundance across the continent. This is not true for neotropical migrants.

Different patterns in the relationship between average

Table 1. Analysis of covariance testing for differences among four avian taxa in the relationship between the logarithm of average abundance and demographic fragmentation as measured by the fractal dimension of the semivariance.

Taxon	*Number species*	*Number migrants*	*Estimated slope*	*Standard error*	*t-test for* H_o: b = 0
Icterinae	15	4	0.05	0.05	1.09
Parulinae	43	42	−0.29	0.12	−2.51**
Tyrannidae	19	15	−0.28	0.16	−1.68*
Vireonidae	9	8	0.06	0.19	0.30

*$p < 0.10$, df = 78, *MSE* = 0.77.
**$p < 0.05$, d = 78, *MSE* = 0.77.

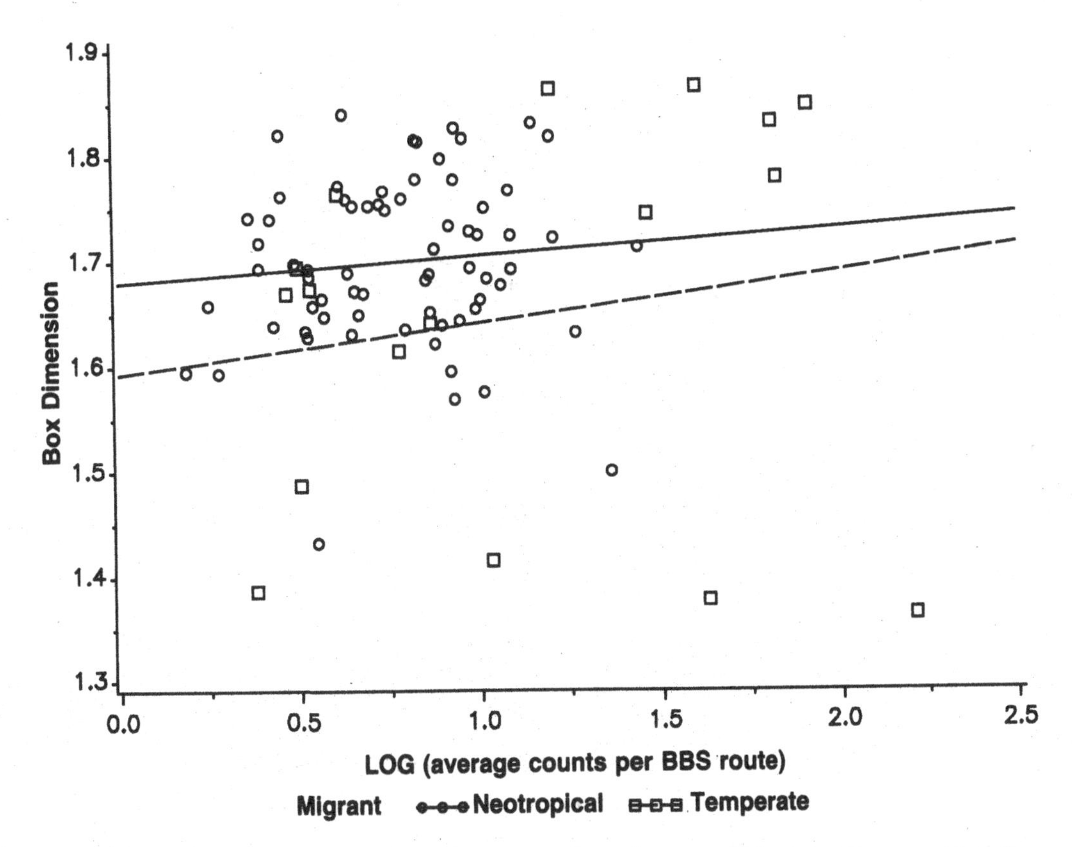

Figure 2. Relationship between the logarithm of average abundance and degree of areographic fragmentation measured by the box dimension for neotropical migrants and residents during the breeding season. Lines were estimated by least squares regression.

abundance, demographic fragmentation, and areographic fragmentation for birds of different taxa that have different migratory strategies suggest different processes generating them. The significant negative relationship between demographic fragmentation and abundance for neotropical migrants suggests that, for these species, processes that influence patterns of abundance affect the degree to which populations become fragmented and isolated across geographic space. For resident species, patterns of areographic and demographic fragmentation are independent of average abundance. Instead, areographic fragmentation increases with decreasing demographic fragmentation; that is, species with relatively smooth changes in abundance over space tend to have more jagged or convoluted range edges.

Although there may be other minor differences, the major ecological difference between the two groups of species is their migratory behavior. Neotropical migrants are exposed to different sets of ecological conditions between breeding and nonbreeding seasons, and during migration. Hence, population abundance among these species may respond to changes in ecological conditions in either North America during the breeding season or in the neotropics during the nonbreeding season. In contrast, populations of resident species are affected only by changes in ecological conditions on the North American continent. It is not immediately evident why these differences in migratory status should correspond to the different patterns shown in this study.

Abundances of neotropical migrants are significantly lower on average than of resident species (B. A. Maurer, unpublished data). This suggests that neotropical migrants will tend to be more sensitive to extinction than residents because, all else being equal, they tend to maintain lower population sizes across their geographic range. Lower population size increases the likelihood of extinction of local populations (see Pimm 1991 for a review of current ideas on extinction of local populations). This would translate into a higher rate of loss of local populations in neotropical migrants than in residents, ultimately driving continental populations of migrants towards extinction. Greater sensitivity of local populations to extinction may be one of the reasons why demographic fragmentation is related to average abundance in neotropical migrants but not in residents.

The patterns that we have documented suggest that it might be possible to use demographic fragmentation as an indicator of population decline for neotropical migrant species. If demographic fragmentation can be shown to increase when populations of a species decrease, then monitoring programs could be designed to detect changes in demographic fragmentation of species over time. We suggest that the analytical procedures described in this paper be used to monitor the geo-

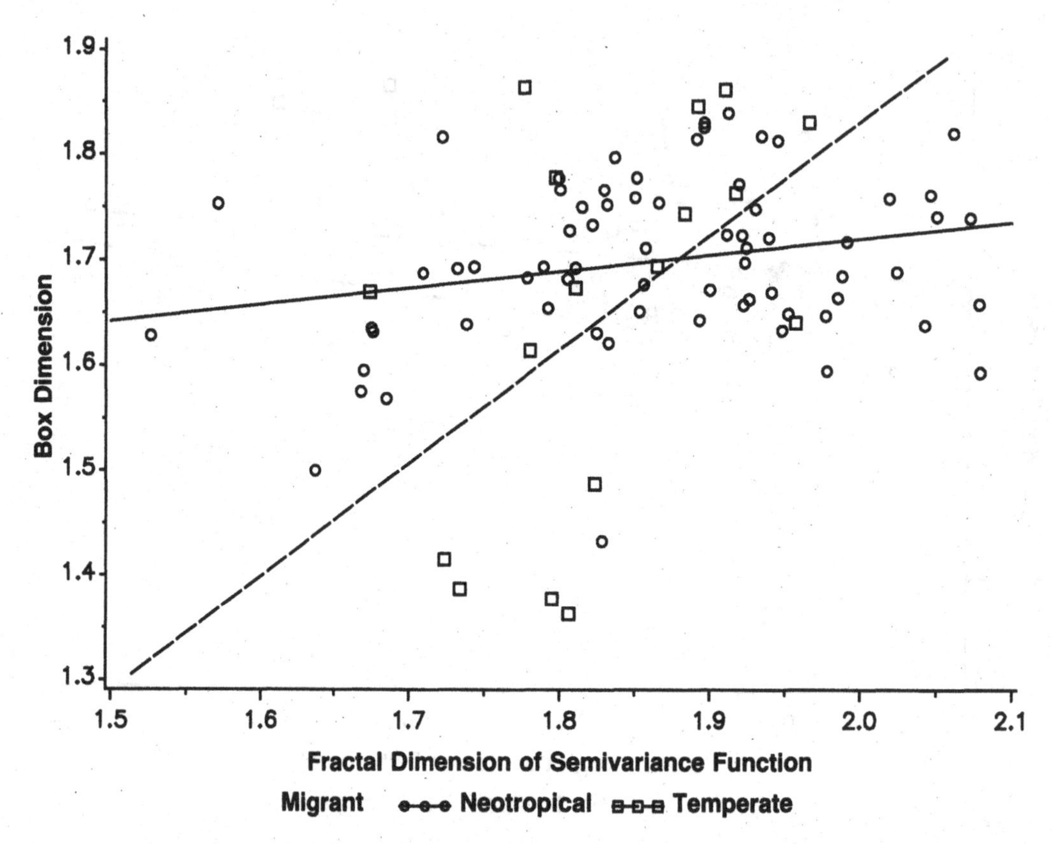

Figure 3. Relationship between areographic fragmentation and demographic fragmentation for neotropical migrants and residents during the breeding season. Note that areographic fragmentation decreases with increasing demographic fragmentation for residents but not for neotropical migrants. Lines were estimated by least squares regression.

graphic structure of populations as well as their abundance. Populations of a species may increase in one portion of the geographic range while declining in another. The BBS has been designed to allow the monitoring of populations within political boundaries and within "physiographic regions." Neither of these ways of categorizing samples is likely to correspond to natural spatial units to which populations of species respond. In fact, most species probably respond independently of other species to ecological factors across their geographic ranges. In our laboratory, we are investigating the possibility of using data sets such as the BBS to monitor changes in the spatial structure of geographic populations of species.

Our results also imply that current concern by conservationists and wildlife managers (see Finch 1991; Hagan & Johnston 1992) over populations of neotropical migrant birds is justified. The patterns of geographic range fragmentation that result from interactions of local populations of these species with their habitats both in North America and in the neotropics are distinctively different from those of related species that do not migrate to the neotropics. Neotropical migrants also have low populations compared to residents. These factors combined suggest that neotropical migrants may be more sensitive to environmental changes induced by human activities than are resident species. We caution that it is not likely that monitoring programs will see drastic changes in these populations over the course of a few years. The mechanisms leading to population decline across a species' geographic range may take decades, or even centuries, to have significant effects. If such is the case, it is important to maintain long-term monitoring programs such as the BBS over periods of time that exceed the normal lifetime of bureaucratic programs, and to continue to analyze the data obtained through such efforts to identify changes in both abundance across geographic ranges and in spatial patterns of abundance.

Conclusions

The species of neotropical migrants we studied appear to be at greater risk of extinction than are residents. A substantial amount of the threat to these species comes from habitat loss and increased fragmentation of once relatively continuous habitat on the North American breeding grounds. This does not discount the potential importance of loss of wintering habitat to some species due to forest fragmentation in the neotropics. Indeed, it is likely that the population declines documented for these species within recent years can be attributed to the cumulative effects of changes in both nearctic and neotropical habitats (Robbins et al. 1989*b*, Finch 1991). Although management efforts in North America aimed at improving breeding habitat will not necessarily solve the problem of population decline, we suggest that such management efforts will likely slow the decline. In addition, management efforts aimed at improving breeding habitat would send a signal to Latin American nations that the governments of North America are committed to solving the problems faced by neotropical migrants on their breeding grounds.

We suggest that management activities concentrate on maintaining current breeding habitat and increasing population densities within it rather than on the acquisition of areas not currently protected. This does not mean that agencies responsible for management activities should ignore the acquisition of reserves that are designated specifically to encourage populations of neotropical migrants. We simply suggest that it is important to improve and encourage populations whenever possible in the areas in which they are now found. This might be accomplished, for example, by decreasing timber harvests on large tracts of public forest that currently support healthy populations of neotropical migrants.

Acknowledgments

We wish to thank Sam Droege at the U.S. Fish and Wildlife Service Office of Migratory Bird Management for facilitating our use of the Breeding Bird Survey data base. Dani Montague and Nikkala Pack assisted with the considerable amount of computer analysis required. Marc-André Villard and three anonymous reviewers provided many useful suggestions on a previous draft of this paper. This research was supported by U.S. Environmental Protection Agency grant no. R81-8358-010 and by Brigham Young University.

Literature Cited

Askins, R. A., J. F. Lynch, and R. Greenberg. 1990. Population declines in migratory birds in eastern North America. Current Ornithology **7**:1–57.

Barnsely, M. 1988. Fractals everywhere. Academic Press, New York.

Brown, J. H. 1984. On the relationship between abundance and distribution of species. American Naturalist **124**:255–279.

Burrough, P. A. 1983. Multiscale sources of spatial variation in soil. I. The application of fractal concepts to nested levels of soil variation. Journal of Soil Science **34**:577–597.

Burrough, P. A. 1986. Principles of geographic information systems for land resources assessment. Oxford University Press, Oxford, England.

Collins, S. L. 1983*a*. Geographic variation in habitat structure of the Black-Throated Green Warbler (*Dendroica virens*). Auk **100**:382–389.

Collins, S. L. 1983*b*. Geographic variation in habitat structure for the Wood Warblers in Maine and Minnesota. Oecologia **59**:246–252.

Finch, D. M. 1991. Population ecology, habitat requirements, and conservation of neotropical migratory birds. General Technical Report R-205. USDA Forest Service, Rocky Mountain Forest and Range Experiment Station, Fort Collins, Colorado.

Gates, J. E., and L. W. Gysel. 1978. Avian nest dispersion and fledging success in field-forest ecotones. Ecology **59**:871–883.

Gilpin, M.E. 1988. A comment on Quinn and Hastings: extinction in subdivided habitats. Conservation Biology **2**:290–292.

Goodman, D. 1987. Consideration of stochastic demography in the design and management of biological reserves. Natural Resource Modeling **1**:205–234.

Hagan, J. M., III, and D. W. Johnston. 1992. Ecology and conservation of neotropical migrant landbirds. Smithsonian Institution Press, Washington, D.C.

Hanski, I. 1991. Single-species metapopulation dynamics: concepts, models, and observations. Biological Society of the Linnean Society **42**:17–38.

Hastings, A. 1991. Structured models of metapopulation dynamics. Biological Journal of the Linnean Society **42**:57–71.

James, F. C., R. F. Johnston, N. O. Wamer, G. J. Niemi, and W. J. Boecklen. 1984. The Grinellian niche of the wood thrush. American Naturalist **124**:17–30.

Lynch, J. F., and D. F. Whigham. 1984. Effects of forest fragmentation on breeding bird communities in Maryland, U.S.A. Biological Conservation **28**:287–324.

Mandelbrot, B. B. 1983. The fractal geometry of nature. Freeman, New York, New York.

Maurer, B. A. 1994. Geographical population analysis. Blackwell Scientific Publications, Oxford, England.

Maurer, B. A., and J. H. Brown. 1989. Distributional consequences of spatial variation in local demographic processes. Annales Zoologici Fennici **26**:121–131.

Milne, B. T. 1991. Lessons from applying fractal models to landscape patterns. Pages 199–235 in M. G. Turner and R. H. Gardner, editors. Quantitative methods in landscape ecology. Springer-Verlag, Berlin, Germany.

Milne, B. T. 1992. Spatial aggregation and neutral models in fractal landscapes. American Naturalist **139**:32–57.

Palmer, M. W. 1988. Fractal geometry: a tool for describing spatial patterns of plant communities. Vegetatio **75**:91–102.

Palmer, M. W. 1992. The coexistence of species in fractal landscapes. American Naturalist **139**:375–397.

Pimm, S. L. 1991. The balance of nature? University of Chicago Press, Chicago, Illinois.

Pulliam, H. R. 1988. Sources, sinks, and population regulation. American Naturalist **132**:652–661.

Quinn, J. F., and A. Hastings. 1987. Extinction in subdivided habitats. Conservation Biology **1**:198–208.

Quinn, J. F., and A. Hastings. 1988. Extinction in subdivided habitats: a reply to Gilpin. Conservation Biology **2**:293–296.

Ripley, B. 1981. Spatial statistics. John Wiley & Sons, New York, New York.

Robbins, C. S. 1979. Effect of forest fragmentation on bird populations. Pages 198–212 in R. M. DeGraaf and K. E. Evans, editors. Management of north-central and northeastern forests for nongame birds. General Technical Report NC-51. USDA Forest Service, North Central Forest Experimental Station, St. Paul, Minnesota.

Robbins, C. S., D. Bystrak, and P. H. Geissler. 1986. The breeding bird survey: its first fifteen years, 1965–1979. Resource Publication 157. Fish and Wildlife Service, U.S. Department of the Interior, Washington, D.C.

Robbins, C. S., D. K. Dawson, and B. A. Dowell. 1989*a*. Habitat area requirements of breeding forest birds of the middle Atlantic states. Wildlife Monographs **103**:1–34.

Robbins, C. S., J. R. Sauer, R. S. Greenberg, and S. Droege. 1989*b*. Population declines in North American birds that migrate to the neotropics. Proceeding of the National Academy of Sciences, U.S.A. **86**:7658–7662.

Robertson, G. P. 1987. Geostatistics in ecology: interpolating with known variance. Ecology **68**:744–748.

Shy, E. 1984. Habitat shift and geographical variation in North American tanagers (Thraupinae: *Piranga*). Oecologia **63**:281–285.

Sugihara, G., and R. M. May. 1990. Applications of fractals in ecology. Trends in Ecology and Evolution **5**:79–86.

Temple, S. A., and J. R. Cary. 1988. Modeling dynamics of habitat-interior bird populations in fragmented landscapes. Conservation Biology **2**:340–347.

Terborgh, J. W. 1989. Where have all the birds gone? Princeton University Press, Princeton, New Jersey.

Villard, M.-A., K. Freemark, and G. Merriam. 1992. Metapopulation theory and neotropical migrant birds in temperate forests: an empirical investigation. Pages 474–482 in J. M. Hagan, III, and D. W. Johnston, editors. Ecology and conservation of neotropical migrant landbirds. Smithsonian Institution Press, Washington, DC.

Whitcomb, R. F., C. S. Robbins, J. F. Lynch, B. L. Whitcomb, M. K. Klimkiewicz, and D. Bystrak. 1981. Effects of forest fragmentation on avifauna of the eastern deciduous forest. Pages 125–205 in R. L. Burgess and D. M. Sharpe, editors. Forest island dynamics in man dominated landscapes. Springer-Verlag, New York, New York.

Wilcove D. 1985. Nest predation in forest tracts and the decline of migratory songbirds. Ecology **66**:1211–1214.

Wright, S. J., and S. P. Hubbell. 1983. Stochastic extinction and reserve size: a focal species approach. Oikos **41**:466–476.

Contributed Papers

Area Requirements for the Conservation of Rain Forest Raptors and Game Birds in French Guiana

J. M. THIOLLAY
Laboratoire d'Ecologie
E.N.S. 46, rue d'Ulm
75230 Paris CEDEX 05, France

Abstract: *The distribution and density of diurnal raptors and large game birds were studied in the vast unbroken rain forest of French Guiana to bring some concrete data to the design of a national park. Mainly by mapping the movements of raptors displaying over the canopy and estimating densities on understory strip transects, I tried to assess the species composition of 11 2,500-ha plots of primary forest, then to determine the actual population of a 10,000-ha area including all the major forest habitats. Among 27 primary forest raptors, 7 were recorded in all plots and 4 in only 1 or 2 plots. There was a near doubling of species with a tenfold increase in area, but no 10,000-ha area and few 100,000-ha quadrats would include the full regional set of forest raptors. A 10,000-ha plot included less than 100 pairs of diurnal raptors and only 23 species, 21 of which were represented by a mere 1 to 8 pairs. The three rarest species (harpy eagle, black caracara, and orange-breasted falcon) have extremely low densities and patchy distributions (on average 1 pair for 10,000 to 300,000-ha), but narrow habitat selection or specialized diet are not the sole correlates of rarity. The dispute over the best conservation strategy cannot be settled without a knowledge of the species' ecology that is currently lacking. Local patchiness and low dispersal ability usually argue for single large reserves, whereas some widely scattered species may be better protected in a network of smaller reserves. According to the results and the (still unsupported) minimum size estimates of a viable population, the smallest Guianan rain forest reserve including a complete bird community should be somewhere between 1 and 10 million hectares. Moreover, the large game birds were found to be highly sensitive to hunting pressure even when it occurs*

Paper submitted June 3, 1988; revised manuscript accepted September 27, 1988.

Resumen: *Con el objeto de recoger datos concretos para el diseño de un parque nacional se estudió la distribución y densidad de aves de rapiña diurnas y aves de caza en los bosques pluviales, extensos e intactos de la Guyana Francesa. Se trató de evaluar la composición de especies en 11 parcelas de 2,500 hectareas de bosque primario, trazando los movimientos de las aves de rapiña observados encima del dosel y calculando las densidades debajo del dosel en transectos lineales. Después se determinó la población actual en una area de 10,000 hectareas incluyendo todos los habitats mayores de la selva. De 27 aves rapaces del bosque primario, 7 fueron documentadas en todas las parcelas y 4 en sólo 1–2 parcelas. Se registró una casi duplicación de población de especies al aumentar el área en diez veces, pero ningúna de las áreas de 10 hectáreas y pocos de los cuadrantes de 10 hecatáreas incluyeron todas las especies de aves de rapiñas forestales de la región. En 10,000 hectáreas, se observaron menos de 100 parejas de sólo 23 especies, de las cuales 21 estaban representadas por sólo 1 a 8 parejas. Las tres especies más raras (la aguila arpía, el caracara negro y el halcón de pecho naranjo) tienen densidades extremadamente bajas y distribuciones irregulares (un promedio de una pareja en 10 áreas de 3 × 10 hectáreas), pero ni la selección fina de habitat, ni la dieta especializada son las únicas correlaciones de rareza. La disputa sobre la mejor estrategia conservacionista no puede ser resuelta sin un conocimiento de la ecología de la especie, el cual en estos momentos falta. Irregularidades locales y baja capacidad de dispersión argumentan mayormente el establecimiento de grandes reservas únicas, mientras que especies extensamente dispersas pueden ser mejor protegidas con una red de reservas más pequeñas. De acuerdo a los resultados y las estimaciones (hasta ahora no documentadas) de tamaño mínimo de una población viable, la reserva forestal más pequeña de Guyana incluyendo una comunidad completa de aves debe*

around the protected area. Their seasonal movements make them especially vulnerable.

ser más o menos de 1 a 10 millones de hectareas. Por otra parte, las grandes aves de caza revelaron ser muy sensitivas a la presión de cacería aún cuando ésta occura sólo alrededor del área protegida. Sus movimientos estacionales les hace especialmente vulnerables.

Introduction

Predators are important biological indicators to take into account in the design of an optimal nature reserve. Among them, diurnal raptors (Falconiformes) are especially well suited by virtue of their high diversity, wide distribution, and relative conspicuousness. They may be key elements in the equilibrium of natural communities, where predation pressure is one of the fundamental determinants of population structure, life history traits, and demography (Taylor 1984). In tropical forests, this functional role may have far-reaching effects (see Oniki 1979, 1985). Diurnal raptors include up to 8% of all bird species recorded in any Guianan rain forest site (Thiollay 1986*a*). They are often very sensitive to habitat disturbance (Thiollay 1985*b, c*) and many of them are likely to disappear from isolated forest fragments (Leck 1979; Willis 1974, 1979). While they are one of the most threatened groups among tropical forest birds, their life histories and ecological requirements are still very poorly known (Thiollay 1985*b*).

Raptors seem also to have one of the lowest densities among rain forest birds (Thiollay 1986*a*). Thus, any area encompassing viable populations of the complete set of predatory birds would be likely to include a whole natural community. Accordingly, they may become a basis and a justification for the size of protected forest areas. Large game birds have similar attributes. They also exemplify those rain forest populations with low reproductive rates and turnover, which are very sensitive to hunting pressure.

Various attempts have been made to assess the minimum critical size of forest reserves (Diamond 1975; Diamond & May 1976; Frankel & Soulé 1981; Lovejoy et al. 1984; Wilcox 1984). They often stem from the theory of island biogeography (MacArthur & Wilson 1967; MacCoy 1983; Simberloff 1986). To add some concrete data to the much-debated problem of the conservation of whole rain forest communities (the only functionally acceptable goal), I have tried to assess the spatial distribution and the overall density of every raptor species (Falconiformes) within a large area of rich primary South American rain forest. Such a survey is difficult and the results are tentative, but it is the first published attempt of this kind for a complete raptor community in any lowland primary forest. This habitat being among the richest and most threatened terrestrial ecosystems worldwide, it is imperative to develop appropriate census methods to obtain more basic data on the actual carrying capacity of existing forest reserves for such large endangered organisms.

The next step will be to define the size of a viable population of each species. Apart from theoretical or genetic considerations, we have almost no long-term natural experiments on the minimum size of isolated populations of such predators, except on islands or in temperate habitats. The peculiar demography and dynamic of humid tropical forest species remain to be investigated.

Study Areas

French Guiana is a 90,000 km^2 territory with 95% of its area still covered with subequatorial lowland rain forest (2–5° N). The 80,000 inhabitants and associated clearings are concentrated on a coastal strip and to a much lesser extent along the eastern (Oyapock River) and western (Maroni River) borders. Over 70,000 km^2 are unbroken, humid (2–3 m annual rainfall) primary forest, still unlogged, undisturbed, and without a single inhabitant since the extinction of Indian populations one or two centuries ago.

All study sites are hilly, lowland (under 600 m), evergreen rain forest made up of a mosaic of the following habitats, in decreasing order of frequency:

(1) High stands with closed canopy and open understory on well-drained slopes and plateaus;
(2) Lower, more disturbed areas with discontinuous canopy, dense undergrowth, and vine tangles;
(3) Swampy bottoms with pure stands of *Euterpe oleracea* palms;
(4) Small watercourses under closed canopy; and
(5) Low stands of Myrtacae or Clusiacae on thin soil and rocky outcrops.

Additional, localized habitats include small expanses of dense, uniform bamboolike Graminaceae called "cambrouzes" (on Sw, Ar, No, and Ko sites, see Table 1); large rivers and associated seasonally flooded flats (Ar, Ko, Vo); and high, bare granitic inselbergs, protruding above the surrounding forest (Em, Cr, No, Be, Vo).

Each sample plot (Fig. 1) occurs within a huge ex-

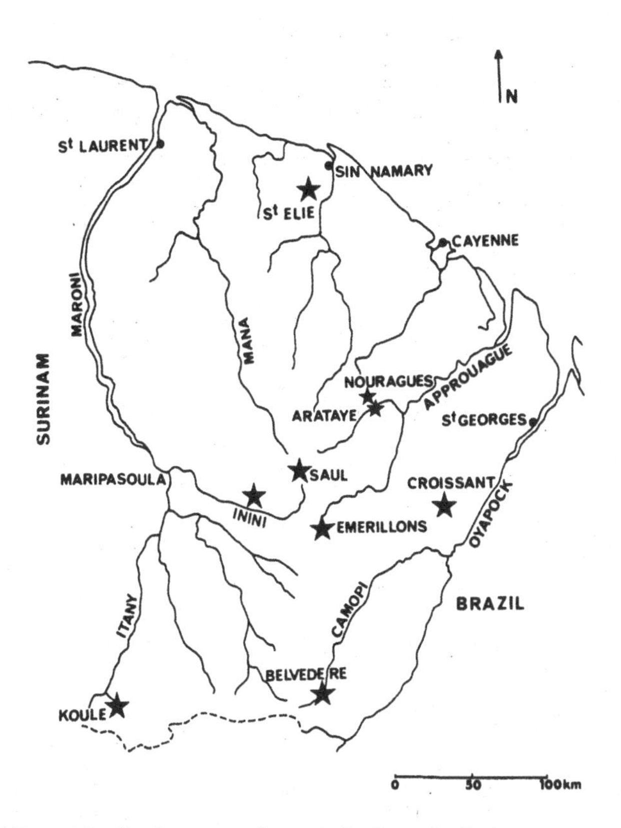

Figure 1. Study areas (stars) in French Guiana, with main rivers and small towns (solid dots). The rain forest is continuous all over the country except a narrow strip of coastal savanna in the north.

panse of unbroken primary forest. Only three of them are moderately hunted (Sw, Se, El), so habitats' flora and fauna are likely to be in a near virgin state. The ten sites in Guiana are spread over all the country and are representative of the natural conditions throughout the biogeographical zone of the Guianas (Amapa state to Guyana). In Table 1, they have been ordered from the most central one (first surveyed) to increasingly more peripheral sites (10 to 190 km) from the central site (see Fig. 1). An eleventh locality (4°40′ N–56°10′ W) has been added in neighboring Surinam, 350 km from the central site in Guiana, where personal records are confirmed by Donahue & Pierson (1982).

The main study area, a 10,000-ha quadrat, including two of the previous plots (Ar and No), was designed to be representative of the country's vegetation cover, i.e., to encompass most forest types including both a rather large river and an inselberg. Thus it is expected to harbor a higher species richness than more uniform forest tracts of similar size. Toward the north or northeast, the nearest zone occasionally hunted was about 5 km away from the limit, the nearest logging area about 30 km, and the nearest human settlement at 50 km. Undisturbed virgin forest extended at least to 75 km in all other directions.

Methods

Census techniques will be described and discussed elsewhere in more depth (Thiollay, in press). They are briefly summarized here.

Survey of Presence/Absence on 2,500-ha Plots

All diurnal raptors and the three biggest game bird species were searched for within eleven 2,500-ha areas of primary rain forest from 1981 to 1986 over a total period of 9 months, always between July and December. Fine weather, full daylight census periods included 1,633 hours in the forest understory, 499 h. from open and dominant lookouts, and 191 h. on rivers. Several methods were used, according to the behavior of the species involved: searching through natural openings and from rocky outcrops for species hunting over the canopy (vultures, *Ictinia*, *Elanoides*, etc.) or displaying in mid to late morning (*Leptodon*, *Harpagus*, *Buteogallus*, *Spizastur*, *Spizaetus*, etc.); surveying cliffs and river banks for habitat specialists (*Falco*, *Daptrius ater*); and recording nonsoaring species (*Leucopternis melanops*, *Harpia*, *Daptrius*, *Micrastur*, game birds) on foot through the understory or listening for *Micrastur* calls at dawn. Trapping, mist-netting, and playback broadcasting brought some additional information. All the species are assumed to be residents, and evidence of territorial or breeding behavior was observed in almost all of them.

Density Estimation on a 10,000-ha Quadrat

This area was the largest tract of forest I could survey within a reasonable time (72 days in 1986–87) and the smallest area likely to include most of the raptor community. It was centered around a high inselberg that provided very convenient lookouts overlooking all the area, except the southern edge, which was crossed by a river providing additional openings. A 600-ha core area, including the inselberg, the field station, and a dense network of trails, was the most intensively surveyed. A 4,200-ha circle, extending 3 km away from the four outmost lookouts of the inselberg, was the maximum area over which the movements of soaring raptors could be followed with 10 × 40 binoculars and an 11–33× telescope. All the canopy over the remaining outer parts was in sight within less than 2.5 km of nine other lookouts (riverbank and tree-fall gaps on slopes). Seventy of the 100 one square km quadrats dividing the area were fully crossed on foot two or more times. The census is likely to be less and less complete and accurate from the inner (600 ha) to the outer parts of the quadrat (see Table 3). Two concurrent methods were used.

First, a mapping method entailed following and drawing on a map the movements of birds displaying or flying over the forest and seen or heard in the understory. Thus, territories, assumed to be the approximate home

Table 1. Species recorded within ten 2,500-ha plots of primary rain forest in French Guiana (Sw = Saül west; Se = Saül east; In = Inini Valley-Montagne Bellevue; Em = Massif des Emerillons; Cr = Pic du Croissant; Ar = Saut Pararé, Arataye River; No = montagne des Nouragues; Be = Mont Belvédère; El = piste de St-Elie; Ko = Koulé Koulé Valley) and one in Surinam: Vo = Voltzberg-Raleigh Falls Reserve.

Species	*Sw*	*Se*	*In*	*Em*	*Cr*	*Ar*	*No*	*Be*	*El*	*Ko*	*Vo*
Greater yellow-headed vulture, *Cathartes melambrotus*	+	+	+	+	+	+	+	+	+	+	+
King vulture, *Sarcoramphus papa*	+	+	+	+	+	+	+	+	+	+	+
Gray-headed kite, *Leptodon cayanensis*	+					+		+	+		
Hook-billed kite, *Chondrohierax uncinatus*		+						+			+
Swallow-tailed kite, *Elanoides forficatus*	+	+	+	+	+	+	+	+	+	+	+
Rufous-thighed kite, *Harpagus diodon*	+			+	+		+	+		+	
Double-toothed kite, *Harpagus bidentatus*	+	+	+	+	+	+	+	+	+	+	+
Plumbeous kite, *Ictinia plumbea*	+	+	+		+	+	+		+	+	+
Tiny hawk, *Accipiter superciliosus*		+					+				+
Bicolored hawk, *Accipiter bicolor*		+		+		+	+	+			+
Gray-bellied hawk, *Accipiter poliogaster*							+				
Black-faced hawk, *Leucopternis melanops*		+	+			+	+	+	+	+	+
White hawk, *Leucopternis albicollis*	+		+	+	+	+	+	+	+	+	+
Great black hawk, *Buteogallus urubitinga*	+			+	+	+	+	+	+	+	+
Crested eagle, *Morphnus guianensis*		+		+	+	+	+	+		+	+
Harpy eagle, *Harpia harpyja*			+	+	+					+	+
Black-and-white hawk eagle, *Spizastur melanoleucus*	+	+	+	+	+	+	+	+	+	+	
Black hawk eagle, *Spizaetus tyrannus*		+	+							+	+
Ornate hawk eagle, *Spizaetus ornatus*	+	+	+	+	+	+	+	+	+	+	+
Red-throated caracara, *Daptrius americanus*	+	+	+	+	+	+	+	+	+	+	+
Black caracara, *Daptrius ater*										+	+
Barred forest falcon, *Micrastur ruficollis*		+		+		+	+	+	+		+
Lined forest falcon, *Micrastur gilvicollis*	+	+	+	+	+	+	+	+	+	+	+
Slaty-backed forest falcon, *Micrastur mirandollei*	+				+	+	+		+		+
Semi-collared forest falcon, *Micrastur semitorquatus*	+	+	+	+		+	+	+	+	+	+
Bat falcon, *Falco rufigularis*									+		+
Orange-breasted falcon, *Falco deiroleucus*					+		+				
Total forest species	15	17	14	16	16	18	21	18	17	18	22

Additional nonforest species recorded: Black vulture, *Coragyps atratus* (Vo) White-tailed hawk, *Buteo albicaudatus* (Vo) Broad-winged hawk, *Buteo platypterus* (migrant, Se) Short-tailed hawk, *Buteo brachyurus* (Be, El) Gray hawk, *Buteo nitidus* (Sw, El, Vo) Slate-colored hawk, *Leucopternis schistacea* (No) Crane hawk, *Geranospiza caerulescens* (Se) Osprey, *Pandion haliaetus* (migrant, Ar, Vo) Laughing falcon, *Herpetotheres cachinnans* (El)

range of a pair, were derived by the Minimum Convex Polygon method (Southwood 1978). The limits of neighboring conspecific territories were determined by repeated simultaneous observations of the pairs involved. Calls of *Micrastur* and group sizes of *Daptrius* were also used to separate different units.

The density of mobile, nonterritorial vultures and kites was estimated by extrapolating to the whole study

Table 2. Density estimates from strip census (number of individuals per 100 ha ± SD) of large game birds on ten 2,500-ha study plots (see Table 1) of primary rain forest in French Guiana with different hunting pressures. d = detection distance maximizing density estimate, calculated from No-site records only.

	Red and green macaw (Ara chloroptera[a] *(d = 100 m)*	*Black curassow* Crax alector *(d = 25 m)*	*Gray-winged trumpeter* Psophia crepitans *(d = 40 m)*
4 sites (In, Em, Be, Ko) No hunting within ≥50 km Census: 557 km	6.05 (±3.32)	8.37 (±1.09)	13.87 (±5.28)
3 sites (Cr, Ar, No) Nearest hunting areas: 3 to 20 km away Census: 673 km	0.81 (±0.90)	1.38 (±0.87)	3.24 (±2.40)
3 sites (Sw, Se, El) Regularly hunted Census: 475 km	0.38 (±0.15)	0.39 (±0.34)	0.64 (±0.60)

[a] *Some scarlet macaws,* Ara macao, *included.*

site the mean number of individuals crossing a fixed 1,000-ha area per hour during the optimum activity period. This was consistent with the maximum number of birds seen at any time over the 10,000-ha area (data in brackets in Table 3).

Then, especially for nonsoaring species, an understory strip transect census was performed. The network of trails was designed to pass within less than 500 m of every point. It also complemented the above mapping method and gave overall results most often consistent with it. The density estimate was $\hat{D} = n/2\,dL$, where n is the number of birds detected within the 2 d-wide strip transect of total length L (517 km) and d is the specific flushing or detection distance from the observer under which a given species almost always flies or calls and thus can be seen.

Results

Species Distribution Between Sites

The sample plots' area (2,500 ha) was designed to be larger than the mean territory size of all the raptor species but the largest ones (*Sarcoramphus, Morphnus, Harpia*). Since a species may well be recorded on a plot that includes only part of its home range and since every species is potentially uniformly distributed over the Guianan forested area, one would expect to meet most, if not all, of the species on each study site. Yet among the regional pool of 27 primary forest species, only 7 were recorded on all 11 plots, 8 species on 7 to 10 plots, and 8 others on 3 to 6 plots, while the last 4 species occurred on 1–2 plots only (Table 1). All three game birds studied (Table 2) were present, at least formerly, on every plot. Consequently, the coefficients of species similarity between two samples, $Ss = 2\,C/A + B$, where C is the number of species found in both communities

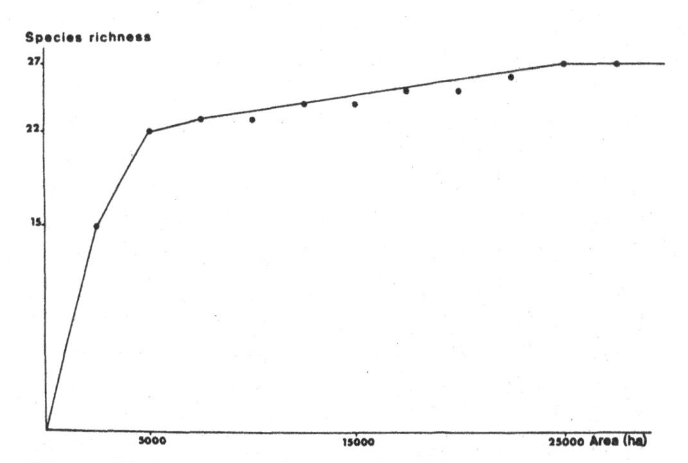

Figure 2. Cumulative number of species recorded within 11 successive 2,500-ha plots of primary rain forest, located at increasing distances (10 to 350 km) from the first plot surveyed. Only forest species are included.

Table 3. Resident population of raptors on the main study site, in number of territorial pairs recorded within three successively larger areas. The lack of precision of most figures comes from the many territories overlapping the limits of the concentric zones or, for the largest area, from some insufficiently known distributions. Home ranges were outlined by the Minimum Convex Polygon method. The overall density estimate $\hat{D}$ (last column), drawn from strip census, is given in number of individuals per 10,000 ha, the birds being recorded along the transects regardless of their age or sex. Only young birds, obviously following adults, have been excluded.

	600 ha	*4,200 ha*	*10,000 ha*	$\hat{D}$
Cathartes melambrotus	<1	≥2	≥4	(19)
Sarcoramphus papa	<1	±1	2	(9)
Leptodon cayanensis	—	—	≥1	3
Elanoides forficatus	<1	≥1	≥2	(≥10)
Harpagus diodon	<1	±1	≥1	4
Harpagus bidentatus	≤1	±3	>4	15
Ictinia plumbea	<1	2	3	(9)
Accipiter superciliosus	1	1	≥1	9
Accipiter bicolor	≤1	±2	≥3	4
Accipiter poliogaster	<1	1	≥1	3
Leucopternis melanops	<1	1	>1	4
Leucopternis albicollis	1	≤4	6	10
Buteogallus urubitinga	1	≤5	8	17
Morphnus guianensis	<1	≤1	≥1	4
Spizastur melanoleucus	<1	≤2	≥3	7
Spizaetus tyrannus	—	—	1	2
Spizaetus ornatus	<1	2	≤4	<13
Daptrius americanus	1	±5	≥12	<199
Micrastur ruficollis	<1	±2	>4	10
Micrastur gilvicollis	±2	>6	>12	72
Micrastur mirandollei	<1	>1	≥3	4
Micrastur semitorquatus	≤1	≤3	≥5	12
Falco deiroleucus	<1	1	1	(2)

of total species number A and B (Able & Noon 1976), are not high despite the large area, the habitat similarity, and the geographical proximity of plots. On all possible pairs of sites, 30% have a coefficient of 60–75% and the remaining 70% a coefficient of 76–82%.

About 80% of the total set of species (regional richness) is reached within the first 5,000 ha but the remaining 20% require an additional area four times larger to be included (Fig. 2). In fact, of the 14–21 species recorded on each 2,500-ha plot, no more than 12–16 were represented by at least one resident pair with most of its home range area included. This fits well the prediction from island biogeography of a near doubling of the species number resulting from a tenfold increase in land area. The data on the main study area lead to a similar conclusion. A 1,000-ha central quadrat included ≥50% of one home range of no more than 11–12 species (Table 3), exactly half the richness of the whole 10,000-ha zone.

This study area itself has more diversified habitats than most randomly chosen forest sites and is consequently more species rich than usual (23 species out of the regional set of 27). Any 10,000-ha area of unbroken primary forest in Guiana is more likely to harbor only

20–21 resident species or fewer. From the known distributions obtained by extensive surveys through the country, few (if any) 100,000-ha areas would include the full set of 27 forest species, whereas most 1,000,000-ha areas would. Thus, as expected from the slope of the species area curve (Fig. 2), the rarest taxa need very large areas to be added to an eventually protected community.

The moderate hunting pressure that may occur on or around some of the study sites seriously influences the density of game birds (Table 2). The three largest species may even completely disappear from the heavily hunted forests. Although more rarely shot, diurnal raptors were also found to be sensitive to hunting pressure in the same areas (Thiollay 1985*a*), perhaps as an indirect consequence of hunting on the prey communities.

Density Estimates on a 10,000-ha Area

Within the central 600-ha area, 21 species were recorded but no more than 5 of them (total = 6 pairs) had at least one nearly complete territory included in the quadrat (Table 3). Within the 4,200-ha circle, the same 21 species were seen, but only about 44 territories from 20 of them were fully included, whereas the randomly drawn limits of the 10,000-ha quadrat encompassed over 83 pairs of 23 species, but probably less than 100.*

The overall density (≤1 pair per 100 ha, all species together) is rather low. In a large area of temperate broad-leaved forest, I found an average 4 pairs of forest-breeding raptors per 100 ha (4 species, Thiollay 1967). Only 2 of our 23 species, *Daptrius americanus* and *Micrastur gilvicollis*, are represented by ≥12 pairs or flocks per 10,000 ha. Similar densities for these two species were recorded on a 6,000-ha area around Saül (Thiollay 1985*a*). Populations sizes of the remaining species range from 1 to 8 pairs per 10,000 ha (see Fig. 3).

The two species most patchily distributed and apparently localized by narrow habitat requirements are the black caracara and the orange-breasted falcon. The caracara is found widely scattered only along medium-size to large rivers. Two randomly chosen perpendicular straight lines through the forested zone of Guiana, crossing near the study area (i.e., 300 km east-west along the 4° N and 360 km north-south on 53° W), would only cross 8 rivers suitable to the species. If each strip is 10 km wide (the minimum spacing between two groups ever found in Guianan rivers), and if one home range of caracara is met at every river crossing (a very optimistic assumption), the resulting density would be one group for 800–900 km^2. The falcon is even rarer. Although it is seen regularly in the coastal zone far from suitable rocks and may breed in tree holes, no breeding pair has ever been found in French Guiana outside a cliff. The old granitic relief provides very few breeding sites. From extensive ground, river, and aerial surveys, we know that the three known pairs or appropriate cliffs nearest to the breeding pair of the study area are located in a straight line at 65 to 78 km. On the often surveyed forested quadrat between 3°30′ to 5° N and 52° to 53°30′ W, i.e. more than 25,000 km^2, no more than 8 pairs or acceptable cliffs have been located. This gives an average density of one pair for over 3,000 km^2.

At the other extreme, the two most abundant species (red-throated caracara and lined forest falcon) are almost continuously distributed. Conversely, for most other species, the proportion of forest seemingly not covered by a territory appears very high (40–80%, Table 4), although this is likely to be an overestimate and the nonterritorial forest may well be a refuge for immature birds.

The territory sizes, as measured here, consistently average several hundred hectares each (Table 4). They are quite similar to those of comparable temperate raptors (Cramp 1980; Géroudet 1965; Thiollay 1967). They are probably related more to the diet or other ecological requirements of the species involved than to body size. For instance, a pair of golden eagle, *Aquila chrysaetos* (mean weights 3–6 kg) forages over an average 50–150 km^2, as against 30–60 km^2 for the smaller Bonelli eagle, *Hieraaetus fasciatus* (1.5–2.5 kg). Medium weight species (600–1,200 g) such as the common buzzard, *Buteo buteo*, have hunting territories of 4–6 km^2 in forested areas, while the more specialized honey buzzard, *Pernis apivorus*, may increase it up to 10–40 km^2. Bird predators have comparatively larger hunting ranges: 4–12 km^2 for the small European sparrowhawk, *Accipiter nisus* (110–280 g) and 30–50 km^2 for the larger goshawk *Accipiter gentilis* (700–1,400 g).

The three large game birds studied (Table 2) have an overall density at least 9 times higher than that of all raptors together. On the main study site (No), the macaw, the curassow, and the trumpeter have estimated densities of 184, 107, and 567 individuals per 10,000 ha, compared with about 3 to 20 for most raptor species. However, on sites further from any hunted area, the respective density of the three game birds is on average 2 to 8 times higher, whereas within hunting zones, it is 2 to 5 times lower. Thus a rather moderate hunting pressure in an otherwise undisturbed primary forest may reduce the overall density of these species from up to 30 birds/100 ha in pristine conditions to about 1/100 ha. Such low populations are probably maintained only by immigration from surrounding areas that are not hunted.

Habitat Selection and Rarity

The rain forest raptor community is divided into several foraging guilds or subcommunities associated with par-

**One pair of an additional species, the Hook-billed kite, was found breeding in March 1989 within this quadrat (previously overlooked or newly established?).*

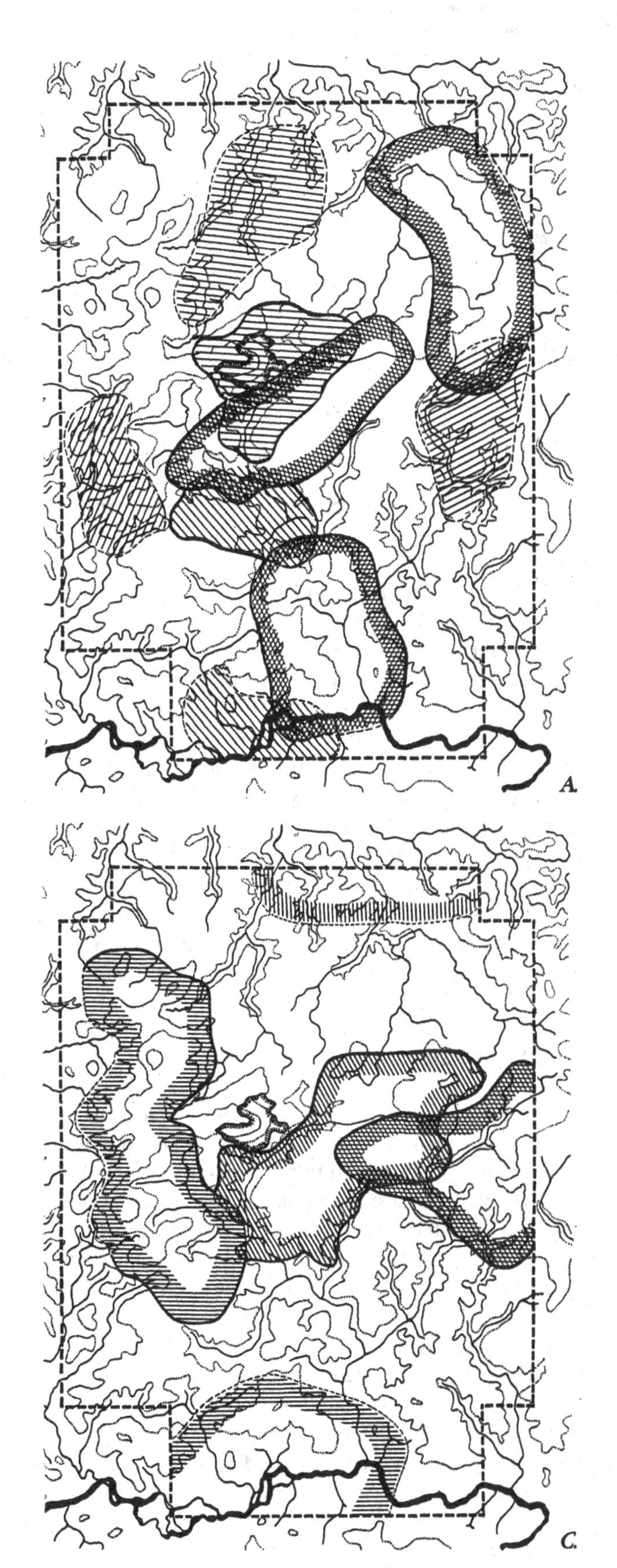

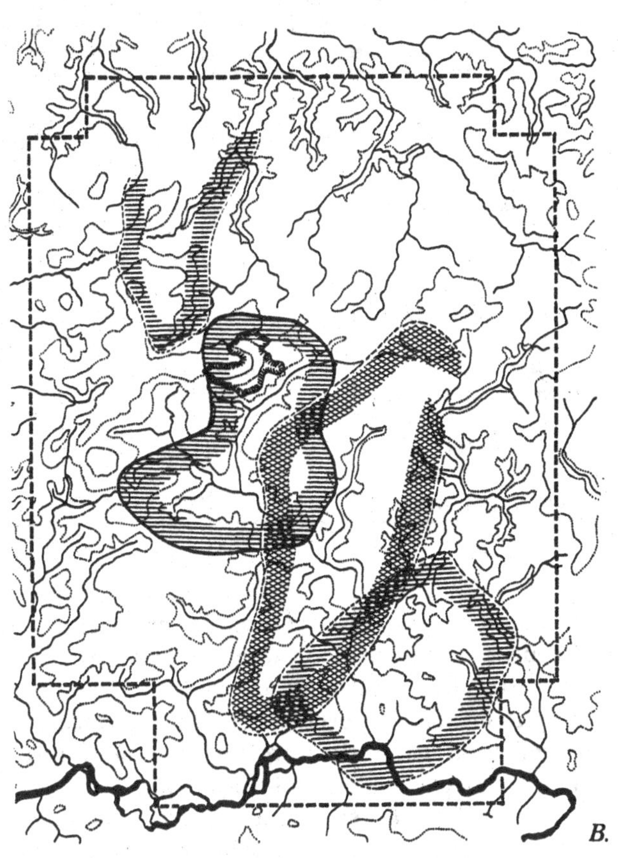

Figure 3. Approximate territories of resident pairs of raptors on the 10,000-ha study area of primary rain forest. Limits include the movements of soaring or displaying birds observed from lookouts above the canopy. Broken lines indicate boundaries that were not enough substantiated or not mapped accurately.

A. Ictinia plumbea (crosshatched)
Leucopternis albicollis (plain hatched)
B. Spizastur melanoleucus (plain hatched)
Morphnus guianensis (crosshatched)
C. Spizaetus ornatus (plain hatched)
Spizaetus tyrannus (crosshatched)

Table 4. Body and territory sizes of selected species. Mean weights are medians of the few data given for males and females by Brown & Amadon (1968). Territories, on the main study site, were mapped by the Minimum Convex Polygon method, then measured on the map to the nearest 50 m (length) or 5 ha (area). Only the two best known territories of each species are mentioned (one for *Spizastur*). The percentage of area not covered by a known territory is calculated from the central 4,200-ha zone most intensively surveyed.

		Territory size		
	Mean weight (g)	*Maximum length (m)*	*Area (ha)*	*Unoccupied area (%)*
Harpagus bidentatus	170–220	3,700	550	60
		3,050	370	
Ictinia plumbea	190–280	5,150	805	79
		3,500	825	
Leucopternis albicollis	600–700	3,300	500	62
		2,750	355	
Buteogallus urubitinga	1,000–1,400	3,450	450	66
		2,850	400	
Spizastur melanoleucus	700–>850	4,250	925	65
Spizaetus ornatus	1,000–1,600	5,800	1,090	43
		6,900	1,375	
Daptrius americanus	600–770	3,000	415	11
		3,250	510	

ticular strata, vegetation structure, or microhabitats. The frequency or distribution pattern of most species is at least partly correlated with their habitat distribution and requirements. However, some species (marked thereafter with an asterisk) are unaccountably rare within their specific widespread forest type.

Thus less than half the species are typical of (although not strictly restricted to) the understory or canopy interior of the high unbroken primary forest (rufous-thighed* and double-toothed kites; Bicolored,* gray-bellied,* and black-faced* hawks; crested and harpy* eagles; black-and-white and ornate hawk eagles; red-throated caracara; and all four forest falcons). Two other species (gray-headed* and hook-billed* kites) are most often found near gaps. Three species are restricted to the upper canopy (swallow-tailed and plumbeous kites, tiny hawk*), while the king and greater yellow-headed vultures search from over the canopy for carcasses in both undergrowth and gaps. Natural openings, such as (a) multiple tree-fall gaps with vine tangles, (b) rocky outcrops, and (c) rivers play an important role. They condition the presence of the white hawk (a, b), great black hawk (a, b, c), black hawk eagle (a), black caracara* (c), bat falcon (c), and orange-breasted falcon* (b). Larger rocky, swampy, or degraded areas occasionally attract the white-tailed, short-tailed, gray, and crane hawks, black vulture, and laughing falcon. Six additional species, as well as three migrants, are commonly recorded along edges, in coastal woodlands or mangroves.

The biggest and probably most threatened species, the harpy eagle, has not been recorded from the main study area or from 55% of the sample plots, although it is widespread in the country and not very shy. It combines three patterns of rarity, common to many raptors: patchy distribution, large home range size (at least several thousand hectares), and spatial segregation from its nearest competitor, the crested eagle. In the two sample sites where both species were found and could be watched from inselbergs, their respective territories did not overlap.

Outside narrow habitat ranges, which account for some of the patchy distributions, there is no persistent ecological correlate of rarity. The most numerous species (*Daptrius*) has a narrow diet (wasp nests), whereas the rarest ones (*Accipiter*) have an apparently broader prey base (birds). Among the four coexisting forest falcons, densities seem to differ much more widely between species than diet (snakes and birds), forest structure, foraging height, or behavior might explain.

Discussion and Conclusion

The rain forest of French Guiana and its fauna are typically Amazonian. Fifty-four species of diurnal raptors have been recorded from this comparatively small country, of which 36 are resident in wooded areas, 11 are purely associated with savannas, marshes, or mangroves, and 7 are migrants or vagrants (Thiollay 1985*d* and unpublished data). Only 27 of them normally occur within large unbroken tracts of primary forest (let alone 9 other occasional species).

Our density estimates are tentative. Classical census methods are inappropriate for most raptors and specific behaviors are so different that a survey of every species always requires two or more different methods. There are some discrepancies between the density estimates based on territory mapping and strip census. Both techniques may lead to underestimations (due to overlooked parts of home ranges and undetected birds) and are currently untestable without comparisons with more specific and more time-consuming methods. However, even if the distribution patterns of territories were quite different or changed seasonally, it is unlikely that the overall densities of resident pairs would be much higher, or that some species would have been consistently overlooked (the three *Accipiter* species are the most easily missed). The aim of this first study is only to emphasize the low specific densities, patchy distributions, and huge size of reserves likely to retain a complete raptor community. These data are seemingly consistent (i.e., within the same order of magnitude) with personal observations in primary rain forests of 15 other Old and New World tropical countries. So the conclusions may have a far-reaching significance that is not thrown into doubt by relatively small-scale local differences.

Diamond (1980) drew attention to the strikingly patchy distribution of many tropical species, mainly on a large geographical scale. Our results illustrate a more local scale patchiness within an apparently uniform habitat, and this may be an equally general rule among tropical forest species. Interspecific competition, predation pressure, or specific susceptibilities to disturbance as well as availability of suitable habitat and food supply are likely to play an important role in distribution patterns through local immigration-extinction equilibria in such a species-rich area, and territories may be expected to shift with time (see also Rafe et al. 1985; Murphy & Wilcox 1986).

Habitat selection and diet specialization are not always obviously involved in patchy distributions. Density and population size are weakly related to body size or habitat availability. A lot of rare species are small-sized and diet or habitat generalists, whereas some common species are bigger and more specialized. Forest interior species, which are likely to be the most sensitive to disturbance and therefore extinction-prone, include both the majority of rare species and the two commonest ones.

The conservation benefit for raptors of a single large vs. several smaller reserves of equivalent size is strongly dependent upon their density, distribution pattern, and other biological characteristics. Distributional patchiness at the geographical scale argues against allocating all the available area to a single large refuge, except if the limits encompass significant populations of every species. However, few small reserves would not harbor populations large enough to have a reasonably low extinction rate. Moreover, at least several forest raptors, even among the most abundant (*Micrastur*), rarely, if ever, cross wide expanses of open habitats. Because of this probable low dispersal ability, small reserves could be efficient only if they were linked by forest corridors. Such a network is often difficult to establish and manage in practice. Nevertheless scattered small reserves would be the easiest way to protect species associated with very localized habitats, such as the orange-breasted falcon on inselbergs.

In most cases, patchy distribution, low density, and still uncertain dispersal ability are the main arguments for a reserve of the largest possible size instead of a fragmented network. However, its minimum size strongly depends on the minimum size of a viable population. This lowest population size, the size likely to ensure a species' long-term survival without immigration from abroad, is still conjectural (Soulé 1986). The often quoted estimate of 5,000 individuals (based on a genetic and mathematical approach), even if it is reduced to 500 or less (Richter-Dyn & Goel 1972), is alarmingly high. It means for raptors 100 to 1,000 breeding pairs, i.e., for most of the species studied here, primary forest areas of about 1 to 10 million hectares. The 1.7 million-ha national park that I have proposed in Southern Guiana (Thiollay 1986*b*) is within the lower range of the above estimate. It was based on a study of bird population and habitat distribution, but also on the availability of unbroken virgin forest without a single inhabitant and outside the hunting range of Indian tribes or gold-diggers.

The conservation of large game birds may pose an even greater challenge. The three species studied here are major seed dispersers, significant food sources for some human populations, and potentially important tourist attractions, being among the easiest big animals to see in the forest. Their natural density is higher than that of most raptors and a reserve as "small" as 10,000 to 100,000 ha may well include a viable population of them all. Unfortunately, they have very low breeding rates (personal unpublished data) and they perform seasonal movements (macaws and curassows) or habitat shifts (trumpeters) of unknown extent, sometimes leading to highly vulnerable concentrations. This, together with their relative tameness and conspicuousness, make them very susceptible to hunting pressure. A decrease of their density was even apparent when hunting occurred only around the protected areas within which they were censused. In French Guiana, as elsewhere, they are severely hunted and disappear from large areas of otherwise undisturbed forest. Therefore, they should have high conservation priority. Their long-term survival depends on effectively protected areas, surrounded by safe buffer zones, based on a knowledge of their food requirements and seasonal movements.

Acknowledgments

This study has been funded by a grant of the French Ministry of Environment (SRETIE). J. L. Dujardin and O. Tostain have been very helpful in the field, as well as F. Thiollay, who also typed the manuscript. The field station of the Nouragues is managed by P. Charlesdominique, director of the Laboratoire ECOTROP, who has also contributed to many aspects of the field maintenance. To all of them, I am very grateful.

Literature Cited

Able, K. P., and B. R. Noon. 1976. Avian community structure along elevational gradients in the northeastern United States. Oecologia (Berlin) **26**:275–295.

Brown, L., and D. Amadon. 1968. Eagles, hawks, and falcons of the world. Country Life Books, Feltham, England.

Cramp, S., editor. 1980. Handbook of the birds of Europe, the Middle East and North Africa. Oxford University Press, Oxford, England.

Diamond, J. M. 1975. The island dilemma: lessons of modern biogeographic studies for the design of natural reserves. Biological Conservation **7**:129–146.

Diamond, J. M. 1980. Patchy distributions of tropical birds. Pages 57–75 in M. Soulé & B. Wilcox, editors. Conservation biology. Sinauer, Sunderland, Massachusetts.

Diamond, J. M., and R. M. May. 1976. Island biogeography and the design of natural reserves. Pages 163–186 in R. M. May, editor. Theoretical ecology. Blackwell Scientific Publications, Oxford, England.

Donahue, P. K., and J. E. Pierson. 1982. Birds of Suriname: an annotated checklist. J. E. Pierson, Harpswell, Maine.

Frankel, O. H., and M. E. Soulé. 1981. Conservation and evolution. Cambridge University Press, Cambridge, England. 327 pp.

Géroudet, P. 1965. Les rapaces diurnes et nocturnes d'Europe. Delachaux & Niestlé. Neuchâtel, Switzerland.

Leck, C. F. 1979. Avian extinction in an isolated tropical wet forest preserve, Ecuador. Auk **96**:343–352.

Lovejoy, T. E., J. M. Rankin, R. O. Bierregaard, K. S. Brown, L. H. Emmon, and M. E. Van der Voort. 1984. Ecosystem decay of Amazon forest remnants. Pages 295–325 in H. Nitecki, editor. Extinctions. University of Chicago Press, Chicago.

MacArthur, R. H., and E. O. Wilson. 1967. The theory of island biogeography. Princeton University Press, Princeton, N.J.

MacCoy, E. E. 1983. The application of island biogeography theory to patches of habitat: how much land is enough? Biological Conservation **25**:53–61.

Murphy, D. D., and B. A. Wilcox. 1986. On island biogeography and conservation. Oikos **47**:385–387.

Oniki, Y. 1979. Is nesting success of birds low in the tropics? Biotropica **11**:60–69.

Oniki, Y. 1985. Why robin eggs are blue and birds build nests: statistical tests for Amazonian birds. Pages 536–545 in P. A. Buckley, M. S. Foster, E. S. Morton, R. S. Ridgely, and F. G. Buckley, editors. Neotropical ornithology. American Ornithologists' Union Monographs 36, A.O.U., Washington, D.C.

Rafe, R. W., M. B. Usher, and R. G. Jefferson. 1985. Birds on reserves: the influence of area and habitat on species richness. Journal of Applied Ecology **22**:327–336.

Richter-Dyn, N., and N. S. Goel, 1972. On the extinction of a colonizing species. Theoretical Population Biology **3**:406–433.

Simberloff, D. 1986. Design of nature reserves. Pages 315–338 in M. B. Usher, editor. Wildlife conservation evaluation. Chapman and Hall, London.

Soulé, M. E., editor. 1986. Conservation biology: the science of scarcity and diversity. Sinauer, Sunderland, Massachusetts. 584 pp.

Southwood, T. R. E. 1978. Ecological methods with particular reference to the study of insect populations, 2nd ed. Chapman and Hall, London. 524 pp.

Taylor, R. J. 1984. Predation. Chapman and Hall, New York. 166 pp.

Thiollay, J. M. 1967. Ecologie d'une population de rapaces diurnes en Lorraine. Terre et Vie **21**:116–183.

Thiollay, J. M. 1985*a*. Raptor community structure of a primary rain forest in French Guiana and effect of human hunting pressure. Raptor Research **18**:117–122.

Thiollay, J. M. 1985*b*. Falconiforms of tropical rain forests: a review. Pages 155–165 in I. Newton and R. D. Chancellor, editors. Conservation studies on raptors. ICBP Technical Publication No. 5, Cambridge, England.

Thiollay, J. M. 1985*c*. Composition of Falconiform communities along successional gradients from primary rain forest to secondary habitats. Pages 181–190 in I. Newton and R. D. Chancellor, editors. Conservation studies on raptors. ICBP Technical Publication No. 5, Cambridge, England.

Thiollay, J. M. 1985*d*. Birds of prey in French Guiana. A preliminary survey. Bull. W.W.G. Bulletin of the World Working Group on Birds of Prey **2**:11–15.

Thiollay, J. M. 1986*a*. Structure comparée du peuplement avien dans trois sites de forêt primaire en Guyane. Revue d'Ecologie (Terre et Vie) **41**:59–105.

Thiollay, J. M. 1986*b*. Projet de Parc National dans le Sud de la Guyane Française. Rapport Ministère de l'Environnement, Paris. 52 pp.

Thiollay, J. M. The census of diurnal raptors in a primary rain forest: comparative methods and species detectability. Journal of Raptor Research. In press.

Wilcox, B. A. 1984. In situ conservation of genetic resources: determinants of minimum area requirements. In J. A. McNeely and K. R. Miller, editors. National parks, conservation, and development. IUCN and Smithsonian Institution Press, Washington, D.C.

Willis, E. O. 1974. Populations and local extinctions of birds on Barro Colorado island, Panama. Ecological monographs **44**:153–169.

Willis, E. O. 1979. The composition of avian communities in remanescent woodlots in southern Brazil. Papeis Avulsos Zoologia (Sao Paulo) **33**:1–25.

Spatial Models and Spotted Owls: Exploring Some Biological Issues Behind Recent Events

SUSAN HARRISON
Division of Environmental Studies
University of California, Davis
Davis, CA 95616, U.S.A.

ANDY STAHL
Sierra Club Legal Defense Fund
705 Second Avenue, Room 203
Seattle, WA 98104, U.S.A.

DANIEL DOAK
Board of Environmental Studies
University of California, Santa Cruz
Santa Cruz, CA 95064, U.S.A.

Few species will ever challenge conservation biology as much as the Northern Spotted Owl, *Strix occidentalis caurina*, has done (Simberloff 1987; Murphy & Noon 1992). The Forest Service and other U.S. agencies, mandated to protect the owl, face formidable pressures to neither under- nor overestimate its needs. Biologists were asked to produce the bottom line: what minimal amount of old-growth forest, in what spatial configuration, would give the owl a high likelihood of survival? This is a very difficult question in strictly biological terms. The owl's extreme sparsity—its distribution of about 3000 pairs over millions of acres in three states—means that existing formulas for "minimum viable population size" are clearly inapplicable. Spatial dynamics—the theoretically challenging issue of how populations are affected by their distribution across an irregular habitat—had to be central to any defensible analysis.

Recently, conservation biologists faced each other in the legal arena, some supporting a proposed plan for the owl and others backing lawsuits to block it. Metapopulation theory and the validity of mathematical modeling became the subjects of court testimony and rulings. Here we describe the biological issues in this unusually science-rich legal debate. We do so from a partisan perspective, as expert witnesses for the litigants, and we invite responses from other viewpoints.

Paper submitted August 10, 1992; revised manuscript accepted July 9, 1993.

The Thomas Strategy

In 1989, the U.S. government created a commission of experts from the Forest Service, the Bureau of Land Management, the Fish and Wildlife Service, and the Park Service to coordinate these agencies' plans for managing the owl. The report of this commission (Thomas et al. 1990) marked a milestone in population viability analysis because of its explicit, quantitative treatment of spatial population dynamics (see below). Development of the commission's report, henceforth referred to as the "Thomas Strategy" after the commission's chairman Jack Ward Thomas, is described in detail by Murphy and Noon (1992).

The Thomas Strategy calls for the preservation of 7.7 million acres of old-growth forest in a system of large and evenly spaced blocks, called "habitat conservation areas," throughout the range of the owl. In Oregon and Washington, each block is to be large enough for at least 20 pair territories and no more than 12 miles from another block. In northern California, where forest fragments are smaller, the standards are at least 10 territories and no more than 7 miles. To permit dispersal between habitat conservation areas, forest between them is to be managed under the "50–11–40 rule," in which each quarter-township (9 miles^2) contains at least 50% forest with trees at least 11 inches in diameter at breast height and with at least 40% canopy closure.

At present, the habitat conservation areas designated by the Thomas Strategy consist of only 30% old-growth forest. The rest is secondary forest that will be allowed to regrow, requiring an estimated 50–100 years. The strategy meanwhile allows about 500,000 acres of old-growth forest outside of the habitat conservation areas to be cut. The strategy thus converts the present landscape of many small, partly connected fragments of old-growth forest into one of fewer, larger blocks in the course of one-half to one century.

Spatial Models and Threshold Effects

Why this particular plan, and how likely is it to succeed? The complexity and long timescale involved in this question make computer simulation modeling an indispensable tool. The Thomas Strategy utilizes two models, the "individual-territory model" and the "territory-cluster model," to address the interaction between the owl's demography and forest fragmentation (see McKelvey et al. 1992; Lamberson et al. 1992 for details of the models). These models use a basic approach developed by Lande (1988) and Doak (1989) and add greater complexity and realism.

Owl demography is linked to fragmentation by making juvenile dispersal success a function of the proportion of the landscape that is forested. A newly fledged juvenile in search of a territory is assumed to visit a limited number of sites, which are either suitable (forested) or unsuitable, and either occupied or vacant. A juvenile goes on to breed only if it finds a suitable vacant territory.

Thresholds for extinction are a key behavior of such models. When the percentage of landscape that is forested reaches a critical lower limit, the owl population collapses, because too few juveniles find territories to balance the mortality of existing territory holders. Extinction thus occurs well before the last old-growth disappears. Lande (1988), Doak (1989), and the Thomas Strategy all suggest that the threshold may lie near 20%, the current percentage of the landscape that is old-growth forest in the Northwest.

The individual-territory models adds further realism by requiring juveniles to encounter mates in order to breed. The result is a second extinction threshold (otherwise known as the Allee effect), a population density below which the owl population collapses because of mate-finding failure.

The individual-territory model is nonspatial; forest exists as a uniform percentage of the landscape, all of which is accessible to every owl. To examine the effects of patch size and spacing, the territory-cluster model of Doak (1989) and the Thomas Strategy portrays the landscape more realistically: forest exists in clusters of territories, corresponding to habitat-conservation areas. Owls disperse more readily within than between clusters. In Doak's model, all clusters are equally distant from one another, but the Thomas Strategy model has an explicit spatial structure: clusters are arranged in a regular fashion across an edgeless plane (a torus).

The territory-cluster model showed that owls would often become extinct in a universe of five territory-clusters, but would persist far better with a cluster size of 10. Gains in persistence are smaller as cluster size increases from 10 to 20. The Thomas Strategy concludes that a minimum of 20 territories per habitat-conservation area offers reasonable security. The Thomas Strategy stresses the consistency of these results with empirical studies of local population extinction.

Models Go to Court

Soon after its release, the Thomas Strategy was adopted by the U.S. Forest Service, manager of 74% of existing spotted owl habitat and 68% of known owl pairs. The Bureau of Land Management, with 22% of known owl pairs, opted instead for its own less restrictive strategy, and also asked the so-called "God Squad" (the Endangered Species Committee) to exempt 44 planned timber sales from the Endangered Species Act. (In early 1992, the God Squad granted 13 exemptions, but these were later declined by the Clinton Administration.)

But when the seemingly more cooperative Forest Service attempted to use the Thomas Strategy, it was promptly thwarted by environmentalist lawsuits. Ruling in *Seattle Audubon Society (SAS) et al. vs. Evans* in March 1991, and again in *SAS et al. vs. Moseley* in May 1992, U.S. District Judge William Dwyer blocked the Forest Service from selling timber under its new "Thomas-based" management plan. Dwyer ruled that this plan carried significant and unacknowledged risks to the owl, and also (in the 1992 ruling) that the Forest Service had neglected to consider the other species inhabiting old-growth forests.

Timber interests, already hostile to the Thomas Strategy, were outraged by Dwyer's rulings. Their calls for congressional intervention and modification of the Endangered Species Act intensified. Conservation biologists who participated in or supported the Thomas Strategy also questioned the "extremism" of the environmentalist litigants.

Weaknesses of the Thomas Strategy

The critics of the Thomas Strategy did not take issue with its general approach, including its use of models. But all models include some tenuous assumptions, and in the intense scrutiny generated by the legal process, some key weaknesses in the Thomas Strategy's analyses were identified.

One concerned the search behavior of juvenile owls. The cluster model portrays juveniles as searching their natal cluster thoroughly; then, if unsuccessful in finding a territory, they move away in linear paths until striking another cluster, which they again search thoroughly, and so on. This is a highly efficient search pattern, and the Thomas Strategy notes that the results depend fairly strongly on it. Unfortunately, details of how juvenile movements really respond to variable landscapes remain unknown.

Another issue was habitat geometry. The cluster model depicts round patches and an edgeless universe. Preliminary work on a model using maps of the actual landscape (Lamberson et al. 1992; McKelvey et al. 1992) shows, not surprisingly, that ragged patches and edge effects greatly reduce the viability of the owl population.

The Thomas Strategy scrupulously points out these and other limitations, but it nonetheless projects a high degree of confidence. Consideration of the above issues alone suggests that some optimism is required. But the litigants also raised an even more basic question: will the owl even survive until the day, 50 to 100 years from now, when its habitat consists of the prescribed array of patches?

The Transition Question

The Thomas Strategy calls for the near-term loss of 500,000 acres of old-growth forest, which will give way to net habitat gain after 50–100 years. In the "worst-case scenario," the report states, the owl population may decline by 50–60% during this time. Implicitly assumed is that the owl population will survive the transition and stabilize once loss of habitat ceases. But it could reach the threshold of either too little habitat or too low density and become extinct before the Thomas Strategy comes into full effect.

This issue could be assessed directly using models in which the landscape is dynamic; some work of this kind has already been performed (Lamberson et al. 1992). But the Forest Service argued that the models were "not yet sufficiently validated" to incorporate the issue into its risk analyses. In fact, the Forest Service did not even acknowledge the transition issue in its 1992 environmental impact statement, which was supposed to disclose all major risks of the proposed plan. Meanwhile, new demographic evidence suggests that the near-term extinction of the owl is a serious possibility.

Dire Demography?

Trends in the owl population are not estimated simply by counting owls, partly because the existence of "floater" owls lacking territories may conceal trends in the territorial breeding population (Thomas et al. 1990). Instead, estimates of survival and fecundity from studies of marked territorial owls are assembled into stage-projection matrices, from which lambda, the yearly rate of population increase, is calculated. Lande (1988), the first to apply these techniques, detected no trends: lambda = 0.96, not significantly different from 1.0. In sensitivity analyses, Lande found that adult survival is the most critical parameter influencing lambda for this long-lived iteroparous species. In the 1990 Thomas Strategy, this analysis is repeated using 3–6 years of data from three populations. One of these, in the Oregon Coast Range, was in sharp decline (lambda = 0.85). The other two, in the Olympic Peninsula and northern California, were in possible decline (lambda = 0.98 and 0.95; not statistically different from 1.0).

In 1991, U.S. Fish and Wildlife Service biostatisticians analyzed 4–7 years of data from over 2000 marked owls in five populations (Anderson & Burnham 1992). They found all populations to be declining at 7.5% to 10% per year, a rate at which about 15 pairs of owls would remain in 50 years. They also found the rate of decline to be accelerating, evidently due to an unsuspected and still unexplained decrease in the survival of adult territorial owls. (Increasing competition with the Barred Owl, an aggressive and more edge-tolerant congener of the Spotted Owl, is one possibility proposed by owl biologists.) These findings were a major basis for Judge Dwyer's 1992 ruling.

The Problem with Monitoring

The Thomas Strategy includes provisions for monitoring and modification (adaptive management). But the litigants argued that, according to the Thomas Strategy's own models, these safeguards won't work. Because of the owl's sparsity, its longevity, and the existence of the pool of "floaters," there will be a long lag between the time when the critical habitat threshold is reached and the time when the resulting population crash becomes detectable. Thus, by the time any trends can be proven, the fate of the population may already be sealed.

The Forest Service's Responses

In the 1992 ruling, Judge Dwyer focused on three sets of issues: (1) risks associated with new information about the owl; (2) risks to species besides the owl; (3) risks causing the refusal of the Bureau of Land Management to participate in the strategy. In March 1993, the Forest Service released a new report (the "Scientific Assessment Team Report") addressing these risks. Most pertinent here is the first set of issues, risks to the owl. The Scientific Assessment Team focuses on the validity of the new evidence for sharp demographic decline. Backed by some owl biologists, the team argues that (1) lambda may have been underestimated because of emigration from the study areas, and (2) some data on absolute *densities* of owls indicate only a moderate decline, roughly equal to the rate of habitat loss (1–3% per year). Thus the team argues that there is no evidence for a precipitous ("threshold effect") collapse of the owl population.

Counter-arguments exist to each of these claims. First, Lande, the Thomas Strategy authors, and Anderson and Burnham all showed that substantial *juvenile* emigration (which does occur), has little influence on estimates of lambda; conversely, substantial *adult* emigration, which would affect lambda, is not evident from any existing data. Second, the use of density data to estimate owl population trends was earlier disavowed by the Thomas Strategy itself for the reasons explained earlier. It can thus be argued that Anderson and Burnham used the best-accepted techniques on the most complete data set available, so that until disproven, their findings stand as strong grounds for doubting the owl's present security.

Summary

As litigants in the recent lawsuits against the Forest Service, we attempted to show that the Northern Spotted Owl may be in rapid decline, that the reason is loss and fragmentation of its habitat, and that the decline may become irreversible once a certain threshold of habitat loss is reached. Thus, on present evidence, any plan entailing further loss of old-growth forest is incompatible with the legal mandate to insure the owl's survival. In response, the Forest Service and its scientific supporters argue that the imminent collapse is not yet proven, and that the Thomas Strategy is a reasonable compromise between opposing interests.

Two fundamental issues emerge from this debate. One is the burden of proof question: if the consequences of the plan are uncertain, should it be allowed or not? (We would argue that the degree of risk is sufficiently high that the law says "not.") The other concerns the role of science in the process. We acknowledge the sincerity of the Thomas Strategy's creators and backers. But we also believe this case has highlighted a dangerous tendency among some conservation biologists to see political "realities" and compromise solutions as a paramount goal. Even if science is never value-free in practice, our aim in cases such as this should be to provide an accurate assessment of biological risks. Weighing these against other concerns and reaching compromises is the job of decision makers. In this case, we feel it would be more honest for the Forest Service to say "the Thomas Strategy is very risky for the owl, but we're going to do it anyway," than to claim that it is a safe strategy.

Acknowledgments

We thank David Ehrenfeld and David Wilcove for their helpful comments.

Literature Cited

Anderson, D. R., and K. P. Burnham. 1992. Demographic analysis of Northern Spotted Owl populations. Recovery Plan for the Northern Spotted Owl, Appendix C. U.S. Fish and Wildlife Service, Portland, Oregon.

Doak, D. 1989. Spotted Owls and old growth logging in the Pacific Northwest. Conservation Biology 3:389–396.

Lamberson, R. H., K. McKelvey, B. R. Noon, and C. Voss. 1992. A dynamic analysis of Northern Spotted Owl viability in a fragmented forest landscape. Conservation Biology 6:505–512.

Lande, R. 1988. Demographic models of the Northern Spotted Owl (*Strix occidentalis caurina*). Oecologia 75:601–607.

McKelvey, K., B. R. Noon, and R. H. Lamberson. 1992. Conservation planning for species occupying fragmented landscapes: The case of the Northern Spotted Owl. Ch. 26 in P. M. Kareiva, J. G. Kingsolver, and R. B. Huey, editors. Biotic interactions and global change. Sinauer, Sunderland, Massachusetts.

Murphy, D. D., and B. R. Noon. 1992. Integrating scientific methods with habitat conservation planning: Reserve design for Northern Spotted Owls. Ecological Applications 2:3–17.

Simberloff, D. 1987. The Spotted Owl fracas: Mixing academic, applied and political ecology. Ecology 68:766–771.

Thomas, J. W., E. D. Forsman, J. B. Lint, E. C. Meslow, B. R. Noon, and J. Verner. 1990. A conservation strategy for the Northern Spotted Owl. Report of the Interagency Scientific Committee to Address the Conservation of the Northern Spotted Owl. U.S. Department of Agriculture Forest Service, Portland, Oregon.

Land Forms and Winter Habitat Refugia in the Conservation of Montane Grasshoppers in Southern Africa

MICHAEL J. SAMWAYS
Department of Zoology and Entomology
University of Natal
P.O. Box 375
Pietermaritzburg
3200
Republic of South Africa

Abstract: *The importance of topography in evaluation of insect conservation is discussed with reference to grasshoppers in the Natal Drakensberg Mountains, South Africa. For these insects, the hilltops act as thermal refugia from winter cold air drainage. Additionally, the increased insolation on the eastern and northern sides of the hilltops, compared with the western and southern sides, attracts the grasshoppers. Crevices in the hill summits provide further microrefugia. Burning reduces grasshopper numbers by half, and although many species are adapted to burn conditions, the reduced grass cover accentuates the thermal influences. Other animals, and plants, also selectively inhabit these hilltops, making them highly significant conservation refugia in a matrix of thermally inhospitable land. Evaluations for biological conservation should take note of the strong influence that land forms may have on plant and animal local distributions.*

Resumen: *La importancia de la topografía en la evaluación de la conservación de insectos es discutida con referencia a los saltamontes de las montañas Natal Drakensberg en Sud Africa. Para estos insectos las cimas de los cerros actúan como refugio térmico de los vientos invernales. Adicionalmente, el mayor aislamiento de los lados este y norte de las cimas de los cerros comparados con los lados oeste y sur, atrae a los saltamontes. Las grietas en la cima de los cerroc proporcionan microrefugios más alejados. Las quemas reducen el número de saltamontes a la mitad y aunque muchas especies están adaptadas a las condiciones de quema, la reducida cubierta de pastos acentúa las influencias térmicas. También otras plantas y animales habitan selectivamente estas cimas haciéndolas refugios altamente significativos para la conservación en una matriz de tierra térmicamente inhóspita. Las evaluaciones para la conservación biológica deben de notar la fuerte influencia que la topografía puede tener en la distribución local de plantas y animales.*

Introduction

Maps, due to the nature of the printed page, are generally two dimensional. Decisions on the acquisition of land for biotic conservation are often based on maps. Although contours may be represented, the land forms are rarely taken into account except for aesthetic reasons or for general conservation of entire montane areas. Ecologically, land forms play a major role in shaping community characteristics (Whittaker 1975; Swanson et al. 1988). Behaviorally, the mesoclimatic patterns (Yoshino 1975) shape the behavior of mobile ectotherms to select microclimatically suitable zones and

Paper submitted November 29, 1989; revised manuscript accepted June 25, 1990.

patches for optimizing life activities (Uvarov 1977; Murphy & Weiss 1988). Scale is important, and for small animals such as insects, the vertical element of the "plantscape" (Duffey 1974; Samways 1989) is additive upon that of the landscape.

The southern African landscape, with its varied and ancient topography (Partridge & Maud 1987), provides an ideal natural laboratory for examining the interaction between land forms and biota. It is especially pertinent in the conservation context, with many groups showing a high level of endemicity. Of these groups, one of the most conspicuous is the Acridoidea. The extensive savannah grasslands, most of which fall within Zonobiome II (Walter 1985), provide an extensive matrix that has given rise to high species richness (Johnsen 1987). Additionally, grasshopper biomass is extremely high, accounting for between 76% and 93% of the aboveground phytophagous insects (Gandar 1982). These, in turn, are resources for vertebrates (Manry 1982; Rowe-Rowe & Lowry 1982; Barnard 1987).

During an initial survey of insect species in a major reserve area, the Natal Drakensberg Mountains, it was apparent that grasshoppers were abundant in midwinter, especially on hills. Here I discuss the possibility that elevated land forms act as island thermal refugia, and I relate this concept to conservation management.

Figure 1. One of the winter refuge hills for grasshoppers in the Natal Drakensberg. Note the shadow on the right (western) side of the summit, where grasshoppers were scarce. This differed from the eastern and northern sides of the summit that received high insolation and supported high numbers of grasshoppers.

Sites, Sampling Methods, and Analysis

Geographical Area and Sites

Sampling took place in the Cathedral Peak research area in the Natal Drakensberg Mountains (29°14′E, 28°59′S). This is a major, virtually pristine conservation area. The vegetation was principally *Themeda triandra* Forsk.–dominated grassland. Data were gathered in July and August 1987, June and July 1988, and June 1989 from bases of hills at 1740 m a.s.l. to summits at 2003 m a.s.l. (Fig. 1). Hillside inclination ranged from 25° to 29°, and transects were between 200 m and 360 m long. Some of the hills were round-topped while others were flat-topped (kopjes) with a basalt wall (krantz) (Fig. 1).

Sampling

This was by visual counts. The ground was too steep and rocky to use enclosed quadrats (Gandar 1982; Monk 1983) or sweep nets. Counts were made of all adult grasshoppers by slowly walking a 1 m × 5 m quadrat (= 1 sample unit) using a 1 m–wide aluminum pronged fork carried in front for 5 m, or alternatively, where the terrain was too rough, in string-demarcated areas. Unlike European conditions, where cool weather can drastically affect this type of sampling, the consistently atmospherically clear South African winter days usually permit constancy of sampling.

Abundance profiles along individual hillsides involved ten stations spaced equally along the transect length. At each station, the sample consisted of ten sample units.

For multivariate analysis, sampling involved stations at the base, midway, and at the summits of hills. Samples were ten sample units of a 1 m × 5 m quadrat visually sampled. Samples were taken from the four compass points and repeated on ten hills. For ANOVA of overall comparative grasshopper abundance, square root–transformed data were used after establishing that p from the Taylor's Power Law transformation procedure was 0.302. Also, as the data showed strong nonlinearity along the transects and zeros were present, Nonlinear Multidimensional Scaling (NMDS) was used in addition to Principal Components Analysis (PCA). These samples were also the data base for drawing a rank-abundance curve.

Ground cover was estimated using a point frame, with

five stakes, repositioned ten times to make up the sample at each site at each quadrant around the ten hills.

Soil surface temperatures were 1 min long taken five times at each site using a Thies Clima hygrothermometer with a Pt100 Resistance thermometer probe inserted 10 mm below the soil surface. Relative humidity was measured with the humidity probe 1.5 m above the ground. Other meteorological data were obtained from the meteorological station in the research area, and from Schulze (1975), Tyson (1968), and Tyson et al. (1976).

Experimental verification was carried out with 300 mm × 100 mm wire-framed net cages each containing six randomly collected grasshoppers, and each translocation was replicated six times.

Effect of burning was determined using individual hills that were burnt across their midline. Sampling was repeated on either side of this line, and ten sample units (1 m × 5 m) at each station made up the sample. Ten equally spaced samples were taken along either side of the burning line.

Results

Species

A total of 19 species was present, with clear dominance by *Dnopherula* near *cruciata* Bolivar (Fig. 2).

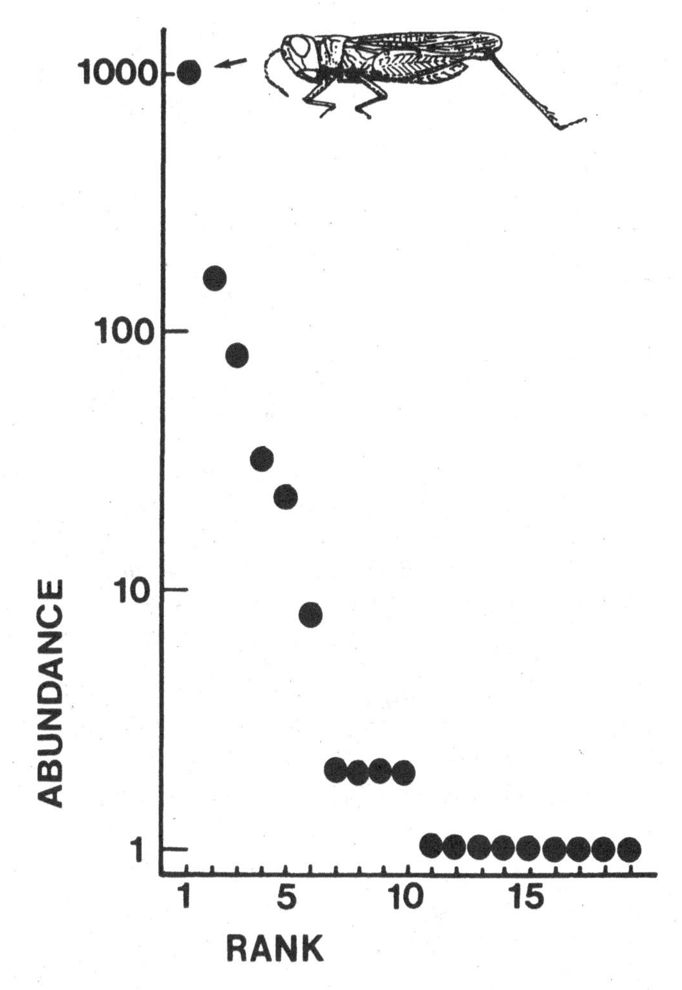

Figure 2. Rank abundance curve of the total grasshopper assemblage used in the multivariate analysis. Species: Rank 1 = Dnopherula *sp. nr* cruciata *Bolivar, 2* = Faureia rosea *Uvarov, 3* = Machaeridia conspersa *Bolivar, 4* = Anthermus granulosus *Stal, 5* = Acorypha ferrifer *Walker, 6* = Acanthacris ruficornis *Fabricius, 7* = ?Vitticatantops *nr* conscitus *Key, 8* = Pseudoarcyptera *sp., 9* = Heteracris herbacea *(Serville), 10* = Phaeocatantops sulphureus *(Walker), 11* = Lentula obtusifrons *Stal, 12* = Pnorisa *sp., 13* = Eyprepocnemis *nr* calceata *(Serville), 14* = Gymnobothrus *sp. A, 15* = Gymnobothrus *sp. B, 16* = Acrida acuminata *Stal, 17* = Anaeolopus socius *(Stal), 18* = Diablepia *sp., 19* = Cantantops sordidus *(Walker).*

Influence of Aspect and Elevation

There were significant differences in square root–transformed mean abundances of grasshoppers from different positions on the hillsides ($F = 69.7$, d.f. = 8.81, $P < 0.00001$) (Fig. 3). The summit (S in Fig. 3) and northern and eastern midslopes (MN and ME) clearly supported larger numbers than the other locations. PCA indicated a fairly homogenous grouping of the low-density locations, with the three highest density sites separating off (Fig. 4A). A Detrended Principal Components Analysis (DPCA) was nonsignificant ($r^2 = 0.062$, d.f. = 2.6, $F = 0.198$, n.s.), suggesting no arching. Nevertheless, as the data were clearly nonlinear (Fig. 3), NMDS was also applied (Fig. 4B). This method tended to group the six highest density locations away from the three lowest (Fig. 4B).

The overall outcome of the analyses suggests three grasshopper density locations: (1) Very high: midslope, east (ME); midslope, north (MN); and summit (S); (2) Low: base, north (BN); base, east (BE); midslope, west (MW); (3) Very low: base, south (BS); base, west (BW); and midslope, south (MS).

Possible Influential Environmental Variables

Direct gradient analysis, in which square root–transformed grasshopper abundance was regressed on various environmental variables, showed that temperature played a significant role in determining the grasshopper distributions across the topographic locations (Table 1). Before sunrise, cold air drainage was a characteristic feature of the mesoclimate (Fig. 5), with a 10° C difference between the air temperature just above the grass at the bottom of the hill and the top. At sunrise, the effect of insolation was rapid (Fig. 6), resulting at times in a 21° C difference in air temperature between the bottom and the top of the hill. At midday, air and ground temperatures differed with aspect by as much as 16° C

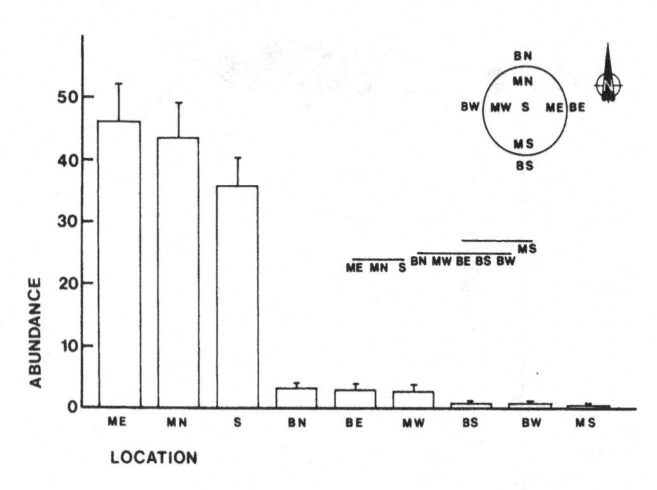

Figure 3. Mean total grasshopper abundances (with positive S.E. bars) at nine locations replicated on ten hills. The locations, illustrated diagrammatically from above in the top right of the figure, are S = summit, MN = midslope (north), BN = base (north), ME = midslope (east), BE = base (east), MS = midslope (south), BS = base (south), MW = midslope (west), BW = base (west). The Scheffe Multiple Range Test (center right of the figure) indicates that three homogenous groups are possible at the 5% level.

when the hill was steep-sided, particularly where there was a krantz (Fig. 7).

Experimental Verification

The translocation results (Fig. 8) illustrated the importance of avoiding cold-air drainage for survival. Mortality was total at the base of hills when the grasshoppers were exposed to air on the surface of the grass. This indicated that grasshoppers survive by either living at or near the summit of hills, or by taking advantage of the insulation effect of the dry grass cover.

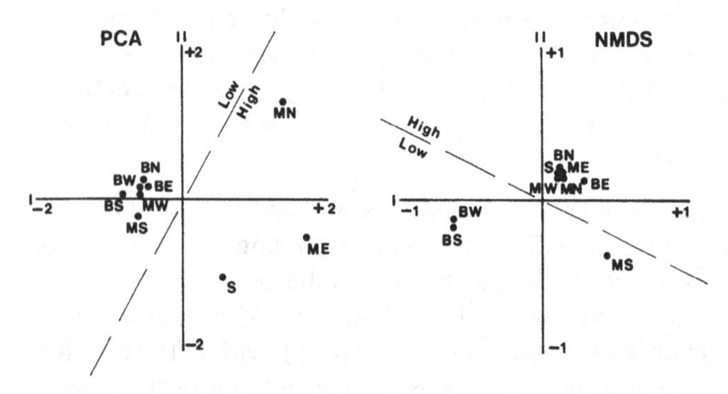

Figure 4. Ordination of the hill sites using Principal Components Analysis (a) and Nonmetric Multiple Dimensional Scaling. PCA clusters the six sites with the lowest density of grasshoppers (a), while NMDS clusters the six sites with the highest density (b).

Table 1. Linear regression of square root–transformed overall grasshopper abundance and several environmental variables across nine topographic locations.

Variable	*Regression coefficient (r)*	*Significance level (P)*
Grass cover (%)	−0.57	0.11
Forb cover (%)	−0.51	0.16
Soil cover (%)	−0.11	0.78
Rock cover (%)	0.44	0.23
Temperature before sunrise (° C)	0.64	0.06
Temperature at sunrise (° C)	0.86	0.002
Temperature at midday (° C)	0.66	0.05
Midday relative humidity (%)	−0.12	0.77

Impact of Vegetation Burning

Burning removes the grass cover, leaving ash on the soil surface. At midday, this results in soil surface temperatures being as much as 10° C higher on the sunny side of hills than on the shady side (Fig. 7). Removal of the grass insulating layer resulted in the night-time temperature inversion (Fig. 5) directly interfacing with the soil. With no refuge, grasshoppers were exposed to low temperatures and suffered total mortality (Fig. 8).

Burning reduced overall grasshopper abundance to about half (Fig. 9). The mid northern and eastern slopes carried the greatest number of individuals, while the southern and western burnt slopes were almost devoid of grasshoppers (Fig. 10). With the exception of the location at the base and at the top of the hill in Figure 9, all mean sizes of populations on burnt versus unburnt ground were significantly different at the 5% level (*t*-test on square root–transformed data).

Figure 5. Nocturnal cold air drainage can be extreme in the Natal Drakensberg Mountains, causing frost on the low-lying ground (arrowed). Visible here only on the burnt ground on the right.

Figure 6. At sunrise the hilltops in the Natal Drakensberg Mountains immediately receive high insolation (a), leading within minutes to a temperature differential of 21° C between the top of the hill (15° C) and the frost-prone bottom of the hill (−6° C) (b).

Exposure

Although the transformed means of ME and S, and MN and S, in Figure 3 are not significantly different (t = 1.30, d.f. = 18, n.s., and t = 1.08, d.f. = 18, n.s.), the grasshopper densities at the summit were consistently lower than on the midslopes (P = 0.0005). It is impossible to determine statistically the reason because botanical composition and microtopography on the summits also differ from lower down. However, the root cause is likely to be exposure, particularly since the Drakensberg is well known for its distinctive wind patterns (Tyson et al. 1976), which at times can be harsh.

Supporting circumstantial evidence for the negative influence of cold winds comes from results from saddles and spurs and from microtopography. Mean grasshopper abundances on the ridge of two saddles and one spur were significantly different (lower) than on the slope (t = 7.8, 12.3, and 9.5, all d.f. = 18, all $P < 0.001$).

Where the krantz had a scree skirt, grasshopper abundances were consistently less on the summit and scree than on the midslope (Fig. 10). But where the krantz was regularly and deeply divided around its perimeter, grasshopper abundances were high in the clefts (Fig. 11).

Figure 7. Soil temperatures (in °C) (10 mm below the surface) at about midday, 10 August 1987, on the slopes and summit of a hill which was burnt (a) on the northern round to the southwestern side, but unburnt (b) and grassy from northeast round to south. N.B., the bumps in the slope are schematic exaggerations of the perennial root system of grass tussocks.

Discussion

Species

Insects of the Drakensberg Mountains are little known. The large number of uncertain species and one uncer-

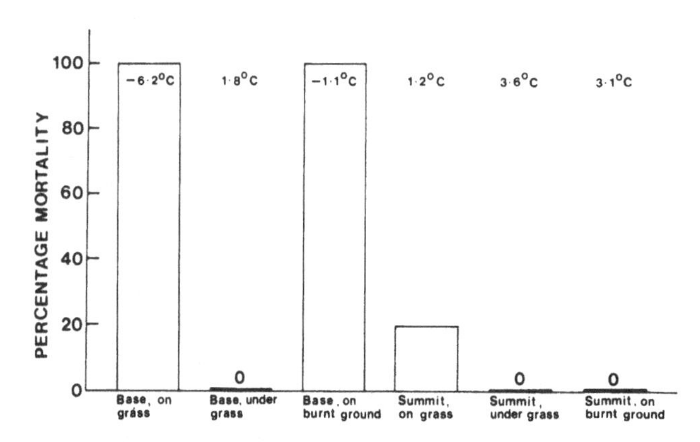

Figure 8. Average mortality percentage of grasshoppers when translocated from a thermally salubrious frost-free location to other topographic locations on hillsides in the Natal Drakensberg Mountains.

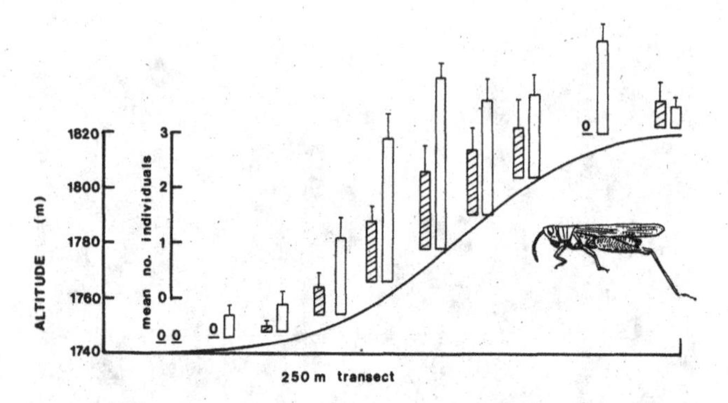

Figure 9. Mean (with positive S.E. bars) number of grasshoppers on burnt ground in 5 m^2 guadrats on burnt (shaded) and unburnt (unshaded) grassland up a hillside in the Natal Drakensberg Mountains.

tain genus in Figure 2 bear this out. Southern Africa is well known for its high level of biotic endemicity, which for species in the Tettigonioidea reaches 75% (Ragge 1974) and in Acridoidea is 47% (Johnsen 1987). Within the Drakensberg, which supports many Gondwana relics, this figure is likely to be higher. Conservation should therefore aim to conserve as many habitats as possible to retain this high number of endemics.

Conservation Evaluation and Topography

The results here have shown that topography plays a major role in determining grasshopper distributions and that land forms need to be taken into account for conservation decisions. In effect, these small hills act like island refugia in a matrix of thermally inhospitable land. Although mountain refugia have been described for small mammals (Brown 1971), the reason for the outcome is different from that of these grasshoppers. The

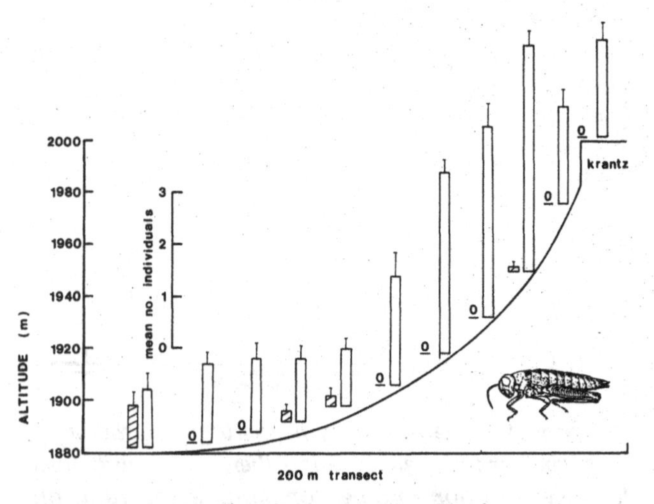

Figure 10. Mean (with positive S.E. bars) number of grasshoppers on burnt ground in 5 m^2 quadrats on the southwest (shaded) and northeast (unshaded) sides of a hill in the Natal Drakensberg Mountains.

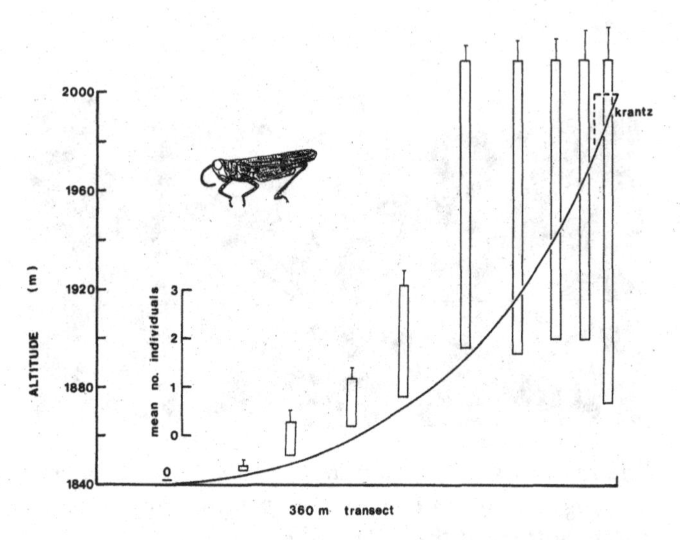

Figure 11. Mean (with positive S.E. bars) number of grasshoppers in 5 m^2 quadrats on a highly insolated east-facing, grassy slope rising up to a sheltered cleft in the krantz of a hill in the Natal Drakensberg Mountains.

mammal distributions have arisen through vicariance events, when the surrounding landscape was eroded. The grasshoppers, although distributed in the Drakensberg by megavicariance events, have mesodistributions apparently shaped by mesoclimatic events.

Land Forms and Mesoclimatic Influences

Acridids are well known to favor certain elevational zones (Claridge & Singrao 1978), and topography is known to influence grasshopper daily and seasonal movements (reviewed by Uvarov, 1977). However, these studies report summer results in the northern hemisphere, which is quite different from the winter, southern hemisphere situation.

In the Drakensberg Mountains, cloud cover and precipitation are low, with sunshine reaching 90% (Tyson et al. 1976). At these altitudes (= 1900 m a.s.l.), the mean daily midwinter air temperature is moderate at 10° C, but grass minimum temperatures (ave. = −3° C) are generally considerably lower than air minimum temperatures (ave. = 5° C) (Killick 1963). Nocturnal temperature inversions are a common feature of mountain valleys (Yoshino 1975), and may drop as much as 10° C in 100 m (Pedgley 1982). With often windless, clear nights, such inversions are a distinct feature of the Drakensberg (Tyson 1968), and function in addition to diurnal air temperature fluctuations, which may be as much as 25° C even on flat ground.

The regression analyses indicated that these mesoclimatic features, rather than vegetation type or percentage cover, were responsible for these grasshopper distributions. The translocation experiments verified this by illustrating that grasshoppers can suffer total mortality from cold air drainage. They can avoid the temperature inversion by selecting the thermal belt up the hillside.

Influence of Aspect

Aspect has a major effect upon insolation. Schulze (1975) showed that for 30° slopes in midwinter, the incoming radiation is nine times greater on a north-facing slope than on a south-facing one, while a horizontal surface is slightly over halfway between. A krantz can exaggerate this effect (Fig. 1).

The influence of insolation is highly significant (Table 1). Both the early morning insolation on the hilltops (Fig. 6) and the midday temperatures are important. Grasshoppers use the hills not only to escape the nighttime temperatures, but to maximize insolation. Presumably a disadvantage is increased exposure to the occasional severe wind patterns characteristic of this area (Tyson et al. 1968). Avoidance of climatic exposure is apparently the reason for exceptionally high grasshopper numbers in the clefts of hill summits (Fig. 11).

Influence of Burning

Grass burning, which simulates lightning strikes to reduce bush encroachment (Tainton & Mentis 1984), reduced overall grasshopper abundance (Fig. 9). The reasons for this may be complex, but certainly the translocation experiments (Fig. 8) indicated that removal of grass cover enhances the mortal effect of cold air drainage. Fire also initially deprives grasshoppers of food, although Gandar (1982) found that some species survived by eating ash. Indeed, melanism can be relatively common, and provides crypsis on burnt ground (Chapman 1952). Also, elevated postburn soil temperatures, through increased insolation, can allow nymphs to emerge earlier than usual, which, coupled with more nutritious vegetative regrowth, progressively increases grasshopper populations (Warren et al. 1987). Additionally, profuse vegetation regrowth can also cause enhanced maturation of females, leading to an increase in numbers of eggs laid, and the increased postburn insolation of the soil favors simultaneous hatching of eggs, leading to relatively stable year-to-year population levels (Warren et al. 1987).

Acridids also show behavioral adaptations to fire. They immediately fly away in front or to the side of a fire, whereas flightless individuals may find shelter under rocks (Rice 1932). These rocks may also give protection from large predators, which include birds (Manry 1982; Barnard 1987; Johnson et al. 1987), and mammals (Rowe-Rowe & Lowry 1982), but possibly not from smaller ones such as spiders and ants (Belovsky et al. 1990). These hill summits function as an elevated launching site for rapid dispersal, and the increased predominance of boulders and crevices provides protection for the less mobile species. In short, acridids are well adapted to this hilly, fire-prone grassland ecotope. Further, Jago (1973) has pointed out that the African grasslands, particularly in the equatorial region, are a species dynamo, where there is much genetic plasticity and species diversity (Dirsch 1965).

Conservation Directives

It is well known that grasshoppers are indicators of grassland history and quality (Bei-Bienko 1970). Further, these Drakensberg grasshoppers are well adapted, behaviorally and physiologically, to taking advantage of hilltops as refugia. Additionally, they are well adapted to fire, which is a naturally frequent phenomenon in the area. As such, they nicely illustrate the significance of topography in determining mesodistributions. But hilltops also facilitate other activities such as mate meeting in butterflies and other insects (Alcock 1987), and they function as lookout points for birds of prey. Further, Killick (1963) has emphasized their interesting botanical significance as refugia. Conservation evaluation, therefore, should consider the third (vertical) dimension of the landscape in addition to the two horizontal ones normally used in delimiting geographic areas for conservation.

Acknowledgments

The Natal Parks Board and the Ukhahlamba Management Committee permitted access to the sites, and the University of Natal Research Fund and the Foundation for Research Development provided financial support. Dr. John P. Grunshaw of the Overseas Development and Natural Resources Research Institute, London, kindly identified the species, and Mrs. Ann Best and Mrs. Myriam Preston processed the manuscript.

Literature Cited

Alcock, J. 1987. Leks and hilltopping in insects. Journal of Natural History **21**:319–328

Barnard, P. 1987. Foraging site selection by three raptors in relation to grassland burning in a montane habitat. African Journal of Ecology **25**:35–45.

Bei-Bienko, G. Y. 1970. Orthopteroid insects (Orthopteroidea) of the national park areas near Kursk and their significance as indices of the local landscape. Zhurnal Obshchei Biologii **31**:30–46. (In Russian; English translation in the Department of Zoology and Entomology, University of Natal, Pietermaritzburg.)

Belovsky, G. E., J. B. Slade, and B. A. Stockhoff. 1990. Susceptibility to predation for different grasshoppers: an experimental study. Ecology **71**:624–634.

Brown, J. 1971. Mammals on mountaintops: nonequilibrium insular biogeography. American Naturalist **105**:467–478.

Chapman, K. 1952. Ecological strategies on solitary Acrididae in England and South Africa. Journal of the Entomological Society of Southern Africa **15**:165–203.

Claridge, M. F., and J. S. Singhrao. 1978. Diversity and altitudinal distribution of grasshoppers (Acridoidea) on a Mediterranean mountain. Journal of Biogeography **5**:239–250.

Dirsch, V. M. 1965. The African genera of Acridoidea. Cambridge University Press, Cambridge and London, England.

Duffey, E. 1974. Nature reserves and wildlife. Heinemann, London, England.

Gandar, M. V. 1982. The dynamics and trophic ecology of grasshoppers (Acridoidea) in a South African savanna. Oecologia (Berlin) **54**:370–378.

Jago, N. D. 1973. The genesis and nature of tropical forest and savanna grasshopper faunas, with special reference to Africa. Pages 187–196 in B. J. Meggers, E. S. Ayensu, and D. Duckworth, editors. Tropical forest ecosystems in Africa and South America: a comparative review. Smithsonian Institution Press, Washington, D.C.

Johnsen, P. 1987. The status of the South African Acridoidea s.l. (Orthoptera: Caelifera). Pages 293–295 in B. Baccetti, editor. Evolutionary biology of orthopteroid insects. Ellis Horwood, Chichester, U.K.

Johnson, C. G., L. A. Nickerson, and M. J. Bechard. 1987. Grasshopper consumption and summer flocks of nonbreeding Swainson's Hawks. The Condor **89**:676–678.

Killick, D. J. B. 1963. An account of the plant ecology of the Cathedral Peak area of the Natal Drakensberg. Botanical Survey Memoir No. 34. Botanical Research Institute, Department of Agricultural Technical Services, Pretoria, South Africa.

Manry, D. E. 1982. Habitat use by foraging Bald Ibises *Geronticus calvus* in western Natal. South African Journal of Wildlife Research **12**:85–93.

Monk, K. A. 1983. Morphological variation in relation to habitat in some British grasshoppers. Journal of Natural History **17**:75–85.

Murphy, D. D., and S. B. Weiss. 1988. A long-term monitoring plan for a threatened butterfly. Conservation Biology **2**:367–374.

Partridge, T. C., and R. R. Maud. 1987. Geomorphic evolution of southern Africa since the Mesozoic. South African Journal of Geology **90**:179–208.

Pedgley, D. 1982. Windborne pests and diseases: meteorology of airborne organisms. Ellis Horwood, Chichester.

Ragge, D. R. 1974. Tettigonioidea. Pages 37–38 in W. G. H. Coaton, editor. Status of the taxonomy of the Hexapoda of southern Africa. Entomology Memoirs, Department of Agriculture, Union of South Africa. Volume 38. Department of Agriculture, Pretoria, South Africa.

Rice, L. A. 1932. The effect of fire on the prairie animal communities. Ecology **13**:392–401.

Rowe-Rowe, D. T., and P. B. Lowry. 1982. Influence of fire on small mammal populations in the Natal Drakensberg. South African Journal of Wildlife Research **12**:130–139.

Samways, M. J. 1989. Insect conservation and landscape ecology: a case-history of bush crickets (Tettigoniidae) in southern France. Environmental Conservation **16**:217–226.

Schulze, R. E. 1975. Incoming radiation fluxes on sloping terrain: a general model for use in southern Africa. Agrochemophysica **7**:55–60.

Swanson, F. J., T. K. Kratz, N. Caine, and R. G. Woodmansee. 1988. Landform effects on ecosystem patterns and processes. BioScience **38**:92–98.

Tainton, N. M., and M. T. Mentis. 1984. Fire in grassland. Pages 115–147 in P. de V. Booysen and N. M. Tainton, editors. Ecological effects of fire in South African ecosystems. Springer-Verlag, Berlin, East Germany.

Tyson, P. D. 1968. Nocturnal local winds in a Drakensberg valley. South African Geographical Journal **50**:15–32.

Tyson, P. D., R. A. Preston-Whyte, and R. E. Schulze. 1976. The climate of the Drakensberg. Report No. 31. Natal Town and Regional Planning Commission, Pietermaritzburg, South Africa.

Uvarov, B. P. 1977. Grasshoppers and locusts. Volume 2. Centre for Overseas Pest Research, London, England.

Walter, H. 1985. Vegetation of the Earth and ecological systems of the geo-biosphere. 3d edition. Springer-Verlag, Berlin, East Germany.

Warren, S. D., C. J. Scifres, and P. D. Teel. 1987. Response of grassland arthropods to burning: a review. Agriculture Ecosystems and Environment **19**:105–130.

Whittaker, R. H. 1975. Communities and ecosystems. 2nd edition. Macmillan, New York.

Yoshino, M. M. 1975. Climate in a small area. University of Tokyo Press, Tokyo, Japan.

Response of Early Successional Vertebrates to Historic Changes in Land Use

JOHN A. LITVAITIS
Wildlife Program
Department of Natural Resources
University of New Hampshire
Durham, NH 03824, U.S.A.

Abstract: *Unlike other regions of North America, forested habitats in New England have increased substantially in the past 100 years. The proportion of land in New Hampshire covered by forests was 47% in 1880 and 87% in 1980. This increase was largely the result of a region-wide abandonment of farms and the subsequent colonization of these lands by second-growth forests. I examined the sequence of farm abandonment, forest colonization, and forest maturation that occurred in New Hampshire in relation to changes in the abundance and distribution of a group of forest mammals and birds that have undergone substantial declines. A modeled pattern of secondary succession resulted in the availability of approximately 195,000 ha of early seral habitats (10–25 years after abandonment) from 1905 to 1940. These habitats then matured into closed-canopy forests by about 1960. Concurrent to the loss of early successional habitats, populations of New England cottontails (*Sylvilagus transitionalis*) decreased from an apparent continuous distribution throughout 60% of New Hampshire to fragmented populations that occupy less than 20% of the state. Bobcats (*Felis rufus*) responded functionally (*S. transitionalis *in diet: 1951–1954 = 43%, 1961–1964 = 10%) and numerically (mean annual harvest of bobcats: 1951–1954 = 350, 1965–1969 = 36) to changes in cottontail abundance. Eighteen of 26 species of migratory passerines that nest in the forests of northern New England also declined during the period their populations were monitored (1934–1987). Eight (44%) of the species that declined are associated with early successional habitats, and these species consistently exhibited population declines during the 1950s. The reduction of early successional species may be extended in space and time by current land uses that fragment and isolate patches of habitat. Ownership patterns of forest lands in New England (ex-*

Paper submitted March 16, 1992; revised manuscript accepted September 8, 1993.

Respuesta de vertebrados sucesionales tempranos a cambios históricos en el uso de la tierra

Resumen: *A diferencia de otras regiones en Norte América, los hábitats boscosos en Nueva Inglattera se han incrementado substancialmente en los últimos 100 años. La proporción de tierra cubierta por bosques en New Hampshire fue del 47% en 1880 y del 87% en 1980. Este incremento fue, en gran medida, el resultado de un abandono a nivel regional de granjas y la subsecuente colonización de estas tierras por un crecimiento secundario de bosques. Yo examiné la secuencia del abandono de las granjas, la colonización de los bosques y la maduración de los bosques que ha ocurrido en New Hampshire en relación con los cambios en la abundancia y distribución de un grupo de mamíferos y pájaros que han sufrido declinaciones substanciales. Un patrón modelado de sucesión secundaria resultó en la disponibilidad de apróximadamente 195,000 ha de hábitats serales tempranos (10–25 años después del abandono) entre 1905 y 1940. Subsecuentemente, hacia 1960, estos hábitats, maduraron hacia bosques de canopeo cerrado. Conjuntamente con la pérdida de hábitats sucesionales tempranos, las poblaciones de conejos (*Sylvilagus transitionalis*) de Nueva Inglaterra decrecieron de una distribución aparentemente contínua a través del 60% de New Hampshire, a una población fragmentada que ocupa <20% del estado. Los "bobcats" (*Felis rufus*) respondieron funcionalmente (*S. transitionalis *en dieta: 1951–54 = 43% versus 1961–64 = 10%) y numéricamente (media anual de la captura de Linces: 1951–54 = 350, 1965–69 = 36) a cambios en la abundancia de conejos. Dieciocho de las 26 especies de paserinidos migratorios que anidan en los bosques del norte de Nueva Inglaterra también declinaron durante el periodo en que la población fue monitoreada (1934–87). Ocho (44%) de las especies que declinaron están asociadas con hábitats sucesionales tempranos, y estas especies consistentemente exhibieron declinaciones du-*

cluding Maine) reveal 88% private ownership with an average holding of 10 ha. This suggests that large tracts of early successional habitats will be restricted to industrial and state/national forests. Although even-aged management of a portion of these forests may be perceived as incompatible with area-sensitive and interior species, clustering of clearcuts and maintaining large tracts of mature habitats could sustain diverse populations of forest vertebrates.

rante los años 50. La reducción de especies sucesionales tempranas puede ser extendida (en espacio y tiempo) por los usos actuales de la tierra que fragmentan y aislan los patches de hábitat. Los patrones de tenencia de tierra de bosques en Nueva Inglaterra (excluyendo Maine, 88% en manos privadas con una media de tenencia de 10 ha) sugieren que grandes extensiones de hábitats sucesionales tempranos van a estar restringidos a usos industriales y reservas estatales/nacionales. Aunque un manejo con stands de la misma edad de porciónes de estos bosques puede ser percibido como incompatible con especies sensibles al área y con especies interiores, el agrupamiento de áreas taladas y el mantenimiento de grandes extensiones de hábitats maduros pueden sustentar poblaciones diversas de vertebrados del bosque.

Introduction

Land-use patterns can have profound effects on the abundance, distribution, and diversity of terrestrial vertebrates (Burgess & Sharpe 1981; Saunders et al. 1991). For example, conversion of forests to agricultural lands or housing developments may isolate and substantially reduce remaining populations of forest-dwelling species (Ambuel & Temple 1983; Robbins et al. 1989). Decline of neotropical migratory landbirds that nest in the forests of eastern North America (Askins et al. 1990; Finch 1991; Hagan & Johnston 1992) is suspected to result from reduction and fragmentation of forested habitats (Wilcove 1985; Wilcove et al. 1986). Loss of continuous forests has also been implicated in local and regional extinctions of some mammals (Matthiae & Stearns 1981; Henderson et al. 1985).

Unlike other portions of North America, forests habitats in New England have increased. This increase was in response to land-use changes during the last century (Irland 1982; Brooks 1989). In New Hampshire, for example, the proportion of land that was forested was reduced from an estimated 95% at the time of European colonization to 47% coverage by 1880 (Harper 1918). Agriculture resulted in large-scale clearing of forests in New England during the Eighteenth and Nineteenth Centuries (Harper 1918; Irland 1982). By the mid-1800s, however, farmers in New England were unable to compete with more productive farms in the Midwest (Irland 1982). As a result, farms were abandoned throughout northern New England during the late 1800s and the first half of this century (Black 1950). Much of this land was subsequently colonized by second-growth forests (Irland 1982), and within 100 years, 87% of New Hampshire was forested (Fig. 1; Frieswyk & Malley 1985*a*). A similar pattern of land-use change and reforestation also occurred in Vermont and southern Maine (Harper 1918; Irland 1982).

One reaction to such a change in landscape composition could be an increase in the abundance and distribution of forest-dependent vertebrates. However, the numeric response of a particular species to reforestation may be restricted to a specific seral stage or forest age-class (Johnston & Odum 1956; Titterington et al. 1979; Helle 1985). Some species may become abundant as regenerating forests reach the age that provides them with plenty of food and cover and then decline as the forest matures. This sequence was used to examine changes in the abundance of a group of forest mammals and birds that have undergone substantial declines in New England during the past three decades.

New England Cottontails

Since the 1960s, the New England cottontail (*Sylvilagus transitionalis*) has declined substantially throughout its range (Chapman & Stauffer 1981). In response to that decline, the U.S. Fish & Wildlife Service has listed this species as a candidate for threatened or endangered sta-

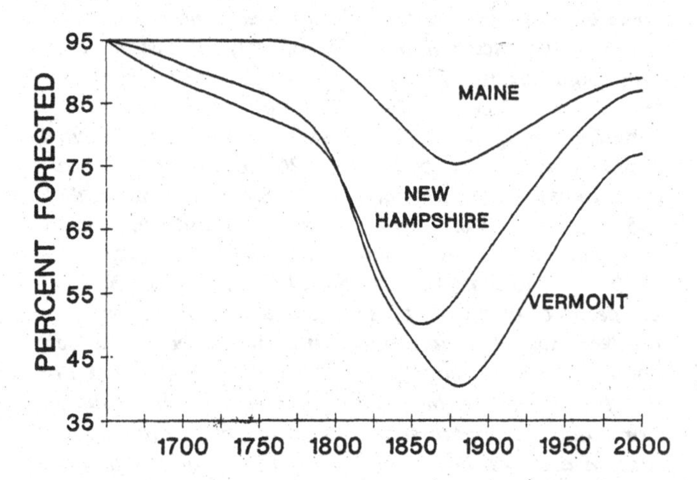

*Figure 1. Forest area in three New England states from 1650 to the present. Estimates from 1650 to 1910 were taken from Harper (1918), and the remaining values were linearly extrapolated using current estimates (Maine: Powell & Dickson 1984; New Hampshire: Frieswyk & Malley 1983*a*; Vermont: Frieswyk & Malley 1983*b*).*

tus (Federal Register, Vol. 54, No. 4:553–579, January 6, 1989). Previous researchers speculated that the decline of this lagomorph was a response to habitat alteration or to competitive interactions with expanding populations of eastern cottontails (*S. floridanus*) (Linkkila 1971; Chapman & Morgan 1973). In New Hampshire, the distribution of New England cottontails decreased from approximately 60% of the state in 1950 (Stevens 1950, illustrated in Jackson 1973) to fragmented populations that currently occupy less than 20% of the state (Fig. 2). Habitat change seems to be a more likely explanation of the decline of New England cottontails, because eastern cottontails did not colonize New Hampshire until the late 1960s or early 1970s (Jackson 1973) and they have been restricted to the extreme southern portion of the state since arriving (T. Walski, New Hampshire Fish & Game Department, personal communication; Litvaitis, unpublished data).

To evaluate the effects of historical land use, I first examined the current distribution of New England cottontails in New Hampshire by sampling habitats considered suitable (Chapman 1975). Lagomorphs were inventoried in 53 sites during the winters of 1990 and 1991. Of these, 39 contained New England cottontails (Litvaitis & Barbour, unpublished data). Occupied sites ranged from 0.2 to approximately 15 ha, and most were dominated by shrubs or seedlings/saplings and were associated with idle agricultural lands, powerline corridors, highway islands, and the edges of industrial parks. Collectively, occupied sites were characterized as disturbance patches where secondary succession had progressed approximately 10–25 years. The historic availability of such habitats was then estimated by examining the pattern of farm abandonment in New Hampshire.

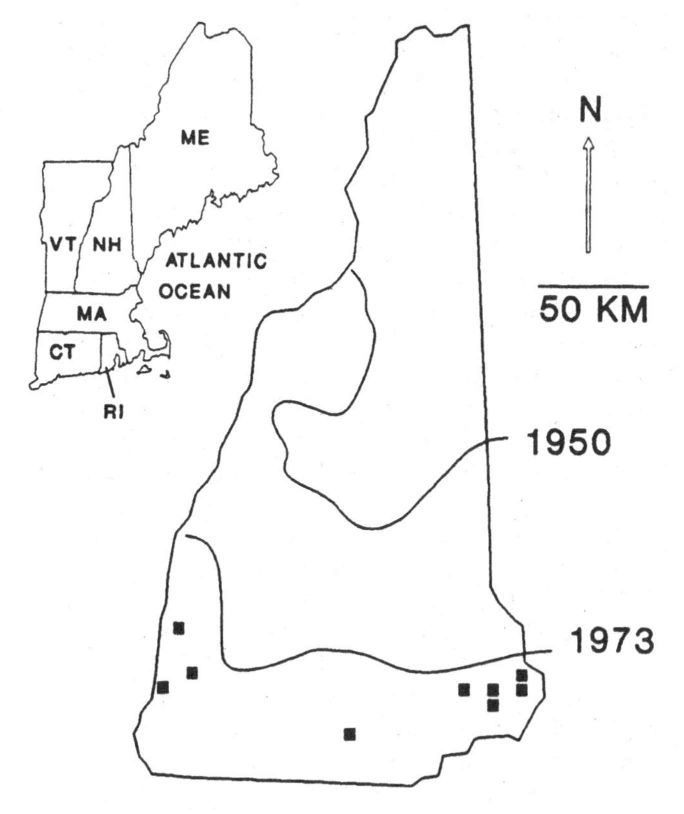

Figure 2. Northern range limits of New England cottontails in New Hampshire during 1950 (Stevens 1950) and 1973 (Jackson 1973). Squares indicate approximate location of current remnant populations (Walski, personal communication; Litvaitis, unpublished data). Insert map of New England includes Maine (ME), New Hampshire (NH), Vermont (VT), Connecticut (CT), and Rhode Island (RI).

From 1880 to 1930, the amount of cropland and nonwooded pastures in the state declined by 650,000 ha (mean rate of abandonment = 13,000 ha/yr, Black 1950:152). Although some of this land was converted to other uses, most was left idle and colonized by second-growth forests (Irland 1982). The availability of habitats suitable to New England cottontails was estimated using the average abandonment rate of 13,000 ha/yr, and modeled habitats were considered occupied by cottontails from 10 to 25 years after abandonment. The initial habitats became available in 1890, and 13,000 ha were recruited every year until 1940. Because such habitat would be suitable for approximately 15 years, 195,000 ha of early successional habitat accumulated from 1905 to 1940 (Fig. 3). Modeled habitats matured by around 1960 into age classes that had limited understory foliage (Aber 1979) and were no longer suitable to New England cottontails (Fig. 3). This rapid decline of early successional habitats is supported by statewide forest inventories. Specifically, the amount of forested land in New Hampshire with trees having an average diameter of less than 13 cm diameter at breast height (dbh) de-

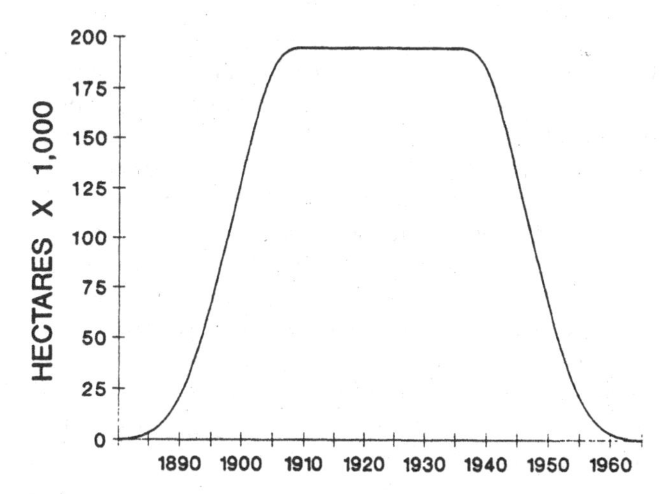

Figure 3. Modeled recruitment of early successional habitat (10–25 years after disturbance) resulting from the annual abandonment of 13,000 ha of cropland and nonwooded pastures in New Hampshire during 1880–1930.

clined (845,425 ha in 1960; 481,860 ha in 1973; and 131,335 ha in 1983), while more mature forests (mean dbh ≥ 13 cm) have increased (1,141,375 ha in 1960; 1,012,875 ha in 1973; 1,816,880 ha in 1983—R. Brooks, U.S. Forest Service, personal communication).

The habitats resulting from farm abandonment do not represent all of the habitats available to New England cottontails. Cottontails probably occupied dense understory vegetation associated with forest gaps, regenerating stands that colonized natural and human-induced disturbances, stream corridors, and shrub-dominated wetlands. However, the modeled chronology of forest maturation coincided with the decline of this species in New Hampshire, suggesting that the widespread loss of early successional habitats had a profound effect. In addition, the north-south range contraction (Fig. 2) may be explained by a more precipitous loss of agricultural lands in the northern counties of the state (Black 1950: 155).

Bobcats

The large-scale increase and subsequent decline of early successional plant communities may have influenced other vertebrates associated with this habitat. New England cottontails were a major prey of bobcats (*Felis rufus*) in New Hampshire during the 1950s (Litvaitis et al. 1984). During 1951–1954, remains of New England cottontails were identified in 43% of bobcat carcasses examined that contained prey (n = 87; Stevens & Litvaitis, unpublished data). This dropped to 10% in carcasses collected during 1961–1965 (n = 108). Bounty and fur buyer records also indicated that the average annual harvest of bobcats in the state declined from 350 during 1951–1954 to 36 during 1965–1969 (New Hampshire Fish & Game Department, unpublished data; Fig. 4). Although trapper harvest rates can be influenced by commercial demand (Erickson 1981), the value of bobcat pelts from New Hampshire did not exceed $10 until 1971, and the bounty program on this species was not terminated until 1973. Therefore, it is unlikely that the change in annual harvests was a result of declining trapper efforts. Expanding populations of coyotes (*Canis latrans*) also have been implicated in the decline of bobcat populations (Litvaitis & Harrison 1989). Coyotes were not common in New Hampshire until the mid or late 1970s (New Hampshire Fish & Game Department, unpublished data), however, and they probably had little influence on bobcats during the period of decline.

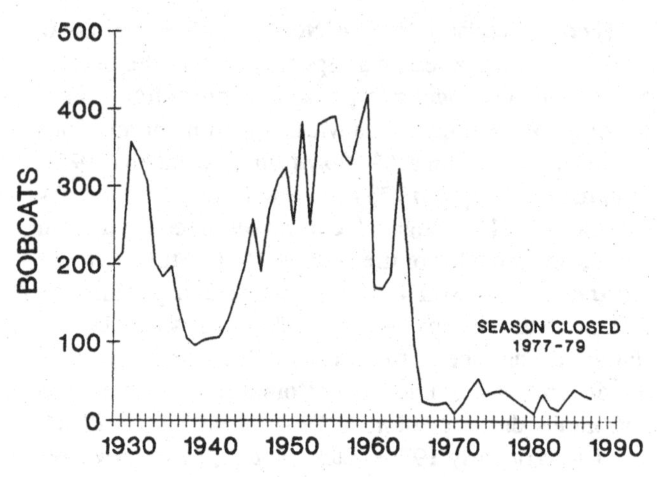

Figure 4. Annual harvests of bobcats in New Hampshire. Values prior to 1966 were obtained from records of statewide bounty payments (Stevens, unpublished data), and those from 1966 to 1987 were obtained from bounty records, surveys of fur buyers, or tallies of registered pelts (New Hampshire Fish & Game Department, unpublished data).

Migratory Passerines

In a recent paper, Hill and Hagan (1991) summarized the population trends of some migratory birds that nest in the forests of New England. These data were collected during spring migrations (1937–1989) at sites in eastern Massachusetts. Because a specific sampling protocol was not followed by the observers involved in this monitoring effort, some variation undoubtedly was introduced to the subsequent data. Also, the effects of suburbanization and urbanization on the migration paths of this region are not known. Nonetheless, the duration of the monitoring period makes this data set unique, and although it was collected in eastern Massachusetts, it may be relevant to trends in populations of conspecifics that nest farther north (such as in New Hampshire). Of 26 species sampled, 18 declined significantly during the 53-year monitoring period. Eight of the species that declined are known to nest in early successional habitats (Least Flycatcher [*Empidonax minimus*], Nashville Warbler [*Vermivora ruficapilla*], Golden-Winged Warbler [*V. chrysoptera*], Chestnut-Sided Warbler [*Dendroica pensylvanica*], Magnolia Warbler [*D. magnolia*], American Redstart [*Setophaga ruticilla*], Wilson's Warbler [*Wilsonia pusilla*], and Canada Warbler [*W. canadensis*]) (DeGraaf & Rudis 1986; Morse 1989). All eight species exhibited declines through the 1950s (Fig. 5), but Wilson's Warblers experienced an unexplained increase during the late 1970s.

Previous researchers have provided several explanations for these declines. Tropical deforestation may have substantially reduced winter habitats (Terborgh 1980; Hutto 1988). Hill and Hagan (1991), however, indicated that winter habitats were not uniform among the species that declined in their study. Deforestation may have actually increased the abundance of habitats used by some of these species (for example, the scrub used

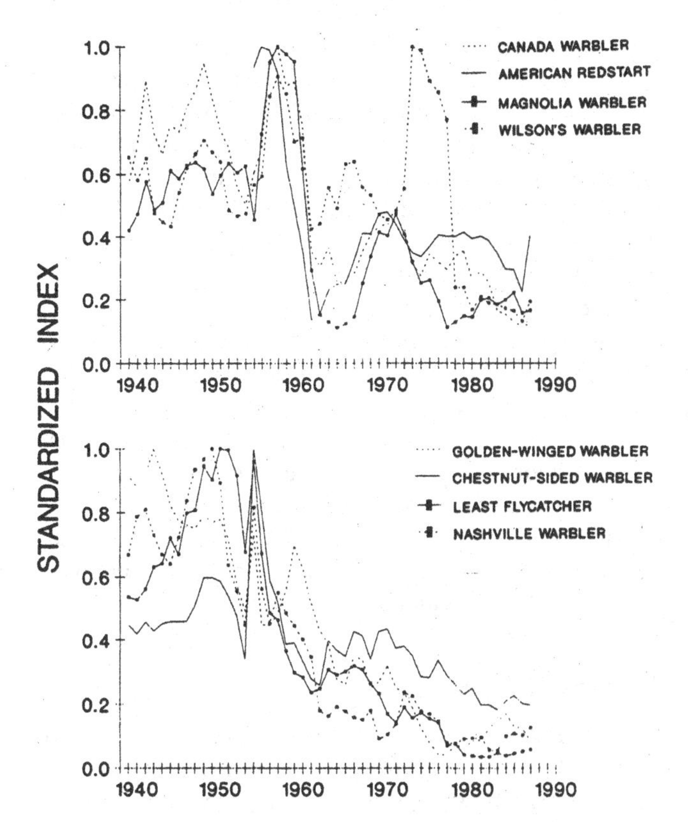

Figure 5. Annual indices of some migrating land birds that nest in the forests of New England. Samples (number observed/hr) were obtained in eastern Massachusetts during spring (May) migrations (Hill & Hagan 1991). Indices were standardized by dividing the highest annual observation rate for a particular species into each annual observation rate for that species.

by wintering Least Flycatchers). Competition with expanding populations of Blue-Winged Warblers (*Vermivora pinus*) has been suggested as a possible cause of the regional decline of Golden-Winged Warblers (Confer & Knapp 1979). Yet Confer and Knapp (1981) and Gill (1980) observed that historic land-use practices provided ideal habitats (brushy old fields) for Golden-Winged Warblers and that the abundance of such habitats has declined markedly. Holmes et al. (1986) also attributed the decline of a local population of Least Flycatchers in northern New Hampshire to forest maturation. Therefore, it seems plausible that a single factor—forest maturation—caused the decline of these passerines. This conclusion is supported by an examination of recent trends (1966–1988) in Breeding Bird Survey data on 55 species of land birds that nest in coastal New England (Witham & Hunter 1992). Of the 18 species that showed significant declines, nine were associated with early successional habits. Changes in land uses (1962–1966 versus 1983–1988) along the survey routes used in this study included decreases in agricultural land (−9%) and early successional habitats (−12%). Nine of the 15 species that showed increases in this survey were associated with closed-canopy forests, further indicating that regional populations were responding to forest maturation.

Historic Abundance and Current Concerns

Although reasonable data exist on the recent patterns of abundance of these species, little is known about their distributions prior to the colonization of New England by Europeans. Anecdotal accounts suggest that some early successional species were relatively rare at that time. For example, New England cottontails were restricted to the southern portion of New Hampshire until the mid-1800s, when they colonized central and northern portions of the state (Silver 1974). Early naturalists also have commented on "population explosions" by several warblers (for example, Chestnut-Sided, Nashville, and Golden-Winged) in response to an increased abundance of early successional habitats during the late 1800s and early 1900s (summarized by Morse 1989: 262–264). According to Bent (1953), the Chestnut-Sided Warbler was so rare that Audubon observed this species only once during his field excursions in the early 1800s. Present-day populations, therefore, may actually be more representative of pre-Columbia densities.

The patterns reported here indicate that the decline in abundance and distribution of some forest vertebrates may be explained by a "successional wave" passing through the forests of northern New England (Fig. 6). The decline of early successional species, however, may be extended in space and time beyond the effects of forest maturation. Current landscape modifications, especially suburbanization, are fragmenting and isolating patches of habitat, thus reducing the viability of local

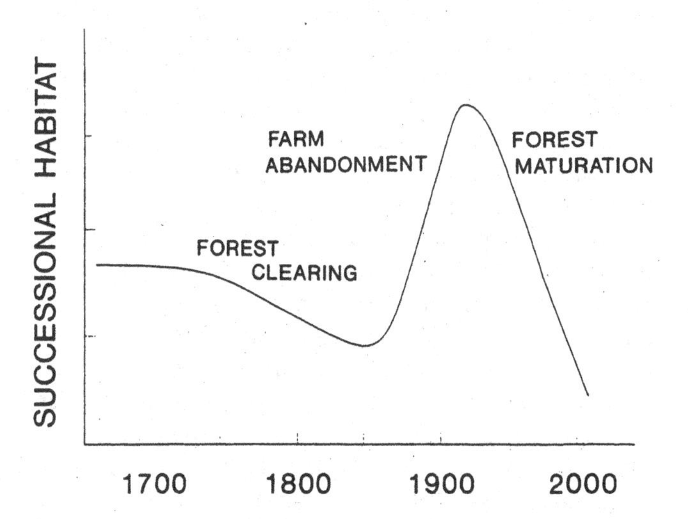

Figure 6. Suggested pattern of events that influenced the abundance of early successional habitat in northern New England from 1650 to the present.

populations. For example, winter survival rates of New England cottontails were significantly lower in habitat patches of less than 2.5 ha (0.35) than in larger patches (0.69; Barbour & Litvaitis 1993). As a result, small, isolated populations of rabbits had high extinction rates (Litvaitis, unpublished data). Habitat fragmentation also has been associated with reduced survival and reproductive success among birds nesting on (such as the Nashville Warbler) or near (such as the Chestnut-Sided Warbler) the ground because of increased activity by predators and nest parasites along habitat edges (Yahner 1988). Wide-ranging species that utilize disjunct patches of early successional habitat, such as bobcats, are hampered by factors that accompany fragmentation, including increased road densities. Collisions with vehicles was the second most common form of mortality (20% of all deaths) among a sample of transmitter-equipped bobcats in Maine (Litvaitis et al. 1987). As a result, relatively large tracts of early successional habitat may be needed to sustain local and regional populations of these species.

Current ownership patterns of forest lands in New England (excluding Maine) reveal 88% private ownership with an average holding of 10 ha and the majority less than 40 ha. These forests will likely continue to mature (Brooks & Birch 1988; Brooks 1989). Also, the trend among landowners is toward subdivisions that are distributed among an increasing number of owners (Brooks & Birch 1988). This trend will likely limit opportunities to provide early seral habitats, except on industrial and public forests.

Even-aged timber management is utilized by timber corporations in northern portions of Vermont, New Hampshire, and Maine, and they will perpetuate early successional habits in these areas (Brenneman & Eubanks 1990). Clear-cutting on state and national forests could be used to manage populations of early successional species, but this may be perceived as contrary to policies regarding the effects of timber management on area-sensitive and forest-interior species (see Norse et al. 1986). Even-aged management, however, can be compatible with efforts to sustain large tracts of mature forest. Clustering clearcuts around selected nuclei rather than dispersing them throughout a forest would reduce the edge and road effects associated with scattered cuts (Franklin & Forman 1987) and would increase the likelihood of sustaining populations of early-sere dependent species. Large areas of mature forest could be maintained with uneven-aged management or no-cut zones that limit disturbance and fragmentation. Although this may be an oversimplification of a complex management issue, it does suggest an opportunity to retain early successional habitats with limited detriment to vertebrates dependent on older and larger tracts of forest habitat.

Historic land use in New England has resulted in even-aged forests that now dominate the landscape. Natural disturbances may increase structural diversity as these forests age, but it is unlikely that such disturbances will provide a variety of seral stages and forest age classes for many decades. Timber harvesting has the potential to increase the structural diversity of these forests over a shorter time period. Management programs that include clear-cutting are not appropriate for the entire region, especially where historic or current land uses and ownership patterns have resulted in diverse forests or where site characteristics (such as topography, soils, watersheds, and species of special concern) are sensitive to disturbance. In suitable areas, even-aged forest management does provide an opportunity to increase or sustain the availability of early successional habitats. Given public attitude towards clearcutting, however, such management policies may be difficult to implement on public lands.

Acknowledgments

I thank Bob Brooks, Dave Capen, Bob Eckert, John Hagan, Mac Hunter, John Kanter, John Lanier, Mariko Yamasaki, and two anonymous reviewers for their thoughtful comments on early drafts of this paper. John Hagan generously provided access to data sets on migratory birds. This is scientific contribution 1793 of the New Hampshire Agricultural Experiment Station.

Literature Cited

Aber, J. D. 1979. Foliage-height profiles and succession in northern hardwood forests. Ecology **60**:18–23.

Ambuel, B., and S. A. Temple. 1983. Area-dependent changes in the bird communities and vegetational southern Wisconsin forests. Ecology **64**:1057–1068.

Askins, R. A., J. F. Lynch, and R. Greenberg. 1990. Population declines in migratory birds in eastern North America. Pages 1–57 in D. M. Power, editor. Current ornithology. Plenum Press, New York.

Barbour, M. S., and J. A. Litraitis. 1993. Niche dimensions of New England cottontails in relation to habitat patch size. Oecologia. In press.

Bent, A. C. 1953. Life histories of North American wood warblers. Bulletin of the United States National Museum **203**:1–734.

Black, J. J. 1950. The rural economy of New England. Harvard University Press, Cambridge, Massachusetts.

Brenneman, R. E., and T. E. Eubanks. 1990. Forest fragmentation in the Northeast—an industry perspective. Pages 23–25 in R. M. DeGraaf and W. M. Healy, compilers. Is forest fragmentation a management issue in the Northeast? General Technical Report NE-140. U.S. Forest Service, Northeast Forest Experiment Station, Radnor, Pennsylvania.

Brooks, R. T. 1989. History and future trends for wildlife and wildlife habitat in northeastern United States. Pages 37–45 in

Timber management and its effects on wildlife. Proceedings of the 1989 Conference on Pennsylvania State Forest Resources Issue conference. Pennsylvania State University, University Park, Pennsylvania.

Brooks, R. T., and T. W. Birch. 1988. Changes in New England forests and forest owners: Implications for wildlife habitat resources and management. Transactions of the North American Wildlife & Natural Resources Conference 53:78–87.

Burgess, R. L., and D. M. Sharpe, editors. 1981. Forest island dynamics in man-dominated landscapes. Springer–Verlag, New York.

Chapman, J. A. 1975. *Sylvilagus transitionalis.* Mammalian Species No.55. American Society of Mammalogists.

Chapman, J. A., and R. P. Morgan. 1973. Systematic status of the cottontail complex in western Maryland and nearby West Virginia. Wildlife Monograph 36:1–54.

Chapman, J. A., and J. R. Stauffer. 1981. The status and distribution of the New England cottontail. Pages 973–983 in K. Myers and C. D. MacInnes, editors. Proceedings of the World lagomorph Conference. University of Guelph, Ontario, Canada.

Confer, J. L., and K. Knapp. 1979. The changing proportion of Blue-Winged and Golden-Winged Warblers in Tompkins County and their habitat selection. Kingbird 29:8–14.

Confer, J. L., and K. Knapp. 1981. Golden-Winged Warblers and Blue-Winged Warblers: The relative success of a habitat specialist and a generalist. Auk 98:108–114.

DeGraaf, R. M., and D. D. Rudis. 1986. New England wildlife: Habitat, natural history, and distribution. General Technical Report NE-108. U.S. Forest Service, Northeast Forest Experiment Station, Broomall, Pennsylvania.

Erickson, D. W. 1981. Furbearer harvest mechanisms: An examination of variables influencing fur harvests in Missouri. Pages 1469–1491 in J. A. Chapman and D. Pursley, editors. Proceedings of the Worldwide Furbearer Conference, Frostburg, Maryland.

Finch, D. M. 1991. Population ecology, habitat requirements, and conservation of neotropical migratory birds. General Technical Report RM–205. U.S. Forest Service, Fort Collins, Colorado.

Franklin, J. F., and R. T. T. Forman. 1987. Creating landscape patterns by forest cutting: Ecological consequences and principles. Landscape Ecology 1:5–18.

Frieswyk, T. S., and A. M. Malley. 1985*a.* Forest statistics for New Hampshire—1973 and 1983. Resource Bulletin NE-88. U.S. Forest Service, Northeast Forest Experiment Station, Broomall, Pennsylvania.

Frieswyk, T. S., and A. M. Malley. 1985*b.* Forest statistics for Vermont—1973 and 1983. Resource Bulletin NE-87. U.S. Forest Service, Northeast Forest Experiment Station, Broomall, Pennsylvania.

Gill, F. B. 1980. Historical aspects of hybridization between Blue-Winged and Golden-Winged Warblers. Auk 97:1–18.

Hagan, J. M., III, and D. W. Johnston, editors. 1992. Ecology and conservation of neotropical migrant landbirds Smithsonian Institution Press, Washington, D.C.

Harper, R. M. 1918. Changes in the forest area of New England in three centuries. Journal of Forestry 16:442–452.

Helle, P. 1985. Habitat selection of breeding birds in relation to forest succession in northeastern Finland. Annales Zoologici Fennici 62:113–123.

Henderson, M. T., G. Merriam, and J. Wegner. 1985. Patchy environments and species survival: Chipmunks in an agricultural mosaic. Biological Conservation 31:95–105.

Hill, N. P., and J. M. Hagan, III. 1991. Population trends of some northeastern North American landbirds: A half-century of data. Wilson Bulletin 103:165–182.

Holmes, R. T., T. W. Sherry, and F. W. Sturgess. 1986. Bird community dynamics in a temperate deciduous forest: Long-term trends at Hubbard Brook. Ecological Monographs 56:201–220.

Hutto, R. L. 1988. Is tropical deforestation responsible for the reported declines in neotropical migrant populations? American Birds 42:375–379.

Irland, L. C. 1982. Wildlands and woodlots—a story of New England's Forests. University Press of New England, Hanover, New Hampshire.

Jackson, S. N. 1973. Distribution of cottontail rabbits (*Sylvilagus* spp.) in northern New England. M. S. thesis. University of Connecticut, Storrs, Connecticut.

Johnston, D. W., and E. P. Odum. 1950. Breeding bird populations in relation to plant succession on the Piedmont of Georgia. Ecology 37:50–62.

Linkkila, T. E. 1971. Influence of habitat upon changes within interspecific Connecticut cottontail populations. M. S. thesis. University of Connecticut, Storrs, Connecticut.

Litvaitis, J. A., and D. J. Harrison. 1989. Bobcat-coyote niche relationships during a period of coyote population increase. Canadian Journal of Zoology 67:1180–1180.

Litvaitis, J. A., C. L. Stevens, and W. W. Mautz. 1984. Age, sex, and weight of bobcats in relation to winter diet. Journal of Wildlife Management 48:632–635.

Litvaitis, J. A., J. T. Major, and J. A. Sherburne. 1987. Influence of season and human-induced mortality on social organization of bobcats (*Felis rufus*) in Maine. Journal of Mammalogy 68:100–106.

Matthiae, P. E., and F. Stearns. 1981. Mammals in forest islands in southeastern Wisconsin. Pages 55–66 in R. L. Burgess and D. M. Sharpe, editors. Forest island dynamics in man-dominated landscapes. Springer-Verlag, New York.

Morse, D. H. 1989. American warblers. Harvard University Press, Cambridge, Massachusetts.

Norse, E. A., K. L. Rosenbaum, D. S. Wilcove, B. A. Wilcox, W. H. Romme, D. W. Johnston, and M. L. Stout. 1986. Conserving biological diversity in our national forests. The Wilderness Society, Washington, D.C.

Powell, D. S., and D. R. Dickson. 1984. Forest statistics for Maine—1971 and 1982. Resource Bulletin NE-81. U.S. Forest Service, Northwest Forest Experiment Station, Broomall, Pennsylvania.

Robbins, C. S., D. K. Dawson, and B. A. Dowell. 1989. Habitat area requirements of breeding birds of the Middle Atlantic states. Wildlife Monograph **103**:1–34.

Saunders, D. A., R. J. Hobbs, and C. R. Margules. 1991. Biological consequences of ecosystem fragmentation: A review. Conservation Biology **5**:18–32.

Silver, H. 1974. A history of New Hampshire game and furbearers. 2nd edition. New Hampshire Fish and Game Department, Concord, New Hampshire.

Sevens, C. L. 1950. Cottontail rabbit distribution in New Hampshire. Unpublished survey. University of New Hampshire, Durham, New Hampshire.

Terborgh, J. W. 1980. The conservation status of neotropical migrants: Present and future. Pages 21–30 in A. Keast and E. S. Morton, editors. Migrant birds in the neotropics. Smithsonian Institution Press, Washington, D.C.

Titterington, R. W., H. S. Crawford, B. N. Burgason. 1979. Songbird responses to commercial clearcutting in Maine spruce-fir forests. Journal of Wildlife Management **43**:602–609.

Wilcove, D. S. 1985. Nest predation in forest tracts and the decline of migratory songbirds. Ecology **66**:1211–1214.

Wilcove, D. S., C. H. McLellan, and A. P. Dobson. 1986. Habitat fragmentation in the temperate zone. Pages 237–256 in M. E. Soulé, editor. Conservation biology. Sinauer Associates, Sunderland, Massachusetts.

Witham, J. W., and M. L. Hunter, Jr. 1992. Population trends of neotropical migrant landbirds in northern coastal New England. Pages 85–95 in J. M. Hagan, III, and D. W. Johnston, editors. Ecology and conservation of neotropical migratory landbirds. Smithsonian Institution Press, Washington, D.C.

Yahner, R. H. 1988. Changes in wildlife communities near edges. Conservation Biology 2:333–339

Pollination in *Dianthus deltoides* (Caryophyllaceae): Effects of Habitat Fragmentation on Visitation and Seed Set

OLA JENNERSTEN
Uppsala University
Department of Zoology
Box 561
S-751 22 Uppsala
Sweden

Abstract: *I analyse the effects of habitat fragmentation on the pollination success of a perennial, butterfly-pollinated, caryophyllaceous herb, the maiden pink,* Dianthus deltoides *L. The study was conducted in July 1986 and July 1987 at two different sites in southwest Sweden, an undisturbed "mainland" site and a fragmented site consisting of "habitat islands" within a heavily utilized agricultural area. The fragmented area had a lower diversity and abundance of both flowering plants and flower-visiting insects.* Dianthus *flowers received fewer visits in the fragmented area than in the mainland area, and the seed set was much lower. Hand pollination increased seed set up to 4.1 times in the fragmented area, but no significant differences were found between hand-pollinated and control flowers at the mainland site. There were no differences between the two sites in standing crop of nectar, ovule number per flower, or seed set of bagged flowers, hand-pollinated flowers, and hand-pollinated fertilized flowers. Thus, the difference in natural seed set between the two sites can be explained by differences in pollinator service.*

Resumen: *Analizé los efectos de fragmentación de habitat sobre el éxito de polinización de la hierba perenne, "caryophyllacedes, Dianthus deltoides L." que es polinada por mariposas. Este estudio fue conducido en julio de 1986 y 1987 en dos diferentes lugares en el suroeste de Suecia, en un lugar pristino del continente y en un lugar fragmentado compuesto de "islas de habitat" dentro una área agricola sumamente utilizada. La área fragmentada tuvo una diversidad más baja y una menor abundancia de plantas florecentes y de insectos tipo visitantes de flores. Las flores "Dianthus" recibieron menos visitas en las áreas de habitat fragmentada que en la area del continente aun no modificado. La siembra tambien era más baja. Polinización por mano aumentó la siembra 4.1 veces más en la área fragmentada, pero niguna diferencia significante fue encontrada entre las flores polinadas por mano y en el control en el lugar del continente. No hubo diferencias en cosechas paradas del néctar, numero de óvule por flor, o la siembra de flores de bolsas, flores polinadas por mano, y las flores fertilizadas y polinadas por mano, entre los dos lugares. Asi, la diferencia entre la siembra natural de los dos lugares puede ser explicado por las diferencias en el servicio de polinización.*

Introduction

Fragmentation of natural habitats, a common feature in today's world, occurs when large expanses of habitat are transformed into a number of smaller patches isolated by an area unlike the original (Wilcove, McLellan, & Dobson 1986). Habitat fragmentation is generally considered to be harmful to animals and plants. Fragmentation and reduction of habitat can reduce species richness and relative abundance of species (MacArthur & Wilson 1967; Simberloff & Wilson 1969; Carleton & Taylor 1983; Diamond 1984). Geographically restricted

Paper submitted 5/5/88; revised manuscript accepted 8/25/88.

species may face a decrease in genetic polymorphism (Lande 1980; Karron 1987) and an increase in inbreeding (Franklin 1980), and populations might be more vulnerable to environmental fluctuations and demographic stochasticity (Shaffer 1981).

Specialized plant-pollinator systems are sensitive to disturbances of any kind, and the pollinator as well as the plant may be influenced if either of the two changes in abundance (Janzen 1974; Gilbert 1980). Thus, pollination systems are useful for analyzing effects of fragmentation. Pollinator behaviour and the number of pollen grains transferred are influenced by the geometry (e.g., size, density, shape) of a plant population (Handel 1983 and references therein). Island populations of plants may have lower seed set than mainland populations of the same species because lower pollinator availability results in pollen limitation (Spears 1987). The high proportion of generalized pollination systems on islands has been interpreted as a result of island-induced reductions in pollinator diversity (Linhart & Feinsinger 1980).

In this study I compare visitation rates by pollinators to, and the seed set of, the butterfly-adapted maiden pink, *Dianthus deltoides* L. (Caryophyllaceae), in a mainland and an island area. The mainland area harbors a continuous population of *Dianthus* and consists of a varied landscape of forests and meadows (here called site A). The island area consists of two habitat fragments with isolated patches of *Dianthus*. The two islands are surrounded by large fields very unlike the original habitat (site B). The aim of this study is to investigate the effects of fragmentation on pollination success, by asking three questions: (1) Are there differences in diversity and abundance of anthophilous insects and flowering plants between the two sites? (2) Are there differences in the reproductive success of *Dianthus* between the two sites? (3) If so, what mechanisms are behind the differences in reproductive success?

Materials and Methods

The Plant

The maiden pink, *Dianthus deltoides*, is a perennial caryophyllaceous herb that grows in sunny habitats on dry sandy soils and is distributed throughout Europe (Clapham, Tutin, & Warburg 1962). Each plant carries one to several flowering shoots, normally with one terminal flower each. The flower is rosy red and has a deep (12–14 mm), tubular, leathery calyx, which has been interpreted as a protection against nectar robbery (Proctor & Yeo 1973). Nectar is secreted from the base of the stamens (Proctor & Yeo 1973), and each flower contains up to 0.9 μl nectar with a sugar content between 15 and 30% (this study). The flower is protandrous, first presenting ten pollen-bearing anthers for approximately two days and, subsequently, a pistil of two twisted styles that protrudes from the flower for, on average, one and a half additional days. Thus, total flower lifetime ranges from two to four days (this study). The flowers close each night, at the study site between 1800 and 0800 h (Greenwich Mean Time + 2h), so they are visited by diurnal insects only. The flowers are butterfly-adapted (Proctor & Yeo 1973) and are visited by butterflies for nectar and by flies for pollen (Knuth 1898; Jennersten 1984; this study). The flower contains about 100 ovules and when fertilized it will develop into a one-celled capsule. The seeds are obovate and about 2 mm long. They eventually shake out of the capsule, so they disperse only for short distances.

Study Site

The study was conducted at two sites in the province of Dalsland, southwestern Sweden. Site A is a dry, sandy meadow of approximately 1 ha. It is situated in a landscape of continuous meadows, abandoned and cultivated fields, and mixed forests mainly of spruce, ash, birch, and aspen. Fields and meadows constitute about 35% of the area in which site A is located. The other site, B, is located in an agricultural district where cultivated fields make up most of the area. Site B consists of two habitat fragments, isolated from nearby forest by approximately 200 meters of cultivated fields, mainly of barley and oats. Thus very few nectar plants are found in the fields. The islands are 80 × 40 m and 50 × 30 m and consist of dry sandy soil with bare rocks. The density of *Dianthus* is roughly equivalent at the two sites but the total population at site A is much larger than the one at site B. The study was performed in July 1986 and July 1987.

Insect and Plant Survey

In 1986 I laid out six transects of 10 m each along an S-shaped line at site A, and five transects of 10 m each per island at site B, to measure insect and plant diversity. I censused each transect five times during one week, recording anthophilous insects (= Lepidoptera and Hymenoptera) observed during a two-minute period within 2.5 m of each site of the transect. All plant species in flower within the transect areas were noted once during the same week, using a three-grade abundance scale (1 = few, 2 = common, 3 = dominant).

Dianthus Visitors

I established four 1 × 2 m plots at site A and five at site B (three on one island and two on the other) around *Dianthus* plants in flower. The plots at site A contained on average 30, 167, 465, and 553 flowers, and at site B 22, 80, 109, 246, and 306 flowers per plot. The plants within these plots were observed during sunny weather for 15-minute sampling intervals between 0900 and 1700 h. The plots at site A were observed for a total of 6.5 hours distributed over four days between July 1–4 and 10.5 hours between July 10–16; at site B the plots were observed for 8.0 hours, distributed over four days between July 1–4, 1986. I identified all flower visitors to the lowest possible taxonomic category as well as the number of flower visits per individual insect within a single plot. I also counted the number of blooming *Dianthus* flowers once a day within each sample plot.

Nectar Measurements

Nectar was collected by inserting 1 μl glass micropipettes into the corolla tube. Standing crop of nectar was measured at site A on July 3, 11, 14, 15, and 16 and at site B on July 4 and 10, 1986. Fifteen flowers were randomly selected on each occasion between 1000 and 1500 h.

Pollination Experiments

In 1986, I marked ten plants with at least three flowering stalks for a pollination experiment at site A only. A flower in receptive female phase on one stalk was hand-pollinated with pollen from other individuals to maximize pollination, a bud of the second stalk was bagged to exclude visiting insects throughout flowering to analyse degree of self-pollination, and the flower of the third stalk was not manipulated, thereby serving as a naturally pollinated control. At both sites I marked ten additional flowers distributed on individual plants in each of the nine sampling plots (see "*Dianthus* Visitors") (a total of 40 at site A and 50 at site B) for subsequent analysis of seed set.

In 1987 pollination experiments were performed simultaneously at both sites, early (July 11–14) and late (July 22–27) in the flowering season of *Dianthus*. The experiments consisted of five treatments performed on 10–25 individual flowers per treatment. In the first treatment one flower per plant was hand-pollinated when in female receptive stage; a second treatment was hand-pollinated and water was subsequently added every second day (30 ml) until the seed capsules were collected; and a third treatment was hand-pollinated, with a liquid fertilizer added (3% "Complesal" N-K-P = 142-205-59 g/l), 30 ml every second day as above. In a fourth treatment insects were excluded by bagging buds with nylon mesh bags, and finally, the fifth group of plants was not manipulated and served as a naturally pollinated control.

Seed capsules were collected when mature but before opening, and the number of developed seeds and undeveloped ovules was later counted using a stereo microscope. Percent fertilization was calculated as the proportion of ovules that developed into seeds. Only capsules not disturbed by predators were considered. I used arcsine-square-root transformed data in statistical analysis using proportions. I used the SAS statistical package (SAS Institute, Inc., 1985) for statistical analysis.

Results

Insect and Plant Diversity

Site A had a higher diversity of both flowering plants and anthophilous insects (Fig. 1) (data from 1986). I found 18 species of flowering plants and 24 species of insects including 10 butterfly and 8 bumblebee species at site A, and a total of 8 plant and 11 insect species including 4 butterflies and 2 bumblebee species at site B. Insect richness at site A was 5.17 encounters per unit transect and time (SD = 2.49, min = 2, max = 13, N = 30), and only 0.60 (SD = 0.67, min = 0, max = 3, N = 50) at site B, a ninefold difference. The mainland site had significantly more insect encounters during all five sampling occasions compared to the island site (Wilcoxon's rank sum test, $p < 0.02$ all times).

Insect Visitation Rates

Dianthus deltoides flowers at site A were visited more often than were those at site B (data from 1986). On average, 11.6% of the flowers were visited every hour at the mainland site (A), and 3.6% were visited at the island site (B) (Table 1). The difference is significant

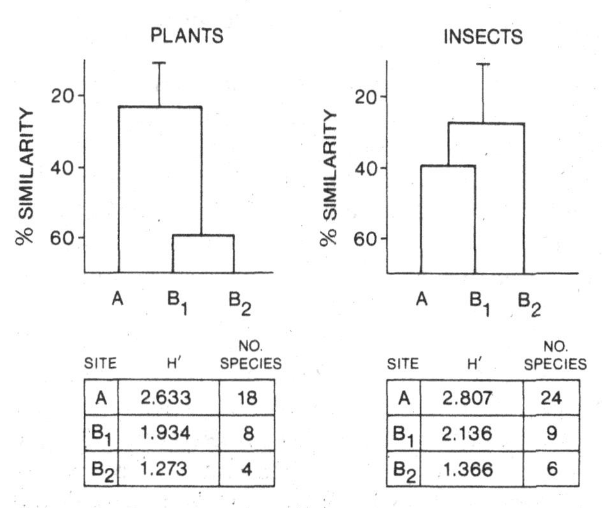

SITE	H'	NO. SPECIES
A	2.633	18
B_1	1.934	8
B_2	1.273	4

SITE	H'	NO. SPECIES
A	2.807	24
B_1	2.136	9
B_2	1.366	6

Figure 1. Proportional similarity (Czekanowski's index), diversity (Shannon-Weiner H'), and the number of species of blooming plants and anthophilous insects (bumblebees and butterflies) in unfragmented (Site A) and fragmented habitat (Site B, data from the two islands are presented, 1 and 2). $PS = \Sigma_i \cdot \min(pi,qi)$, $H' = -\Sigma_i \cdot (pi)(\ln pi)$.

Table 1. Proportion (%) of *Dianthus deltoides* flowers visited per hour at site A (Mainland) and site B (Island). At site A and site B data were collected on July 1–4 (early), and at site A also on July 10–16 (late). Means with different letters are significantly different ($p < 0.05$, Wilcoxon's rank sum test for two samples). Data from 1986.

Site	Mean	SE	Range
All visitors			
Mainland (early)	11.6 a	3.4	1.2–35.1
Mainland (late)	08.8 a	1.9	0.0–20.5
Island (early)	03.6 b	1.8	0.0–16.8
Nectar feeders			
Mainland (early)	04.4 a	1.9	0.0–18.5
Mainland (late)	05.3 a	1.9	0.0–20.5
Island (early)	01.6 b	1.5	0.0–16.8

whether or not pollen feeders are included. Assuming an even eight hours of visitation (few visitors before 0900 h and after 1700 h) for each of three days (the mean flower life), the expected mean number of visits per flower exceeded one at site A (all visitors: 2.78, nectar feeders only: 1.06). At site B expected visitation did not reach one (all visitors: 0.86, nectar feeders: 0.38). The probability (P) of a flower receiving at least one visit during its lifetime may be described by a Poisson distribution if one assumes random visitation, and can be estimated as:

$$P = 1 \cdot 1/e^x$$

where x is the mean number of visits (Straw 1972; Schmitt 1983). Each *Dianthus* flower had a 93.8% probability of being visited at least once at site A (65.2% for nectar feeders) and 57.7% (31.6% for nectar feeders) at site B.

At site A there was no difference in visitation over the season (this assessment was not done at site B): 11.6% of the flowers were visited per hour early in the season, and 8.8% were visited late (Table 1), suggesting that visitation is even throughout the flowering season (during suitable weather). The probability of at least one visit late in the season was 87.9% (all visitors) and 72.0% (nectar feeders only).

Butterflies, the most frequent *Dianthus* visitors, contributed 49.1% (N = 112) of all visits to *Dianthus* flowers. The two hesperiid species, *Ochlodes venata* (Brem. & Grey) and *Thymelicus lineola* (Ochs.), were the most frequent nectar-feeding visitors. Dipteran visitors contributed 46.4% of all visits, with syrphid and anthomyiid flies the dominating groups; both are mainly pollen feeders.

Nectar and Flower Sex Ratio

Standing crop of nectar did not differ between the two sites (ANOVA: DF = 1, F = 0.32, $p > 0.5$). Overall, female stage flowers had significantly more nectar than male flowers (Female: mean = 0.14 μl, SE = 0.02, N = 64; Male: mean = 0.02 μl, SE = 0.01, N = 50; ANOVA: DF = 1, F = 21.51, $p < 0.001$), but there was no overall difference in standing crop of nectar between early- and late-blooming flowers (ANOVA: DF = 2, F = 0.01, $p > 0.9$).

The sex ratio of *Dianthus* flowers did not differ between site A (75.4% males) and B (69.2% males) early in the flowering period when observations were made at both sites ($X^2 = 4.0$, $p > 0.05$). However, the sex ratio changed over the course of the season; male-stage flowers dominated early in the season and female-stage flowers dominated late in the season (Fig. 2) (data from site A).

Seed Set

Significantly more seeds were produced per flower at site A compared to flowers at site B in both 1986 and 1987, but the number of available ovules did not differ between the two sites (Table 2). Thus, significantly more ovules developed into seeds at the mainland site (A). The result was consistent between subgroups; each subgroup at site A produced more seeds than any one of the subgroups at site B (data from 1986) (Fig. 3).

In neither 1986 or 1987 did hand-pollination increase seed set when compared to seed set of control flowers at site A (Figs. 4, 5). At site B, however, hand-pollination increased natural seed set 2.1 times early in the season and 4.1 times late in the season in 1987 (not done in 1986) (Fig. 5). Neither water nor fertilizer increased seed set of hand-pollinated flowers (Fig. 5). Seed set following the three hand-pollinated treatments (hand-pollination, hand-pollination plus water, hand-pollination plus fertilizer) did not differ between the two sites (Fig. 5).

Overall seed set of bagged flowers was lower than seed set of the other treatments, and seed set of bagged

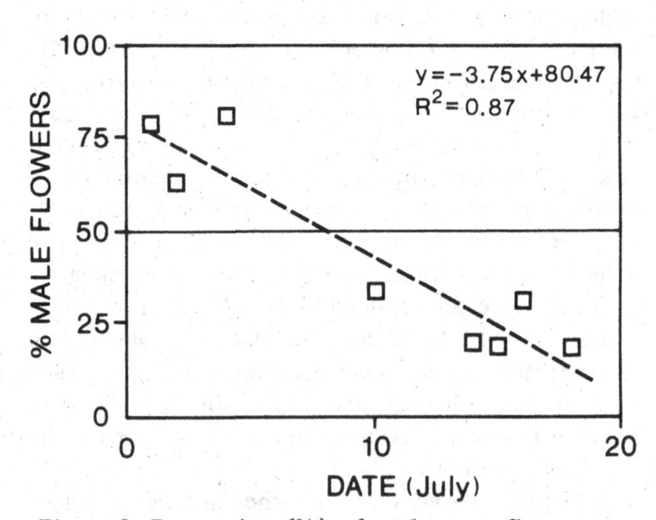

Figure 2. Proportion (%) of male stage flowers during the flowering period of Dianthus deltoides.

Table 2. Mean number of seeds and ovules per capsule in naturally pollinated *Dianthus deltoides* from site A (Mainland) and site B (Island). *** = p < 0.001, NS = p > 0.05 (ANOVA).

		Seeds				
		Mean	*SE*	*Range*	*N*	
1986	Mainland	65.34	2.94	13–95	36	DF = 1
	Island	27.11	2.38	00–67	50	F = 106.04***
1987	Mainland	52.22	3.92	01–101	36	DF = 1
	Island	23.65	4.66	00–93	34	F = 20.72***
		Ovules				
		Mean	*SE*	*Range*	*N*	
1986	Mainland	120.06	2.78	75–152	36	DF = 1
	Island	113.09	2.40	83–158	50	F = 3.59 NS
1987	Mainland	097.06	3.70	17–127	36	DF = 1
	Island	099.97	1.75	73–118	34	F = 0.49 NS

flowers did not differ between site A and site B (Fig. 5). There was, however, a time effect at site B; seed set of bagged flowers early in the season did not differ from the seed set of the other treatments at that time (Fig. 5).

In summary, significant differences between the mainland and the island sites were found for plant and insect diversity, insect visitation rates to *Dianthus*, number of seeds, and percentage of ovules developed into seeds. There were no differences between the sites for standing crop of nectar, ovule number per flower, and seed set following hand pollination (Table 3).

Discussion

The fragmented area exhibited a much lower diversity of flowering plants and anthophilous insects than did the continuous area. This is possibly due to the low percentage of original habitats in the agricultural area, hence the availability of nectar and larval food plants for butterflies as well as nest sites for bumblebees. Furthermore, the residual areas have shallow soils and bare rocks (therefore they are not cultivated), so they are easily affected by drought. "Cornucopian species" (sensu Mosquin 1971) may, however, temporarily attract large numbers of insects to these habitat islands. For example, when *Sedum telephium* began to flower numerous honey bees and bumblebees appeared (personal observation). The interisland area is not a complete barrier to insects but it probably requires rich nectar sources to attract bees to the isolated islands. The

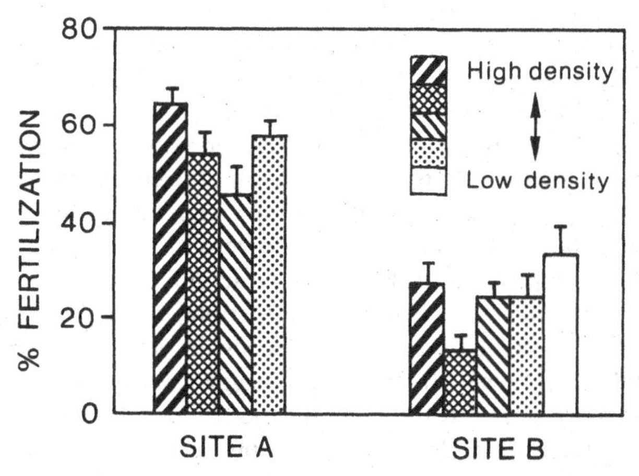

Figure 3. Proportion of ovules that developed into seed in Dianthus deltoides. *Standard error of mean is indicated by a vertical line. Four sample plots from unfragmented habitat (site A) and five plots from fragmented habitat (site B) sorted after descending* Dianthus *flower density within sites. One-way ANOVA; DF = 8, F = 10.99, p < 0.001.*

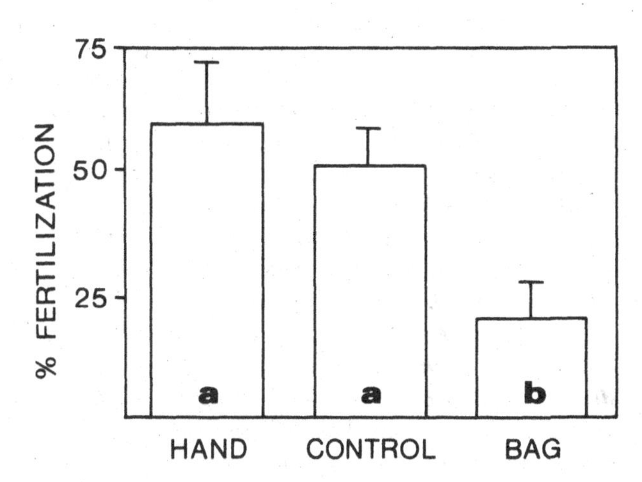

Figure 4. Proportion of ovules that developed into seed (% fertilization) in Dianthus deltoides, *following pollination experiment in 1986 at site A only. Ten plants with at least three stalks were treated in one of three ways: one flower was hand-pollinated (HAND), one was bagged with nylon-mesh net (BAG), and one was left without treatment (CONTROL). Vertical line indicates standard error of mean. Bars with the same letter are not significantly different (Duncan's multiple range test [ANOVA; DF = 2, F = 3.75, p < 0.0398]).*

SOURCE	DF	F	p
MODEL (1987)	19	9.70	0.0001
Site	1	5.05	0.0253
Experiment	4	26.22	0.0001
Date	1	24.63	0.0001
Site × Experiment	4	7.69	0.0001
Site × Date	1	0.06	0.8060
Experiment × Date	4	3.88	0.0044
Site × Experiment × Date	4	1.47	0.2127

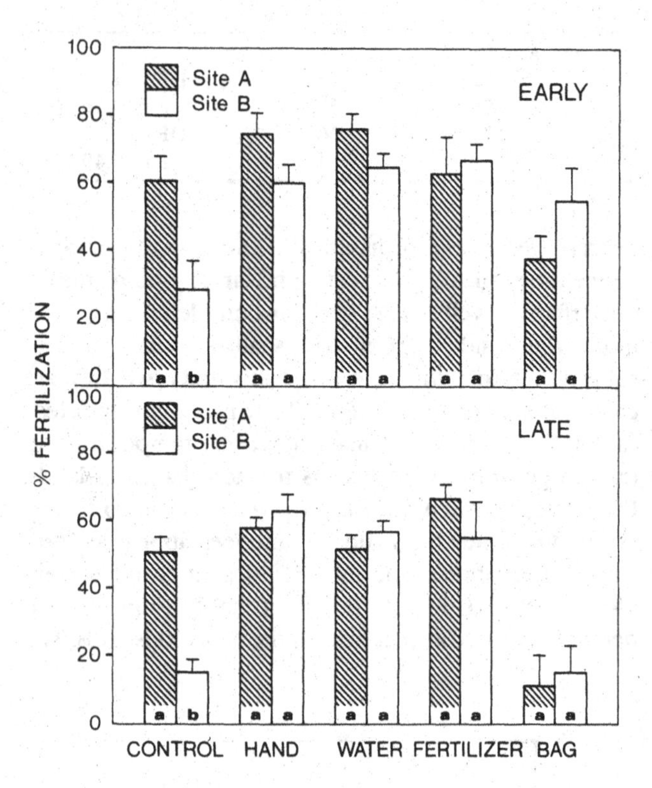

Figure 5. Proportion of ovules that developed into seed (% fertilization) in Dianthus deltoides, *following different experimental treatments in 1987. Plants were hand-pollinated (HAND), hand-pollinated and given water (WATER), hand-pollinated and given liquid fertilizer (FERTILIZER), bagged with nylon-mesh nets (BAG), and not manipulated to provide natural pollinated controls (CONTROL), both at the undisturbed mainland site (A, hatched bars) and at the fragmented island site (B, open bars) early (July 11–14) as well as late (July 22–27) in the flowering season of* Dianthus. *Vertical line indicates standard error of mean. Bars with the same letter are not significantly different (between sites) (Duncan's multiple range test), three-way factorial ANOVA (type III SS):*

diversity and abundance of insects at the island site are therefore typically low, but with high local and temporal variation. This result agrees with Spears (1987), who found that insect abundance on islands varied widely within and between islands, but that on the mainland (Florida) insect abundance was more constant.

Table 3. The ratio of means (site A/site B) for number of visits, standing crop of nectar, number of seeds, number of ovules, and percentage fertilization of naturally pollinated flowers. Statistical tests performed on base data (not on ratios) are Wilcoxon's rank sum test for two samples (number of visits) and ANOVA (nectar and seed set), *** = $p < 0.001$, * = $p < 0.05$, NS = $p > 0.05$.

Source	*Ratio 1986*		*Ratio 1987*	
All visits	1.63	*	—	—
Nectar visits	2.06	*	—	—
Nectar	0.83	NS	—	—
Seeds	2.41	***	2.21	***
Ovules	1.06	NS	0.97	NS
Fertilization	1.71	***	2.36	***

The high proportion of butterflies visiting *Dianthus deltoides* suggests that *Dianthus* is butterfly-pollinated. Several butterfly species carry *Dianthus* pollen grains and furthermore are able to carry more than enough conspecific pollen grains to fertilize all of the ovules of a *Dianthus* flower (Jennersten 1984).

The seed set of *Dianthus* flowers at the mainland site was much higher than at the island site. Since flower sex ratio (availability of pollen), ovule number, and seed set following hand pollination and bagging of flowers did not differ between the two sites, ecological factors probably cause the differences in seed set of control flowers. Furthermore, since hand-pollination increased seed set at the island site, but not at the mainland site, pollen limitation is probably the factor behind the observed pattern (see Bierzychudek 1981). Reduced pollen flow can be explained by reduced pollinator visitation and/or effectiveness (Levin & Kerster 1974; Spears 1987). The low visitation rate by insects is therefore the probable mechanism behind the low seed set of *Dianthus* in the fragmented area, since pollen dispersal in *Dianthus* is dependent on insects.

Reproductive success may also be resource-limited, and full development of ovules may thus not occur (Stephenson 1981; Lee & Bazzaz 1982; Willson & Burley 1983). Differences in soil quality could therefore account for the differences in seed set between the two sites. However, watering and adding of fertilizer did not increase fertilization values at either site, indicating similar reproductive capacity. Hence, resource limitation is not a probable explanation for these differences.

Dianthus is clearly self-compatible, since bagged plants produced seeds (selfed seeds do germinate; unpublished data). The seed set of bagged flowers and the seed set of control flowers did not differ at the island site, suggesting that most seeds produced in the fragmented area result from self-pollination. A protandrous plant like *Dianthus* growing in areas with low pollinator visitation is more likely to produce selfed seeds because self-pollen will remain on the anthers when the flower becomes receptive. In areas with high visitation rates self-pollen will be carried away before receptivity. A high level of selfing will result in reduced genetic variation with time (Darwin 1876; Lee & Hartgerink 1986; Stephenson & Winsor 1986). The island plants might therefore lose genetic diversity, resulting in expression of deleterious alleles that interfere with the reproductive capacity or make the plant more vulnerable to stresses such as disease and climatic change (Ledig 1986). A few long-distance pollinator movements may, however, greatly increase the probability of outcrossing (Schmitt 1983). Thus a few visits by butterflies, acknowledged as potential long-distance pollinators (Schmitt 1980; Courtney, Hill, & Westerman 1982), may greatly increase the probability for outcrossing. All "mainland" flowers had a high probability of being visited, and thus were ensured of high potential for outcrossing and reproduction. Low reproductive success may therefore be one mechanism that causes a lowering in species diversity in fragmented and isolated areas. Studies on seed germination and seedling performance are needed to test for fitness differences of the island and the mainland *Dianthus* population.

Acknowledgments

I thank Johan Åberg, Kajsa Högfeldt, and Catharina Lehman for assistance in the field; Gustav Härd for letting us work on his property; Astrid Ulfstrand for illustrations; Douglass Morse, Sharon Kinsman, and Torbjörn Ebenhard for criticism on earlier drafts; and Eric Menges and Stefan Ås for general discussions. This study was made possible through grants from the Swedish Environmental Protection Board (SNV) and a postdoctoral fellowship from the Swedish Natural Science Research Council (NFR) at the Graduate Program in Ecology and Evolutionary Biology, Brown University, Rhode Island, U.S.A.

Literature Cited

Bierzychudek, P. 1981. Pollinator limitation of plant reproductive effort. *American Naturalist* **117**:838–840.

Carleton, T. J., and S. J. Taylor. 1983. The structure and composition of a wooded urban ravine system. *Canadian Journal of Botany* **61**:1392–1401.

Clapman, A. R., T. G. Tutin, and E. F. Warburg. 1962. *Flora of the British Isles.* Cambridge University Press, Cambridge, England.

Courtney, S. P., S. P. Hill, and A. Westerman. 1982. Pollen carried for long periods by butterflies. *Oikos* **38**:260–263.

Darwin, C. 1876. *On the Effects of Cross- and Self-fertilization in the Vegetable Kingdom.* Murray, London, England.

Diamond, J. M. 1984. "Normal" extinctions of isolated populations. Pages 191–246 *in* M. H. Nitecki, editor. *Extinctions.* University of Chicago Press, Chicago, Illinois.

Franklin, I. R. 1980. Evolutionary change in small populations. Pages 134–150 *in* M. E. Soulé and B. A. Wilcox, editors. *Conservation Biology.* Sinauer, Sunderland, Massachusetts.

Gilbert, L. E. 1980. Food web organization and conservation of neotropical diversity. Pages 11–34 *in* M. E. Soulé and B. A. Wilcox, editors. *Conservation Biology.* Sinauer, Sunderland, Massachusetts.

Handel, S. N. 1983. Pollination ecology, plant population structure, and gene flow. Pages 163–211 *in* L. Real, editor. *Pollination Biology.* Academic Press, London, England.

Janzen, D. H. 1974. The deflowering of Central America. *Natural History* **83**:48–53.

Jennersten, O. 1984. Flower visits and pollination efficiency of some North European butterflies. *Oecologia (Berlin)* **63**:80–89.

Karron, J. D. 1987. A comparison of levels of genetic polymorphism and self-compatibility in geographically restricted and widespread plant congeners. *Evolutionary Ecology* **1**:47–58.

Knuth, P. 1898. *Handbuch der Blutenbiologie.* Verlag von W. Engelmann, Leipzig, German Democratic Republic.

Lande, R. 1980. Genetic variation and phenotypic evolution during allopatric speciation. *American Naturalist* **116**:463–479.

Ledig, F. T. 1986. Heterozygosity, heterosis, and fitness in outbreeding plants. Pages 77–104 *in* M. E. Soulé, editor. *Conservation Biology.* Sinauer, Sunderland, Massachusetts.

Lee, T. D., and F. A. Bazzaz. 1982. Regulation of fruit maturation pattern in an annual legume, *Cassia fasciculata. Ecology* **63**:1374–1388.

Lee, T. D., and A. P. Hartgerink. 1986. Pollination intensity, fruit maturation pattern, and offspring quality in *Cassia fasciculata* (Leguminosae). Pages 417–422 *in* D. L. Mulcahy, editor. *Pollen biotechnology and ecology.* Springer Verlag, New York.

Levin, D. A., and H. W. Kerster. 1974. Gene flow in seed plants. *Evolutionary Biology* **7**:139–220.

Linhart, Y. B., and P. Feinsinger. 1980. Plant-hummingbird interaction effect of island size and degree of specialization on pollination. *Journal of Ecology* **68**:745–760.

MacArthur, R. H., and E. O. Wilson. 1967. *The Theory of Island Biogeography.* Princeton University Press, Princeton, New Jersey.

Mosquin, T. 1971. Competition for pollinators as a stimulus for evolution of flowering time. *Oikos* **22**:398–402.

Proctor, M., and P. Yeo. 1973. *The Pollination of Flowers.* Collins, London, England.

SAS Institute, Inc. 1985. *SAS User's Guide: Statistics.* Version 5 edition. SAS Institute, Inc., Cary, North Carolina.

Schmitt, J. 1980. Pollinator foraging behaviour and gene dispersal in *Senecio* (Compositae). *Evolution* **34**:934–943.

Schmitt, J. 1983. Density-dependent pollinator foraging, flowering phenology, and temporal pollen dispersal pattern in *Linanthus bicolor. Evolution* **37**:1247–1257.

Shaffer, M. L. 1981. Minimum population sizes for species conservation. *Bioscience* **31**:131–134.

Simberloff, D. S., and E. O. Wilson. 1969. Experimental zoogeography on empty islands. *Ecology* **50**:278–295.

Spears, E. E., Jr. 1987. Island and mainland pollination ecology of *Centrosema virginianum* and *Opuntia stricta. Journal of Ecology* **75**:351–362.

Stephenson, A. G. 1981. Flower and fruit abortion: Proximate causes and ultimate functions. *Annual Review of Ecology and Systematics* **12**:253–279.

Stephenson, A. G., and J. A. Winsor. 1986. *Lotus corniculatus* regulates offspring quality through selective fruit abortion. *Evolution* **40**:453–458.

Straw, R. M. 1972. A Markov model for pollinator constancy and competition. *American Naturalist* **106**:597–620.

Wilcove, D. S., C. H. McLellan, and A. P. Dobson. 1986. Habitat fragmentation in the temperate zone. Pages 237–256 *in* M. E. Soulé, editor. *Conservation Biology.* Sinauer, Sunderland, Massachusetts.

Willson, M. F., and N. Burley. 1983. *Mate Choice in Plants: Tactics, Mechanisms, and Consequences.* Princeton University Press, Princeton, New Jersey.

Forest Fragmentation and Alien Plant Invasion of Central Indiana Old-Growth Forests

TIMOTHY S. BROTHERS

Department of Geography
Indiana University
Indianapolis, IN 46202, U.S.A.

ARTHUR SPINGARN

Indiana Department of Natural Resources
Division of Nature Preserves
Indianapolis, IN 46204, U.S.A.*

Abstract: *We assessed alien plant invasion of old-growth forest islands in rural central Indiana by means of paired sample sites along "warm" (south and west) edges and "cool" (north and east) edges in each of seven forest remnants. In general, the fragments appeared resistant to alien invasion; alien species richness and frequency dropped sharply inward from forest edges, and forest interiors were relatively free of aliens. However, aliens were more diverse and more frequent on warm edges than on cool ones. The main factor limiting invasion is probably low light availability in forest interiors, though limited dispersal and low disturbance levels may also be important. We suggest that invasion is discouraged by the edge response of the forests themselves: development of a dense wall of bordering vegetation that reduces interior light levels and wind speeds. The most successful invaders are mostly escaped ornamentals or other species not commonly found in adjoining fields. They can be expected to become more prominent in the future, but their spread at the landscape scale may be slowed by the "oceans" of cropland surrounding these forest islands.*

Resumen: *Hemos evaluado la invasión por plantas foráneas de islas de bosques de antiguo crecimiento en la región rural de Indiana central por medio de sitios de muestreo apareados a lo largo de bordes "cálidos" (sur y oeste) y bordes "frescos" (norte y este) en cada uno de siete remanentes boscosos. En general, los fragmentos parecieron resistentes a la invasión foránea; la riqueza y frecuencia de especies foráneas decayó marcadamente hacia el interior del borde de los bosques, y el mismo se encontraba relativamente libre de especies foráneas. Sin embargo, las especies foráneas eran más diversas y más frecuentes en los bordes cálidos que en los bordes frescos. El principal factor que limita la invasión es, probablemente, la escasés de iluminación en el interior de los bosques; de todos modos, una dispersión limitada y bajos niveles de perturbación pueden también ser importantes. Sugerimos que la invasión es desanimada por la respuesta de los mismos bordes de los bosques: el desarrollo de una densa pared de vegetación fronteriza que reduce los niveles de iluminación interior y la velocidad de los vientos. Los invasores más eficaces son predominantemente ornamentales u ostras especies que comunmente no se encuentran en los campos adyacentes. Se puede esperar que estas especies se hagan más predominantes en el futuro, pero su diseminación al nivel del paisage puede ser retardada por los "océanos" de tierras de cultivo que circundan estas islas de bosques.*

* *Present address: U.S. Environmental Protection Agency, Region III, Philadelphia, PA 19107, U.S.A.*
Paper submitted June 21, 1990; revised manuscript accepted March 8, 1991.

Introduction

A quarter-century ago, John Curtis pointed out that agricultural clearing had fragmented the once-extensive forests of the Midwestern United States into small, scattered islands in a sea of farmland (Curtis 1956). Curtis suggested that the biotas of these forest remnants had been altered not only by direct human disturbances such as logging and grazing but by the indirect effects of fragmentation itself, including increased isolation, decreased size, and habitat changes along forest margins. Forest fragmentation has now become the subject of intensive research. One approach, inspired by the equilibrium theory of island biogeography, has been to examine whether reductions of forest area and increasing forest isolation (so-called "island" effects) have caused local extinctions in forest islands (Helliwell 1976; Weaver & Kellman 1981; Middleton & Merriam 1983; Harris 1984; Wilcove 1987). A second approach, closely related to the first, has been to investigate whether microenvironmental changes caused by creation of forest edges ("edge" effects) have led to shifts in species composition at forest margins (Ranney et al. 1981; Whitney & Runkle 1981; Lovejoy et al. 1986; Alverson et al. 1988; Yahner 1988; Pursell 1989).

This work has emphasized loss of native forest species rather than introduction of aliens. However, forest islands, unlike true islands, are usually surrounded by a sea of potential alien invaders. Forest fragmentation could encourage alien invasions for at least two reasons. First, fragmentation increases the ratio of nonforest to forest and of forest edge to interior (Hill 1985); these changes should increase both the abundance of aliens in the landscape and their average proximity to the remaining forest fragments. Second, microenvironmental changes at forest edges could provide points of entry for aliens, because these are predominantly species of open, disturbed habitats rather than closed forests. The extent of alien invasion might be expected to differ, however, with edge aspect: south and west edges should be warmer and drier and have higher light levels than north and east edges (Wales 1967). (Both of these effects could also favor weedy native species, but we exclude natives from consideration here.)

We examined whether forest fragmentation has encouraged alien plant invasion of old-growth forest remnants in central Indiana. ("Old growth" forests were defined as forests never subject to major human disturbance.) Instead of examining "whole-island" patterns, we focused on local patterns at the ecotone between field and forest edge. The following questions guided our research:

1. How do the abundance and diversity of alien plants vary from forest edge into the interior?
2. Do patterns of invasion differ with edge aspect? In particular, are aliens more diverse or abundant on south and west edges than north and east edges?
3. What environmental factors control alien invasions at forest edges?

Research Design

Forest Selection

The study was restricted to rural areas of the Central Till Plain natural region of Indiana (Homoya et al. 1985). This area, originally dominated by beech-maple forest (Braun 1950), is now largely devoted to agriculture; the remaining upland forests are mostly small woodlots surrounded by fields. The few remaining old-growth stands were identified from information provided by the Indiana Natural Heritage Program.

The following criteria were established for forest selection:

1. Only upland mesic or flatwoods forests would be studied; dry forests and bottomland forests were excluded.
2. Forests must be minimally disturbed. All forests on the Central Till Plain probably have some history of human disturbance, but we chose only forests with no signs of intensive logging, grazing, or clearing. Three of the forests are in fact nature preserves and the others are of similar natural value.
3. Forests must have at least one acceptable "warm" (south or west) edge and one "cool" (north or east) edge, both bordered by cropland. "Acceptable" edges were straight, at least 150 m long (a length dictated by sampling design, as explained below), and had no more than 5 m of bordering woody secondary growth. The age of the edges could not be controlled, but examination of topographic maps and observation of old fences along many edges indicated that none was recently created.

We ultimately selected seven forests, ranging from 8 ha to 23 ha in area. One cool and one warm edge was sampled at each forest. Most edges were bordered by corn or soybean fields; several were fallow and one was planted in clover. Sugar maple (*Acer saccharum*) and beech (*Fagus grandifolia*) dominated the four mesic forest stands. Diversity was higher in the three flatwoods stands, where dominants included sugar maple, basswood (*Tilia americana*), white ash (*Fraxinus americana*), and a mix of oak (*Quercus*) and hickory (*Carya*) species.

Sampling Procedure

At each forest, one study site was randomly located along each edge at least 50 m from either end. In two instances, however, sites were shifted or interrupted to avoid seasonal ponds; this brought one site within about 30 m of the nearest lateral edge.

Study sites consisted of five 50 m belt transects, each 4 m wide and divided into ten 4 m × 5 m quadrats (Fig. 1). Transects were parallel to the edge (defined as the average center line of the trunks of the outermost mature trees) and centered at 2 m outside the edge and 2 m, 8 m, 20 m, and 50 m inside it. (These are referred to hereafter as the exterior, or −2 m, transect and the +2 m, +8 m, +20 m, and +50 m transects.) Exterior transects were beneath the forest canopy overhang but commonly extended across the plow line into the adjacent field. The position of the innermost (+50 m) transect was suggested by Pursell's recent demonstration of an edge response in native understory herbs as much as 40 m from the forest margin at Davis Woods, one of the forests we studied (Pursell 1989). In addition to the transects, two 10 m × 50 m forest overstory plots (termed "edge" and "interior" plots) were overlaid on this transect grid, paralleling the edge and centered at 5 m and 50 m inside it (Fig. 1).

During two sampling periods (June–early July and late August–September 1989) we noted the frequency of all alien species, as well as the frequency of tree fall mounds, canopy gaps, cut stumps, and human-created trails, in all transects. Alien species were defined as species not native to the Central Till Plain natural region, as determined from the *Flora of Indiana* (Deam 1940) and *Plants of the Chicago Region* (Swink & Wilhelm 1979). Doubtful cases were treated as aliens. For example, *Chenopodium album* may include populations of native and European origin; *Campsis radicans* has probably spread northward from southern Indiana since European settlement; and *Solanum carolinense* has apparently migrated into Indiana from the southeastern United States. Alien crop plants were not counted unless they were volunteers. Unknown species were collected for later identification. Of approximately 250 collections, 36 could not be identified to species, but only 10–15 of these (mostly grasses and species of *Polygonum* and *Rumex*) are possible aliens, and their inclusion would not materially change our results.

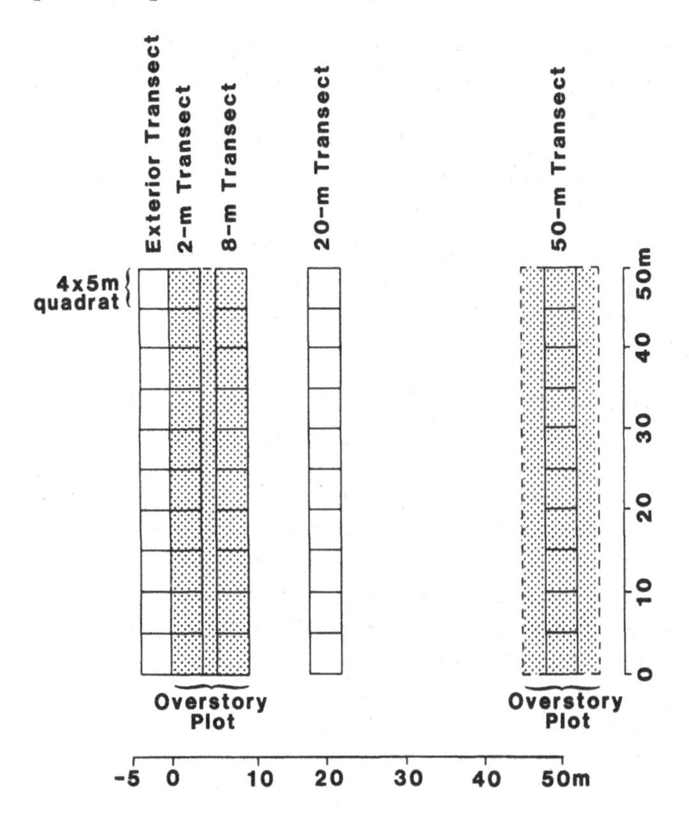

Figure 1. Design of sample plots. Forest overstory plots are stippled.

In the second round of sampling, we recorded the density of all saplings (2.5–10 cm dbh) and the density and basal area of all trees (>10 cm dbh) in the forest overstory plots.

During September 2–18, 1989, microenvironmental measurements were taken at four of the seven forests, selected to provide two edges in each cardinal direction (eight edges total). Measurements and instrumentation were as follows: (1) photosynthetically active radiation at 30 cm above ground level (Licor model 185B radiometer/photometer and model 190SB quantum sensor); (2) air temperature, relative humidity, and dew point at 30 cm height (Psychron model 566 aspirating psychrometer); and (3) soil temperature at 5 cm depth (Atkins model 38642-T thermocouple thermometer).

All measurements were taken under clear skies with scattered clouds, two to three days after a rain. Radiation measurements were made between 10:00 A.M. and 2:00 P.M. EST and temperature and humidity measurements between 2:00 and 3:00 P.M. Measurements were made on both edges of a given forest during the same day; we randomized the edge at which measurements were begun, subject to the condition that two sets of measurements begin on warm edges and two on cool edges. Within transects, measurements were made in stratified random fashion. Temperature and humidity were measured at one location in the 20–25 m quadrat on each transect (a total of five readings per site) and radiation measurements in every quadrat (10 readings per transect, or 70 per site). Environmental measurements (one set of temperature/humidity readings and five radiation readings) were made at the same time in the adjacent field, about 30 paces from the forest edge. Here soil temperature was measured at 5 cm depth as before, but air temperature and relative humidity were measured at approximately 1.5 m above ground level and radiation at approximately 2 m to assess conditions above the crop canopy where possible.

Results

Species Composition

Fifty-eight alien plant species were identified on the transects. Most frequently encountered were dandelion

(*Taraxacum officinale*), multiflora rose (*Rosa multiflora*), goosefoot (*Chenopodium album*), black nightshade (*Solanum nigrum*), and trumpet creeper (*Campsis radicans*) (Table 1). Most of the aliens are common weeds of annual crops, but they also include escaped ornamentals (e.g., *Euonymus fortunei*, *Ligustrum vulgare*, and *Berberis thunbergii*), pasture weeds (e.g., *Rosa multiflora*) and weeds of broad habitat distributions (e.g., *Taraxacum officinale*). All but five species are herbs; no alien trees and only five alien shrubs and vines were encountered.

Patterns Across the Field/Forest Ecotone

Microenvironment changed markedly at the field/forest ecotone. The most dramatic change was in light intensity, which characteristically dropped to less than 1% of full sun at the +2 m transect (Fig. 2). Light levels were low even on exterior (−2 m) transects, which were shaded by the overhanging forest canopy; median light levels on five of the eight exterior transects were less than 2% of full sun. However, light intensity varied by several orders of magnitude within transects because of sunflecks. Temperature fell and humidity rose across the ecotone; two meters inside the forest, mean air temperature was 1.8° C lower, soil temperature 3.7° C lower, and relative humidity 9% higher than in the adjacent field. All of these microenvironmental changes occurred within a few meters from the forest edge; there were no consistent trends in understory environment beyond the 8 m transect.

Overstory density was significantly greater at forest edges than in forest interiors. Combined tree-sapling density was 1399 (±374 SE) stems/ha in edge plots, compared to 877 (±98 SE) stems/ha in interior plots (paired t-test; $t = 4.82$, $p < .001$). Along most edges, trees were obviously asymmetrical, leaning toward the edge and branching more profusely on that side. These canopy responses created a noticeable wall of vegetation at most edges.

Patterns of alien species also changed sharply at the field-forest boundary. Mean alien species richness declined from 11.1 species on exterior (−2 m) transects to 1.5 species at 8 m and remained low beyond (Fig. 3a). Of the 58 alien species encountered during this study, 37 were found only on exterior transects; six others failed to extend beyond the +2 m transects. Frequency of aliens (the percentage of transect quadrats occupied) fell from 86% on exterior transects to 22% at 8 m (Fig. 3b). Fifty meters into the forest, about one quadrat in ten contained aliens.

These data show that the individual frequencies of most species dropped abruptly at or near the forest edge. For example, *Taraxacum officinale*, the single most abundant species encountered, had a frequency of 57% on exterior (−2 m) transects but only 4% at +8 m (Fig. 4a). Most other species dropped out even more rapidly (e.g., *Chenopodium album*, Fig. 4b). However, a few aliens exhibited flatter response curves; most prominent of these were *Rosa multiflora* and *Solanum nigrum* (Fig. 4c–d). Two others, *Alliaria officinalis* and *Euonymus fortunei*, were at least as frequent on interior transects as exterior ones (Table 1).

Differences in the exterior and interior frequencies of individual species were evaluated by chi-square tests, employing a correction for continuity (Siegel 1956). All but 3 of the 20 most common species were significantly more frequent on exterior transects than interior transects at $p < .001$. Of the latter, *Rosa multiflora* was also more frequent on exterior transects, but at a lower significance level ($p < .05$); *Alliaria officinalis* and *Euonymus fortunei* showed no significant preference.

Although herbaceous species dominated the alien vegetation of forest margins, woody species became relatively prominent among the sparse aliens of forest interiors. The mean species richness of alien herbs (annuals, biennials, and perennials combined) fell from 10.1 on exterior transects to 0.7 at 50 m, whereas woody species richness declined only from 1.1 to 0.4. Patterns of alien frequency were similar.

Both light intensity and alien abundance decreased sharply across forest margins, but we found no close relationship between median light level and either alien frequency or alien abundance. Spearman rank correlations, calculated separately for each of the four forests where environmental measurements were made, were nonsignificant in all but one case (Ogden Woods: alien richness vs. light intensity, $r_s = 0.59$; $p < .05$). Correlations remained nonsignificant when all four forests were combined. We did not attempt correlations for individual species because zero frequencies on interior transects produced many tied ranks.

Aspect Patterns

Alien species richness was greater on south and west edges than north and east edges (Fig. 3a). This difference was not significant if exterior transects were included (paired t-test; $t = 1.44$; $p > .05$), but it was highly significant if exterior transects were removed to reduce the effect of variation in agricultural treatment of adjoining fields ($t = 3.65$; $p < .01$). Overall alien frequency was also greater on south and west aspects (Fig. 3b). This difference was highly significant whether or not exterior transects were included (paired t-test [all transects]; $t = 4.99$; $p < .01$).

Many individual species showed apparent aspect preferences (Table 1), largely for "warm" south and west aspects. However, only eight species were present at enough sites to justify statistical tests of aspect differ-

Table 1. Presence and frequency of alien species encountered on transects. Presence is the total number of edges (out of 14) at which each species occurred. Frequency is the percentage of transect quadrats in which each species occurred. Of 700 quadrats sampled, 140 were on exterior transects and 560 on interior transects; 350 were on north or east edges and 350 on south or west edges.

			Frequency			
Species	*Habit*[a]	*Presence*	*Exterior transects*	*Interior transects*	*N/E edges*	*S/W edges*
Taraxacum officinale	P	14	57%	7%	12%	22%
Rosa multiflora	W	9	16	9	7	14
Chenopodium album	A	8	29	3	7	9
Solanum nigrum	A	8	15	4	9	3
Campsis radicans	W	5	19	3	1	11
Poa compressa	P	6	29	0	0	12[b]
Daucus carota	B	7	26	1	tr[c]	11[b]
Lactuca serriola	A	8	19	2	2	9
Setaria faberii	A	5	26	tr	5	6
Stellaria media	A	8	24	1	8	3[b]
Leonurus cardiaca	P	2	14	2	4	5
Rumex obtusifolius	P	5	19	0	4	3
Poa pratensis	P	4	14	0	3	3
Phleum pratense	P	4	14	0	1	5
Plantago major	P	5	12	0	2	3
Digitaria sanguinalis	A	2	11	0	0	5
Triticum aestivum	A	3	8	1	2	2
Euonymus fortunei	W	2	2	2	tr	4
Alliaria officinalis	B	4	1	2	0	4
Glechoma hederacea	P	1	7	tr	3	0
Setaria lutescens	A	3	9	0	0	3
Sida spinosa	A	3	8	0	0	3
Dactylis glomerata	P	2	7	0	3	tr
Polygonum persicaria	A	2	6	0	3	0
Capsella bursa-pastoris	A	1	1	tr	3	0
Abutilon theophrasti	A	3	6	0	1	1
Barbarea vulgaris	B	3	4	tr	1	1
Trifolium pratense	P	2	6	0	2	0
Medicago sativa	P	1	5	0	2	0
Rumex crispus	P	4	4	0	1	1
Bromus inermis	P	2	4	0	tr	1
Arctium minus	B	2	4	0	2	0
Solanum carolinense	P	2	4	0	1	1
Medicago lupulina	A	1	4	0	2	0
Amaranthus hybridus	A	1	4	0	2	0
Cirsium arvense	P	2	3	tr	0	1
Agropyron repens	P	3	4	0	1	1
Portulaca oleracea	A	1	4	0	0	1
Allium vineale	P	1	4	0	0	1
Prunella vulgaris	P	2	3	0	0	1
Commelina communis	A	2	1	tr	0	1
Physalis pubescens	A	2	3	0	1	0
Potentilla norvegica	B	2	2	0	1	0
Verbascum blattaria	B	2	2	0	0	1
Galinsoga ciliata	A	1	1	0	1	0
Pastinaca sativa	B	1	1	0	0	1
Berberis thunbergii	W	1	1	tr	0	1
Veronica officinalis	P	1	1	0	0	tr
Tragopogon dubius	B	1	1	0	0	tr
Ipomoea hederacea	A	1	1	0	0	tr
Atriplex patula	A	1	1	0	0	tr
Bromus tectorum	A	1	1	0	0	tr
Festuca arundinacea	P	1	1	0	0	tr
Plantago lanceolata	P	1	1	0	0	tr
Echinochloa crus-galli	A	1	1	0	tr	0
Setaria viridis	A	1	0	tr	tr	0
Nepeta cataria	P	1	1	0	0	tr
Ligustrum vulgare	W	1	0	tr	0	tr

[a] A = annual, B = binennial, P = herbaceous perennial, W = woody perennial. Growth habit was determined from Gleason (1963), U.S. Department of Agriculture (1971), and Muenscher (1980).
[b] Indicates significant ($p < .05$) aspect difference, as measured by the randomization test for matched pairs (Siegel 1956).
[c] trace

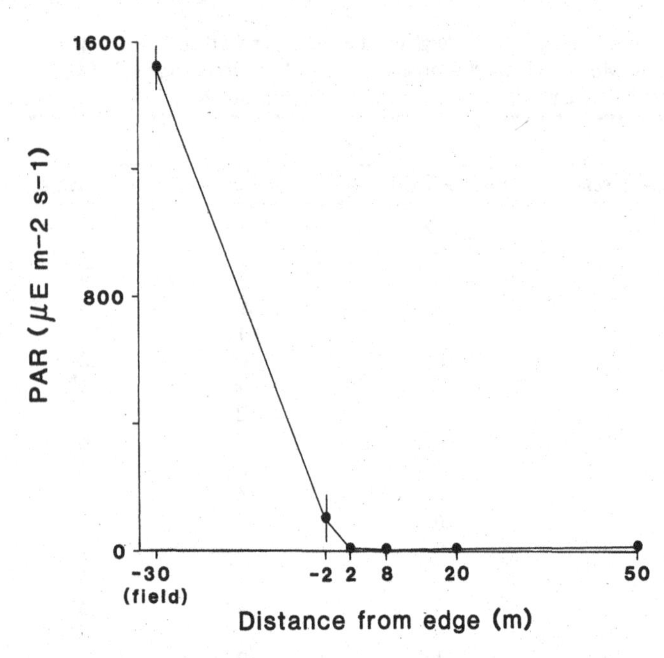

Figure 2. Mean intensity (±SE) of photosynthetically active radiation, by transect position. Each value is the mean of four median light levels (see text). Standard errors for the five interior transects, too small to show, were 3.2 (+2 m), 2.4 (+8 m), 1.9 (+20 m), and 8.8 (+50 m).

ences. Of these, two species (*Poa compressa* and *Daucus carota*) had significant preferences ($p < .05$) for warm aspects and one (*Stellaria media*) a significant preference for cool aspects, as judged by a two-tailed randomization test for matched pairs (Siegel 1956).

Despite these aspect differences in alien abundance, we found no consistent environmental differences across aspects. Overstory density in edge plots was higher on south and west aspects than on north and east, but this difference was not significant (S/W: 1494 ± 156 stems/ha; N/E: 1302 ± 130 stems/ha; paired t-test; $t = 1.33$, $p > .05$).

Disturbance

Nineteen percent of the quadrats on forest-interior transects crossed canopy gaps, tree fall mounds, or trails. However, a chi-square test showed no significant association between the presence of aliens and these combined disturbances (chi-square test; $\chi^2 = 1.50$; $p > .05$). Chi-square results remained nonsignificant even when edge effects were minimized by considering only the two innermost (+20 m and +50 m) transects ($\chi^2 = 2.32$; $p > .05$).

Discussion

The remaining islands of mesic old-growth forest in central Indiana are surrounded by alien agroecosystems, but they have not been swamped by alien species. Most of the aliens at forest margins have moved no more than a meter or two into forest interiors, and none can be considered common relative to native understory species. Aliens are in fact less successful than our frequency data suggest: most aliens encountered in interior transects were small, isolated, nonreproducing plants. *Taraxacum officinale,* the most common alien, was never observed in flower or fruit on interior transects.

Even the smallest forest remnants studied here have thus retained "interior" habitats, considered from the standpoint of alien invasion. This conclusion is consistent with Levenson's (1981) estimate that mesic Midwestern forest islands must remain larger than about 3.8 ha to sustain forest-interior communities. We know of no other studies specifically concerned with alien plant invasions at midwestern forest edges, but data from Gysel (1951) indicate that most aliens in the forest-farm mosaic of southern Michigan are unable to invade mature forests. These results suggest that, given freedom from major human alteration, midwestern mesic forest remnants can exclude most of the presently available alien plants from forest interiors.

The most obvious potential limit to alien establishment in these Indiana forests is low light availability. Late-summer radiation levels near the forest floor, measured at midday under generally clear skies, were generally less than 20 $E\mu\ m^{-2}\ s^{-1}$, or less than 1% of full sun. Such radiation levels are below the compensation points of many native woodland herbs (Sparling 1967) and probably well below the compensation points of most agricultural weeds in surrounding fields (Blackman & Wilson 1951; Grime 1965). The lower abundance of aliens on north and east edges is consistent with this hypothesis, in that light levels should be lowest on north edges. Though alien patterns were poorly correlated with our light measurements, we suspect that the correlations would improve if light availability were monitored during an entire growing season.

The invasion resistance of these forests is probably enhanced by their low level of disturbance. Disturbance can encourage alien invasion of forests by reducing competition for light and other resources, by triggering germination of light-sensitive seeds, and by creating the patches of bare mineral soil many weeds require for establishment (Cross 1975; Harper 1977:133; Lorence & Sussman 1986). We were therefore surprised to find that aliens were not significantly more abundant in tree fall gaps and along trails than they were elsewhere. One explanation may be that individual tree falls and trails do not admit enough light to support germination and growth of the available aliens. In a recent study of factors limiting alien invasion of Michigan forests, Cid-Benevento (1987) showed that artificial disturbance of the forest floor encouraged germination of *Chenopodium album,* but all of the resulting seedlings died

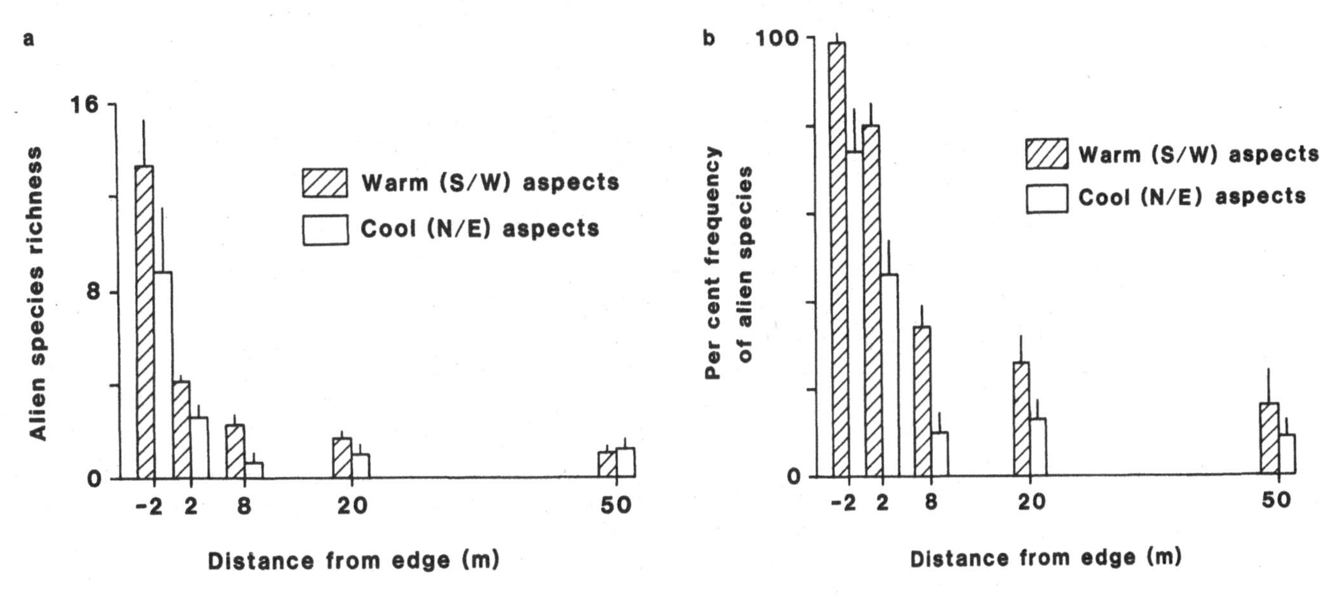

Figure 3. (a) Mean alien species richness and (b) mean frequency of transect segments containing aliens, summarized by transect position and aspect. Vertical lines above bars are standard errors. Sample size = 7 for each combination of transect position and aspect.

without flowering, apparently because light remained inadequate beneath the closed canopy.

On the other hand, aliens may simply be unable to reach suitable disturbed sites. Many crop weeds apparently move only short distances without human assistance, persisting by dispersal in time (long-lived seed) rather than dispersal in space (Holzner et al. 1982). For example, Hume and Archibold (1986) recently showed that natural seed dispersal from a weedy pasture to an adjacent cultivated field in Saskatchewan was essentially limited to the first 7 m beyond the pasture edge, despite the great variety of dispersal mechanisms among the pasture weeds. Dispersal into the forests studied here may be even more limited, at least for wind-dispersed species, because of low wind speeds beneath the canopy. The most prominent aliens of deep interior transects (*Rosa multiflora* and *Solanum nigrum*) are animal-dispersed. Cavers et al. (1979) have suggested that *Alliaria officinalis* is also animal-dispersed, via sticky seeds.

In summary, it seems likely that most of the available aliens are prevented from invading these forests by some combination of low light availability, lack of disturbed substrate, and poor dispersal. Assessment of the relative importances of these factors for individual species will require more detailed study.

It is noteworthy that the most successful alien invaders are not weeds of nearby croplands but plants from other kinds of disturbed sites, such as gardens and pastures. Some, such as *Taraxacum officinale,* would also be expected in croplands, but most are woody perennials (e.g., *Rosa multiflora, Campsis radicans, Euonymus fortunei, Berberis thunbergii,* and *Ligustrum vulgare*) that are poorly adapted to annual cultivation. For these species, the forests are perhaps akin to true islands, accessible only by long-distance dispersal from other forests unless roads or fence rows bridge the gap. As Janzen (1983) has noted, forest fragmentation may in fact help protect forests against invasion if the surrounding cropland is a barrier to shade-tolerant aliens. Firm conclusions about this effect will require examination of the landscape-scale dispersal patterns of the invaders.

We suggest that forest fragmentation does produce another protective response: development of a dense wall of marginal vegetation that armors the forest against invasion by reducing interior light levels and wind speeds. This "self-armoring" effect has been described in other forest-edge studies (Wales 1967; Ranney et al. 1981), but apparently its significance for invasion has not been noted. Aliens themselves contribute to the development of the vegetation wall, creating a negative invasion feedback. Of course, new edges would at first be unarmored and perhaps highly vulnerable to alien invasion. Kapos (1989) found that microclimatic changes extended at least 40 m inward from freshly created edges in Amazonian forest fragments, some 30 m deeper than the effects measured here. The point is that forests may contain species that help repair this damage and maintain the integrity of the community as a whole. Holland and Olson (1989) have recently emphasized the importance of such native "repair plants" in limiting alien invasion of Southern Hemisphere forests.

Our results offer hope for preserve management, because they suggest that small, islandlike forest preserves are not automatically vulnerable to invasion. Nevertheless, we do not believe that alien invasions have reached equilibrium. In Indiana, as elsewhere, alien introduc-

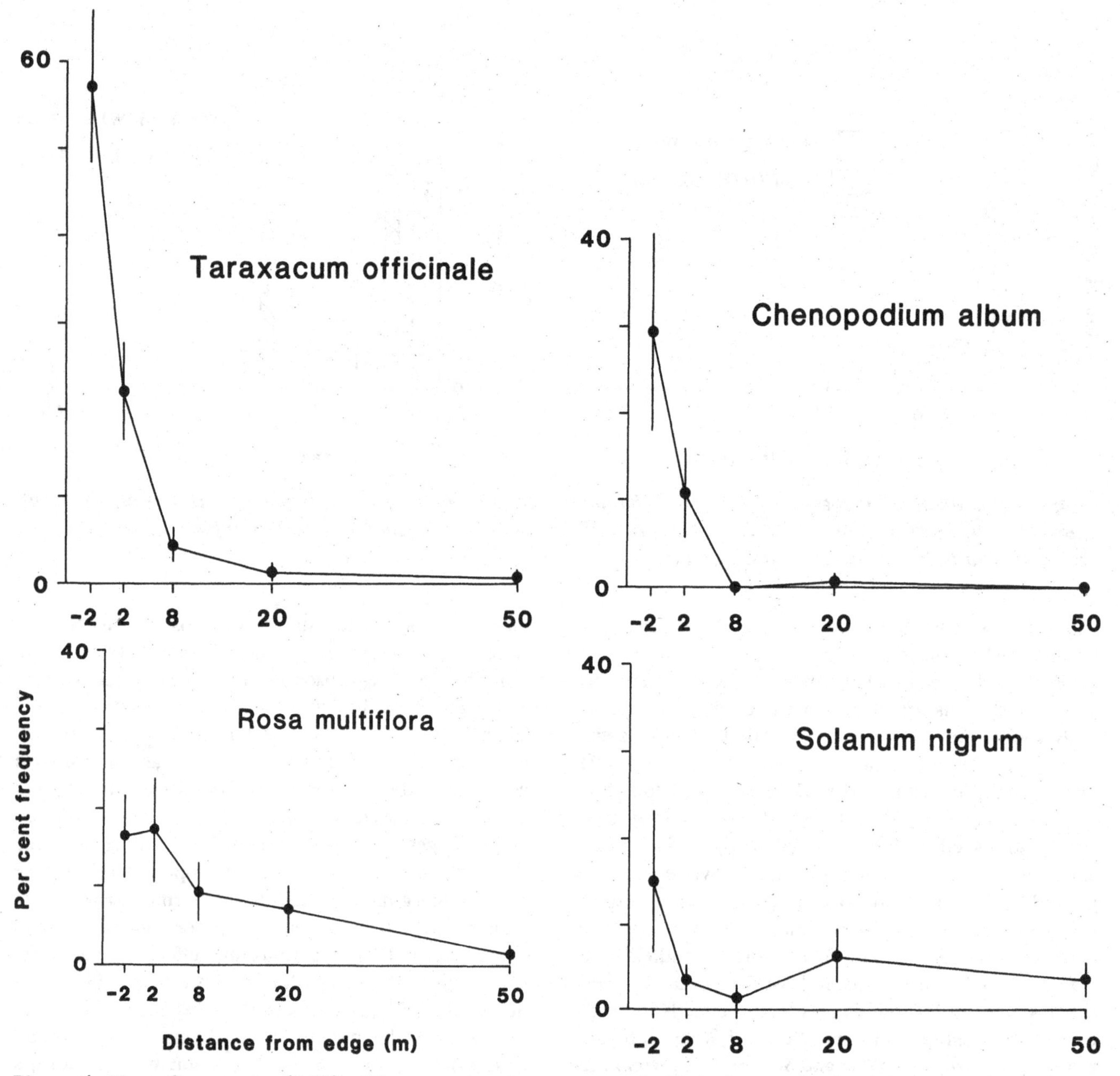

Figure 4. Mean frequencies (±SE) of four commonly encountered aliens, by transect position. Sample size = 14 at each position.

tions have been heavily weighted toward pioneer species (Sauer 1988:89). Future introductions may include shade-tolerant species with a better chance of invading closed forest. Garlic mustard (*Alliaria officinale*) is a case in point. Unknown in Indiana in 1940 (Deam 1940), it is now spreading rapidly into shaded habitats throughout the state (Cavers et al. 1979; Swink & Wilhelm 1979). *Lonicera japonica* and *Lonicera tatarica*, both prominent invaders of urban forest preserves in Indianapolis, may also reach these rural preserves in time. Even the alien heliophytes of adjacent croplands are a latent threat: their seeds may accumulate in these forests as a long-lived seed bank, awaiting only some major natural or human-caused disturbance to spring to life. Finally, other kinds of forests may be more vulnerable to invasion than the mesic forests studied here: for example, dry upland woods, where light availability is greater, or bottomland woods subject to recurrent fluvial disturbance. Future research should examine both the detailed mechanisms of alien invasion and broad patterns of alien movement in a variety of fragmented forest landscapes.

Acknowledgments

This research was supported by a grant from the Indiana Academy of Science. The Indiana Division of Nature Preserves, the Indiana chapter of the Nature Conservancy, Purdue University, and several landowners gave access to study sites. Lisa Kennedy provided excellent field assistance, and Kevin Mickey drafted one of the illustrations. Eric Menges, Mike Homoya, Cloyce Hedge, and two anonymous reviewers suggested improvements in the manuscript.

Literature Cited

Alverson, W. S., D. M. Waller, and S. L. Solheim. 1988. Forests too deer: edge effects in northern Wisconsin. Conservation Biology **2**:348–358.

Blackman, G. E., and G. L. Wilson. 1951. Physiological and ecological studies in the analysis of plant environment VI. The constancy for different species of a logarithmic relationship between net assimilation rate and light intensity and its ecological significance. Annals of Botany, N.S. **57**:63–94.

Braun, E. L. 1950. Deciduous forests of eastern North America. Hafner Publishing Co., New York.

Cavers, P. B., M. I. Heagy, and R. F. Kokron. 1979. The biology of Canadian weeds. 35. *Alliaria petiolata* (M. Bieb.) Cavara and Grande. Canadian Journal of Plant Science **59**:217–229.

Cid-Benevento, C. R. 1987. Distributional limits of old-field and woodland annual herbs: the relative importance of seed availability and interference from herbaceous vegetation. American Midland Naturalist **117**:296–306.

Cross, J. R. 1975. Biological flora of the British Isles. *Rhododendron ponticum* L. Journal of Ecology **63**:345–364.

Curtis, J. T. 1956. The modification of mid-latitude grasslands and forests by man. Pages 721–736 in W. L. Thomas, editor. Man's role in changing the face of the Earth. University of Chicago Press, Chicago, Illinois.

Deam, C. C. 1940. Flora of Indiana. Wm. B. Burford Printing Co., Indianapolis, Indiana.

Gleason, H. A. 1963. The new Britton and Brown illustrated flora of the northeastern United States and adjacent Canada. 3 vols. Hafner Publishing Co. for the New York Botanical Garden, New York.

Grime, J. P. 1965. Shade tolerance in flowering plants. Nature **208**:161–163.

Gysel, L. W. 1951. Borders and openings of beech-maple woodlands in southern Michigan. Journal of Forestry **49**:13–19.

Harper, J. L. 1977. Population biology of plants. Academic Press, London, England.

Harris, L. D. 1984. The fragmented forest: island biogeography theory and the preservation of biotic diversity. University of Chicago Press, Chicago, Illinois.

Helliwell, D. R. 1976. The effects of size and isolation on the conservation value of wooded sites in Britain. Journal of Biogeography **3**:407–416.

Hill, D. B. 1985. Forest fragmentation and its implications in central New York. Forest Ecology and Management **12**:113–128.

Holland, P., and S. Olson. 1989. Introduced *versus* native plants in austral forests. Progress in Physical Geography **13**:260–293.

Holzner, W., I. Hayashi, and J. Glauninger. 1982. Reproductive strategy of annual agrestals. Pages 111–121 in W. Holzner and M. Numata, editors. Biology and ecology of weeds. Dr. W. Junk, The Hague, Netherlands.

Homoya, M. A., D. B. Abrell, J. R. Aldrich, and T. W. Post. 1985. The natural regions of Indiana. Proceedings of the Indiana Academy of Science **94**:245–268.

Hume, L., and O. W. Archibold. 1986. The influence of a weedy habitat on the seed bank of an adjacent cultivated field. Canadian Journal of Botany **64**:1879–1883.

Janzen, D. H. 1983. No park is an island: increase in interference from outside as park size decreases. Oikos **41**:402–410.

Kapos, V. 1989. Effects of isolation on the water status of forest patches in the Brazilian Amazon. Journal of Tropical Ecology **5**:173–185.

Levenson, J. B. 1981. Woodlots as biogeographic islands in southeastern Wisconsin. Pages 13–39 in R. L. Burgess and D. M. Sharpe, editors. Forest island dynamics in man-dominated landscapes. Springer-Verlag, New York.

Lorence, D. H., and R. W. Sussman. 1986. Exotic species invasion into Mauritius wet forest remnants. Journal of Tropical Ecology **2**:147–162.

Lovejoy, T. E., et al. 1986. Edge and other effects of isolation on Amazon forest fragments. Pages 257–285 in M. Soulé, editor. Conservation biology: the science of scarcity and diversity. Sinauer Associates, Sunderland, Massachusetts.

Middleton, J., and G. Merriam. 1983. Distribution of woodland species in farmland woods. Journal of Applied Ecology **20**:625–644.

Muenscher, W. C. 1980. Weeds. 2nd edition. Comstock, Ithaca, New York.

Pursell, F. A. 1989. Long-term change in tree populations and the distribution of plants in the edge of woodlands in eastern Indiana. Purdue University, Lafayette, Indiana. M.S. thesis.

Ranney, J. W., M. C. Bruner, and J. B. Levenson. 1981. The importance of edge in the structure and dynamics of forest islands. Pages 67–95 in R. L. Burgess and D. M. Sharpe, editors. Forest island dynamics in man-dominated landscapes. Springer-Verlag, New York.

Sauer, J. D. 1988. Plant migration: the dynamics of geographic patterning in seed plant species. University of California Press, Berkeley, California.

Siegel, S. 1956. Nonparametric statistics for the behavioral sciences. McGraw-Hill, New York.

Sparling, J. H. 1967. Assimilation rates of some woodland herbs in Ontario. Botanical Gazette **128**:160–168.

Swink, F., and G. Wilhelm. 1979. Plants of the Chicago region. Morton Arboretum, Lisle, Illinois.

U.S. Department of Agriculture. 1971. Common weeds of the United States. Dover, New York.

Wales, B. A. 1967. Climate, microclimate and vegetation relationships on north and south forest boundaries in New Jersey. William L. Hutcheson Memorial Forest Bulletin **2**:1–60.

Weaver, M., and M. Kellman. 1981. The effects of forest fragmentation on woodlot tree biotas in southern Ontario. Journal of Biogeography **8**:199–210.

Whitney, G. G., and J. R. Runkle. 1981. Edge versus age effects in the development of a beech-maple forest. Oikos **37**:377–381.

Wilcove, D. S. 1987. From fragmentation to extinction. Natural Areas Journal **7**:23–29.

Yahner, R. H. 1988. Changes in wildlife communities near edges. Conservation Biology **2**:333–339.

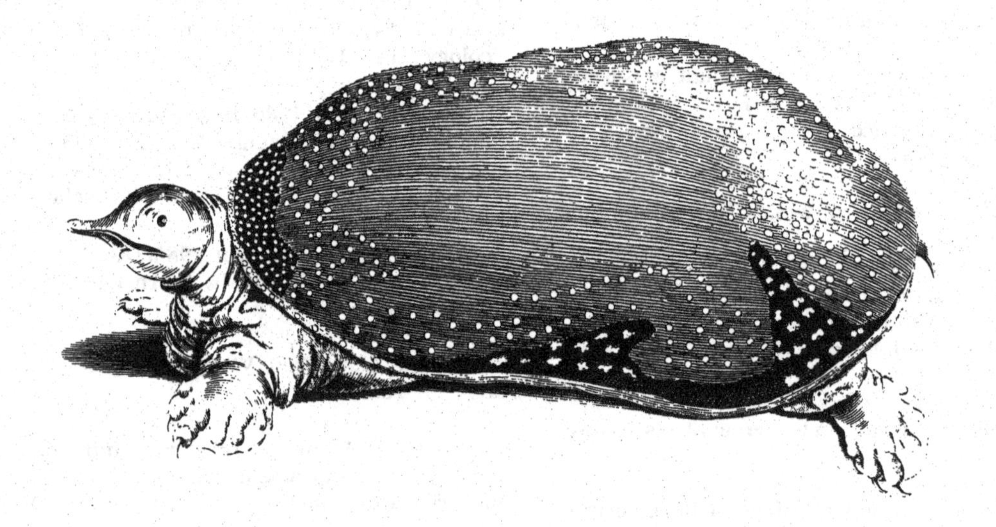

Trends in Landscape Heterogeneity Along the Borders of Great Smoky Mountains National Park

JONATHAN P. AMBROSE*
U.S. National Park Service
Cooperative Studies Unit
Institute of Ecology
The University of Georgia
Athens, GA 30602, U.S.A.

SUSAN P. BRATTON
U.S. National Park Service
Cooperative Studies Unit
Institute of Ecology
The University of Georgia
Athens, GA 30602, U.S.A.

Abstract: *Border areas of Great Smoky Mountains National Park were examined in a photogrammetric study of temporal trends in landscape heterogeneity. Analysis of data from maps and aerial photographs showed that from 1940 to 1978, most areas outside the park borders experienced an increase in forest patch density and cleared patch density, a decrease in forest patch size, and little change in percentage of forest cover. Areas inside the park borders showed an increase in forest patch size and percentage of forest cover, and a decrease in density of forested and cleared patches.*

Significant differences in spatial continuity of vegetation existed between areas inside and outside the park borders at the time of the park's establishment. Land use changes since 1940 have increased differences in landscape heterogeneity between park and nonpark lands. The areas of greatest change have generally been those in closest proximity to the legal park border, where relatively sharp gradients in patch size, patch density, and forest cover now exist. Differences in landscape heterogeneity between adjacent lands outside the park have decreased, as have differences between adjacent park lands.

Local variations in land use history have influenced the nature and intensity of boundary effects along the park border. Different sections of the park border vary greatly in both

* *Correspondence should be sent to this author. Present address: Georgia Department of Natural Resources, Freshwater Wetlands and Heritage Inventory Program, Rt. 2, Box 119D, Social Circle, GA 30279.*

Paper submitted 10/3/89; revised manuscript accepted 10/16/89.

Resumen: *Las áreas al borde del Parque Nacional Great Smoky Montains fueron examinados en un estudio fotogramétrico temporal en la betereogeneidad de paisajes. El análisis de los datos de mapas y fotografías aéreas mostraron que entre 1940 y 1978, la mayoría de las áreas que se encuentran fuera de los límites del parque mostraron 1) un incremento en la densidad de fragmentos de tipos bosques, 2) un incremento en la densidad de fragmentos deforestados, 3) una disminución en el tamaño de fragmentos de tipos bosques, y 4) un cambio mínimo del porcentaje de la cobertura de bosque. Las áreas dentro de los límites del Parque mostraron un incremento en 1) el tamaño de los fragmentos de bosques, 2) el porcentage del tipo de covertura de bosque, y 3) una disminución de la densidad de fragmentos de bosque y deforestados.*

Diferencias significativas en la continuidad espacial de la vegetación existó entre las áreas dentro y fuera de los límites del parque cuando éstos fueron establecidos. Desde 1940, los cambios en el uso de la tierra y betereogeneidad del paisaje ban aumentado entre el área del parque y el área fuera del parque. Las áreas con el mayor cambio ban sido generalmente las más próximas a los límites legales del parque, donde abora existen gradientes relativamente marcados en el tamaño y densidad de los fragmentos, y la covertura del bosque. Diferencias en la betereogeneidad del paisaje entre tierras adjacentes fuera del parque ba disminuido tanto como las diferencias entre tierras adjacentes al parque.

Variaciones locales en la historia del uso de la tierra ban influenciado la naturaleza e intensidad del efecto de borde a lo largo de los límites del parque. Secciones diferentes de

spatial patterns and temporal trends in landscape heterogeneity. The results of the study point to the limited utility of the island biogeographic model in predicting management problems in this park.

los límites del parque varían grandemente en las tendencias de los patrones espaciales y temporales de la hetereogeneidad del paisaje. Los resultados de este estudio señalan el uso limitado del modelo de la biogeografía de islas en predecir problemas de manejo en este parque.

Introduction

One of the most significant recent developments in ecology has been an attempt to understand the spatially dynamic, mosaic nature of ecosystems. Ecologists have long been aware of the importance of change and spatial patterning in natural systems, but the emphasis has historically been on successional development of equilibrium communities. More recently, researchers have turned their attention to disturbance-related phenomena and the spatial heterogeneity of communities. The "patch dynamics" concept presented by Pickett and Thompson (1978) and the similar "shifting mosaic" concept developed by Bormann and Likens (1979) stimulated investigations of interactions within and between relatively discrete units in complex landscapes. The conceptual development of landscape ecology (Forman & Godron 1986) brought together ideas for geography and ecology and emphasized the need to understand the structural and functional interrelationships of spatially heterogeneous combinations of ecosystems. The importance of spatial pattern and temporal change in landscape units has been demonstrated at a variety of scales (Wiens 1976; Pickett & White 1985; Whitcomb et al. 1981; Rudis & Ek 1981; Opdam et al. 1983).

Recent discussions of the patchy nature of landscapes have emphasized the importance of ecotones and corridors as major integrative structural components. Forman and Godron (1981) pointed out that boundaries between landscape elements are relatively distinct. Patch edges or ecotones serve as landscape boundaries and may greatly influence patch properties. Wiens et al. (1985) adovcated a conceptual framework of landscape ecology that focuses on the functional dynamics of landscape boundaries. They stressed the need to look at landscape structures as composites of their constituent units, and to examine boundary phenomena in terms of the spatial dynamics of individuals and populations. They stated further that to understand landscape-level patterns and processes, researchers must understand how boundaries influence exchanges of materials and energy between landscape elements and how these exchanges in turn affect boundary characteristics.

Boundaries are important in both natural and managed landscapes. But boundaries generated by humans are often particularly sharp and functionally significant. Agricultural practices produce abrupt transitions between patches of cropland and adjacent natural communities, affecting the population dynamics of species in both types of habitat patch (Middleton & Merriam 1981). Urbanization causes further disruptions in the distribution patterns of natural communities, often leading to the development of characteristic urban assemblages of plants and animals (Crowe 1979; Whitney & Adams 1980). Roads and utility corridors can isolate populations of forest interior species (Oxley et al. 1974). Where agricultural, industrial, and residential land use zones are in close proximity to relatively undisturbed natural habitats, a complex pattern of communities of different species composition, different successional stages, and different human influences is created.

One of the most important areas for applied research into landscape boundary phenomena may be the field of conservation biology. Evaluations of the effectiveness of conservation areas in preserving habitat and species diversity require assessments of the effects of activities and development outside these preserves on interior communities. Current global rates of habitat destruction and species extinction underscore the need to develop methods for evaluating existing nature preserves and planning future preserves.

The realization that most conservation areas are destined to become isolated fragments of formerly widespread natural habitats has resulted in the application of island biogeographic theory in predicting species extinction rates and area-diversity relationships for nature preserves (Wilcox 1980). However, the relevance of island biogeographic theory to continental nature preserves depends upon the existence of an isolating zone between the preserve and outside areas. Analysis of boundary phenomena at the borders of preserves is necessary to delineate the actual or effective size of these areas for a given species or group of species.

The objective of this study was to determine whether significant changes in landscape heterogeneity have occurred along the borders of Great Smoky Mountains National Park since its establishment. A photogrammetric survey of park borders was conducted to investigate changes in the size and spatial distribution of forested and cleared patches from 1940 to 1978. Data from aerial photographs and timber cruise maps were analyzed to assess changes in the spatial heterogeneity of vegetation inside and outside the park. These trends were compared between sections of the park border, and were interpreted in the context of land use histories of the different border sections. Information from the photogrammetric survey was used to describe and predict the development of ecological boundaries near the park borders.

The Study Area

The Great Smoky Mountains, part of the Appalachian Mountain system, comprise a 115 km segment of the Unaka range, which forms the boundary between the states of Tennessee and North Carolina. The Smokies are separated from the rest of the Unakas by the Pigeon River drainage on the northeast and the Little Tennessee River drainage on the southwest. Great Smoky Mountains National Park (GRSM) contains approximately 209,000 ha of the main divide, lower spur ridges, and foothills of the Great Smoky Mountains.

Establishment of GRSM in the 1930s was an unprecedented event in the history of the National Park Service. This ambitious project required the purchase of over 6,600 individual tracts of land by state and federal governments (Campbell 1960). These tracts included Cherokee Indian reservation lands, private homes and farms, mining properties, and timber company holdings. Other national parks established to that point had been created primarily from lands already owned by the federal government, especially national forest lands; these federal lands were located in areas of low population density. The Smokies, however, contained several small communities and a large number of isolated homesites.

Between 70 and 80 percent of the park's lands have been logged, farmed, or grazed (Pyle 1985). Although some logging of the high-elevation spruce-fir forests occurred, most long-term human disturbance was confined to the lower elevations. Most of these areas, located along the periphery of the park, are presently undergoing succession to mature low-elevation forests. Areas outside the park have experienced significant changes in land usage, including a general decline in agricultral activities and an increase in development for tourism and residential use (Carpenter 1982).

Methods

Information for this survey came primarily from two sources: aerial photographs from the park archives and U.S.G.S. topographic maps. The archives contained black-and-white photographs of park lands from 1939 and 1953 and color photographs from 1978–1982. Unfortunately, the 1939 photographs had so much distortion and variation in scale that they were useless for the purposes of this project. Copies of prepark timber cruise maps prepared by the Tennessee Valley Authority in 1940 were obtained from the park archives. These maps showed the distribution of major forest types and cleared patches in the vicinity of the present-day park borders in the late 1930s and were accurate enough to use in the survey.

Eight areas around the park's periphery were chosen for the survey (Fig. 1). These areas are listed in Table 1, with brief descriptions of their location and land uses. For ease of sampling, relatively straight sections of the border were chosen. For each of the eight park border sections sampled, areas of forested and cleared land were mapped from the aerial photographs to 7.5-minute U.S.G.S. topographic quadrangles. In some cases, the original editions of these quadrangles were used, and information on forested and cleared patches was verified from the aerial photographs. The area covered by this mapping effort included a 3 km wide band on both sides of the legal park boundary.

To measure vegetational continuity from the maps, the border sections were sampled using line intercepts. A set of parallel transect lines drawn on transparent acetate sheets was superimposed over the border areas. These transect lines, spaced 1 cm apart (corresponding to 240 m on the ground), were positioned perpendicularly to the park border. The number of transect lines in each area varied with the length of the boundary section sampled (total = 248 transects; mean = 31; range = 18 to 48).

For each transect, the amount of forested and cleared land was measured within each kilometer-wide section from 3 km within the park (zone -3) to 3 km outside the park (zone 3). The width of each forested or cleared patch encountered by the transect line was recorded, as well as the number of patches of each type, the maximum patch size for each type, and the total percentage of forest cover for that section of the transect. More than 10,500 measurements of forested and cleared patches were made for the eight surveyed sections.

The Friedman test, a nonparametric test of means, was used to determine the statistical significance of block means. In this statistical analysis, year and border zone served as the classification variables. Questions evaluated by the statistical test were: (1) have there been significant changes in the size and spatial distribution of forested and cleared patches within the designated border zones since the late 1930s; and (2) how has the relationship between park lands and outside areas changed since the late 1930s, with respect to vegetational continuity? The Friedman test evaluated the contribution of year and zone interactions to the variation in each of the five variables measured. A statistical significance level of 0.05 was used in each individual test.

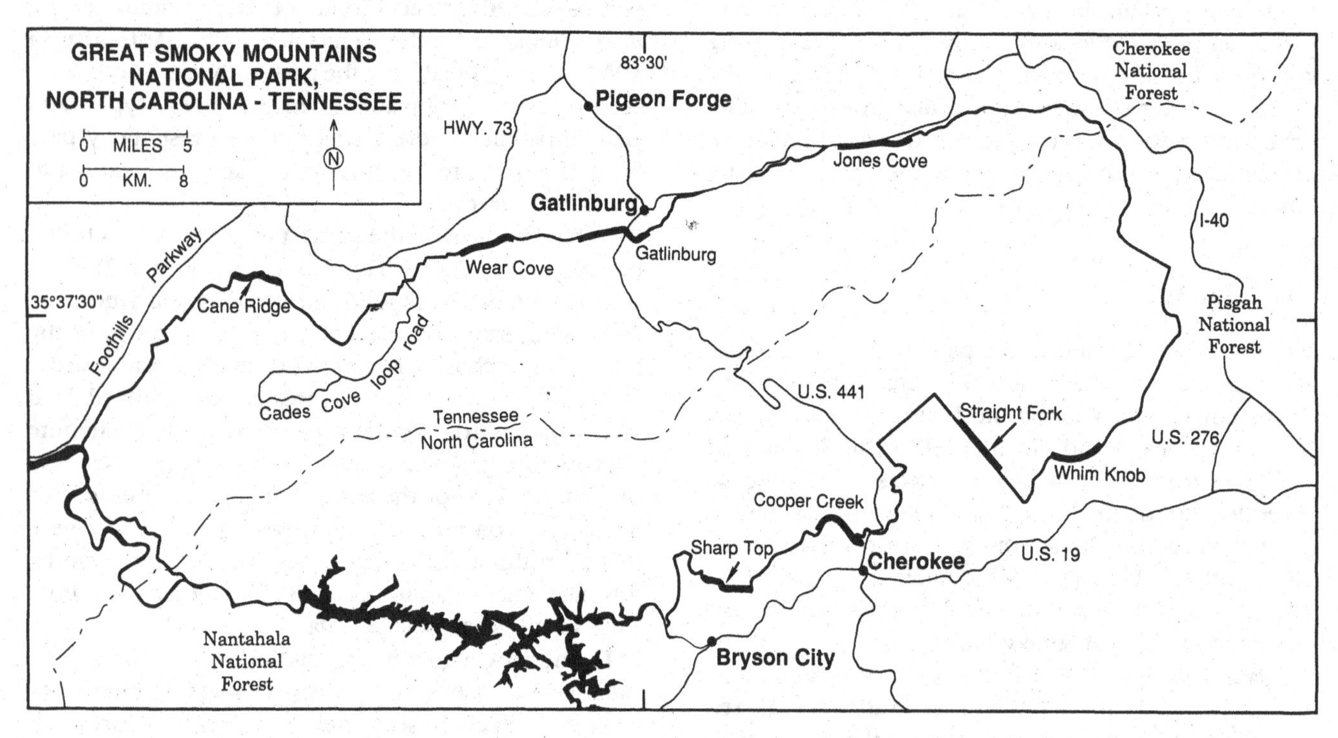

Figure 1. Locations of park border sections sampled in the photogrammetric survey.

Results

Within the park borders, percentage of forest cover and maximum forest patch width increased from 1940 to 1978, while forest patch density, cleared patch density, and maximum cleared patch width decreased (Fig. 2). In contrast, areas outside the park borders showed an increase in forest patch density and cleared patch density, a decrease in forest patch width and cleared patch width, and little change in percentage of forest cover.

These overall trends do not reflect the temporal patterns of forest continuity in every section of the park border, however. Analysis of trends for individual sections of the border indicated that some nonpark areas underwent a brief period of forest patch consolidation (as measured by an increase in forest patch size and total forest cover, and a decrease in patch density) after the park's establishment. Nonpark lands in the Wear Cove section, for example, showed a significant decrease in patch density and an increase in total forest cover from 1940 to 1953. This trend was reversed from 1953 to 1978, indicating a change in land use intensity resulting in forest patch fragmentation. A similar trend was seen in the Cooper Creek section.

Results of the Friedman test of differences across time periods for each zone in each of the eight border sections revealed that the Gatlinburg and Jones Cove sections showed significant differences ($p < .05$) between mean values for 1940, 1953, and 1978. The Gatlinburg section showed a significant increase in forest patch density in zones 1, 2, and 3, a significant increase in cleared patch density and maximum cleared patch width in zones 1 and 2, and a significant decrease in maximum forest patch width in zones 1 and 2 (Fig. 3). The Whim Knob section showed a consistent trend similar to that seen in the Gatlinburg section, though changes in this section were statistically insignificant.

The Jones Cove section showed a trend in forest continuity directly opposite to that of the Gatlinburg area. In this section, all six zones experienced a significant increase in total forest cover and maximum forest patch width, and a decrease in cleared patch density. In addition, zones -3, -2, -1, 1, and 3 showed a significant decrease in maximum cleared patch width, and all zones except zone -2 showed a significant decline in forest patch density (Fig. 3). Thus, this border section has undergone a consistent pattern of forest patch consolidation both outside and inside the park boundaries since 1940. This pattern was also seen in the Cane Ridge section, though the changes were statistically insignificant.

Friedman F-tests of between-zone differences within each time period revealed that changes in forest patch size and distribution have resulted in an intensification of differences across the park border. Table 2 indicates the significance of between-zone differences in forest

Table 1. Locations and descriptions of adjacent land uses for eight sections of the Great Smoky Mountains National Park border.

Boundary section	*Approximate location on U.S.G.S. topographic maps (7.5-minute series)*	*Current land use outside park border*	*Length of park border sampled*
Cane Ridge	Cane Ridge to Ace Gap, Kinzel Springs quadrangle	Private homes and farms; timber lots	5.0 km
Cooper Creek	Cooper Creek to Oconaluftee River, Smokemont quadrangle	Cherokee Indian reservation; tourist developments; private homes and farms	6.5 km
Gatlinburg	Gatlinburg quadrangle (all)	Private homes; rental properties; tourist and recreational developments	12.0 km
Jones Cove	Jones Cove quadrangle (all)	Private homes and farms; tourist and recreational developments	11.5 km
Sharp Top	Sharp Top to Thomas Ridge, Bryson City quadrangle	Private homes and farms; tourist developments	8.5 km
Straight Fork	Straight Fork to Selma Creek, Bunches Bald quadrangle	Cherokee Indian reservation; private homes and farms; timber lots	7.0 km
Wear Cove	Little Roundtop to Little Brier Gap, Wear Cove quadrangle	Private homes and farms; timber lots; tourist developments	7.0 km
Whim Knob	Whim Knob to Little Bald Knob, Bunches Bald and Dellwood quadrangles	Private homes and farms; timber lots	4.5 km

patch density; all other variables measured in this survey showed a similar pattern. The F_r values from these tests indicate that secondary succession inside the park and increasing development outside the park have produced a sharpening of differences between adjacent lands in the border areas. This is true even in the Jones Cove area, where forest patch consolidation has occurred both inside and outside the park borders. Patch consolidation has occurred more rapidly inside the park than in areas just outside the borders; the net result has been an intensification of the gradient in patch size and density across the park border from 1940 to 1978.

Discussion

Results of the photogrammetric survey indicate that, for all variables measured, significant differences existed between lands inside and outside the borders at the time of the park's establishment. Areas just inside the new park borders had more continuous forest cover and less evidence of human habitation. The gradient in disturbance (as measured by patch size and density) across

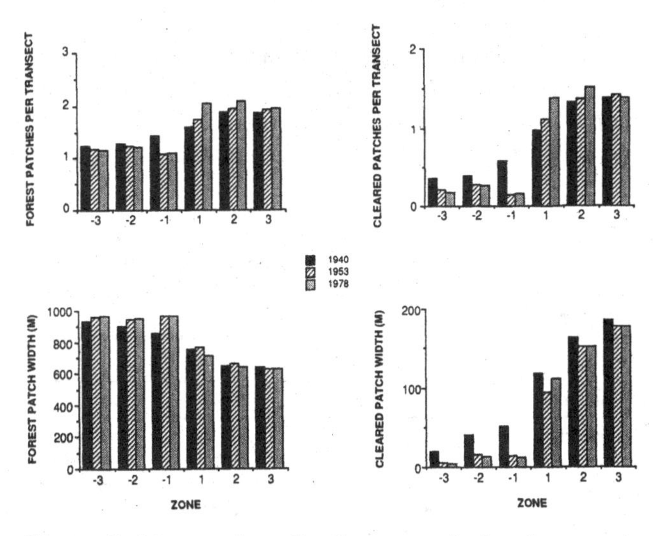

Figure 2. Mean values for forest patch density, cleared patch density, maximum forest patch width, and maximum cleared patch width in eight sections of the Great Smoky Mountains National Park border.

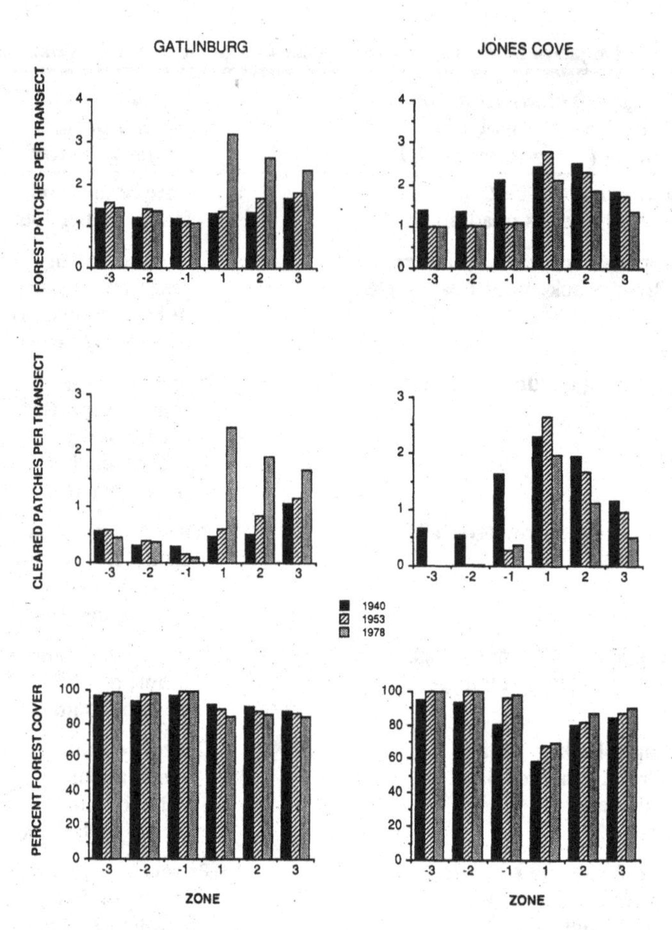

Figure 3. Mean values for forest patch density, cleared patch density, and percentage of forest cover in the Gatlinburg and Jones Cove border sections.

the park border was continuous, however. Values for total forest cover, density of forested and cleared patches, and size of the respective patches described smooth curves along the sampling transects from 3 km within the park to 3 km outside the park.

Patterns of human-caused disturbance along the border of the park have been greatly influenced by topography. Prior to the park's establishment, clearings along the periphery of the Smokies were produced primarily by farming and logging activities. In the low foothills, these cleared areas formed large, convex patches; at higher elevations, the clearings closely followed the narrow valley bottoms, producing a dendritic pattern of elongate, irregularly shaped disturbance patches. Until large-scale logging operations began, few large clearings existed in the steep-sloped central Smokies.

After establishment of the park, the elongate cleared patches at higher elevations became chains of small, disconnected clearings, and then patches of young successional forest. At lower elevations inside the park, the wider valleys underwent succession to large, even-aged stands usually dominated by *Liriodendron tulipifera*. In low-elevation sites outside the park, some consolidation of forest patches occurred between the late 1930s and the early 1950s, resulting in a decrease in the number of forest and cleared patches and an increase in the size of forest patches. In most areas, however, this trend was reversed from the 1950s to the late 1970s, as development for tourism and residential use increased.

The overall trend in vegetational continuity since 1940 has been an intensification of the gradient in patch size, patch density, and total forest cover across the park border. Differences between adjacent zones within the park have decreased, as secondary succession has

Table 2. F values for between-zone comparisons of forest patch density to eight sections of the Great Smoky Mountains National Park border.

	Year		
Section	*1940*	*1953*	*1978*
Cane Ridge	23.19***	25.51***	21.46***
Cooper Creek	8.12 (n.s.)	14.79*	21.20***
Gatlinburg	8.49 (n.s.)	19.46**	52.73***
Jones Cove	38.27***	98.96***	52.61***
Sharp Top	54.14***	74.59***	60.90***
Straight Fork	40.10***	21.34***	41.66***
Wear Cove	10.00 (n.s.)	14.06*	13.50*
Whim Knob	6.40 (n.s.)	9.98 (n.s.)	15.24**

*Tabled values are Friedman test statistic (F_r) with df = 5; * = $p < .05$, ** = $p < .01$, and *** = $p < .001$.*

brought most areas within the border to an equivalent state of landscape heterogeneity. Similarly, areas outside the park are becoming more alike with respect to overall landscape heterogeneity. In most cases, the nonpark areas that have undergone the greatest change in patch size and density have been those closest to the border. Those areas have seen an increase in the number of forest and cleared patches, a decrease in maximum forest patch size, and a slight decrease in maximum cleared patch size. Overall changes in total forest cover since the park's establishment have been very small, both inside and outside the borders.

The two border sections that showed distinctly opposite trends for nonpark lands in this study were the Gatlinburg and Jones Cove areas. These opposite trends are related to the different land use histories of these two areas. While Jones Cove is primarily an agricultural area, Gatlinburg is an example of a "gateway community," characterized by rapid, essentially unplanned growth and an economy based on tourism. Construction of buildings, roads, and utility corridors in this border section has resulted in the fragmentation of large forested tracts. Since land immediately adjacent to the park is in highest demand for vacation home site development, this zone has received the greatest impact on vegetational continuity.

While the number of forest patches and cleared patches has increased significantly in the Gatlinburg area, forest patch width has decreased, and maximum cleared patch width has stabilized at about 100 m. At the same time, little change in total forest cover has occurred (less than 10% change since 1940 in all zones). This indicates that development in the Gatlinburg area has not, in general, produced a number of large cleared areas in the landscape, but has resulted in the creation of a substantial amount of new forest edge.

The effects of this increase in forest edge are more subtle, and potentially more serious, than those caused by large, isolated clearings in the surroundings landscape. The creation of new forest edge affects areas well interior to the path of the cut, by increasing lateral wind and light penetration, altering temperature fields, and increasing evapotranspiration. This produces more xeric conditions in the vicinity of the edge, and promotes the growth of shade-intolerant plants over that of more shade-tolerant forest interior species. These effects may extend as far as 15–30 m inward from the original cut area (Ranney et al. 1981), greatly reducing the area of forest interior habitat. Effects on the reproductive success of forest-interior avifauna may extend much further inward from the patch edge, perhaps as much as 300–600 m (Wilcove et al. 1986).

Levenson (1976) estimated that patches of eastern deciduous forest less than 2.3 ha in size consisted almost entirely of edge, and that a minimum of about 3.8 ha was necessary to maintain a viable forest interior environment. For a circular patch, this would indicate a diameter of 220 m. The average forest patch width in areas closest to the park (zone 1) in the Gatlinburg section was 264 m in 1978. Assuming a minimal 15–30 m edge effect, the average forest patch in zone 1 would consist of 20% to 40% edge habitat. These forest patches probably have little or no suitable nesting habitat for forest-interior songbirds.

Another impact from the creation of more forest edge is the establishment of new corridors for invasive plant exotics such as *Lonicera japonica, Microstegium vimineum,* and *Albizzia julibrissin.* These species may be particularly troublesome where young successional forests within the park border lie adjacent to areas of diffuse disturbance (Ambrose 1987). Thus park lands near Gatlinburg, especially those near roadways, have a relatively high risk of being influenced by exotic plants from areas outside the park.

Only about 25 km east of Gatlinburg, the Jones Cove area shows a completely different trend in landscape heterogeneity. In this section, lands outside the park have been consolidated into larger tracts of unbroken forest cover since 1940. The Jones Cove section showed a significant increase (18% in zone 1) in total forest cover for areas outside the park from 1940 to 1978. Information from aerial photographs indicated that much of this forest patch consolidation was due to secondary succession occurring on abandoned small pastures, crop patches, and logging sites.

Until recently, land use outside the park in the Jones Cove area consisted almost entirely of truck and pasture crops, livestock farms, orchards, and small-scale logging operations. Abandonment of the small agricultural plots since 1940 may be due primarily to changes in agricultural methods; with a heavier reliance today on large farm machinery, cultivation of small isolated fields is no longer cost-efficient or convenient. A general decline in small-scale logging, changes in logging methods (less clearcutting of steep slopes), and an increase in residential use has resulted in the reforestation of some logged-over areas. Nevertheless, large cleared patches remain in areas near the park borders, creating abrupt ecotones between forested park lands and crop land or pasture land.

Within the last decade, the Jones Cove area has experienced a rapid increase in development of lands for residential use and tourism. This development has not, in general, produced a large number of new cleared areas outside the park. Instead, areas already cleared have been converted to other uses, with little change in the total forest cover. Areas outside the park in this section of the border were in better condition (in terms of spatial continuity of vegetation) in 1978 than in 1940. However, improved road access from Interstate Highway 40 and the completed Foothills Parkway will probably stimulate further development in this area. If so, a reversal of the trend toward forest patch consolidation will occur, and expected impacts on the park's biotic communities in this area will be similar to those predicted for the Gatlinburg section.

The Cane Ridge section showed a trend in landscape heterogeneity similar to that seen in Jones Cove. This area has a land use history much like that of Jones Cove; it has also experienced a recent increase in development for commercial and residential uses, so present trends in vegetational continuity outside the park may soon be reversed. This area's proximity to the Foothills Parkway seems to be a major factor in the recent expansion of development. The Sharp Top border section, near Bryson City, North Carolina, showed a response intermediate to that of the Gatlinburg and Jones Cove sections, with evidence of forest fragmentation close to the park, and forest consolidation in areas further away from the park border.

Conclusions

As stated previously, principles from island biogeography have been used to develop design criteria for nature preserves (Diamond 1975; Wilcox 1980), despite a lack of evidence to support the concept of continental parks as islands. One of the greatest problems in applying the island model to continental nature preserves is resolving the question of whether legal borders are congruent with ecological boundaries. As several studies have indicated, conservation problems often arise for the very reason that legal and biotic boundaries are not congruent (Newmark 1985). Secondly, the tendency to focus attention on the park as a whole, and on the exterior borders only, may obscure the importance of ecological boundaries within the park borders. Internal boundaries that subdivide areas within the park may be more important than those along the legal borders. For some forest interior species, nearly any cleared patch of substantial size may constitute an ecological boundary.

White et al. (1985) discussed the application of preserve design criteria based on the "island model" to the preservation of vascular plant species in GRSM. They concluded that island biogeographic theory, while conceptually powerful, was not very useful when applied to the park as a whole. Design criteria specifying optimum size, shape, alignment, and degree of fragmentation had to be evaluated with respect to many other factors, including topographic diversity, species endemism, genetic source areas, and disturbance patterns.

In Great Smoky Mountains National Park, low-elevation areas just inside the park borders have been, and will continue to be, the sites of most human-caused disturbance. Completion of a planned peripheral trail system intended to divert hikers from the Appalachian Trail will increase diffuse human impacts near the park borders. More importantly, expansion of the Foothills Parkway will stimulate further development outside the park borders, causing additional loss of habitat for forest interior species. At the same time, the roadway may serve as a barrier to dispersal of the more mobile park organisms. In short, while there is little chance of GRSM becoming an "island park" in the near future, impacts on species including loss of habitat, direct mortality, and loss of dispersal routes are likely to become much more important in peripheral areas of the park in the coming years.

The nature and intensity of boundary effects at the borders of nature preserves depend not only upon present patterns of land use outside the preserve, but on the entire history of land use changes both within and outside the borders. Because land use histories vary among lands at the borders of nature preserves, different sections of the boundary may vary greatly, both in the intensity of boundary effects and in temporal trends in these effects. For this reason, simple models based on uniform trends of disturbance along the legal border will be inadequate for assessing or predicting the relative isolation of preserve biota. Differences in the nature of boundary effects along and within preserve borders should be investigated to develop management schemes that are adaptable to local conditions and trends.

To accurately assess the effects of land use changes, periodic disturbance, and successional processes on preserve communities, long-term monitoring at a variety of spatial scales is recommended. Photogrammetric surveys of preserve border areas can be used to detect trends in landscape heterogeneity and to locate areas in which these trends indicate a need for concern. Field studies within representative border sections are also necessary, to relate these broad-scale patterns of spatial heterogeneity to patterns and processes observable on the ground. In this way, general trends in boundary effects can be assessed in terms of impacts on local plant and animal communities and on the preserve as a whole.

Acknowledgments

This research was partially funded by the Man and the Biosphere Program of UNESCO, through contact number 18-892 from the United States Forest Service. The National Park Service Cooperative Studies Unit, University of Georgia, furnished additional funds and equipment for the study. The staff of Uplands Field Research Laboratory, Great Smoky Mountains National Park, provided access to maps and other research materials. Special thanks go to research coordinator John Piene, biologist Peter White, archivist Kathleen Manscill, and biological technician Charlotte Pyle for their assistance in this regard. Frank Golley and Peter White gave valuable advice and criticisms during the course of the project.

Literature Cited

Ambrose, J. P. 1987. Dynamics of ecological boundary phenomena along the borders of Great Smoky Mountains National Park. NPS-CPSU Technical Report 34, Institute of Ecology, University of Georgia, Athens, Georgia.

Bormann, F. H., and G. E. Likens. 1979. Pattern and process in a forested ecosystem. Springer-Verlag, New York.

Campbell, C. 1960. Birth of a national park in the Great Smoky Mountains. University of Tennessee Press, Knoxville, Tennessee.

Carpenter, D. E. 1982. Impacts and influences on the Great Smoky Mountains National Park: an annotated bibliography with a discussion and review of selected findings, recommendations, and conclusions. U.S. Department of the Interior, National Park Service, Research/Resources Management Report SER-64.

Crowe, T. M. 1979. Lots of weeds: insular biogeography of vacant urban lots. Journal of Biogeography **6**:169–181.

Diamond, J. M. 1975. The island dilemma: lessons of modern geographical studies for the design of nature preserves. Biological Conservation **7**:129–146.

Forman, R. T. T., and M. Godron. 1981. Patches and structural components for a landscape ecology. Bioscience **31**:733–740.

Forman, R. T. T., and M. Godron. 1986. Landscape ecology. John Wiley and Sons, New York.

Levenson, J. B. 1976. Forested woodlots as biogeographic islands in an urban-agricultural matrix. Ph.D. dissertation. University of Wisconsin, Milwaukee, Wisconsin.

Middleton, J., and G. Merriam. 1981. Woodland mice in a farmland mosaic. Journal of Applied Ecology **18**:703–710.

Newmark, W. D. 1985. Legal and biotic boundaries of western North American national parks: a problem of congruence. Biological Conservation **33**:197–208.

Opdam, P., H. van Dam, S. M. ten Houte de Lange, J. T. R. Kalkoven, F. Kraght, and A. H. P. Stumpel. 1983. A comparative study of spatial patterns in a landscape. Annual Report 1982, Research Institute of Nature Management, Leersum, Netherlands.

Oxley, D. J., M. B. Fenton, and G. R. Carmody. 1974. The effects of roads on populations of small mammals. Journal of Applied Ecology **11**:51–59.

Pickett, S. T. A., and P. S. White. 1985. The ecology of natural disturbance and patch dynamics. Academic Press, Orlando, Florida.

Pyle, C. 1985. Vegetation disturbance history of Great Smoky Mountains National Park: an analysis of archival maps and records. USDI, NPS-SER Research/Resources Management Report SER-77.

Ranney, J. W., M. C. Bruner, and J. B. Levenson. 1981. The importance of edge in the structure and dynamics of forest islands. Pages 67–95 in R. L. Burgess and D. M. Sharpe, editors. Forest island dynamics in man-dominated landscapes. Springer-Verlag, New York.

Rudis, V. A., and A. R. Ek. 1981. Optimization of forest island spatial patterns: methodology for analysis of landscape pattern. Pages 241–256 in R. L. Burgess and D. M. Sharpe, editors. Forest island dynamics in man-dominated landscapes. Springer-Verlag, New York.

Whitcomb, R. F., C. S. Robbins, J. F. Lynch, B. L. Whitcomb, M. K. Klimkievitz, and D. Bystrak. 1981. Effects of forest fragmentation on avifauna of the eastern deciduous forest. Pages 125–206 in R. L. Burgess and D. M. Sharpe, editors. Forest island dynamics in man-dominated landscapes. Springer-Verlag, New York.

White, P. S., R. I. Miller, and S. P. Bratton. 1985. Island biogeography and preserve design: preserving the vascular plants of Great Smoky Mountains National Park. Natural Areas Journal **3**:4–13.

Whitney, G. G., and S. D. Adams. 1980. Man as a maker of new plant communities. Journal of Applied Ecology **17**:431–448.

Wiens, J. A. 1976. Population responses to patchy environments. Annual Review of Ecological Systematics **7**:81–120.

Wiens, J. A., C. S. Crawford, and J. R. Gosz. 1985. Boundary dynamics: a conceptual framework for studying landscape ecosystem. Oikos **45**:421–427.

Wilcove, D. S., C. H. McClellan, and A. Dobson. 1986. Habitat fragmentation in the temperate zone. Pages 237–256 in M. Soulé, editor. Conservation biology: the science of scarcity and diversity. Sinauer Associates, Sunderland, Massachusetts.

Wilcox, B. A. 1980. Insular ecology and conservation. Pages 95–117 in M. E. Soulé and B. A. Wilcox, editors. Conservation biology: an evolutionary-ecology perspective. Sinauer Associates, Sunderland, Massachusetts.

An Ecological Evaluation of Proposed New Conservation Areas in Idaho: Evaluating Proposed Idaho National Parks

R. GERALD WRIGHT

National Park Service Cooperative Unit
Department of Fisheries and Wildlife Resources
University of Idaho
Moscow, ID 83843, U.S.A.

JAMES G. MACCRACKEN

College of Forestry
Wildlife and Range Sciences
University of Idaho
Moscow, ID 83843, U.S.A.

JOEL HALL

Department of Fisheries and Wildlife Resources
University of Idaho
Moscow, ID 83843, U.S.A.

Abstract: *Four areas in the state of Idaho, U.S.A., have been proposed by various interest groups to be designated as national parks. The four areas average 220,000 ha and contain important biological, scenic, recreational, and geological resources. However, the biological resources that would be protected by these proposals have received relatively little consideration. We used the U.S. Fish and Wildlife Service's Gap Analysis Project databases to evaluate the vegetation types contained in each proposal. The scale of analysis was an ecoregion with the proposals falling within three of six ecoregions that encompass Idaho. Databases included vegetation type, land ownership, and land protection status, which were analyzed using a geographic information system. Vegetation types were used as surrogates for information on the distribution of other biological resources (i.e., biodiversity). A conservation strategy was evaluated that would preserve at least 10% of each vegetation type in an ecoregion. Only 15% of the vegetation types in the three ecoregions were protected under this criterion. The four pro-*

Paper submitted August 24, 1992; revised manuscript accepted July 12, 1993.

Una evaluación ecológica de nuevas áreas de conservación propuestas en Idaho: Evaluando parques nacionales propuestos en Idaho

Resumen: *Cuatro nuevas áreas en el estado de Idaho, U.S.A., han sido propuestas, por distintos grupos intereresados en el tema, para ser designadas como Parques Nacionales. Las cuatro áreas promedian 220.000 ha y contienen recursos biológicos, escénicos, recreacionales y geológicos importantes. Sin embargo, los recursos biológicos que serían protegidos por esas propuestas han recibido poca atención. Nosotros usamos la base de datos del "Gap Analysis Project" del "U.S. Fish and Wildlife Services" para evaluar los tipos de vegetación contenidos en cada propuesta. La escala de análisis fue la de eco-región con las propuestas contenidas en tres de las seis eco-regiones que abarcan Idaho. La base de datos incluyó tipos de vegetación, tenencia de tierras y el estado de protección de las tierras, tales datos fueron analizados usando sistemas de información geográficos. Los tipos de vegetación fueron usados como substitutos de información sobre la distribución de otros recursos biológicos (i.e., biodiversidad). Se evaluó una estrategia de conservación que preservaría por lo menos el 10% de cada tipo de vegetación en cada ecoregión. Sólo el 15% de los tipos de vegetación en*

posals added little to this figure because 67–78% of the vegetation types found in the proposals were already protected elsewhere in each ecoregion. Between 16–30% of the land areas in the proposals provided protection for vegetation types that were previously unprotected. We increased the size of each proposal by including vegetation types in a 15 km zone around each proposal in an attempt to enhance protection. None of the proposals could be feasibly modified to reach the 10% protection goal for all of the vegetation types. However, the protection provided by each proposal could be enhanced to varying degrees with the addition of relatively few hectares to their land area. Although national parks throughout the world play an important role in the conservation of biodiversity, this attribute is often accidental, and as our analysis showed, more attention needs to be devoted to biological data in the selection and design of new parks.

tres ecoregiones se encontraban protegidos actualmente bajo este criterio. Las cuatro propuestas contribuyeron poco a esta cifra debido a que entre un 67–78% de los tipos de vegetación encontrados en las propuestas se hallaban ya protegidos en otras áreas dentro de cada ecoregión. Entre un 16–30% de las áreas en las propuestas proveía protección a tipos de vegetación que no estaban protegidos en el pasado. Nosotros incrementamos el tamaño de cada propuesta incluyendo tipos de vegetación en una zona de 15 km alrededor de cada propuesta a los efectos de mejorar la protección. Ninguno de las propuestas pudo ser modificada feasiblemente para alcanzar la meta de un 10% de protección para todos los tipos de vegetación. Sin embargo, la protección provista por cada propuesta podría ser mejorada en grados diversos mediante la adición de unas (relativamente) pocas hectareas a sus áreas territoriales. Si bien los parques nacionales alrededor del mundo fuegan un rol importante en la conservación de la biodiversidad, este atributo generalmente es accidental y como muestra nuestro análisis se debe prestar más atención a los dátos biológicas en la selección y diseño de nuevos parques.

Introduction

The conservation of biological diversity will require a reserve system that is efficient in terms of maximizing resource representation within a framework of limited opportunities to devote lands to such a purpose. In most areas of the world, "national parks," or more specifically Protected Area categories I, II, and IV of the Commission on National Parks and Protected Areas of the International Union for the Conservation of Nature and Natural Resources (IUCN 1984), provide the greatest protection for biodiversity.

The concept that parks should represent the range of regional biophysical diversity has been advocated as a strategy for all parks (IUCN 1980), for biosphere reserves (Gregg & McGean 1985), and for units of the United States (U.S.), National Park Service (NPS) (National Park Service 1972, 1990). However, the attainment of that goal has been elusive. In the U.S., Crumpacker et al. (1988) found that 24% of Kuchler's (1964) potential natural vegetation types were not adequately represented on federal and Indian lands. Furthermore, on NPS lands they found that representation of 33% of those vegetation types was lacking.

There are two fundamental reasons for this situation. The first is that many land use classifications including national parks can only be designated through national legislation. As a result, factors such as opportunity, aesthetics, administrative costs, local economic impacts, surrounding land uses, and political support are often more important in park establishment than resource considerations and system-wide plans (Wright 1984; Conservation Foundation 1985; Margules 1989; MacCracken & O'Laughlin 1992). As a result, the boundaries for parks rarely coincide with the landscape elements necessary to maintain ecological processes (Noss 1987*a*; Wright 1992).

Second, the development of procedures to meet conservation goals has been limited. Most have been theoretical in nature and rarely applied in real world situations. For example, the NPS identified major physiographic and biologic zones and accompanying resources with the goal of having each represented within the NPS system (National Park Service 1972, 1990). However, this information has seldom been used to guide park establishment (Mackintosh 1991).

Klubnikin (1979) analyzed the distribution of vegetation types in California, U.S.A., relative to the location of parks and preserves and found that most areas tended to preserve only one or two vegetation types. Noss and Harris (1986) and Harris and Gallagher (1989) described a reserve system based on connecting landscape elements using coastal corridors and inland riparian habitats. Australian park planners and ecologists have used automated mapping systems and vegetation cover data to develop conservation strategies for large areas (Bolton & Specht 1983; Purdie et al. 1986; Pressey & Nicholls 1991). Dinerstein and Wikramanayoke (1993) developed an untested strategy, based on conservation potential and potential threats, to prioritize the establishment of reserves in several countries in the Indo-Pacific region. Kirkpatrick (1983) used plant species distribution data to prioritize the selection of natural reserves in Tasmania.

The purpose of this paper is to illustrate a method used to address the contribution of specified areas in Idaho to the conservation of biodiversity. All of the areas analyzed have recently been proposed for national park designation and presumably, along with other areas, would form the basis for a biodiversity reserve system.

The data for each area were analyzed at a regional scale in relation to a specific conservation goal, land ownership–resource protection, and existing land areas that already contributed to meeting the conservation goal. The specific objectives of this study were to: (1) summarize the vegetation types contained in each proposal and the associated ecoregion; (2) contrast the level of resource protection provided by the proposals with that needed to achieve the conservation goal; and (3) modify the proposals if necessary to attempt to reach the conservation goal.

Background

The NPS administers four units in Idaho. All but one are relatively small, and all make little contribution to the preservation of the biodiversity in their associated ecoregions. MacCracken and O'Laughlin (1992) identified and described four other areas that have been proposed for national park status and have significant public support (Fig. 1). Each area is recognized for its important aesthetic, geologic, biological and cultural resources. The following is a brief description of each proposal.

Craters of the Moon National Park and Preserve would be created by expanding the existing 21,686 ha monument approximately six-fold. This area includes significant examples of plains volcanism (Greeley 1977) and supports sagebrush–wheatgrass (*Artemisia* spp.–*Agropyron* spp.) vegetation. This proposal and four alternatives were the subject of a NPS study (NPS 1989).

The Hells Canyon National Park–Chief Joseph National Preserve–Snake River Breaks National Recreation Area is a 123,200 ha proposal centered around the current Hells Canyon National Recreation Area and includes lands in Idaho, Oregon, and Washington (Hells Canyon Preservation Council 1992). Hells Canyon is the deepest river carved gorge in North America. Due to extensive elevation gradients and diverse topographic features the area supports many different ecosystems as well as significant archaeological resources. At the time of our analysis, the available data were limited to the Idaho portions of this proposal.

Since 1911 many attempts have been made to designate a national park centered around the Sawtooth Mountains in central Idaho. The 410,764 ha Sawtooth National Park and National Recreation Area proposal was the result of an NPS study in the early 1970s (NPS 1975). Resources include high elevation rugged mountains, spectacular scenery, and abundant wildlife.

Informal proposals have been made for a national park of 128,324 ha in the Owyhee canyonlands of southwestern Idaho (Crosby 1991). These lands have been recommended for wilderness designation (Bureau of Land Management 1989, 1991). The Owyhee river and its tributaries dissect a high elevation desert plateau resulting in steep, isolated, and very scenic canyons. The area supports sagebrush–wheatgrass vegetation as well as riparian plant communities and an isolated population of mountain sheep (*Ovis canadensis californicus*).

Methods

Four different databases, discussed below, defined ecoregions, vegetation types, land ownerships, and land status. The data were stored, organized, and analyzed using a geographic information system (GIS).

Natural ecoregions as defined by Omernik (1986) were our primary scale of analysis. Ecoregions are considered to be objectively defined geographic units whose ecosystem properties are relatively homogeneous in terms of climate, topography, and vegetation (Bailey 1976). Idaho contains six ecoregions and the park proposals fell into three of those (Fig. 1). Craters of the Moon and the Owyhee canyonlands are in the Snake River Basin/High Desert Ecoregion, the Hells Canyon area is in the Blue Mountains Ecoregion, and the Sawtooth Mountains are in the Northern Rockies Ecoregion.

The existing vegetation of Idaho was mapped at a scale of 1:500,000. This effort used data from several sources to describe the 26 vegetation types that occurred in the three ecoregions that encompassed the four proposals (Appendix 1) (Scott et al. 1993).

Discrete vegetation types are often used as surrogates for information on the distribution of plant and animal species because they are more easily derived and more widely available (Pressey 1990). We assumed that by preserving a sufficient land area of a representative set of vegetation types the protection of a majority of all species would be assured. Some authors have indeed suggested that preserving representative plant communities could protect up to 85% of all species. However, this figure is probably overly optimistic and our strategy essentially represents a "coarse filter" that must be complemented by a species-specific strategy to conserve uncommon species (Noss 1987*a*).

A land ownership database was developed by the Idaho Department of Water Resources (IDWR) at a scale of 1:100,000. Ownerships by all federal and state agencies were combined into one class resulting in two broad categories of land ownership, private and public. We assumed that policy changes in land use designed to enhance biodiversity protection would involve similar actions on all public lands.

Land status maps for Idaho were produced at a 1:100,000 scale by IDWR and three classes were identified: protected public lands, unprotected public lands, and unprotected private lands (Scott et al. 1993). Pro-

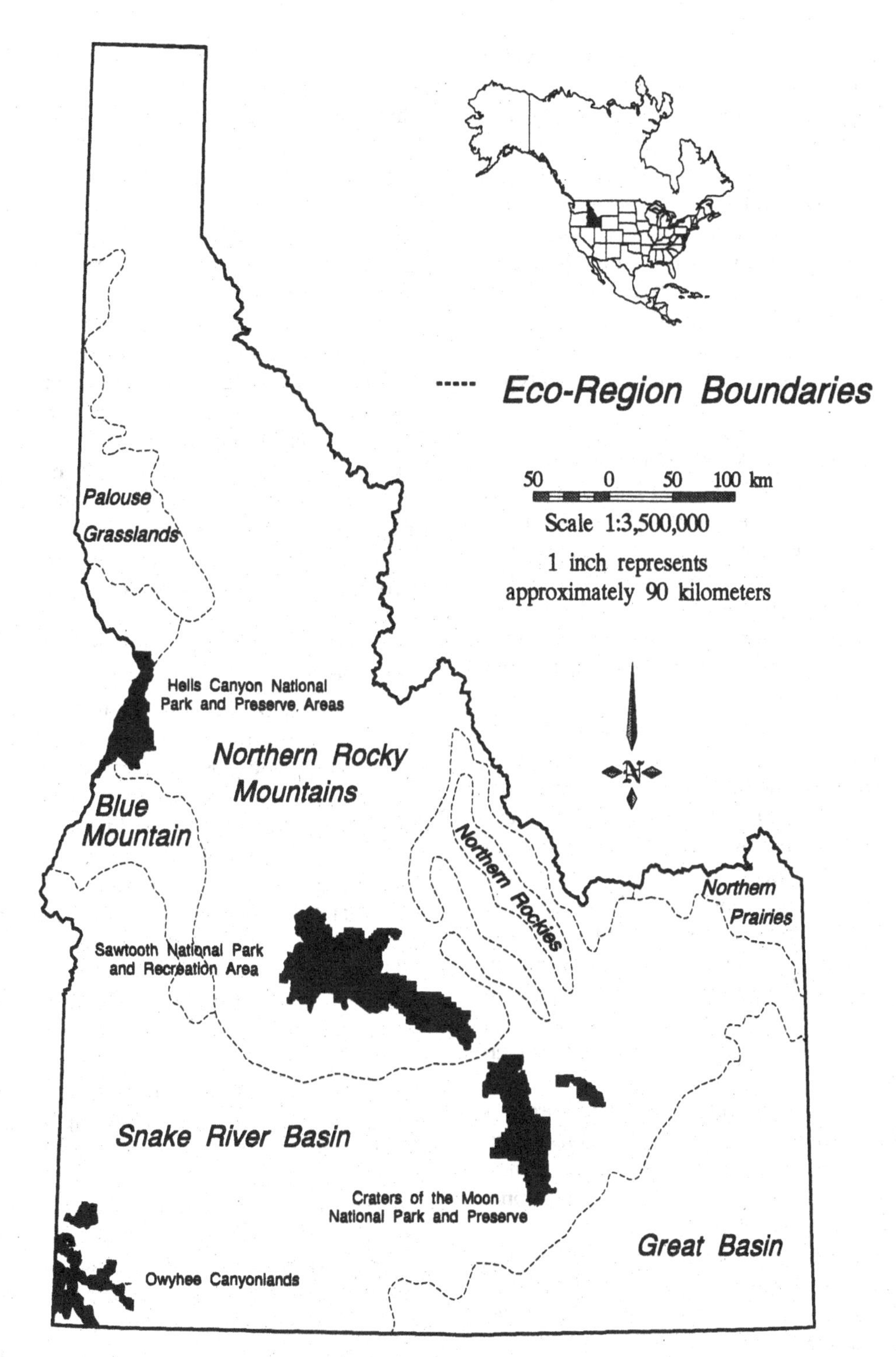

Figure 1. Map of the four national park proposals in Idaho and the boundaries of the ecoregions in which they are located.

tected public lands included those in which most resource extraction (e.g., mining and timber harvest), development (e.g., extensive road systems and facilities), and habitat manipulation (e.g., artificial alteration of native plant communities) is prohibited by law or agency policy. Unprotected public lands were those where consumptive resource use and development are permitted according to various laws and regulations.

All privately owned lands were assumed to have few legal mandates to protect them and were generally unavailable as part of a biodiversity reserve system. This assumption did not take into account the fact that a small percentage of private lands in Idaho are owned and protected by private conservation foundations.

Within each ecoregion, we evaluated a conservation strategy designed to preserve each vegetation type in proportion to its occurrence. Criteria for determining the size of individual reserves include the maintenance of self-sustaining populations of particular species and how many species the area can support (Diamond 1986). Guidelines for the amount of land that should be devoted to a reserve system have been less conclusive, although the International Union for the Conservation of Nature and Natural Resources recommended that 10% of a nation's land area be protected (Miller 1984). Our goal in this analysis was to protect a minimum of 10% of the land area occupied by each vegetation type in each ecoregion. We assumed that all vegetation types had equal importance.

Analyses followed the steps described below.

(1) Land ownership, vegetation, and land status data layers were organized and defined based on the extent of the three ecoregions encompassing the four national park proposals. Only data for the Idaho portion of those ecoregions were available.
(2) The percentage of land area occupied by each vegetation type, land ownership, and land status was calculated for each ecoregion.
(3) Maps of the four proposals were georeferenced and digitized.
(4) The percentage of land area occupied by each vegetation type within the proposals was calculated.
(5) The land area needed by each vegetation type in each ecoregion to meet the 10% goal was calculated. This figure was compared with the area occupied by each vegetation type in a proposal and with areas occupied by each vegetation type already in protected land status. This comparison allowed an assessment of the level of protection afforded a vegetation type and the extent to which a proposal added to that protection.
(6) To evaluate potential modifications to the proposals, a 15 km zone was established around each proposal. Within the zone we plotted the spatial extent of all vegetation types requiring additional protection in an ecoregion.
(7) The areas occupied by vegetation types occurring closest to the proposals that were large enough to meet the conservation strategy were identified as potential additions to the proposals.

Results

Approximately 83% of the lands in the Snake River Basin–High Desert Ecoregion were publicly owned. Protected lands comprised 6.6% of the total land area. Twenty-two vegetation types were present in the ecoregion but only three had at least 10% of their land area in protected status. There were 11 vegetation types that had less than 1% of their land area protected.

Most of the lands (78%) that occurred in the Craters of the Moon proposal were occupied by the sagebrush–lava fields mosaic vegetation type (Big sagebrush) which was already adequately protected in the ecoregion (Table 1). The proposal provided no additional protection for 16 types that were inadequately protected, and made a relatively minor contribution to two others. The proposal substantially increased the protection given to the juniper/sagebrush/bitterbrush woodlands and annual grasslands.

The Owyhee Canyonlands proposal contained six vegetation types (Table 1). It made important contributions to the protection of three vegetation types: montane shrubfields, juniper/sagebrush/bitterbrush woodlands, and sagebrush mosaics. Two of the dominant vegetation types in the proposal, the sagebrush lava–fields mosaic and canyon grasslands were adequately protected elsewhere in the ecoregion.

Publicly owned lands comprised 60% of the Idaho portion of the Blue Mountains Ecoregion. Almost 7% of the total land area is protected. Twelve vegetation types were present in this area and three of these were unprotected. The Hells Canyon proposal did not substantially increase protection for many vegetation types, since 76% of the area was occupied by three types (Subalpine, ponderosa/grass, and canyon grasslands), which were already adequately protected (Table 2). Conversely, the proposal provided no additional protection to five vegetation types that were not adequately protected, and only a relatively minor contribution to two others. The proposal did substantially increase the protection given to the grand fir forest and montane shrubfield types.

Publicly owned lands comprise 88% of the Northern Rocky Mountains Ecoregion with most managed by the U.S. Forest Service. Twenty-four potential vegetation types were found in the Idaho portion of this ecoregion. Because of the large amount of designated wilderness in this ecoregion, a significant proportion of the lands were protected (22.4%). Twelve of the vegetation types had at least 10% of their lands in protected status while four types received no protection.

The majority of lands in the Sawtooth Mountains proposal (77%) supported vegetation types that were already adequately protected in the ecoregion. The proposal provided no additional protection to eight vegetation types that were not adequately protected (Table 3). It substantially increased the protection given to the mixed grand fir, mountain sagebrush, and shrub-steppe types.

It was not possible to modify any of the proposals to protect fully all of the unprotected vegetation types that occurred in the ecoregions using the 15 km limit. In most cases, only a few vegetation types in the zone met the conservation goal. These included shrub-steppe and perennial grasslands in the modified Craters proposal, grand fir, ponderosa pine, and montane shrubfields, in

Table 1. Vegetation protection in the Snake River Basin Ecoregion (Craters Expansion [CEP] and Owyhee Canyonlands proposal [OCP]).

		Vegetation Protection (%)			
*Vegetation type**	*Total area (ha)*	*Current*	*Additional with CEP*	*Additional with OCP*	*Total*
Alpine	3026	0	0	0	0
Subalpine	1995	0	0	0	0
Lodgepole pine	869	0	0	0	0
Grand fir	2748	0	0	0	0
Douglas fir	70,338	0.9	0	0	0.9
Mixed lodgepole	3198	0	0	0	0
Ponderosa pine	12,706	0	0	0	0
Mixed grand fir	13,981	0.3	0	0	0.3
Lodgepole/aspen	1155	0	0	0	0
Ponderosa/grass	41,664	1.6	0	0	1.6
Mountain sagebrush	99,215	1.9	0.4	0	2.3
Montane shrubs	91,404	0	0	5.3	5.3
Woodlands	179,639	6.7	7.8	17.0	31.5
Shrub–steppe	801,178	1.7	0.6	0	2.3
Sagebrush	762,387	6.1	0	7.3	13.4
Big sagebrush	2,138,069	11.6	8.2	0.9	20.7
Salt desert	406,688	4.6	0	0.001	4.6
Canyon grassland	132,143	22.9	0	13.0	35.9
Riparian	23,668	8.2	0	0	8.2
Marshlands	11,629	49.9	0	0	49.9
Annual grassland	681,321	0.8	3.7	0	4.5
Perennial grassland	443,755	1.7	1.3	0	3.0
Total	5,949,033	6.6	3.8	2.2	12.6

**For a detailed description of vegetation types see Appendix 1.*

the Hells Canyon proposal, montane shrubfields in the Owyhees, and mixed grand fir and shrub-steppe in the Sawtooths.

Discussion

Three of the four proposals have very similar characteristics in terms of land area occupied by vegetation types currently in protected status and their contributions to the conservation goal. The exception to this was the Owyhee Canyonlands where 67% of the lands in the proposal were occupied by two vegetation types not adequately protected in the ecoregion. The majority of the vegetation types contained in the other three proposals were already adequately protected elsewhere in their respective ecoregions. In the Snake River Basin this protection is provided by Craters of the Moon National Monument and the Idaho National Engineering Laboratory (253,186 ha combined). In the Blue Mountains and Northern Rockies Ecoregions protection is provided by 1.6 million ha of designated wilderness. The proposals did, however, make significant advances toward the goal for some vegetation types. For example, montane shrubfields in the Snake River Basin currently lack any protection, and the Owyhee proposal would bring 5.3% of this type into protected status (Table 1).

Table 2. Vegetation protection in the Blue Mountains Ecoregion (Hells Canyon Proposal [HCP]).

		Vegetation Protection (%)		
Vegetation type	*Total area (ha)*	*Current*	*Additional with HCP*	*Total*
Subalpine	14,552	19.2	18.7	39.9
Subalpine forests	3751	0	0	0
Grand fir	92,763	4.3	3.2	7.5
Douglas fir	8821	4.3	1.8	6.1
Ponderosa pine	268,111	1.5	1.5	3.0
Ponderosa/grass	113,411	11.9	11.6	23.5
Mountain sagebrush	12,344	0	0	0
Montane shrubs	85,817	0.5	7.4	7.9
Shrub-steppe	16,706	0.3	0	0.3
Big sagebrush	43,973	0.1	0	0.1
Canyon grassland	111,469	24.6	24.5	49.2
Riparian	4912	0	0	0
Total	776,630	6.8	7.3	14.1

Combining data across ecoregions would result in a different view of conservation needs. For example, while no alpine communities are protected in the Snake River Basin, 25% of the lands in this vegetation type are protected in the Northern Rockies. However, efforts to protect representative samples of national environments must be scaled to a relatively homogeneous geographical unit. We feel the ecoregion is the most appropriate unit of spatial analysis and has been used in other studies and management programs to define gaps in the protection of biological and physiographic resources (NPS 1972; Austin & Margules 1986).

The use of buffer zones to expand the protection of

Table 3. Vegetation protection in the Northern Rockies Ecoregion (Sawtooth Proposal [SP]).

Vegetation type	Total area (ha)	Vegetation Protection (%)		
		Current	Additional with SP	Total
Alpine	80,585	24.5	46.3	70.8
Subalpine	735,758	30.2	17.8	48.0
Subalpine forests	467,874	53.0	6.3	59.3
Mountain hemlock	136,732	28.2	0	28.2
Lodgepole pine	337,508	28.6	1.6	30.2
Montane forests	494,369	1.8	0	1.8
Western redcedar	1,242,104	6.9	0	6.9
Grand fir	340,040	13.5	0	13.5
Douglas fir	1,328,461	36.1	3.7	39.8
Mixed lodgepole	493,647	39.1	10.0	49.1
Ponderosa pine	929,872	26.4	0.001	26.4
Mixed grand fir	146,886	2.1	4.2	6.3
Ponderosa/grass	145,049	21.1	0	21.1
Mountain sagebrush	508,047	9.0	14.4	23.4
Montane shrubs	301,034	21.3	0	21.3
Woodlands	399	0	0	0
Shrub-steppe	328,687	2.3	3.5	5.8
Sagebrush	43,205	0	0	0
Big sagebrush	65,233	0.4	0.1	0.5
Salt desert	36,775	4.9	0	4.9
Canyon grassland	41,024	3.2	0	3.2
Riparian	34,027	19.5	33.8	53.3
Annual grassland	460	0	0	0
Perennial grassland	3617	0	0	0
Total	8,196,802	22.4	5.0	27.4

natural reserves has been widely advocated (Oldfield 1988; Mwalyosi 1991). In general, our analyses suggest that simply expanding the proposals did not contribute greatly to the preservation of biodiversity. This was because many of the lands surrounding the proposals were occupied by the same vegetation types that were found in the proposals. Inadequately protected vegetation types were relatively uncommon and widely dispersed in their respective ecoregions. For example, nearly half of the vegetation types in the Snake River Basin occupied only 1.3% of the land area. They also occurred in a small number of distinct areas and thus the probability of their occurring in the 15 km zone was low. This suggests that protecting 10% of some vegetation types in some regions may be not realistic without focusing on those rare types and expanding from those.

All of the modified proposals could be reduced in size without diminishing their ability to meet the conservation goal by excluding portions of vegetation types that are adequately protected by the original proposal or elsewhere in the ecoregion. Since the total area occupied by a proposal is likely to be a source of controversy, it would be advantageous to make those additional modifications. However, we did not consider this option in our analyses because these types of refinements are best conducted by agency planners and administrators.

Although the boundaries of the ecoregions used in this analysis extend beyond the borders of Idaho, we feel that, as in many areas of the world, political boundaries will continue to play an important role in conservation planning efforts (Van Riet & Cooks 1990). For example, Idaho is the only western state that does not have a national park headquartered within its borders. This situation is one of the driving forces behind the national park proposals analyzed in this study, and may be a significant factor leading to the establishment of a new park (MacCracken & O'Laughlin 1992).

It could be argued that by combining the proposals (e.g., Craters of the Moon and Owyhee Canyonlands) resource protection would be enhanced (Table 1, last column). While this is partially true, all of the proposals are highly controversial and the probability of more than one park being established in the near future is low.

The approach outlined in this paper provides an objective basis for setting priorities among various proposals based on their contribution to the protection of biological resources and has wide application to large-scale reserve design throughout the U.S.A. and in many other countries. The methodology would be of value to planners, decision makers, and interest groups as it appears unlikely that more than one proposal will be seriously considered in the near future. This analysis also illustrates that the planning process for new parks needs to be an interdisciplinary effort. The proposals were all based primarily on visual and recreational resources as well as dissatisfaction with current management programs affecting the resources. The proposals would probably do an adequate job of protecting scenic and recreational attributes of each area. However, the relatively small contribution these proposals made toward conserving the biological elements in their respective ecoregions reflects the need for more attention to ecological criteria in proposal selection and design. This point raises considerations similar to those used in arguments over the value of habitat corridors (e.g., Noss 1987*b*; Simberloff et al. 1993). That is, if limited funds and political capital are used to set aside lands for a national park that contributes little to biodiversity protection, the potential to add other biologically diverse lands may be compromised.

This exercise has illustrated the benefits of using resource data in conjunction with a GIS to help define the boundaries of proposed conservation areas (i.e., proposal modification). It has also shown some of the limitations of this technique. The approach we took was the converse of the gap analysis methodology (Scott et al. 1993) which uses overlays of resource data and land status to identify zones of unique habitat assemblages or high species richness that are currently unprotected. Gap Analysis then targets these areas as those where the greatest resource protection can be achieved. By using preselected areas, we were constrained to the data within and adjacent to each proposal. As a result we were less likely (i.e., only by coincidence) to encounter unique resources or regions of high species richness.

However, in some instances, proposal modifications may capture those important areas.

To use the boundaries defined in this manner, extensive ground surveys will be required to verify the accuracy of the vegetation mapping and to help identify natural boundaries which in turn can be tied in with established section lines to provide an accurate cartographic description of the area. Other data layers that would help refine the natural boundaries would be animal species distribution and various topographic features (e.g., elevation and aspect) (DeVelice et al. 1988).

The failure of the proposals to significantly add to the conservation of biological diversity in Idaho was due as much to the resources found in each proposal as it was the large extent of currently protected public lands. With more than 4,000,000 ha of roadless, undeveloped public lands in Idaho (MacCracken et al. 1993), the potential to expand the system is great.

Acknowledgments

This study was funded in part by the U.S. National Park Service, Pacific Northwest Region, and the GAP Analysis Program of the U.S. Fish and Wildlife Service. However, the views expressed in this paper are solely those of the authors and do not necessarily represent those of the supporting agencies. The manuscript benefited greatly from comments by Blair Csuti, Mike Jennings, Reed Noss, Mike Scott, and an anonymous reviewer. Steve Dodson assisted with the graphics. This article is contribution #695 from the University of Idaho Forestry and Wildlife Resources Experiment Station.

Literature Cited

Austin, M. P., and C. R. Margules. 1986. Assessing representativeness. Pages 45–67 in M. B. Usher, editor. Wildlife conservation evaluation. Chapman and Hall, London.

Bailey, R. G. 1976. Description of ecoregions of the United States. Forest Service U.S. Department of Agriculture.

Bolton, M. P., and R. L. Specht. 1983. A method for selecting nature conservation reserves. Australian National Parks and Wildlife Service Occasional Paper No. 8. Canberra, Australia.

Bureau of Land Management. 1989. Owyhee Canyonlands wilderness environmental impact statement. U.S. Department Interior, Bureau of Land Management. Boise District Office, Idaho.

Bureau of Land Management. 1991. Idaho wilderness study report. Vol I. U.S. Department Interior Bureau Land Management. Washington, D.C. BLM-ID-PT-91-018-4333.

Conservation Foundation. 1985. National parks for a new generation. The Conservation Foundation, Washington, D.C.

Crosby, J., ed. 1991. Desert notes. The Committee for Idaho's High Desert, Boise, Idaho.

Crumpacker, D. W., S. W. Hodge, D. Friedley, and W. P. Gregg. 1988. A preliminary assessment of the status of major terrestrial and wetland ecosystems on federal and Indian lands in the United States. Conservation Biology 2:103–115.

DeVelice, R. L., J. W. DeVelice, and G. N. Park. 1988. Gradient analysis in nature reserve design: A New Zealand experiment. Conservation Biology 2:206–217.

Diamond, J. 1986. The design of nature reserve systems for Indonesian New Guinea. Pages 485–503 in M. E. Soulé, editor. Conservation Biology. Sinauer Associates, Sunderland, Massachusetts.

Dinerstein, E., and E. D. Wikramanayake. 1993. Beyond "hotspots": How to prioritize investments to conserve biodiversity in the Indo-Pacific Region. Conservation Biology 7:53–65.

Greeley, R. 1977. Basaltic plains volcanism. Pages 23–44 in R. Greeley and J. S. King, editors. Volcanism of the eastern Snake River Plain, Idaho. National Aeronautics and Space Administration, Washington, D.C.

Gregg, W. B., and B. McGean. 1985. Biosphere reserves: Their history and their promise. Orion 4:50–51.

Harris, L. D., and P. B. Gallagher. 1989. New initiatives for wildlife conservation: The need for movement corridors. Pages 11–34 in G. Mackintosh, editor. Preserving communities and corridors. Defenders of Wildlife, Washington, D.C.

Hells Canyon Preservation Council. 1992. Information package for the proposed Hells Canyon/Chief Joseph National Park and Preserve. Hells Canyon Preservation Council, Joseph, Oregon.

International Union for the Conservation of Nature. 1980. World Conservation Strategy. IUCN. Gland, Switzerland.

International Union for the Conservation of Nature. 1984. Categories, objectives and criteria for protected areas. Pages 47–53 in J. A. McNeely and K. R. Miller, editors. National parks, conservation, and development: The role of protected areas in sustaining society. Smithsonian Institution Press, Washington, D.C.

Kirkpatrick, J. B. 1983. An iterative method for establishing priorities for the selection of natural reserves: An example from Tasmania. Biological Conservation 56:317–328.

Klubnikin, K. 1979. An analysis of the distribution of park and preserve systems relative to vegetation types in California. Master of science thesis, California State University, Fullerton, California.

Kuchler, A. W. 1964. Potential natural vegetation of the conterminous United States. Special. Publication 36, American Geographic Society, New York.

MacCracken, J. G., and J. O'Laughlin. 1992. A national park in Idaho? Proposal and possibilities. Idaho Forest, Wildlife and Range Policy Analysis Group Report 7. University of Idaho, Moscow, Idaho.

MacCracken, J. G., T. Merrill, and J. O'Laughlin. 1993. Analysis of Idaho roadless areas and wilderness. Idaho Forest, Wildlife and Range Policy Analysis Group Report 10. University of Idaho, Moscow, Idaho.

Mackintosh, B. 1991. The national parks: Shaping the system. National Park Service, Washington, D.C.

Margules, C. R. 1989. Introduction to some Australian developments in conservation evaluation. Biological Conservation **50**:1–10.

Miller, K. R. 1984. The Bali Action Plan: A framework for the future of protected areas. Pages 756–764 in J. A. McNeely and K. R. Miller, editors. National parks, conservation, and development. Smithsonian Institution Press, Washington, D.C.

Mwalyosi, R. B. B. 1991. Ecological evaluation for wildlife corridors and buffer zones for Lake Manyara National Park, Tanzania and its immediate environment. Biological Conservation **57**:171–186.

National Park Service. 1972. Part Two of the National Park System Plan: Natural History. U.S. Department Interior, National Park Service, Washington, D.C.

National Park Service. 1975. Study report and plan for the proposed Sawtooth National Park–National Recreation Area, Idaho. U.S. Department of Interior, National Park Service, Washington, D.C.

National Park Service. 1989. Reconnaissance survey, expansion of Craters of the Moon National Monument. U.S. Department Interior, National Park Service, Denver Service Center.

National Park Service. 1990. Resource topics for parklands: Criteria for parks. U.S. Department Interior, National Park Service, Washington, D.C.

Noss, R. F. 1987*a*. From plant communities to landscapes in conservation inventories: A look at The Nature Conservancy. Biological Conservation **41**:11–37.

Noss, R. F. 1987*b*. Corridors in real landscapes: A reply to Simberloff and Cox. Conservation Biology **1**:159–164.

Noss, R. F., and L. D. Harris. 1986. Nodes, networks, and MUMS: Preserving diversity at all scales. Environmental Management **10**:294–309.

Oldfield, S. 1988. Buffer zone management in tropical moist forests. International Union for the Conservation of Nature and Natural Resources Tropical Forestry Paper No. 5. pp. 1–49.

Omernik, J. M. 1986. Ecoregions of the United States. Map. U.S. Environmental Protection Agency, Corvallis, Oregon.

Pressey, R. L. 1990. Reserve selection in New South Wales: Where to from here? Australian Zoologist **26**:70–75.

Pressey, R. L., and A. O. Nicholls. 1991. Reserve selection in the western division of New South Wales: Development of a new procedure based on land system mapping. Pages 98–105 in C. R. Margules and M. P. Austin, editors. Nature conservation: Cost effective biological surveys and data analysis. International Specialized Book Services Incorporated, Portland, Oregon.

Purdie, R. W., R. Blick, and M. P. Bolton. 1986. Selection of a conservation reserve network in the Mulga Biogeographic Region of southwestern Queensland, Australia. Biological Conservation **38**:369–384.

Scott, J. M., B. Csuti, J. J. Jacobi, and J. E. Estes. 1987. Species richness: A geographic approach to protecting future biological diversity. BioScience **37**:782–788.

Scott, J. M., F. Davis, B. Csuti, R. Noss, B. Butterfield, S. Caicco, C. Groves, J. Ulliman, H. Anderson, and R. G. Wright. 1993. Gap analysis: A geographic approach to protection of biological diversity. Wildlife Monograph **123**:1–41.

Simberloff, D., J. A. Farr, J. Cox, and D. W. Mehlman. 1992. Movement corridors: Conservation bargains or poor investments? Conservation Biology **6**:493–504.

Van Riet, W. F., and J. Cooks. 1990. Ecological planning proposal for Kruger National Park. Environmental Management **14**:349–358.

Wright, R. G. 1984. Wildlife resources in creating the new Alaskan parks and preserves. Environmental Management **8**:121–124.

Wright, R. G. 1992. Wildlife research and management in the National Parks. University of Illinois Press, Urbana, Illinois.

Appendix 1

Description of the vegetation types used in the analyses.

Vegetation code	*Description*
Alpine	Grass and sedge communities of *Poa alpina, carex* sp. and *Lazula parviflora*
Subalpine	Parklands including subalpine fir (*Abies lasiocarpa*), mountain hemlock (*Tsuga heterophylla*), whitebark pine (*Pinus albicaulis*), and Engelmann spruce (*Picea engelmannii*)
Subalpine forests	Subalpine fir, Douglas fir (*Pseudotsuga douglasii*), lodgepole pine (*Pinus contorta*), and aspen (*Populus tremuloides*)
Mountain hemlock	Mountain hemlock and subalpine fir
Lodgepole pine	Lodgepole pine with aspen and subalpine fir
Montane forests	Western larch (*Larix occidentalis*) and Douglas fir
Western redcedar	Western redcedar (*Thuja plicata*), with western hemlock, grand fir (*Abies grandis*), Douglas fir, and combinations thereof
Grand fir	Mixed forests of grand fir and Douglas fir
Douglas fir	Douglas fir dominated stands sometimes mixed with lodgepole pine
Mixed lodgepole	Lodgepole pine with mixed conifer, Douglas fir, aspen, and brushfields
Ponderosa pine	Ponderosa pine (*Pinus ponderosa*), forest mixed with Douglas fir, and lodgepole pine
Mixed grand fir	Douglas fir, limber pine (*Pinus flexilis*), and aspen with a mosaic of various mountain shrub species
Lodgepole/aspen	Lodgepole pine with aspen and mountain shrubs
Ponderosa/grass	Ponderosa pine/grasslands with some sagebrush (*Artemesia* sp.) and Douglas fir
Mountain sagebrush	Mountain sagebrush transition zones with Ponderosa pine and Douglas fir

Appendix continued

Vegetation code	*Description*
Montane shrubs	Montane shrubfields with buffalo-berry ***Sheperdia canadensis***, cinquefoil ***Potentilla*** sp., and vetch ***Astragulus alpinus***
Woodlands	Foothills and plains woodlands with western juniper (***Juniperus occidentalis***), sagebrush sp., bitterbrush (***Persia trdidentata***), and greasewood (***Atriplex*** sp.)
Shrub-steppe	Shrub-steppe grasslands with several species of sagebrush, and bitterbrush
Sagebrush	Mosaics of various sagebrush species
Big Sagebrush	Big and low sagebrush mosaics

Appendix continued

Vegetation code	*Description*
Salt desert	Salt desert shrub
Canyon grassland	Canyon grassland and shrub communities
Riparian	Riparian and wetlands including cottonwood and willow floodplain types
Marshlands	Marshes with rushes ***Juncus*** sp., sedges ***Carex*** sp., and cattails ***Typha latifolia***
Annual grassland	Annual grasses including cheatgrass (***Bromus tectorum***) and foxtail fescue (***Festuca megalura***)
Perennial grasslands	Perennial bunchgrasses including bluebunch wheatgrass ***Agropyron spicatum*** and Idaho fescue ***Festuca idahoensis*** grassland

A Comparison of Direct and Environmental Domain Approaches to Planning Reservation of Forest Higher Plant Communities and Species in Tasmania

J. B. KIRKPATRICK

Department of Geography and Environmental Studies
University of Tasmania
GPO Box 252C
Hobart, Tasmania 7001, Australia

M. J. BROWN

Forestry Commission, Tasmania
Surrey House
Macquarie Street
Hobart, Tasmania 7000, Australia

Abstract: *The planning of representative systems of nature conservation reserves can be based on a wide variety of criteria. The ready availability of data on the physical attributes of the environment, and the patchiness of biological data, have made reservation planning based on environmental classifications an attractive option for decision makers. We developed an environmental classification based on ecologically relevant variables and used it to plan a forest reserve system for Tasmania. We then used biological distributional data and the same targets and procedures to choose a forest reserve system. The analyses based on the environmental classification selected the same areas as equivalent analyses based on biological distributional data to a greater degree than could be expected by chance. Many rare species and communities missed selection by environmental classificatory units (environmental domains), however, and the proportions of ranges of selected taxa varied widely. Conversely, environmental domains were missed by a reservation strategy based only on biological data. These domains might reflect some gaps in the biological data. A reservation planning approach based on both biological data and do-*

Paper submitted September 30, 1992; revised manuscript accepted July 16, 1993.

Una comparación de propuestas directas y de dominio ambiental en la planificación de reservas de bosques de las comunidades y especies de plantas altas en Tasmania

Resumen: *La planificación de sistemas representativos de reservas de conservacion de la naturaleza puede basarse en una amplia variedad de criterios. La disponibilidad concreta de datos sobre atributos físicos del medio ambiente y la fragmentación de datos biológicos, han hecho de la planificación de las reservas basadas en clasificaciones ambientales una opoción atractiva para los tomadores de decisiones. Nosotros desarrollamos una clasificación ambiental basada en variables ecológicas relevantes y la utilizamos para planificar un sitema de reserva para Tasmania. Luego utilizamos datos de distribuciones biológicas y los mismos objetivos y procedimientos para elegir un sistema de reserva forestal. Los análisis basados en la clasificación ambiental seleccionaron las mismas áreas que los análisis equivalentes basados en datos de distribuciones biológicas en un grado mayor que sería esperado por azar. Sin embargo, muchas especies raras y comunidades fueron omitidas en la selección basada en la clasificación de unidades ambientales (dominios ambientales) y la proporción de los rangos de los taxones seleccionados varió ampliamente. En forma op*

mains may produce better results than either used in isolation.

uesta, los dominios ambientales fueron omitidos por la estrategia de reservaciones basada sólamente en datos biológicos. Estos dominios reflejarían algunos discontinuidades en los datos biológicos. Una propuesta de planificación basada en datos biológicos y dominios puede producir mejores resultados que cualquiera de estos criterios utilizado independientemente.

Introduction

With increasing conflict over the future of our remaining natural lands and an increasing loss of natural systems to development, there is a need for rapid and cost-effective means for selecting reserves in systems that will maximize biodiversity conservation (Margules & Austin 1990). Many procedures for achieving this end have been proposed (see Kirkpatrick 1983; Game & Peterken 1984; Margules et al. 1988; Pressey & Nicholls 1989; Rebelo & Siegfried 1992). The collection of biotic distribution data is time consuming and expensive, making approaches based on classifications of readily available environmental data very attractive (see Mackey et al. 1988). Such environmental domain analyses (EDAs) have not yet been tested for conservation efficacy against approaches based on biological distributional data. In this paper we compare the results of iterative selections of Tasmanian forest areas for reserves, using known distributions of higher plant species and communities, with the results of selections based on environmental domains.

Methods

Introduction

We undertook the comparisons using the standard Australian Metric Grid of 10 × 10–km squares in order to provide a uniform degree of resolution. Ecological, physiographic, and climatic data are all characteristically scale-dependent (see Levin 1992), but the resolution we have chosen is sufficiently fine to provide a reasonable number of cells (714) for statistical comparison. At the same time it is coarse enough to compensate for varying intensities of data collection.

The data sets we used in the analyses consisted of direct biological records in one case and environmental domains (Mackey et al. 1988) in the other. Each of these data sets is explained in detail below. With both data sets, we undertook two sets of analyses. In the first set we assumed that there were no existing reserves (no existing reserves (NER) analysis) and sought reservation up to various proportional targets. In the second set we had the same targets for reservation but achieved these by adding to existing secure reserves.

Analyses Based on Biological Data

ATTRIBUTES

The attributes used in the various biological analyses were those for which good statewide distributional data were available. They were:

(1) *Forest vegetation mapping units.* We mapped land clearance boundaries from AUSLIG 1:250,000 Landsat 5 MSS cloud-free scenes for 1988 through early 1989. The interpreter (J. B. Kirkpatrick) was unable to distinguish old pine plantations from natural vegetation and assumed that private land that was recently disturbed would be converted to pasture or cropland. This source of error was insignificant because the period between logging of forest and its conversion to pasture or cropland is very brief, as is the persistence of an ambiguous signal from regenerating logged forest. The former source of error was eliminated by the superimposition of the land-clearance map on the unpublished *Forests of Tasmania* map of the Tasmanian Forestry Commission, in which only areas that appeared as natural forest on the latter map and uncleared land on the former map were accepted as natural forest. Natural forest was not taken to include silviculturally regenerated forest.

We divided the area of the 1988–1989 natural forest into 39 mapping units defined by dominant species using the 1:100,000 maps assembled for the Working Group for Forest Conservation (1990), the 1:500,000 vegetation map of Kirkpatrick and Dickinson (1984) and tall forest map of the Forestry Commission (1988), 1:25,000 forest-type maps (Forestry Commission, Tasmania) and vegetation maps (Parks, Wildlife, and Heritage, unpublished data), 1:100,000 maps of the distribution of *Athrotaxis cupressoides* (Cullen & Kirkpatrick, unpublished data; this study), *A. selaginoides* (Brown 1988), *Lagarostrobos franklinii* (Peterson 1990), *Nothofagus gunnii* (Robertson & Duncan 1991), and our field knowledge and that of F. Duncan (applied at 1:500,000).

Almost all the boundaries between units are ultimately based on field verification of the analysis of panchromatic or colored aerial photographs at scales between 1:15,000 and 1:40,000 or mapping at 1:100,000 from direct aerial surveillance (see Brown 1988). Some boundaries resulted from altitudinal or geological extrapolations based on detailed field observations, such as those of Hogg and Kirkpatrick (1974) and Fensham (1989).

(2) *Endemic species groups.* We used the groups that resulted from our earlier association analyses (Kirkpatrick & Brown 1984*a*, 1984*b*) of the distributions of Tasmanian endemic higher plants (Brown et al. 1983), amended appropriately to take into account changes in taxonomic perception and distributional knowledge.

(3) *Rare, threatened, poorly-reserved, and unreserved species.* Species listed as endangered, vulnerable, or rare nationally (Briggs & Leigh 1988) and/or endangered, vulnerable, unreserved, or poorly-reserved in Tasmania (Kirkpatrick et al. 1991) were included by us in the appropriate analyses if they were known to occur in or to be closely associated with forest in at least part of their range. We obtained distributional data from herbarium records; from phytosociological data bases for rainforest (Jarman et al. 1984) and other wet forests (Kirkpatrick et al. 1988), sedgelands (Jarman et al. 1988), grassy ecosystems (Kirkpatrick et al. 1988), wetlands (Kirkpatrick & Harwood 1983), heathlands (Kirkpatrick 1977), dry eucalypt forest (Duncan & Brown 1985), alpine vegetation (Kirkpatrick 1986), vegetation with *Callitris* (Harris & Kirkpatrick 1991), coastal ecosystems (Harris & Kirkpatrick, personal communication), and *Sphagnum* bogs (Whinam et al. 1989) from non-electronic records held by Parks, Wildlife, and Heritage; from records held by the Forestry Commission; from other observations from reliable observers; and from the literature.

(4) *Poorly-reserved and unreserved communities.* We included those forest communities for which only point data were available if they were either poorly reserved or unreserved. These communities have been defined on the basis of their floristics and/or dominance and structure of their strata; they are listed in Kirkpatrick (1991).

FORM OF DATA ENTRY

Species and point community data were entered as presence/absence for each of the 714 10 × 10–km grid squares. The use of presence/absence by grid square rather than the number of records per grid square was necessitated by the marked variability in the intensity of collection of herbarium specimens that has occurred in different parts of the state. We also recorded the most secure land tenure in which each species or point community occurred within each square in the database. We entered the individual mapping unit data for each grid square as both hectares in existing reserves and total hectares. We entered an endemic species group as present in a square if that square contained occurrences of at least two species in the group.

NO EXISTING RESERVES ANALYSES

Our NER analyses assumed that there was no forest reservation in Tasmania and searched for minimum area solutions at various thresholds for the reservation of biotic values within the remaining natural forest. We used an iterative technique similar to that of Kirkpatrick (1983). Our targets for representation in the reserve system were set, in different analyses, at 5%, 10%, 20% and 30% of the areas of each mapping unit and the occurrences of each of the other variables. In the first stage of the analysis, we gave each grid square a score based on its biological content. For any particular square, this score consisted of the total of the areas (ha) of all forest mapping units. This number was added to the number of other entities in the grid square multiplied by 1000. The relative weightings of the different input units affect the order of selection—that is, the priorities—but do not affect the total proportion of each attribute in the national reserve system.

We chose the square with the highest score for reservation. The entities reserved to or above the threshold by this selection were then excluded from the second stage of the analysis, which involved the recalculation of values for all squares. The above sequence of procedures was then repeated until all entities attained the target.

EXTENSION ANALYSES

Our extension analyses aimed for the same percentage reservation targets for biological entities as listed above for the NER analyses. We excluded entities from the analyses, however, if they already attained these targets within the existing forest reserve system. We also excluded the endemic species groups because of the difficulty of attributing a particular tenure type to their occurrence in mixed-tenure grid squares. The problem of grid squares with mixed tenure was addressed by editing out species or point communities that were known to occur only in reserves and by adjusting the area values for the map units to exclude areas in reserves. We used the reserved area and total area for each attribute to calculate a target percentage for unreserved land that would enable the attainment of the selected overall threshold target. We used the same iterative procedure described under the NER analyses for the selection of squares for reservation.

Environmental Domain Analysis

The attributes we used to derive the environmental domain classification were measures of geology, altitude, climate, and solar radiation available on the ArcInfo geographic information system of the Tasmanian Forestry Commission. These are the variables that are known to most influence biotic patterning within Tasmania (Kirkpatrick 1991).

The climate variables (eight temperature and eight precipitation) were derived by BIOCLIM version 2.0 (see Busby 1986). BIOCLIM is a program that produces a set of estimates of 16 climatic variables at any specified latitude, longitude, and altitude, based upon fitted surfaces of precipitation and monthly mean maximum and minimum temperature, which are estimated from data from existing meteorological stations. The program normally provides estimated data at six-minute intervals. We converted these data to the 10 × 10–km grid. A similar process was followed to assign a set of three solar radiation variables to each cell, using the program CLOUDY (Fleming et al. 1987).

A preliminary analysis of the resulting 19 variables showed that many of them were autocorrelated. Accordingly, a principal components analysis (GENSTAT 5 Committee 1987) was used to select those variables that best explained the data set as a whole.

Two solar radiation, four temperature, and three precipitation variables were selected on the basis of this analysis:

December total solar radiation,
June total solar radiation,
mean temperature of the coolest month,
annual temperature range,
mean temperature of the wettest quarter,
mean temperature of the driest quarter,
mean annual rainfall,
rainfall of the driest month, and
seasonality of rainfall.

We determined the proportions of each grid square occupied by the following 12 geological classes from geological maps:

Tertiary basalt,
Cambrian basic igneous,
Jurassic dolerite,
Devonian and Cambrian granite,
Ordovician limestone,
Mathinna sediments,
Triassic sediments,
Permian sediments,
Cambrian and Precambrian sediments,
Tertiary sediments,
Quaternary sediments, and
Ordovician conglomerates and Precambrian metamorphics.

This ordering follows an approximate fertility gradient, as determined for us independently by field geologists and soil scientists, who also aggregated the multitude of mapped Tasmanian geological types into these 12 groups.

Altitude was divided into 300-m zones from sea level to 1500 m. The proportions of each altitudinal class in each grid cell were calculated from topographic data.

The values for each of the 714 cells for each of the 12 geology groups, five altitude zones, two solar radiation variables, four temperature variables, and three precipitation variables were used to define environmental domains for Tasmania using a hierarchical average-link classification in the statistical package GENSTAT 5 (1987). The similarities for the variables used Formula 1 for the geology groups and altitude zones and Formula 2 for the climate and solar radiation variables.

$$1 - |x^i - x^i| \, weight = 1; \text{ unless } x^i = x^j = 0, \text{ then } weight = 0 \text{ range.} \quad (1)$$

$$1 - (x^i - x^i)^2 \, weight = 1 \text{ (range)} \quad (2)$$

The classification can be stopped at any level of division. We wanted our domain classification to have a similar number of classes to our number of forest-mapping units, yet to reflect our knowledge of variation in the physical environment of Tasmania. The classification that we used contained 68 classes at 89% similarity.

Each grid square was attributed to one of the domains. Therefore, overall prioritization using the iterative process was inappropriate, but grid squares were selected for each domain using the area of natural forest. Thus, for each domain, the grid square with the most forest was selected first. Not all forest was needed in the final grid squares selected at each level in each analysis. The hectares required to attain the target were calculated in these cases. We note that the greatest area criterion, when combined with the patterning of grid squares and of forest, does not logically result in the selection of the largest patch of forest that represents a particular domain. For example, a large area of contiguous forest may be split between several grid squares, but most of it could have been contained in one, given a different grid.

Comparison of Biotic and Domain Analyses

We assessed the areas indicated for reservation in the eight domain analyses for their coverage of the biological attributes. We found it necessary in practice to assume that all values contained in a square were covered. Thus, our results give a slightly optimistic picture of the

Table 1. Observed and expected values and significance of the deviation (Fishers exact test) at the 1% (**) level for coincidence between grid squares selected in the domain analysis and the equivalent biotic data for no existing reserves and extension analyses.*

	5%	*10%*	*20%*	*30%*
No Existing Reserves				
Observed	8	15	40	79
Expected	4	7	18	35
Significance	**	**	**	**
Extension Analysis				
Observed	5	3	15	38
Expected	2	3	9	20
Significance	**	N.S.	**	**

* *For a target of 30%, assuming no existing reserves, there were 79 grid squares that were selected in both the biotic and domain analyses. If both of these sets of selections had been randomly distributed, we would have expected a coincidence of 35 grid squares.*

utility of the domain analyses. We calculated the percentage representation of attributes in the grid squares selected in each of the NER and extension domain analyses. In the cases of the extension analyses, we added these values to the percentage representation in existing reserves. These results are presented in the form of box plots that show the median, 25, and 75 percentiles and the range of values. We used Fishers Exact Test to calculate the probability of coincidences between selected grid squares for each pair of comparable domain and biotic analyses.

Results and Discussion

In general, there was a highly statistically significant level of correspondence between the grid squares selected for the domain analyses and those selected in the equivalent biological analyses (Table 1).

The NER domain analyses captured many of the attributes included in the biological analyses (or vice versa), with success increasing dramatically with increased area selected (Table 2). The variability of the attribute capture rate is illustrated in Figure 1. The attributes missed by the 30% NER domain analysis include seven point-location attributes that may occur within the selected squares but for which we may have less than totally complete distributional data, and thirteen point-location attributes that were definitely missed by the selection. The species, communities, mapping units, and endemic groups that were missed were all genuine absences. The domain extension analyses also failed to include many of the biotic attributes (Figure 1, Table 2). The species and communities that were missed or poorly represented in either or both analyses characteristically have very limited distributions in Tasmania.

The fact that the biotic attributes that were missed by the domain analyses usually have relatively restricted distributions decreases the probability of their capture using the criterion of greatest natural forest area. A greater number of domain classes might have served to capture more attributes. However, many of the taxa and communities that were missed have distributions that relate to climatic and disturbance history, not to the contemporary physical environment, so their capture would be a matter of chance unless all domains occupied one grid square each. The various communities with *Callitris* well illustrate the effects of disturbance history in that their present localized occurrences are largely the result of the chance workings of fire history. These communities are relatively widespread on Flinders Island, yet absent from environmentally similar country on the adjacent Tasmanian mainland simply because people were absent from Flinders during all except the last 200 of the last few thousand years (Harris, personal communication). Species such as *Velleia paradoxa* and *Lepidium pseudotasmanicum* have present distributions that reflect the chance of ground disturbance not corresponding with heavy grazing pressure for any sustained period over the last 200 years. They have a wide potential climatic and edaphic environmental range, but they occur sporadically and unpredictably within its bounds. A large proportion of the biotic attributes missed by the 30% domain-extension analysis

Table 2. Nonrepresentation, under representation, and adequate representation in domain analyses.*

	5%			*10%*			*20%*			*30%*		
	0	*−*	*+*	*0*	*−*	*+*	*0*	*−*	*+*	*0*	*−*	*+*
No Existing Reserves												
Species	44	5	64	41	15	57	32	31	50	20	44	49
Communities	38	3	56	33	17	47	26	32	39	20	41	36
Mapping Units	11	8	28	10	9	28	6	15	25	5	18	24
Endemic Groups	5	2	17	5	6	13	3	7	14	3	10	11
Extension Analysis												
Species	35	1	77	29	6	78	26	15	72	17	22	74
Communities	29	1	67	25	11	61	16	21	60	12	39	46
Mapping Units	3	4	40	3	8	36	2	8	36	2	9	36
Endemic Groups	4	0	20	4	0	20	3	1	20	2	3	19

* *0 = not represented; − = under represented; + = over represented.*

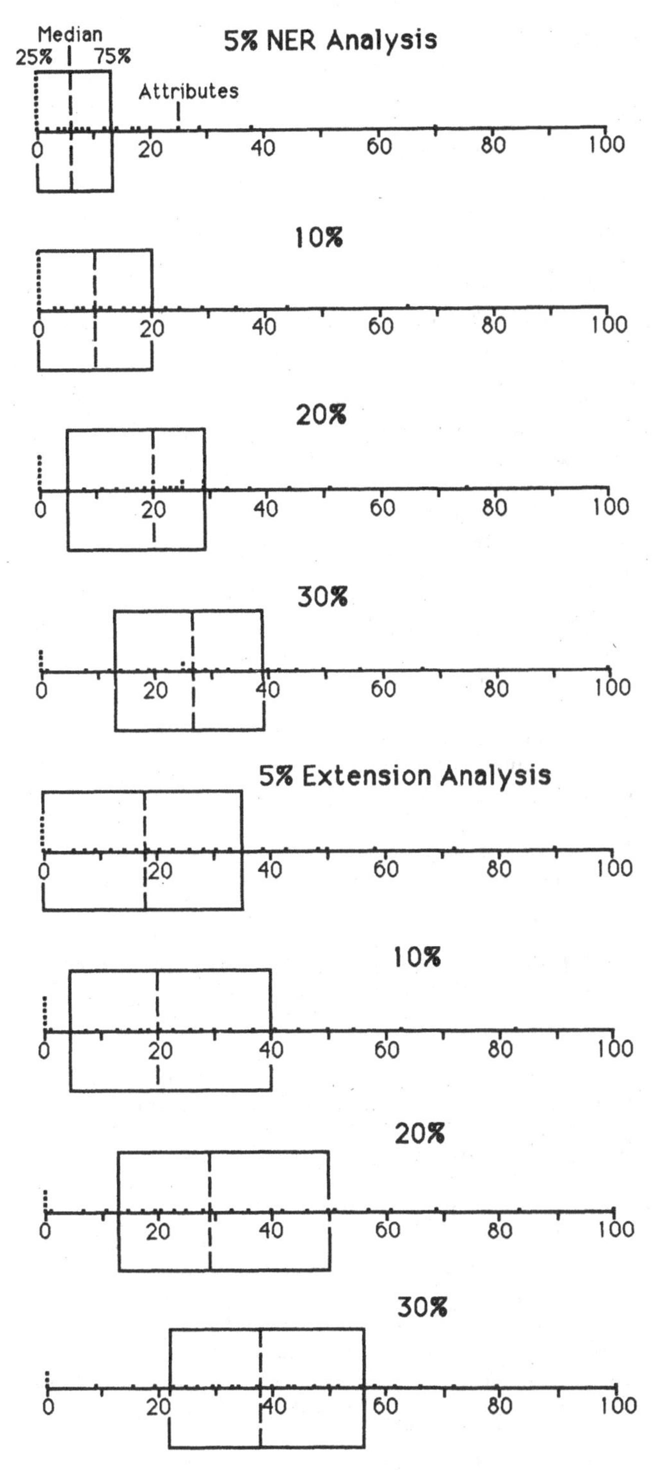

Figure 1. Box plots showing the percentage reservation of individual biotic entities within the grid squares selected for the various domain analyses. NER denotes the analyses that plan reservation assuming no existing reserves. The extension analyses attain reservation targets that include existing reserved areas. Each dot represents 10 biotic entities. The boxes include the 25 to 75 percentiles. The vertical line is the median.

have had their ranges dramatically reduced by the aggregate of land clearance decisions by individual farmers. Their present distributions are therefore unpredictable in environmental terms. The importance of nonedaphic and nonclimatic factors in controlling biotic distribution patterns is well established in the literature (see Stocker 1988; Solem & McKenzie 1991).

The converse of the failure of the domain analyses to encompass all biotic attributes is the failure of the biotic analyses to adequately encompass all domains. The areas selected in the biotic 30% extension analysis, when aggregated with the present reserve system, did not cover any of two domains. Nine other domains had less than 10% of their forest area covered. These were northwestern, western, and coastal domains. The poor representation of the western and northwestern domains probably relates to the lack of significance of rainfall variables in influencing the distributions of higher plant species and communities where precipitation exceeds 1600 mm per annum. The coastal domains contain very small areas of forest so were selected only where distinctly coastal forest communities were present, which was not the case for the two under-represented coastal domains. The lack of representation of domain 57 (three grid cells) in the biotic analyses and the poor representation of domain 64 (inland southeast with only three grid cells) suggests either that the environmental factors that differentiate these domains do not result in restricted target plant species or community occurrences, or that ground surveys in these areas are inadequate.

We conclude that domain analysis, by itself, is not an appropriate tool for the planning of reserves that encompass the full range of biodiversity. Reservations selected on the basis of environmental domains are likely to represent the more widespread biotic attributes in a reasonably satisfactory manner, but they will not capture many of the rarest species and communities. Thus, information on the existence and distribution of such taxa needs to be a major influence on any planning process that purports to have the aim of retaining biodiversity. However, it would seem to be prudent to ensure that the full range of environmental, as well as known biotic, variation is included within any reserve system, because our biotic knowledge is partial. If our experience with vascular plants and plant communities is any guide, we could expect that all common invertebrates and nonvascular plants would be covered by this approach, whereas a reliance on plant species and vegetation variables will not necessarily encompass all less-common species. Thus, we suggest that a complementary or hybrid systematic approach that plans reservation on the basis of known biotic values (as with our biotic analyses), together with a domain analysis, is most likely to be successful in representing the range of biodiversity.

Acknowledgments

We thank the Resource Assessment Commission for permission to publish this work. This paper would not have been possible without the contributions of K. Williams, K. Bridle, A. North, and R. Wardman, and all of the field researchers whose data were made available.

Literature Cited

Briggs, J. D., and J. H. Leigh. 1988. Rare or threatened Australian plants. Special Publication 14. Australian National Parks and Wildlife Service, Canberra, Australia.

Brown, M. J. 1988. Distribution and conservation of King Billy pine. Forestry Commission, Tasmania, Australia.

Brown, M. J., and H. J. Bayly-Stark. 1979. Vegetation of Maria Island. Technical Report 79/1. National Parks and Wildlife Service, Tasmania, Australia.

Brown, M. J., J. B. Kirkpatrick, and A. Moscal. 1983. An atlas of Tasmania's endemic flora. Tasmanian Conservation Trust, Hobart, Tasmania, Australia.

Busby, J. R. 1986. A biogeoclimatic analysis of *Nothofagus cunninghamii* (Hook.) Oerst. in south eastern Australia. Australian Journal of Ecology **11**:1–7.

Duncan, F., and M. J. Brown. 1985. Dry sclerophyll vegetation in Tasmania. Wildlife Division Technical Report 85/1. National Parks and Wildlife Service, Tasmania, Australia.

Fensham, R. J. 1989. The pre-European vegetation of the Midlands, Tasmania: A floristic and historical analysis of vegetation patterns. Journal of Biogeography **16**:29–45.

Fleming, P. M., M. P. Austin, and A. O. Nicholls. 1987. Notes on a radiation index for use in studies of aspect on radiation climate (including details of the SUNDAY and CLOUDY computer packages to investigate solar radiation on slopes). Technical memorandum. CSIRO, Institute of Biological Resources, Division of Water Resources Research, Canberra, Australia.

Forestry Commission. 1988. Very tall eucalypt forests on public land 1:50 000. Forestry Commission, Tasmania, Australia.

Game, M., and G. F. Peterken. 1984. Nature reserve selection strategies in the woodlands of central Lincolnshire, England, Biological Conservation **29**:157–181.

GENSTAT 5 Committee. 1987. Genstat 5, reference manual. Clarendon Press, Oxford, England.

Harris, S., and J. B. Kirkpatrick. 1991. The community ecology of vegetation with *Callitris* in Tasmania. Pages 179–190 in M. R. Banks, S. J. Smith, A. E. Orchard, and G. Kantvilas, editors. Aspects of Tasmanian botany. Royal Society of Tasmania, Hobart, Tasmania, Australia.

Hogg, A., and J. B. Kirkpatrick. 1974. The phytosociology and synecology of some southern Tasmanian eucalypt forests and woodlands. Journal of Biogeography **1**:227–245.

Jarman, S. J., M. J. Brown, and G. Kantvilas. 1984. Rainforest in Tasmania. National Parks and Wildlife Service, Hobart, Tasmania, Australia.

Jarman, S. J., G. Kantvilas, and M. J. Brown. 1988. Buttongrass moorland in Tasmania. Research Report No. 2. Tasmanian Forest Research Council, Inc., Tasmania, Australia.

Kirkpatrick, J. B. 1977. The disappearing heath. Tasmanian Conservation Trust, Inc., Hobart, Tasmania, Australia.

Kirkpatrick, J. B. 1983. An iterative method for establishing priorities for the selection of nature reserves: An example from Tasmania. Biological Conservation **25**:127–134.

Kirkpatrick, J. B. 1986. Conservation of plant species, alliances and associations of the treeless high country of Tasmania. Biological Conservation **37**:43–57.

Kirkpatrick, J. B., editor. 1991. Tasmanian native bush: A management handbook. Tasmanian Environment Centre, Hobart, Tasmania, Australia.

Kirkpatrick, J. B., and M. J. Brown. 1984*a*. Numerical analysis of higher plant endemism in Tasmania. Botanical Journal of the Linnaean Society of London **88**:165–183.

Kirkpatrick, J. B., and M. J. Brown. 1984*b*. The palaeogeographic significance of local endemism in Tasmanian higher plants. Search **15**:112–113.

Kirkpatrick, J. B., and K. J. M. Dickinson. 1984. Vegetation map of Tasmania 1:500,000. Forestry Commission, Tasmania, Australia.

Kirkpatrick, J. B., and C. E. Harwood. 1983. Plant communities of Tasmanian wetlands. Australian Journal of Botany **31**:437–451.

Kirkpatrick, J. B., L. Gilfedder, and R. J. Fensham. 1988. City parks and cemeteries: Tasmania's remnant grasslands and grassy woodlands. Tasmanian Conservation Trust, Inc., Hobart, Tasmania, Australia.

Kirkpatrick, J. B., R. J. Peacock, P. J. Cullen, and M. G. Neyland. 1988. The wet eucalypt forests of Tasmania. Tasmanian Conservation Trust Inc., Hobart, Tasmania, Australia.

Kirkpatrick, J. B., L. Gilfedder, J. Hickie, and S. Harris. 1991. Reservation and conservation status of Tasmanian native higher plants. Wildlife Division Scientific Report 91/2. Department of Parks, Wildlife and Heritage, Tasmania, Australia.

Levin, S. A. 1992. The problem of pattern and scale in ecology. Ecology **73**:1943–1967.

Mackey, B. G., H. A. Nix, M. F. Hutchinson, J. P. MacMahon, and P. M. Fleming. 1988. Assessing representativeness of places for conservation, reservation and heritage listing. Environmental Management **12**:501–514.

Margules, C. R., and M. P. Austin, editors. 1990. Cost effective biological surveys and data analysis. Commonwealth Scientific and Industrial Research Organization, Melbourne, Australia.

Margules, C. R., A. O. Nicholls, and R. L. Pressey. 1988. Selecting networks of reserves to maximize biological diversity. Biological Conservation **43**:63–76.

Peterson, M. J. 1990. Distribution and conservation of Huon pine. Forestry Commission, Tasmania, Australia.

Pressey, R. L., and O. A. Nicholls. 1989. Application of a numerical algorithm to the selection of reserves in semi-arid New South Wales. Biological Conservation **50**:263–278.

Rebelo, A. G., and W. R. Siegfried. 1992. Where should nature reserves be located in the Cape Floristic Region, South Africa? Models for the spatial configuration of reserve network aimed at maximizing the protection of floral diversity. Conservation Biology **6**:243–252.

Robertson, D. I., and F. Duncan. 1991. Distribution and conservation of deciduous beech. Forestry Commission, Tasmania, Australia.

Solem, A., and N. L. McKenzie. 1991. The composition of land snail assemblages in Kimberley rainforests. Pages 247–263 in N. L. McKenzie, R. B. Johnston and P. G. Kendrick, editors. Kimberley rainforests. Surrey Beatty and Sons, New South Wales, Australia.

Stocker, G. L. 1988. Tree species diversity in rainforest—establishment and maintenance. Proceedings of the Ecological Society of Australia **15**:39–47.

Whinam, J., S. Eberhard, J. B. Kirkpatrick, and A. Moscal. 1989. Ecology and conservation of Tasmanian *Sphagnum* peatlands. Tasmanian Conservation Trust Inc., Hobart, Tasmania, Australia.

Working Group for Forest Conservation. 1990. Recommended areas for protection of rainforest, wet eucalypt and dry sclerophyll forest in Tasmania. Forestry Commission, Tasmania, Australia.

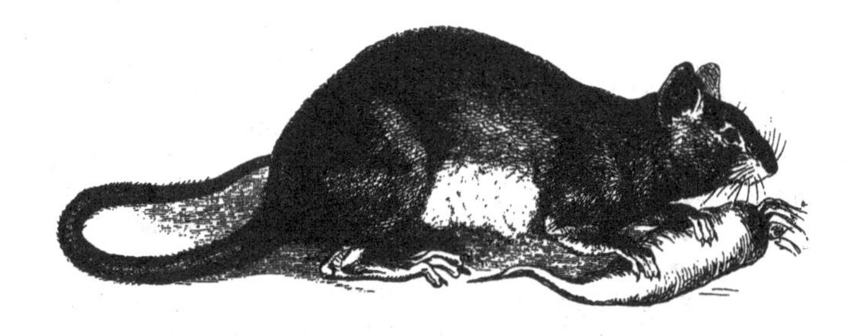

Modeling Effects of Land Management in the Brazilian Amazonian Settlement of Rondônia

VIRGINIA H. DALE
ROBERT V. O'NEILL
Environmental Sciences Division
Oak Ridge National Laboratory
Oak Ridge, TN 37831, U.S.A.

FRANK SOUTHWORTH
Energy Division
Oak Ridge National Laboratory
Oak Ridge, TN 37831, U.S.A.

MARCOS PEDLOWSKI*
Environmental Sciences Division
Oak Ridge National Laboratory
Oak Ridge, TN 37831, U.S.A.

Abstract: *The socioeconomic and ecological aspects of land-use change are interrelated, especially in the Brazilian Amazon, where immigrants are rapidly cutting the forest to establish farms. A computer simulation has been developed that projects land-use changes, carbon release, and the time a family can remain on a farm lot as a function of initial soil and vegetation conditions, market and road infrastructure, and decision variables. The model simulates extremes of land-use practices and typical land-use conditions for central Rondônia, Brazil. Typical land-use practices in Rondônia are based on published accounts of farmers' activities. The most severe practices are defined based on extreme land-management conditions along the Transamazon Highway. The best land-use practices are innovative farming techniques that use a diversity of perennial crops. Model projections using the typical land-use scenario produce changes in land cleared, carbon release, and farmer turnover rates rep-*

* *Current address: Department of Urban Affairs and Planning, Virginia Polytechnic Institute and State University, Blacksburg, VA 24061, U.S.A.*
Paper submitted April 10, 1992; revised manuscript accepted July 21, 1993.

Modelando los efectos del manejo de tierras en los asentamientos de la Amazonia Brasilera de Rondônia

Resumen: *Los aspectos socio-economicós y ecológicos del cambio en el uso de la tiera están inter-relacionados, especialmente en la Amazonía Brasilera dónde los inmigrantes estan cortando rápidamente la selva para establecer terrenos agrícolas. Se ha desarrollado una simulación que proyecta los cambios en el uso de la tierra, las emisiones de carbono y el tiempo que una familia puede quedarse en un terreno agrícola en función de las condiciones iniciales del suelo y la vegetación, el mercado y la infraestructura de rutas y las variables de decisión. El modelo simula tanto condiciones extremas como típicas en las prácticas del uso de la tierra en Rondônia central, Brasil. Las prácticas típicas del uso de la tierra en Rondônia, están basadas en actividades de los agricultores que se encuentran publicadas. Las prácticas más severas son definidas en base a condiciones extremas de manejo de la tierra a lo largo de la ruta Transamasónica. Las mejores prácticas de uso de la tierra son técnicas de cultivo innovadoras que utilizan una diversidad de cultivos peremnes. Las proyecciones del modelo que utiliza el caso de uso de la tierra típico produce cambios en tierras desmon*

resentative of central Rondônia, based on interviews with 87 farmers. Results from both the model projections and the interview data show that farmers who have been on the lots a decade have cleared about half of the lot. The most striking result is the similarity in the typical and worst-case scenarios over the 40-year projection period in the percentage of land cleared and carbon released and in the number of families on the farm lots and abandoning the lots over time. Model results from the best-case scenario compare well to the three farmers out of 87 interviewed who practice innovative farming. In both cases, farmers who have been on their land an average of eight years have deforested less than 20% of the lot. Spatial indices of fractal dimension and contagion are used to quantify the effects of the different land-management practices. Changes in the fractal index over the 40 years of projected land management demonstrate the decrease in complexity as a forested region becomes dominated by agriculture for the typical and worst-case scenarios. Low values of the contagion index indicates that a heterogeneous mix of land uses is retained with the best-case scenario. These results suggest that crop diversity and nontraditional techniques provide both a social and environmental improvement over the other scenarios: the people are able to maintain themselves on the land, less carbon is released, and the land maintains a mix of habitat types. The model results illustrate that both social and environmental effects of land-management practices need to be considered.

tadas, emisiones de carbono y tasas de recambio de los agricultores que son representativas de Rondônia central, basadas en los datos de encuestas llevadas a cabo entre ochenta y siete agricultores. Tanto los resultados de las proyecciones del modelo como de las encuestas muestran que los agricultores que han estado en el lote por una década han desmontado alrededor de la mitad del mismo. El resultado más sorprendente es la similitud entre el caso típico y el peor en el período de cuarenta años de la proyección con respecto al porcentaje de tierra desmontada y carbono emitido y el número de familias que permanecían en los lotes de terrenos agrícolas y aquellas que los abandonaban, a través del tiempo. El resultado del modelo para el mejor caso se ajusta bien a tres, de los ochenta y siete agricultores encuestados, que practican agricultura innovadora. En ambos casos, los agricultores que han estado en su tierra un promedio de ocho años han deforestado menos del 20% del lote. Indices espaciales de dimensión fractal y de contagio son utilizados para cuantificar el efecto de las diferentes prácticas de manejo de la tierra. Cambios en el índice de dimensión fractal en el período proyectado de 40 años de manejo de la tierra, demuestra el decrecimiento en complejidad en la medida en que la región forestada es dominada por agricultura para el caso típico y para el peor caso. Bajos valores del índice de contagio indican que una mezcla heterogénea de usos de la tierra es retenida en el mejor de los casos. Estos resultados sugieren que, comparado con los otros casos, la diversidad del cultivo y las técnicas no tradicionales proveen un ambiente social y ambiental de mejoras: los pobladores son capaces de mantenerse por ellos mismos en la tierra, una menor cantidad de carbono es emitida y las tierras mantienen una mezcla de tipos de hábitat. Los resultados del modelo ilustran que se deben considerar tanto los efectos sociales como los del ambiente en las prácticas de manejo de la tierra.

Introduction

There are two aspects to balancing the social and environmental effects of land-use management. First, landscape characteristics need to be identified that promote biodiversity and support key species (see Franklin et al. 1981; Wright & Hubbell 1983; Lehmkuhl 1985; Dawson et al. 1986). Example landscape characteristics include presence of large trees, decomposing logs, or a mix of habitats critical for the complete lifecycle of key species. The second aspect of balanced land-management policies is that land uses must support human needs. For much of the world, humans require that the land provide a livelihood. Thus, balanced land management must support people in an ecologically responsible manner.

Nowhere is the need to support people in an ecologically responsible manner more pressing than in the Brazilian Amazon. The region has the largest tract of moist tropical rain forest in the world—more than 3.3 million km^2 (Molofosky et al. 1985). Within the past two decades, the forests of Brazil have undergone widespread clearing. The Brazilian state of Rondônia, comprised of 243,000 km^2 located in the western Amazon Basin, is the most extreme example of rampant deforestation. Analysis of AVHRR satellite images shows a five-fold increase in the deforested area in the last decade, with more than 8000 km^2 of the forest eliminated by 1980, increasing to 28,000 km^2 by 1985, and reaching 41,000 km^2 by 1987 (Malingreau & Tucker 1988; Stone et al. 1991). The 1987 estimates of total deforested area range from 16% to 19% of the state (Booth 1989). More recent analysis using high-resolution landsat imagery shows an increase in deforestation from 6281 km^2 in 1978 to 23,998 km^2 in 1988 (Skole & Tucker 1993). Whatever the actual rate, it is clear that forests are being rapidly cleared in Rondônia.

The widespread deforestation in Rondônia results primarily from immigrants clearing forests for farming. The paving of roads is largely responsible for the influx of people into the state (Fearnside 1983). In 1979, only 1434 km of permanent roads existed in Rondônia (IBGE 1979*a*, 1979*b*). By 1988, 25,324 km of roads had been established (DER 1988), an increase of over 1600%. Deforestation occurs not only along the road, but also on adjacent areas as the road permits access to land for farming.

For the Amazon, the landscape features that promote biodiversity are largely associated with the tropical forest. The ecological effects of large-scale forest loss include disruption of hydrological regimes (Shukla et al. 1990), degradation of soil (Hecht 1981; Buschbacher et al. 1988), and changes in greenhouse gases in the atmosphere, which may induce climate change (Houghton et al. 1983; Post et al. 1990; Dale et al. 1991). Deforestation has a direct impact on vegetation patterns and animal movement. The size, spatial distribution, and contact among different plots of residual natural vegetation can be crucial for a species' survival (Bierregaard 1990; Saunders et al. 1991). Prominent effects of newly created edges include abiotic changes—such as temperature, relative humidity, light, and exposure to wind—and different levels of biological change—such as elevated tree mortality, increased leaf fall, and enhanced survival of insectivorous species at increased densities (Lovejoy et al. 1986).

Colonization of the Brazilian Amazon reflected various economic, social, and political intentions (Foresta 1991). From an economic perspective, the abundant natural resources of the Amazon could provide a livelihood for the growing population. From a social perspective, there was a need for the government to relieve social tension, especially in the populated coastal cities (Machado 1991). From a political perspective, the Brazilian government desired to establish Portuguese-speaking citizens along the borders of the country.

Unfortunately, the social and economic goals were not attained. Most of the migrants to the Amazon moved from the south or southwest part of Brazil rather than from overpopulated cities (Pedlowski & Dale 1992). Furthermore, much of the land is being used in such a manner that long-term support of a family is questionable. Land use follows a typical pattern, although site quality and the farmers' cultural background influence farming methods (Coy 1987; Millikan 1988). The colonists usually cut down the forest and burn the slash. During the first four years, annual crops such as rice, corn, beans, or manioc and perennial crops such as coffee, cocoa, and rubber are planted. The fields are burned every year to reduce weeds. Production of these crops becomes infeasible as soil quality decreases and as crops become more susceptible to pests and disease. The steady decline of agriculture following clearing forces many farmers to plant pasture grasses and raise a small number of cattle (see Tucker et al. 1984, 1986; Duncan et al. 1990; Frohn et al. 1990). Under current practices there is no evidence that the land sustains cattle ranching or any other type of farming on fully cleared land, beyond 6 to 8 years. At that time, the farmer either cuts more forest and begins the land degradation process anew, or abandons the land and moves elsewhere.

Thus, there is a pressing need to identify an ecologically responsible way for people to support themselves in the Brazilian Amazon. One alternative is the use of extractive reserves within which rubber, Brazil nuts, and other forest products are collected over a 200-ha to 300-ha area of which about 2 ha are in agricultural crops and 6 ha are in pasture (Schwartzman 1989). Another alternative is agroforestry, which involves planting a combination of agricultural crops and forest products. Other farmers are expanding their sources of income by establishing fish ponds or bee hives from which honey is collected.

Evaluation of alternative land-management practices involves long-term and broad-scale effects on environmental, social, and economic conditions. Such evaluations are difficult because implementing land-management practices is costly and evaluating the effects requires long-term observations for large areas.

Therefore, we have developed a computer model that projects important aspects of the social and ecological consequences of human migration and subsequent land-use changes. The model is currently applied to Ouro Preto in the central part of Rondônia, where colonization schemes over the past 20 years have attempted to take advantage of the relatively good soils. In this paper, we show theoretically how extremes of land-use practices affect human turnover rates, carbon depletion, and deforestation compared to a typical land-use scenario.

Three land-use cases are projected with the model in an attempt to bracket socially extreme conditions. The three scenarios were chosen to reflect the extremes of land management by farmers settling along the Transamazon Highway (the "worst" case), of typical colonists in central Rondônia, and of the few farmers who had no income from milk or cattle (the "best" case). Land-management policies in Rondônia attempted to prevent the high turnover of land owners experienced along the Transamazon Highway (Frohn et al. 1990; Foresta 1991), as has been documented by Moran (1981) and Fearnside (1980, 1984, 1986). Comparing the worst-case and typical land-use scenarios serves to compare the experiences of the farmers with the most drastic land-use practices along the Transamazon Highway with those in central Rondônia. This paper illustrates that the model can evaluate land-management strategies that address the dual objectives of meeting the socioeconomic needs of people while retaining unique landscape characteristics necessary to promote biodiversity.

Model of Land-Use Change

The Dynamic Ecological–Land Tenure Analysis model (DELTA) evaluates ecological effects at the aggregate, region-wide level (300,000 ha and larger) based on the dynamics of land use at the level of the farm lot and the individual farmer (lots average 101 ha) (Southworth et al. 1991). By tracking the history of each lot and the colonists, the model projects single-lot and aggregate land-use patterns reflective of the social and ecological aspects of human settlement. The DELTA model is an exploratory tool that can be used to estimate the importance of components of specific land-management practices (Southworth et al. 1991). The model is described in Dale et al. (1993*a* and 1993*b*) and the Appendix.

Scenarios Considered

The worst-case scenario simulates situations where colonists plant only annual crops or pasture and clear 20 ha of forested land each year for the first 14 years of ownership. None of the land is returned to fallow or used for agroforestry. There is an increasing probability of the colonists leaving the lot up to year 5, after which the colonists are simulated to leave the plots with a 100% probability. Because there is a constant stream of new colonists, the abandoned lots can be reoccupied, and more land cleared for agriculture or pasture. When all of the forests are cleared from a lot and the soil fertility is depleted, a lot is no longer selected by new colonists. In that case, vegetation recovery and carbon accumulation occurs in the lot in the ensuing years. This case is, admittedly, an extreme of the land-use practices in the Amazon, but it meets our intent to bracket the typical practices with excessive conditions.

The typical case simulates land-use activities that occur in central Rondônia (Leite & Furley 1985; Coy 1987; Millikan 1988). Farmers tend to clear an increasing amount of the forest up to year three and stop most of the clearing by year 7, at which time about half of the land is cleared. There is a "50%" law specifying that only half of a lot may be cleared by any one farmer. The Rondônia Department of Environment does attempt enforcement, but there are ways around the law. For example, sequential ownership allows more clearing of a lot. The model simulates the typical situation in which farmers plant annual and perennial crops during the first years and pasture in subsequent years, and then abandon parts of the lots to fallow vegetation beginning about year 4.

The best-case scenario simulates an innovative system. The farmers have a higher probability of staying on the lot (a 0.10 probability of a move per year for the first 10 years), although soil degradation can act to override that probability. Five ha are cleared every year for the first four years, with no clearing thereafter. For the first two years, all of the land is planted into annual crops or pasture, but in years three and four 50% of that land is turned into agroforestry or other nontraditional uses of the land; by year 5 all of the cleared land on the lot is in a diversity of uses.

All cases have similar initial conditions. Two hundred and ninety-four lots are considered with an average lot size of 101 ha, ranging from 53 to 120 ha (based on the Ouro Preto region). Lots are overlaid on the pattern of natural vegetation and soils in central Rondônia (digitized from 1:1,000,000 maps obtained from a radar-image study of the Amazon; RadamBrasil 1978*a*, 1978*b*) and the 1988 primary and secondary road network (digitized from DER 1988). Distances to the major market along primary and secondary roads are calculated in the geographic information system. Maps produced are the result of a single simulation, but the variability of the projected land cleared and carbon released based on 40 replications are compared for the three scenarios.

Colonist families are introduced into the area in a regular pattern: 70 families the first year; 30 in years 2 to 6; 20 in year 7; and 10 per year for every year thereafter. In the simulation, colonists choose an appropriate lot and farm it until conditions cause them to move elsewhere. After leaving a lot, half of the farmers stay in the 294-lot region, and the others move away.

Results

Three scenarios project different amounts of forest clearing (Fig. 1). Almost all of the forest is cleared by year 10 in the worst-case scenario and by year 18 in the typical land-use scenario. This large-scale clearing reflects not only the clearing done by individual farmers, but also the sequential ownership of most lots. Because the best-case scenario projects little turnover in lot ownership, forest clearance approaches 40% and appears to stabilize.

Maps of the deforestation illustrate spatial differences

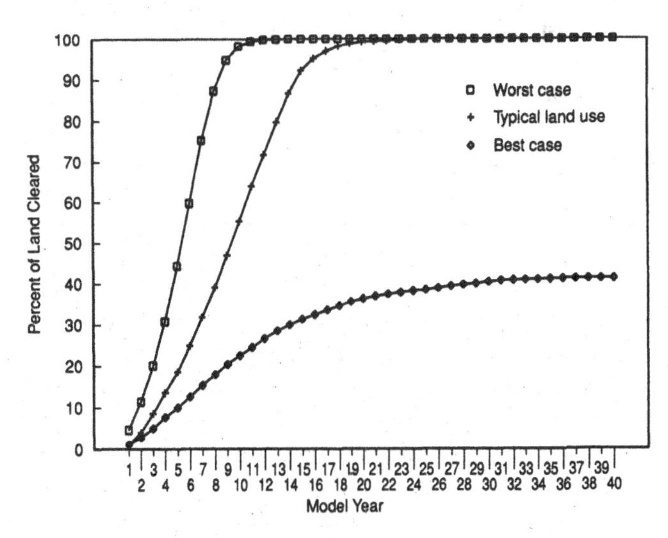

Figure 1. Land cleared over time for worst-case, best-case, and typical land-use scenarios.

of the three scenarios (Fig. 2). For the typical land use, complete deforestation does not occur until year 20, but every parcel has had some clearing by year 10. For the best-case scenario, in contrast, some lots have up to 20% of the forest remaining at year 40 (which is so similar to the map for year 20 that it is not illustrated here). Throughout the 40-year projected time, there is a mix of land uses.

The variability in percentage land cleared and carbon released is higher for the best case than for the other two scenarios (Table 1). The worst-case and typical land-use scenarios project that all of the lots are cleared and almost two-thirds of the carbon is released with little variation, because most of the farmers are following similar practices. In contrast, the best-case scenario allows more variability in the amount of land cleared and carbon released.

The differences in the projected maps of deforestation over time for the three cases is captured by spatial indices (Fig. 3). The contagion index depicts the extent to which land uses are aggregated or clumped (O'Neill et al. 1988) and has a value of 1.0 when the area is dominated by a single land use. Therefore, all scenarios are initiated with a contagion value of 1.0 when the area is completely covered by the undisturbed tropical moist forest. The contagion index declines sharply for all cases as the tropical forest is cut. For the worst and typical land-use cases, it again approaches 1.0 at year 10 and year 20, respectively, when the area becomes dominated by the cleared land. The contagion index remains at about 0.4 for the best land-use scenario, which reflects the mix of forested and cleared land that occurs in this case. The dominance index indicates the degree to which one land use dominates the landscape, and in this case dominance follows the same patterns as the contagion index.

The fractal index measures the complexity in shapes of the land-use patches (Krummel et al. 1987). The original forested perimeter has a fractal value of 1.0, which increases as the land is cleared. Minimum complexity occurs at years 10 and 20 for the worst and typical land-use scenarios, when the land is totally cleared. Thereafter, vegetation recovery induces an increase in the fractal value because of gradual reforestation. For the best case, the fractal increases up to year 25 and remains high, which reflects the stability of the dissected pattern of the landscape after year 25. The higher value of the best case compared to the other two scenarios is due to the more complex patterns generated.

The amount of carbon released from the cleared lots is a function of the different land-use scenarios (Fig. 4). For the worst case, the carbon release quickly rises to more than 60% of the carbon originally in the soils and vegetation. The slight recovery thereafter is due to the secondary vegetation growing back on abandoned lands. The typical land use is projected to cause a slower

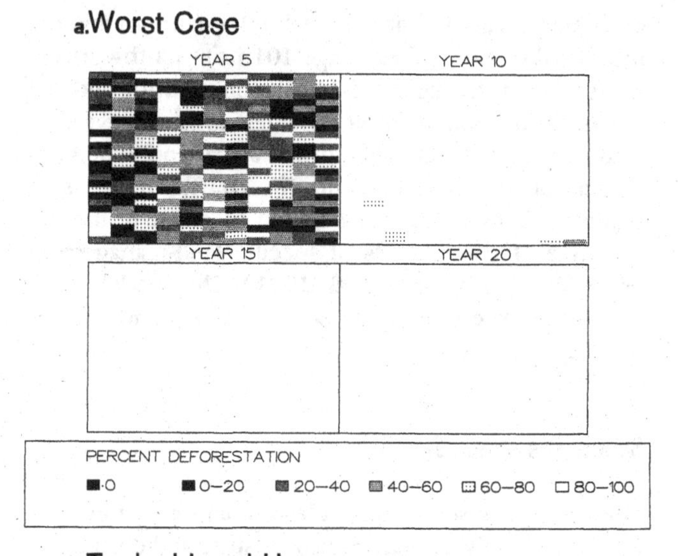

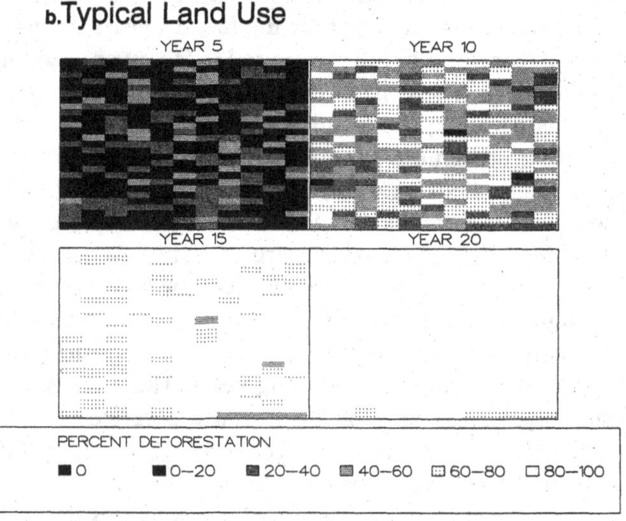

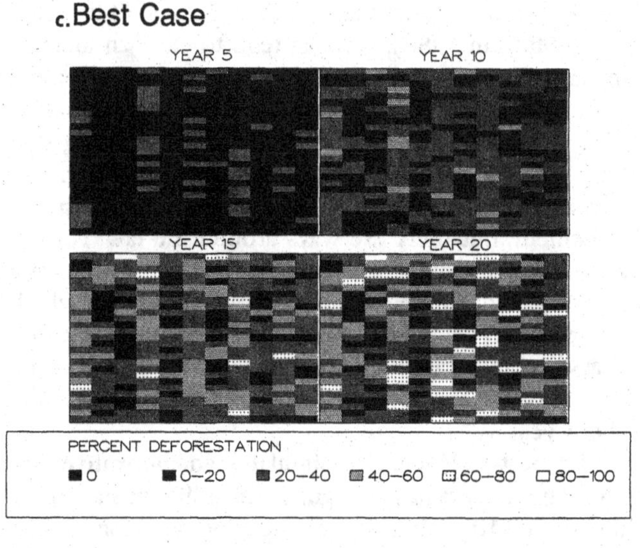

Figure 2. Projected map of the percentage of deforestation for the (a) worst-case, (b) typical land-use, and (c) best-case scenarios for years 5, 10, 15, and 20. For later years, the deforestation maps are similar to the maps produced for year 20 for each scenario.

Table 1. Mean and variance of 40 simulation replications.

Model Scenario	*Percentage Land Cleared* *Mean*	*Percentage Land Cleared* *Standard Deviation*	*Percentage Carbon Released* *Mean*	*Percentage Carbon Released* *Standard Deviation*
Worst Case	99.9	0.0945	55.1	0.217
Typical Case	99.9	0.105	63.6	0.374
Best Case	41.9	1.33	27.2	0.804

increase in carbon release that stabilizes at about 55% of the original carbon by model year 27. The best-case scenario projects a slow but steady increase in the amount of carbon released that has attained 30% of the original carbon by model year 40.

The number of families on the lots over time shows that only the best-case scenario is able to support farmers for the full 40-year projection (Fig. 5). Both the worst case and typical land-use scenarios simulate an increase in the number of families on the lots up to the time at which more than 65% of the land has been cleared and soil quality is greatly reduced by the overuse of the land. At model year 6 for the worst case and year 10 for the typical case, the number of families on the lots begins to decline. No families can be supported for either of these two scenarios after years 14 and 22, respectively.

More families leave the lots under the worst-case and typical-case scenarios (Fig. 5). Only the best-case scenario projects no farmers leaving the area until model year 14, at which time more than 294 families (more than one family per lot) have attempted to enter the system. The average tenure on the lot increases linearly to 33 years for the best-case scenario, in contrast to the worst and typical land-use cases, where the average tenure time never exceeds four years. The constant increase in tenure time for the best-case scenario reflects the constant immigration into the area. The low tenure times for the typical and worst-case scenarios are due to the high clearance rate of the initial colonists and to soil degradation that results from annual crops and pasture.

Discussion

On the basis of recent statistical and satellite data, plus interviews with the farmers (Pedlowski & Dale 1992), the simulated typical land-use conditions appear to reflect quite well land-use changes in central Rondônia (Table 2). Both the simulation projections and interview data suggest that farmers who have been on the land for 10 years have deforested almost half the lot. Satellite imagery also illustrates that the lots are being cleared at a very rapid rate (Stone et al. 1991).

Simulated patterns of land use can show great differences between the worst and best case in the amount of land cleared, carbon released, and number of families supported on the land. The worst-case scenario projects rapid clearing of the forest, large amounts of carbon being released, and the farmers being supported on the land for less than a decade. In contrast, the best-case scenario projects only about 40% of the land being cleared, less than half the amount of carbon being released, and families supporting themselves over the 40 years.

The most striking result is the similarity of the typical-case and worst-case scenarios, which suggests that the migrants' experiences in Rondônia do not greatly differ from those of the extreme farmers along the Transamazon Highway (assuming the soils were as good as the Rondônian soils). This is of concern because a primary goal of the colonization programs in Rondônia was to make improvements over the programs developed for the Transamazon.

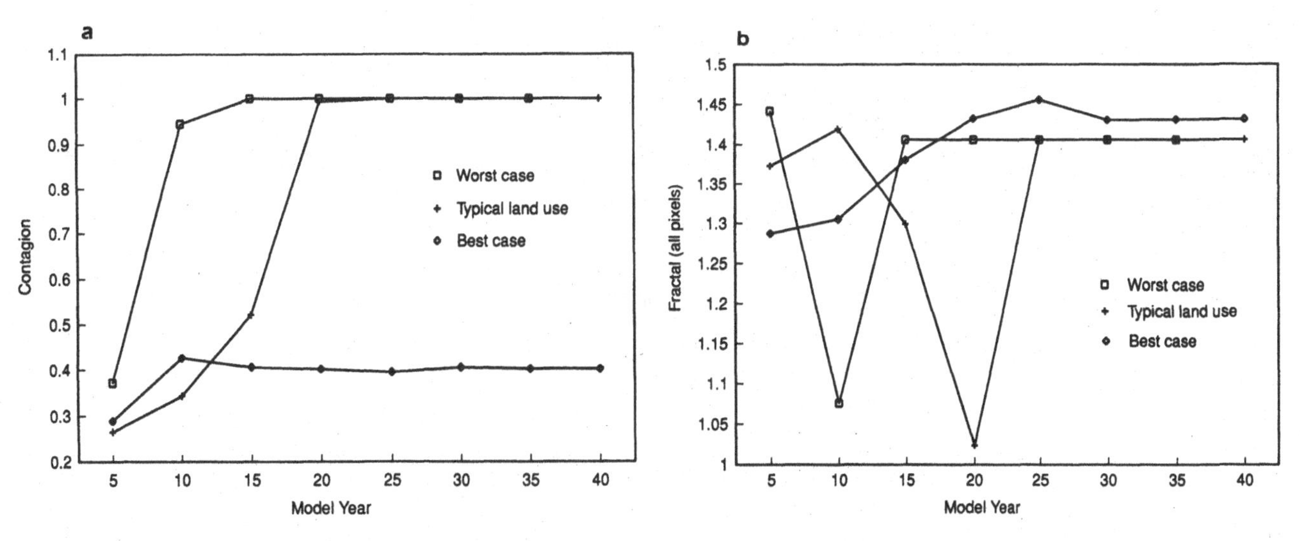

Figure 3. (a) Percentage contagion and (b) fractal index over time for worst-case, best-case, and typical land-use scenarios.

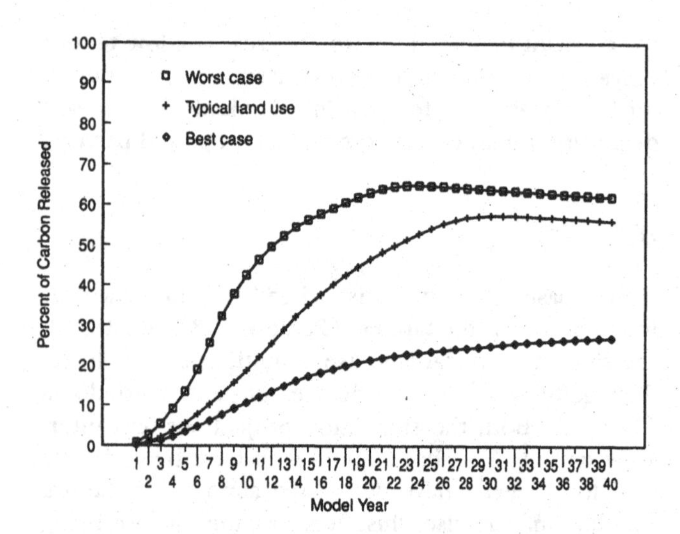

Figure 4. Carbon release over time for worst-case, best-case, and typical land-use scenarios.

The best-case scenario was designed to examine the long-term implications of farmers employing innovative techniques. The simulations suggest that scenario is better than the other cases from both a social and an environmental perspective: the people are able to maintain themselves on the land, less carbon is released, and the land retains a mix of habitat types. These best-case results relate directly to the three farmers in Rondônia out of 87 interviewed who used innovative farming practices and had no income from milk production (Table 2). These select farmers had a higher income from perennial and annual crops than the typical farmers, probably because they grew a higher diversity of crops and had frequent visits from the agricultural extension agents. Furthermore, these farmers employed innovative techniques such as agroforestry, establishing bee hives for honey collection, and digging ponds to grow fish. The value of modeling the best-case scenario became apparent during a conversation with a farmer who supported his family for 20 years by a variety of innovative land-management approaches. After having the modeling approach explained to him, the farmer asked, "What can your research do for me?" Our response was, "The research demonstrates the long-term benefits of innovative land practices."

We were also concerned with quantifying the effects of management alternatives using spatial indices. The abrupt drop and recovery of the fractal index signifies a change in complexity of land-use patterns from dominance by one land use to another. For both the worst-case and typical land-use case, the change was from a forested area to a completely cleared area. The low values for the contagion index indicate a retention of heterogenous land uses. Benefits of retaining a mix of land uses on the landscape include providing for both habitat diversity and movement corridors. For example, the movements of white-lipped peccaries (which are major seed dispersers) are greatly influenced by roads. Thus, it would be useful to further explore the benefits that patches of mixed forest and clearing may provide compared to a totally cleared area. The properties of land-use mixes may be crucial for developing guidelines for balancing socioeconomic and ecological effects of land-use change.

Recognizing the extremely difficult socioeconomic problems that exist in the area, the value of this exercise is to show that there are great differences in the socioeconomic and environmental impact of specific land-management options. Common land-use practices are doing poorly on grounds of both population sustainability and ecological considerations. The model provides the opportunity to examine the impact of land-

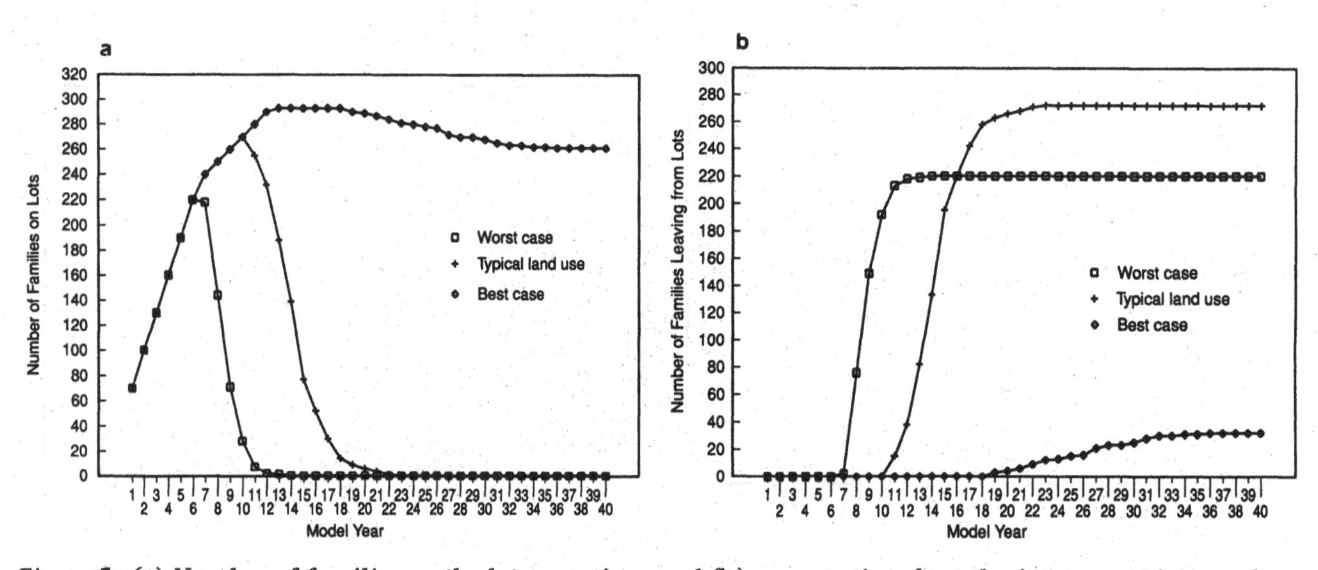

Figure 5. (a) Number of families on the lots over time, and (b) average time for colonists to remain on a lot over time, for worst-case, best-case, and typical land-use scenarios.

Table 2. Comparison of farming activities for two groups of farms in Ouro Preto Do Oeste, Rondônia.*

	Typical Farms (n = 84)		*Best-Case Farms*** (n = 3)		
	Average	*S.E.*	*Average*	*S.E.*	*t-test*
Distance to Ouro Preto (km)	38.7	2.10	54.0	6.6	$p < 0.001$
Time on Lot (yrs)	9.43	0.64	8.67	1.19	ns
Area (ha)					
Deforested	44.21	3.88	18.80	2.55	$p < 0.001$
Perennial	6.08	0.70	7.20	1.13	ns
Annual	5.03	0.45	4.00	0.65	$p < 0.001$
Pasture	30.98	3.72	7.60	1.99	$p < 0.001$
Income for Year (U.S. dollars for Nov. 1990)					
Perennial	2221.0	442.3	5585.7	1586.1	$p < 0.005$
Annual	806.7	189.1	3451.6	1603.3	$p < 0.10$
Milk	708.6	130.2	0.0	0.0	$p < 0.001$
Cattle	544.9	124.0	0.0	0.0	$p < 0.001$
Pigs	18.51	7.94	48.33	39.46	ns
Chickens	8.29	2.80	17.67	14.42	ns

* *Based on interviews summarized in Pedlowski & Dale 1992.*
** *Innovative farming practices; no income from cows.*

management options by projecting the repercussions of specific decisions.

The model results suggest that land-management practices in areas like the Amazon need to consider both socioeconomic and ecologic factors. The exercise of evaluating a scenario can bring many aspects of land management into focus. Such interdisciplinary planning is of critical importance for long-term management success.

Acknowledgment

Barbara Jackson of Oak Ridge National Laboratory ran the spatial analysis program using the model results; this program was developed under U.S. Environmental Protection Agency funding for the Landscape Ecology Group of the Environmental Monitoring and Assessment Program (Interagency Agreement DW89934921-01). Aaron Rosen wrote the user interface for the DELTA model and created the link between the DELTA model and the spatial analysis program. Reviews of the manuscript by M. G. Turner, R. H. Gardner, and an anonymous reviewer were most helpful. Research was sponsored by the Director's R&D Fund of the Oak Ridge National Laboratory and by the Carbon Dioxide Research Program, Atmospheric and Climatic Change Division, Office of Health and Environmental Research, U.S. Department of Energy, under contract DE-AC95–84OR21400, with Martin Marietta Energy Systems, Inc. This is publication 4239, Environmental Sciences Division of the Oak Ridge National Laboratory.

Literature Cited

Bierregaard, R. O. Jr. 1990. Avian communities in the understory of Amazonian forest fragments. Pages 333–343 in A. Keast, editor. Biogeography and ecology of forest bird communities. SPB Academic Publishing, The Hague, The Netherlands.

Booth, W. 1989. Monitoring the fate of forests from space. Science **243**:1428–1429.

Buschbacher, R., C. Uhl, and E. A. S. Serrao. 1988. Abandoned pastures in eastern Amazonia. II. Nutrient stocks in the soil and vegetation. Journal of Ecology **76**:682–699.

Coy, M. 1987. Rondônia: Frente pioneira e programa Polonoroteste. O processo de diferenciação sócio-econômica na periferia e os limites do planejamento público. Tubinguen Geographhische Studien **95**:253–270.

Dale, V. H., R. A. Houghton, and C. A. S. Hall. 1991. Estimating the effects of land use change on global atmospheric CO_2 concentrations. Canadian Journal of Forestry Research **21**:87–90.

Dale, V. H., F. Southworth, R. V. O'Neill, and A. Rosen. 1993*a*. Simulating spatial patterns of land-use change in Rondônia, Brazil. Pages 29–56 in R. H. Gardner, editor. Some mathematical questions in biology. American Mathematical Society, Providence, Rhode Island.

Dale, V. H., R. V. O'Neill, M. A. Pedlowski, and F. Southworth. 1993*b*. Causes and effects of land use change in central Rondônia, Brazil. Photogrammetric Engineering & Remote Sensing **59**:997–1005.

Dawson, W. R., J. D. Ligon, J. R. Murphy, J. P. Myers, D. Simberloff, and J. Verner. 1986. Report 7 of the Advisory Panel on the Spotted Owl. Audobon Conservation. National Audobon Society, New York.

DER (Departmento De Estradas De Rodagem, Rondônia). 1988. Estado De Rondônia. Map, scale 1:1,000,000. Porto Vehlo, Brazil.

Duncan, E. M. J., D. F. M. Mcgregor, and N. A. Q. Vieira. 1990. Pasture development in the cleared forest land in northern Amazonia. The Geographical Journal **15**:283–296.

Fearnside, P. M. 1980. Land use allocation of the Transamazon Highway colonists of Brazil and its relation to human carrying capacity. Pages 114–138 in Barbira-Scazzocchio (ed.). Land, People, and Planning in Contemporary Amazonia. University of Cambridge Centre of Latin American Studies Occasional Paper No. 3, Cambridge, U.K. 313 pp.

Fearnside, P. M. 1983. Land use trends in the Brazilian Amazon region as factors in accelerating deforestation. Environmental Conservation **19**:141–148.

Fearnside, P. M. 1984. Land clearing behavior in small farmer settlement schemes in the Brazilian Amazon and its relation to human carrying capacity. Pages 255–271 in A. C. Chadwick and S. L. Sutton, editors. Tropical rain forests: The Leeds Symposium. Leeds Philosophical and Literary Society, Leeds, England.

Fearnside, P. M. 1986. Human carrying capacity of the Brazilian rainforest. Columbia University Press, New York.

Foresta, R. 1991. The limits of providence: Amazon conservation in the age of development. University of Florida Press, Gainesville, Florida.

Franklin, J. F., K. Cromack, Jr., W. Denison, A. McKee, C. Maser, J. Sedell, F. Swanson, and G. Juday. 1981. Ecological characteristics of old-growth Douglas-fir forests. General Technical Report PNW-118. United States Forest Service, Portland, Oregon.

Frohn, R. C., V. H. Dale, and B. D. Jimenez. 1990. Colonization, road development and deforestation in the Brazilian Amazon Basin of Rondônia. ORNL/TM-11470. Oak Ridge National Laboratory, Oak Ridge, Tennessee.

Hecht, S. B. 1981. Deforestation in the Amazon Basin: Magnitude, dynamics, and soil resource effects. Studies in Third World Societies **13**:61–110.

Houghton, R. A., J. E. Hobbie, J. M. Melillo, B. Moore, B. J. Peterson, G. R. Shaver, and G. M. Woodwell. 1983. Changes in the carbon content of terrestrial biota and soils between 1860 and 1980: Net release of CO_2 to the atmosphere. Ecological Monographs **53**:235–262.

IBGE (Instituto Brasileiro de Geografia e Estatistica). 1979*a*. Porto Velho. Map, scale 1:1,000,000. Rio de Janeiro, Brazil.

IBGE (Instituto Brasileiro de Geografia e Estatistica). 1979*b*. Guapore. Map, scale 1:1,000,000. Rio de Janeiro, Brazil.

Krummel, J. R., R. H. Gardner, G. Suhihara, and R. V. O'Neill. 1987. Landscape patterns in a disturbed environment. Oikos **48**:321–324.

Lehmkuhl, J. F. 1985. Determining size and dispersion of minimum viable populations for land management planning and species conservation. Environmental Management **8**:167–176.

Leite, L. L., and P. A. Furley. 1985. Land development in the Brazilian Amazon with particular reference to Rondônia and the Ouro Preto colonization project. J. Hemming, editor. Change in the Amazon basin. Manchester University Press, Manchester, England.

Lovejoy, T. E., R. O. Bierregaard, Jr., and A. B. Rylands. 1986. Edge and other effects of isolation on Amazon forest fragments. Pages 275–285 in M. E. Soulé, editor. Conservation biology: The science of scarcity and diversity. Sinauer Associates, Sunderland, Massachusetts.

Machado, L. 1991. Frontera agricola en la Amazonia Brasilena. Pages 395–427 in V. I. Coloquio de Geografia Geral, Madrid, Spain.

Malingrequ, J. P., and C. J. Tucker. 1988. Large-scale deforestation in the southeastern Amazon basin of Brazil. Ambio **17(1)**:49–55.

Millikan, B. H. 1988. The dialectics of devastation: Tropical deforestation, land degradation, and society in Rondônia, Brazil. M.A. thesis. University of California, Berkeley, California.

Molofosky, J., C. A. S. Hall, and N. Myers. 1985. A comparison of tropical forest surveys. U.S. Department of Energy, Carbon Dioxide Research Program TR032, Washington, D.C.

Moran, E. F. 1981. Developing the Amazon. Indiana University Press, Bloomington, Indiana.

O'Neill, R. V., J. R. Krummel, R. H. Gardner, G. Sugihara, B. Jackson, D. L. DeAngelis, B. T. Milne, M. G. Turner, B. Zygmnuht, S. W. Christensen, V. H. Dale, and R. L. Graham. 1988. Indices of landscape pattern. Landscape Ecology **1**:153–162.

Pedlowski, M. A., and V. H. Dale. 1992. Land use practices in Ouro Preto Do Oeste, Rondônia, Brazil. ORNL-TM 3850. Oak Ridge National Laboratory, Oak Ridge, Tennessee.

Post, W. M., T.-H. Peng, W. Emanuel, A. W. King, V. H. Dale, and D. L. DeAngelis. 1990. The global carbon cycle. American Scientist **78(4)**:310–326.

RadamBrasil. 1978*a*. Vol. 16. Porto Velho, Brazil.

RadamBrasil. 1978b. Vol. 19. Guapore, Brazil.

Saunders, D. A., R. J. Hobbs, and C. R. Margules. 1991. Biological consequences of ecosystem fragmentation: A review. Conservation Biology **5(1)**:18–32.

Schwartzman, S. 1989. Extractive reserves: The rubber tappers' strategy for sustainable use of the Amazon rainforest. Pages 150–165 in J. O. Browder, editor. Fragile lands of Latin America: Strategies of sustainable development. Westview Press, Boulder, Colorado.

Shukla, J., C. Nobre, and P. Sellers. 1990. Amazon deforestation and climate change. Science **247**:1322–1325.

Skole, D., and C. Tucker. 1993. Tropical deforestation and habitat fragmentation in the Amazon: Satellite data from 1978–1988. Science **260**:1905–1910.

Southworth, F., V. H. Dale, and R. V. O'Neill. 1991. Contrasting patterns of land use in Rondônia, Brazil: Simulating the effects of carbon release. International Social Science Journal **130**:681–698.

Stone, T. A., F. Brown, and G. M. Woodwell. 1991. Estimation, by remote sensing, of deforestation in Central Rondônia, Brazil. Forest Ecology and Management 38:291–304.

Tucker, C. J., B. N. Holben, and T. E. Goff. 1984. Intensive forest clearing in Rondônia, Brazil, as detected by satellite remote sensing. Remote Sensing Environment 15:255–261.

Tucker, C. J., J. R. G. Townsend, T. E. Goff, and B. N. Holben. 1986. Continental and global scale remote sensing of land cover. Pages 221–241 in J. R. Trabalka and D. E. Reichle, editors. The changing carbon cycle: A global analysis. Springer-Verlag, New York.

Wright, J. S., and S. P. Hubbell. 1983. Stochastic extinction and reserve size: A focal species approach. Oikos 41:466–476.

Appendix

Model Overview, Inputs and Outputs

The Dynamic Ecological–Land Tenure Analysis model (DELTA) is written in Fortran and runs on an IBM or clone personal computer with a math coprocessor. A flow-chart (Fig. A) shows the major steps in the program. The spatial characteristics of each lot (size, soils, vegetation, transportation systems, and distances to market) are input from spatial data bases (in this case from digitized maps using the ARC/INFO geographic information system). The number of immigrants is set by the user. Lot attraction variables are calculated from the spatial characteristics and used to determine which lots are selected by new migrants and by farmers that move between lots. Colonists leaving a lot either purchase a new lot, emigrate from the area, or become part of the local labor pool of landless workers. The mean amount of land cleared on each lot and the particular land uses are specified by the user or set from values typical for the region. A distribution based on the mean is then used to assign the proportion cleared for each lot in each year. Agricultural lots can be coalesced into pasture when a minimum area has been cleared. Lots can also be fragmented such that two or more families occupy a lot that was originally designated for one family. Every year the histories of each farmer and each lot are updated, and land quality values are recalculated. The user has the option of printing detailed information or lot summaries for each year. The model is usually run for about 50 years (although the time period could be extended) requiring 8 min for 40 replications.

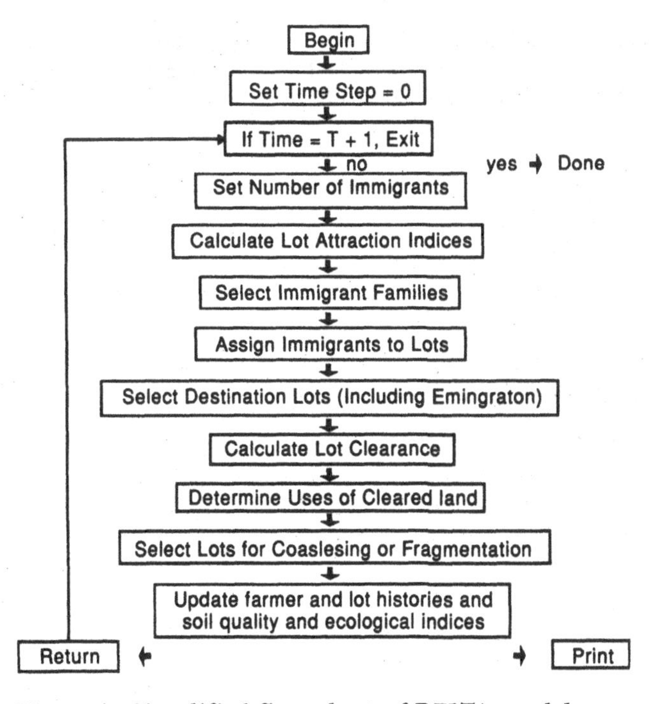

Figure A. Simplified flow chart of DELTA model.

The nonspatial parameters include information about the size of the area, lot occupancy decisions (such as the proportion of a lot's area to be cleared each year), conditions for coalescing lots into pastures, criteria for changes over time in a tenant's mix of land uses on a lot, and choice variables that deal with conditions under which colonists may move to a new lot. A user interface to the model requests values for each parameter and assists the user by providing example inputs. Simulation results are based on initial conditions, highway building patterns, population immigration and emigration conditions, and land-management practices set by the user.

Model outputs include land-use changes by year, maps, and spatial statistics. The land-use changes include average length of tenure on the land, land-use type, percentage deforestation, and loss of carbon in soils and vegetation. A time series of deforestation maps and aggregate spatial statistics (such as dominance, contagion, and fractal dimension) record the effects of a particular settlement strategy on the spatio-temporal pattern of forest clearing and, in turn, the results of this land development on carbon release.

Settlement Diffusion

In each model year, a number of immigrant families are introduced to the region and distributed among the available land locations or "lots." The availability of a lot depends upon a number of factors, such as government programs that offer land under specific conditions. For the study reported here, lot boundaries were defined from a blueprint of the government land-use plan for Ouro Preto in central Rondônia and entered into the Arc/Info GIS (ESRI 1990). Where no such data exist, lots may be selected from a grid of points distributed evenly over the landscape or defined by boundaries evident in remote sensing data.

For each available lot, i, in the system, we calculate a relative lot attractiveness index of the form

$$A_i = a_1X_{i1} + a_2X_{i2} + \ldots + a_kX_{ik}, \tag{1}$$

where X_{ik} is the value of attribute k measured on lot i and where $a1, \ldots, aK$ are attribute weights specified by the user or derived on the basis of empirical evidence. The lot attractiveness attributes, X, may themselves be composite variables requiring prior computation. In the Rondônia case study, they include lot size, three soil quality indices of agricultural suitability (for annual crops, perennials, and pasture), and distance to the nearest local market via primary (paved) and local (unpaved, feeder) roads. A particularly interesting attribute is length of current lot tenure, as this affects both the lot's productivity and its mix of land uses. Both length of current lot tenure and usable lot size are updated each time period within DELTA.

The probability of a tenant selecting a specific lot i is based on a destination selection model of the form

$$P_i = \frac{\exp A_i}{\sum_j \exp A_j} \tag{2}$$

for $j = \{1, 2, \ldots, i, \ldots, J\}$ available lots (see Southworth [1981] for example). The relative choice probabilities, P_i (which sum to 1.0 when taken over all available lots), are then used with a random number selection process to allocate lots to colonists. In this manner, a less attractive lot has some chance of being taken early in the settlement process—a common occurrence because potential colonists, no matter what the land-use situation, rarely have complete information at their disposal.

An exogenous input to the model is the probability of a colonist moving after a specific number of years on a lot. For example, there may be a 10% chance, on the average, of a move during any of the first three years of occupancy, rising to 30% in years 4 through 7, and up to 90% by year 10. A stochastic process selects the colonists that move annually (except during the first year). In the Rondônia example, colonists with longer time on a lot are more likely to move. Such movements could also be made to depend on current soil quality.

If a colonist is selected to move, then the choice of his destination lot is modeled using Equation 2. When available land becomes scarce, or if evidence suggests that some proportion of colonists emigrate from the region, then a test is made to see if one or more colonists are (randomly) allocated to be emigrants during a particular year. Currently, this test is based upon the probability of a colonist leaving the area after a given period of time, or on a measure of land pressure (as a function of the number of immigrants divided by the number of available lots).

Land-Use Change

Currently, DELTA accepts as input a set of expected land-use parameters that determine the combination of land uses to be found on a lot over time. For the Rondônia case study, the mix of land uses on a lot includes undeveloped forest as well as some combination of annual cropping, perennial cropping, animal grazing, and fallow. For example, in the first year of tenure, a family may grow only annual, subsistence crops. In subsequent years they may plant perennial crops. A colonist may also put some cleared and previously cultivated land into grazing.

Carbon content of the soil and vegetation are assumed to affect land-use practices on the lot. The probability of abandoning a lot increases as the land quality declines (as measured by the decline in carbon content). Millikan (1988) notes that illness and chance events influence lot abandonment, but these are not included in the model at this time.

Carbon Release

Ecological indicators that are currently being projected are the amount of deforestation and the loss of carbon in vegetation and soils. Carbon release is based on a piece-wise continuous curve of the carbon per hectare under pristine conditions, farming, pasture, and abandonment, which is derived from the approach of Houghton et al. (1983). For the Rondônia example, the pristine forest is estimated to have 293 Mg/ha of aboveground tree biomass (Brown et al. 1993). Multiplying by Fearnside's (1992) value of 1.43 for the ratio between total and aboveground live biomass gives an estimate of 419 Mg/ha for biomass or 209 Mg carbon per hectare, since approximately half the biomass is composed of carbon. This estimate for carbon is multiplied by the number of hectares in the lot to obtain an initial carbon value for each lot. With farming or pasture use, the initial carbon undergoes a negative exponential decline. For this case study, a carbon loss rate per hectare of $e^{-0.19}$ was obtained by assuming that long-term farms or pasture would be in place 15 years, at which time 10 Mg/ha of carbon remained on the lot. Once a lot is abandoned, if it is not immediately reoccupied, its carbon content is then assumed to improve gradually as vegetation regenerates itself slowly on the typically overworked open land. A constant linear rate of carbon return is used to model this effect. The slope of 0.32 is used if it is assumed that it takes 500 years for the carbon content of the forest to recover fully (Uhl et al. 1988).

If a lot is reoccupied, the negative exponential decline in carbon begins from the amount of carbon on the lot at that time. Because some carbon is always retained in the soil, the carbon depletion is not allowed to go below 10% of the initial carbon value.

Papers Referenced only in Appendix

Brown, I. F., L. A. Martinelle, W. W. Thomas, M. Z. Moreira, C. A. C. Ferreira, and R. A. Victoria. 1993. Studies of a southwestern Amazonian Forest: Uncertainty in estimates of biomass. Ecological Applications.

ESRI (Environmental Systems Research Institute). 1990. Arc-Info Users Guide 380 New York Street, Redlands, California.

Fearnside, P. 1992. Forest biomass in Brazilian Amazonia: Comments the estimate by Brown and Lugo. Intercenia 17(1):19–27.

Southworth, F. 1981. Calibration of multinomial logit models of mode and destination choice. Transportation Research 154:315–325.

Uhl, C., R. Buschbacher, and E. A. S. Serrao. 1988. Abandoned pastures in eastern Amazonia. I. Patterns of plant succession. Journal of Ecology 76:663–681.

References

Alverson, W. S., D. M. Waller, and S. L. Solheim. 1988. Forests too deer: edge effects in northern Wisconsin. Conservation Biology **2(4)**:348–358.

Ambrose, J. P. and S. P. Bratton. 1990. Trends in landscape heterogeneity along the borders of Great Smoky Mountains National Park. Conservation Biology **4(2)**:135–143.

Blake, J. G. 1991. Nested subsets and the distribution of birds on isolated woodlots. Conservation Biology **5(1)**:58–66.

Bleich, V. C., J. D. Wehausen, and S. A. Holl. 1990. Desert-dwelling mountain sheep: conservation implications of a naturally fragmented distribution. Conservation Biology **4(4)**:383–390.

Brothers, T. S. and A. Spingharn. 1992. Forest fragmentation and alien plant invasion of central Indiana old-growth forests. Conservation Biology **6(1)**:91–100.

Dale, V. H., R. V. O'Neill, F. Southworth, and M. Pedlowski. 1994. Modeling effects of land management in the Brazillian amazonian settlement of Rondônia. Conservation Biology **8(1)**:196–206.

Fahrig, L. and G. Merriam. 1994. Conservation of fragmented populations. Conservation Biology **8(1)**:50–59.

Gilpin, M. E. 1988. A comment on Quinn and Hastings: extinction in subdivided habitats. Conservation Biology **2(3)**:290–292.

Harrison, S., A. Stahl, and D. Doak. 1993. Spatial models and spotted owls: exploring some biological issues behind recent events. Conservation Biology **7(4)**:950–953.

Jennersten, O. 1988. Pollination in *Dianthus deltoides* (Caryophyllaceae): effects of habitat fragmentation on visitation and seed set. Conservation Biology **2(4)**:359–366.

Karr, J. R. 1990. Avian survival rates and the extinction process on Barro Colorado Island, Panama. Conservation Biology **4(4)**:391–397.

Kattan, G. H., H. Alvarez-López, and M. Giraldo. 1994. Forest fragmentation and bird extinctions: San Antonio eighty years later. Conservation Biology **8(1)**:138–146.

Kirkpatrick, J. B. and M. J. Brown. 1994. A comparison of direct and environmental domain approached to planning reservation of forest higher plant communities and species in Tasmania. Conservation Biology **8(1)**:217–224.

Lindenmayer, D. B. and H. A. Nix. 1993. Ecological principles for the design of wildlife corridors. Conservation Biology **7(3)**:627–630.

Litvaitis, J. A. 1993. Response of early successional vertebrates to historic changes in land use. Conservation Biology **7(4)**:866–873.

Maurer, B. A. and S. G. Heywood. 1993. Geographic range fragmentation and abundance in neotropical migratory birds. Conservation Biology **7(3)**:501–509.

Mladenoff, D. J. and F. Stearns. 1993. Eastern hemlock regeneration and deer browsing in the northern Great Lakes region: a reexamination and model simulation. Conservation Biology **7(4)**:889–900.

Noss, R. F. 1987. Corridors in real landscapes: a reply to Simberloff and Cox. Conservation Biology **1(2)**:159–164.

Paton, P. W. C. 1994. The effect of edge on avian nest success: how strong is the evidence? Conservation Biology **8(1)**:17–26.

Patterson, B. D. 1987. The principle of nested subsets and its implications for biological conservation. **1(4)**:323–334.

Quinn, J. F. and A. Hastings. 1987. Extinction in subdivided habitats. Conservation Biology **1(3)**:198–208.

Quinn, J. F. and A. Hastings. 1988. Extinction in subdivided habitats: a reply to Gilpin. Conservation Biology **2(3)**:293–296.

Samways, M. J. 1990. Land forms and winter habitat refugia in the conservation of montane grasshoppers in southern Africa. Conservation Biology **4(4)**:375–382.

Saunders, D. A., R. J. Hobbs, and C. R. Margules. 1991. Biological consequences of ecosystem fragmentation: a review. Conservation Biology **5(1)**:18–32.

Simberloff, D. and J. Cox. 1987. Consequences and costs of conservation corridors. Conservation Biology **1(1)**:63–71.

Simberloff, D., J. A. Farr, J. Cox, and D. W. Mehlman. 1992. Movement corridors: conservation bargains or poor investments? Conservation Biology **6(4)**:493–504.

Thiollay, J. M. 1989. Area requirements for the conservation of rain forest raptors and game birds in French Guiana. Conservation Biology **3(2)**:128–137.

Wright, R. G., J. G. MacCracken, and J. Hall. 1994. An ecological evaluation of proposed new conservation areas in Idaho: evaluating proposed Idaho national parks. Conservation Biology **8(1)**:207–216.

Yosef, R. 1994. The effects of fencelines on the reproductive success of Loggerhead Shrikes. Conservation Biology **8(1)**:281–285.

9 780865 424531